T0202799

Lecture Notes in Computer Science 14462

Founding Editors

Gerhard Goos
Juris Hartmanis

The series Lecture Notes in Computer Science (LNCS), including its subseries Lecture Notes in Artificial Intelligence (LNAI) and Lecture Notes in Bioinformatics (LNBI), has established itself as a medium for the publication of new developments in computer science and information technology research, teaching, and education.

LNCS enjoys close cooperation with the computer science R & D community, the series counts many renowned academics among its volume editors and paper authors, and collaborates with prestigious societies. Its mission is to serve this international community by providing an invaluable service, mainly focused on the publication of conference and workshop proceedings and postproceedings. LNCS commenced publication in 1973.

Weili Wu · Jianxiong Guo

Editors

Combinatorial Optimization and Applications

16th International Conference, COCOA 2023
Hawaii, HI, USA, December 15–17, 2023
Proceedings, Part II

 Springer

Editors
Weili Wu ⓘ
University of Texas at Dallas
Richardson, TX, USA

Jianxiong Guo ⓘ
Beijing Normal University
Zhuhai, China

ISSN 0302-9743   ISSN 1611-3349 (electronic)
Lecture Notes in Computer Science
ISBN 978-3-031-49613-4   ISBN 978-3-031-49614-1 (eBook)
https://doi.org/10.1007/978-3-031-49614-1

This Springer imprint is published by the registered company Springer Nature Switzerland AG
The registered company address is: Gewerbestrasse 11, 6330 Cham, Switzerland

Paper in this product is recyclable.

Preface

The papers in these proceedings, which consist of two volumes, were presented at the 16th Annual International Conference on Combinatorial Optimization and Applications (COCOA 2023), December 15–17, 2023, in Honolulu, Hawaii, USA. The topics cover most aspects of combinatorial optimization and applications pertaining to computing.

All accepted papers were selected by an international program committee consisting of a large number of scholars from various countries and regions, distributed over the world, including Asia, North America, Europe, and Australia. Each paper was required to submit in double-blind style and was evaluated by at least three reviewers. The decision was made based on those evaluations through a process containing a discussion period.

Authors of selected papers come from the following countries and regions: Canada, China (including Hong Kong and Macau), Romania, Brazil, UK, India, Belgium, Japan, Germany, Israel, and USA. Many of these papers represent reports of continuing research, and it is expected that most of them will appear in a more polished and complete form in scientific journals.

We wish to thank all who have made this meeting possible and successful, the authors for submitting papers, the program committee members for their excellent work in reviewing papers, the sponsors, the local organizers, and Springer for their support and assistance, We are especially grateful to Yi Zhu and Xiao Li who made tremendous efforts in local arrangements and set-up.

December 2023

Weili Wu
Jianxiong Guo

Organization

General Co-chair

Ding-Zhu Du   University of Texas at Dallas, USA

PC Co-chairs

Weili Wu   University of Texas at Dallas, USA
Jianxiong Guo   Beijing Normal University, China

Web Co-chairs

Xiao Li   University of Texas at Dallas, USA
Ke Su   University of Texas at Dallas, USA

Finance Co-chairs

Jing Yuan   University of Texas at Dallas, USA
Smita Ghosh   Santa Clara University, USA

Registration Co-chairs

Xiao Li   University of Texas at Dallas, USA
Garg Priyanshi   University of Texas at Dallas, USA

Local Chair

Yi Zhu   Hawaii Pacific University, USA

Program Committee Members

An Zhang	Hangzhou Dianzi University, China
Andras Farago	University of Texas at Dallas, USA
Annalisa De Bonis	Università degli Studi di Salerno, Italy
Arash Rafiey	Indiana State University, USA
Bin Liu	Ocean University of China, China
Binhai Zhu	Montana State University, USA
Bo Li	Hong Kong Polytechnic University, China
Chenchen Wu	Tianjin University of Technology, China
Chuanwen Luo	Beijing Forestry University, China
Dachuan Xu	Beijing University of Technology, China
Donglei Du	University of New Brunswick, Canada
Fay Zhong	Shanghai Jiao Tong University, China
Guochuan Zhang	Zhejiang University, China
Guohui Lin	University of Alberta, Canada
Habib Ammari	Texas A&M University-Kingsville, USA
Ho-Lin Chen	National Taiwan University, Taiwan
Huaming Zhang	University of Alabama in Huntsville, USA
Jing Yuan	North Texas University, USA
Joong-Lyul Lee	University of North Carolina at Pembroke, USA
Juraj Hromkovic	ETH Zurich, Switzerland
Kazuo Iwama	Kyoto University, Japan
Lidong Wu	University of Texas at Dallas, USA
Ling-Ju Hung	National Taipei University of Business, Taiwan
Louxin Zhang	National University of Singapore, Singapore
Lu Han	Beijing University of Posts and Telecommunications, China
Meghana Satpute	University of Texas at Dallas, USA
Michael Khachay	Krasovsky Institute of Mathematics and Mechanics, Russia
Mihaela Cardei	Florida Atlantic University, USA
Qianping Gu	Simon Fraser University, Canada
Sergey Bereg	University of Texas at Dallas
Sergiy Butenko	Texas A&M University, USA
Shaojie Tang	University of Texas at Dallas, USA
Shengxin Liu	Harbin Institute of Technology (Shenzhen), China
Shuyang Gu	Texas A&M University - Central Texas, USA
Smita Ghosh	Santa Clara University, USA
Ueverton Souza	Universidade Federal Fluminense, Brazil
Viet Hung Nguyen	Université Pierre & Marie Curie, France
Wei Wang	Xian Jiaotong University, China

Xianyue Li	Lanzhou University, China
Yan Shi	University of Wisconsin - Platteville, USA
Yifei Zou	Shandong University, China
Yong Chen	Hangzhou Dianzi University, China
Yongxi Cheng	Xian Jiaotong University, China
Yuqing Zhu	California State University, Los Angeles, USA
Zhao Zhang	Zhejiang Normal University, China
Zhongnan Zhang	Xiamen University, China

Contents – Part II

Optimization and Algorithms

Extreme Graph and Others

Machine Learning, Blockchain and Others

Contents – Part I

Set-Related Optimization

Applied Optimization and Algorithm

Graph Planer and Others

Modeling and Algorithms

Differentiable Discrete Optimization Using Dataless Neural Networks

Sangram K. Jena[1], K. Subramani[1(✉)], and Alvaro Velasquez[2]

[1] LDCSEE, West Virginia University, Morgantown, WV, USA
{sangramkishor.jena,k.subramani}@mail.wvu.edu
[2] Department of Computer Science, University of Colorado Boulder, Boulder, CO, USA
alvaro.velasquez@colorado.edu

Abstract. The area of combinatorial optimization is characterized by the search for optimal combinations of discrete variables that satisfy some set of constraints. Famous problems in this space include maximum satisfiability and maximum independent set. Due to their discrete dynamics, these problems are not differentiable in their natural formulations. In this paper, we explore the counter-intuitive direction of differentiable discrete optimization by leveraging the recently discovered dataless neural networks, which have been used to yield a single differentiable function that is equivalent to the maximum independent set problem. In particular, we leverage the dataless neural networks framework to derive differentiable forms for a variety of **NP-hard** discrete problems and prove the correctness of our derivations. The proposed differentiable forms open up the avenue for continuous differentiable optimization to be brought to bear on classical discrete optimization problems.

1 Introduction

While hard optimization problems have been solved for decades via approximation algorithms and heuristics that leverage convex relaxations, the power of non-convex optimization afforded by the loss function of modern neural networks and the training algorithms based on backpropagation in said learning models may enable more powerful solutions for discrete optimization. This introduces the possibility to replicate the tremendous success of neural networks, as evidenced by Chat-GPT and AlphaGo, within the realm of combinatorial optimization. While the conventional direction would be to leverage large datasets of combinatorial optimization problems from which to derive relevant patterns using neural networks, a more recent approach has emerged that requires no data. In [2], the maximum independent set problem is framed as a single differentiable function such that neural networks and backpropagation can then be adopted to solve them, where only the given instance of the problem is required and no

---

K. Subramani and S. K. Jena—This research was supported in part by the Defense Advanced Research Projects Agency through grant HR001123S0001-FP-004.

additional data is necessary. This approach upends the conventional wisdom of machine learning, whereby a dataset is assumed for refining the parameters of the learning model (e.g., a neural network) in the direction of more accurate predictions. Instead, dataless neural networks can refine their parameters in terms of the internal connections of the network or other factors not dependent on a set of ground-truth data. To illustrate this, consider a conventional neural network f parameterized by weights θ to be trained on some dataset $\{(x_i, y_i)\}$. For example, x_i can be an instance of a discrete optimization problem and y_i can be the values of the optimal solution. The parameters θ are typically updated using backpropagation by minimizing a differentiable loss function $L(x_i, f(x_i; \theta))$ to make the output $f(x_i; \theta)$ of the neural network as close to y_i as possible. Backpropagation updates the parameters in the direction of $\theta := \theta - \alpha \cdot \partial L(x_i, f(x_i; \theta))/\partial \theta$, where α controls the learning rate. The idea behind the dataless neural networks is that there is no data, so what we have as an output of the neural network is simply $f(e_n; \theta) = f(\theta)$, where e_n is the all-ones vector representing a trivial input to the neural network. Thus, instead of attempting to find patterns in some data set, dataless neural networks attempt to find the optimal solution to a given discrete optimization problem by enforcing a certain structure on f and θ.

The rest of this paper is organized as follows: In Sect. 2, we formally define the problems and the notations considered in the paper. Section 3 describes related work in the literature. In Sect. 4, we design a dataless neural network (dNN) for the maximum dissociation set (MDS) problem. Section 5 discusses a dNN for the k-coloring problem. We design a dNN for the maximum cardinality d-distance matching problem in Sect. 6. Finally, we conclude in Sect. 7 by summarizing our results and discussing avenues for future work.

2   Statement of Problems

In this section, we define the problems and some of the notations considered in this paper.

Definition 1. Dissociation set (DS): *Given a graph $G = (V, E)$ and an integer k, does there exist a vertex set $D \subseteq V$ of size at least k such that the degree of each vertex $v \in D$ is at most one in the induced graph $G' = (D, E')$?*

Definition 2. Maximum dissociation set (MDS): *Given a graph $G = (V, E)$, find a dissociation set $D \subseteq V$ of maximum size.*

Definition 3. Minimum 3-path vertex cover (M3PVC): *Given a graph $G = (V, E)$, find a minimum size set $C_3 \subseteq V$ such that each path having three vertices (path of order 3) contains at least one vertex from C_3 in G.*

It is trivial to observe that a set C_3 of vertices of a graph G is a 3-path vertex cover of G, if and only if its complement $D = V \setminus C_3$ is a dissociation set of G.

Definition 4. k-coloring: *Given a graph $G = (V, E)$ and an integer k, does there exist an assignment of at most k colors to the vertex set V such that no two vertices sharing an edge are assigned the same color?*

Definition 5. Maximum cardinality d-distance matching (MCDM): *Given a bipartite graph $G = (S, T, E)$, where the set $S = \{s_1, s_2, \ldots, s_n\}$ is a strictly ordered set and $T = \{t_1, t_2, \ldots, t_r\}$ and a distance $d \in \mathbb{Z}^+$, find a maximum cardinality subset $\mathcal{M} \subseteq E$ in G such that the degree of every vertex of S is at most one in $\mathcal{M}$ and if $s_i t, s_j t \in \mathcal{M}$, then $|j - i| \geq d$.*

An activation function in a neural network transforms the summed weighted input from the node into the node's activation or output for that input. In our design of dataless neural networks, we use a rectified linear activation function, also known as the ReLU activation function. It is a piecewise linear function that outputs the input directly if it is positive; otherwise, it outputs zero, i.e., $\sigma(x) = max(0, x)$. For any positive integer n, $[n] := \{1, 2, \ldots, n\}$. Unless mentioned otherwise, $|\cdot|$ represents the absolute value or modulus.

The principal contributions of this paper are as follows:

1. A differential approach for the MDS problem (see Sect. 4).
2. A differential approach for the k-coloring problem (see Sect. 5).
3. A differential approach for the MCDM problem (see Sect. 6).

3   Related Work

In this section, we discuss the state-of-the-art result related to the neural network (NN) and dataless neural network (dNN) available in the literature. Our discussion for NN and dNN is particularly based on many combinatorial optimization problems (COPs). The most interesting COPs are **NP-hard**. It is well-known that such problems do not have polynomial time efficient algorithms unless some established complexity-theoretic conjectures fail. Although these problems cannot be solved efficiently, they have applications in almost every domain, such as scheduling, routing, telecommunications, planning, transportation, and decision-making processes [3,7]. Researchers have attempted to address **NP-hard** problems with different efficient, approximate solvers [11]. Broadly, these solvers are categorized into heuristic algorithms [1], approximation algorithms [4], and conventional branch-and-bound methods [15]. Such approaches may produce suboptimal solutions. Some of the other well-studied approaches to dealing with **NP-hard** problems use parameterized [5,8,14] and exact exponential algorithmic techniques [9,10].

Yet another approach to address the COPs is to use the concept of machine learning [3,17]. The use of reinforcement learning to automate the search of the heuristics for COPs is discussed in [6,13]. These models require training based on the problems. More specifically, they rely on supervised learning using datasets of the combinatorial structures of interest drawn from some distribution of problem instances. In [2], the authors introduced dNNs for which no data is required for training. By designing a single differentiable function, they captured the well-known combinatorial optimization problem, the maximum independent set (MIS) problem. They also designed a similar dNN structure for the maximum clique (MC) and minimum vertex cover (MVC) problems related to the MIS problem. To prove that their dNN performs on par or outperforms the existing learning-based methods with respect to the solution size, they implemented them both on real and synthetic large-scale graphs.

The literature discusses several powerful heuristic solvers for the MIS problem. One of the heuristic solvers is ReduMIS [11]. It consists of two components. The first component is an iterative implementation of a series of graph reduction techniques. The second component is the use of an evolutionary algorithm. These methods usually involve extensive training of neural networks (NNs) using large graph datasets for which solutions are known. Another method for the MIS problem, similar to the method of dNN for the MIS problem discussed in [2], was developed in [16]. The method discussed in [16] does not require training data, whereas it uses a graph neural network. More specifically, its output is represented by the probability of each node being in the solution. In contrast to the method discussed in [2], it uses a loss function to adjust its parameter that encodes the graph of interest. Furthermore, the approach discussed by Alkhouri et al. [2] uses n trainable parameters where n is the number of vertices in the input graph. However, the number of tunable parameters used by the approach discussed in [16] are large in size. It uses n parameters in its last layer only. In [2], the authors also showed some experimental results by comparing them with the best heuristics available in the literature. They evaluated success by taking the solution size obtained by ReduMIS as a benchmark. They also showed that their experimental results perform as well or outperform the state-of-the-art learning-based methods discussed in [12].

4   Maximum Dissociation Set

In this section, we discuss a dataless neural network (dNN) for the maximum dissociation set (MDS) problem. Our proposed approach is dataless. However, it is a neural network based technique, but not a learning method. Therefore, it is different from supervised, unsupervised, and reinforcement learning. We leverage the construction of the dNN for the MIS problem [2] to design a dNN for the MDS problem.

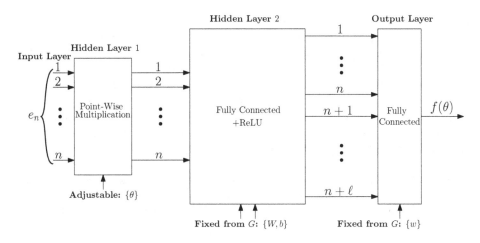

Fig. 1. Block diagram of dNN.

Let $G = (V, E)$ be a graph with n vertices over m edges. Let ℓ be the total number of three paths present in G. With respect to G, we construct a dNN f with trainable parameters $\theta \in [0,1]^n$. The input to the dNN is all-ones vector e_n which does not depend upon any data. The output of the dNN is $f(e_n; \theta) = f(\theta) \in \mathbb{R}$. The dNN for the MDS problem consists of four layers; one input layer, two hidden layers, and one output layer (Fig. 1 represents a block diagram of the proposed network). The input layer e_n is connected with the first hidden layer through an elementwise product of the trainable parameters θ. The first hidden layer is connected to the second hidden layer by the binary matrix $W \in \{0,1\}^{n \times (n+\ell)}$. Observe that the binary matrix is only dependent on G. At the second hidden layer, there exists a bias vector $b \in \{-2, -\frac{3}{4}\}^{n+\ell}$. There is a fully connected weight matrix $w \in \{-1, n\}^{n+\ell}$ in the second hidden layer to the output layer. Note that all the parameters are defined as a function of G as follows:

$$f(e_n; \theta) = f(\theta) = w^T \cdot \sigma((W^T \cdot (e_n \odot \theta)) + b). \tag{1}$$

Here $\odot$ is the element-wise Hadamard product that represents the operation of the first hidden layer of the constructed network. The fully-connected second hidden layer consists of the fixed matrix W and a bias vector b with a ReLU activation function $\sigma(x) = max(0, x)$. The last layer is another fully-connected layer and is expressed in vector w.

On the other hand, we prove that when an MDS $D \subseteq V$ in G is found, $f(\theta)$ attains its minimum value. Therefore, $f(\theta)$ is an equivalent differentiable function of the MDS generated in G. Moreover, D can be constructed from θ as follows. Let $\theta^* = argmin_{\theta \in [0,1]^n} f(\theta)$ be an optimal solution to f. Let $I : [0,1]^n \to 2^V$ be a dissociation set corresponding to θ such that $I(\theta) = \{v \in V \mid \theta_v^* \geq \alpha\}$, for $\alpha > 0$. We show that $|I(\theta^*)| = |D|$. We choose the vertices in the MDS D

in G corresponding to the indices of θ whose value exceeds a threshold (say α). From an input graph $G = (V, E)$, the fixed parameters of f can be constructed as follows: In the binary matrix W, the first $n \times n$ submatrix represents the vertices V of G. Its weights are set equal to the identity matrix I_n (see the 5×5 submatrix in Fig. 2(b) corresponding to the 5 vertices of G in Fig. 2(a)). Furthermore, the remaining ℓ columns of W represent three paths of G and for each three path $\ell_i = (u, v, w)$, the value of $u = v = w = 1$ in the column (see the 5 columns ℓ_1 to ℓ_5 in Fig. 2(b) corresponding to the 5 three paths of G in Fig. 2(a)). For each vertex of G, the corresponding entry of n nodes is $-\frac{3}{4}$ in the biased vector b. For each three path, the corresponding value in the bias vector is set to -2. Finally, the value of -1 is assigned in the entries corresponding to the nodes of G in vector w. For ℓ entries corresponding to the three paths in G, the value is set to n in w. Hence, the parameters W, b, and w are defined as follows:

$$W(i, i) = 1, v_i \in V, i \in [n],$$
$$W(i, n + h) = W(j, n + h) = W(k, n + h) = 1, \forall \ell_h = (v_i, v_j, v_k), h \in [\ell], \tag{2}$$

$$b(i) = -\frac{3}{4}, w(i) = -1, v_i \in V, i \in [n],$$
$$b(n + h) = -2, w(n + h) = n, h \in [\ell]. \tag{3}$$

So, the function in (1) can be rewritten as follows:

$$f(\theta) = -\sum_{v \in V} \sigma\left(\theta_v - \frac{3}{4}\right) + n \cdot \sum_{\substack{uv \in E \\ vw \in E}} \sigma(\theta_u + \theta_v + \theta_w - 2). \tag{4}$$

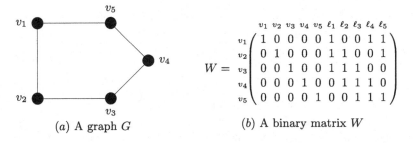

(a) A graph G (b) A binary matrix W

Fig. 2. Representation of a binary matrix W corresponding to G.

An example of the above discussed dNN construction is presented in Fig. 3. The following theorem establishes the relation between an MDS and the minimum value of f in the constructed dNN with respect to a given graph G.

Theorem 1. *Let $G = (V, E)$ be a graph and its corresponding dNN be f. G has an MDS $D \subseteq V$ of size k, if and only if the minimum value of f is $-\frac{k}{4}$.*

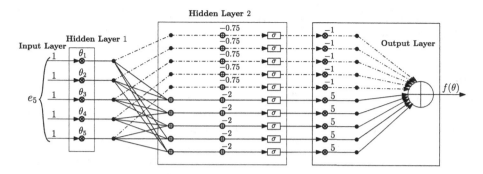

Fig. 3. Construction of dNN f corresponding to the graph in Fig. 2(a) for the MDS problem.

Proof. Let D be an MDS of size k in G. For each $v_i \in V$, set the value of θ_{v_i} as follows: If $v_i \in D$, then set $\theta_{v_i} = 1$. Otherwise, set $\theta_{v_i} = 0$. Consider the output f represented in Fig. 4 for an arbitrary set of three nodes $v_i, v_j, v_k \in D$.

As per the construction of the dNN, the three path values denote the outputs of the preceding nodes in the network. Furthermore, the ith neurons in the first hidden layer are denoted by η_i^1 and the second hidden layer is denoted by η_i^2. Observe that each node v_i, v_j, and v_k contribute $-\frac{1}{4}$ to the output $f(\theta)$. As per the definition of the MDS, there does not exist a three path with the nodes v_i, v_j, and v_k in D. Therefore, the output of $\eta_{n+\ell}^2 = 0$. Moreover, the size of the MDS in G is $|D| = k$. It immediately proves that for an MDS D of size k, $f(\theta) = -\frac{k}{4}$, that is the minimum value of the function.

Conversely, assume that the minimum value of the output function f is $f(\theta) = -\frac{k}{4}$. We construct an MDS D of size k in G from f as follows: From

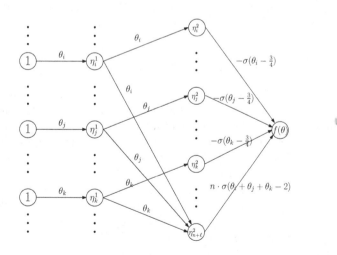

Fig. 4. Output with respect to three arbitrary nodes.

the construction of the dNN, it is clear that, for each set of three nodes v_i, v_j, and v_k such that they form a three path, $\theta_{v_i} + \theta_{v_j} + \theta_{v_k} \leq 2$. Otherwise, f does not achieve its minimum value. To prove this, assume that $\theta_{v_i} + \theta_{v_j} + \theta_{v_k} > 2$. It follows that the neuron $\eta_{n+\ell}^2$ contributes $n \cdot (\theta_u + \theta_v + \theta_k - 2) > 0$ to the output $f(\theta)$. This is a contradiction to the fact that f achieves its minimum value. We can simply assign the value of $\theta_{v_i}, \theta_{v_j}$, or θ_{v_k} as zero and reduce the value of f further. So, it is clear that for any three vertices v_i, v_j, and v_k in G, such that these three vertices create a three path, the θ value of at most two of the three vertices can be 1. Each such vertex, which θ value is one, contributes $-\frac{1}{4}$ to $f(\theta)$ through η_i^2. Furthermore, it contributes a value of 0 to $f(\theta)$ through $\eta_{n+\ell}^2$. That means there are k entries of value 1 in θ. For each entry of θ with value 1, take the corresponding vertex in the MDS D. It is clear that D is an MDS in G of size k.   □

5 k-Coloring

In this section, we first design a dNN for the 3-coloring problem. Next, we extend the discussed dNN of the 3-coloring problem for the k-coloring problem.

Let $G = (V, E)$ be a graph having n vertices over m edges. With respect to G, we construct a dNN f for the 3-coloring problem, similar to the dNN of the MDS problem, with trainable parameters $\theta \in [1,3]^n$. The input to the dNN is all-ones vector e_n which does not depend upon any data. The output of the dNN is $f(e_n; \theta) = f(\theta) \in \mathbb{R}$. The dNN for the 3-coloring problem consists of an input layer, two hidden layers, and one output layer. The input layer e_n is connected with the first hidden layer through an elementwise product of the trainable parameters θ. The first hidden layer is connected to the second hidden layer by the binary matrix $W \in \{0,1\}^{n \times (n+m)}$, which is only depend on G. There is a bias vector $b \in \{1, -\frac{1}{2}\}^{n+m}$ at the second hidden layer. There is also a fully connected weight matrix $w \in \{-1, n\}^{n+m}$ in the second hidden layer to the output layer. Note that all the parameters are defined as a function of G. The output of f is given by (5).

$$f(e_n; \theta) = f(\theta) = w^T \cdot \sigma((W^T \cdot (e_n \odot \theta)) + b). \qquad (5)$$

On the other hand, we prove that when a 3-coloring assignment of V in G is found, $f(\theta)$ attains its minimum value. Therefore, $f(\theta)$ is an equivalent differentiable function of the 3-coloring solution generated in G. Moreover, each vertex of V in G can be assigned a color with respect to θ as follows. Let $\theta^* = argmin_{\theta \in [1,3]^n} f(\theta)$ be an optimal solution to f. Let $I : [1,3]^n \rightarrow 3^V$ be the corresponding 3-coloring with respect to θ such that $I^i(\theta) = \{v \in V \mid \theta_v^* = i\}$ for $i \in \{1, 2, 3\}$. We assign color i to the vertices in $I^i(\theta)$. From an input graph $G = (V, E)$, the fixed parameters of f can be constructed as follows: In the binary matrix W, the first $n \times n$ submatrix represents the vertices V of G. Its weights are set equal to the identity matrix I_n. Furthermore, the remaining m columns of W represent edges of G and for each edge uv, the value of $u = v = 1$ in the column. For each vertex $v \in V$ in G, the corresponding entry of n nodes is $-\frac{1}{2}$ in the biased vector b. For each edge, the corresponding value in the bias vector is set to 1. Finally, the value of -1 is assigned in the entries corresponding to the nodes of G in vector w. For m entries corresponding to the edges in G, the value is set to n in w. So, the function in (5) can be rewritten as follows:

$$f(\theta) = -\sum_{v \in V} \sigma(\theta_v - \frac{1}{2}) + n \cdot \sum_{uv \in E} \sigma((-|\theta_u - \theta_v|) + 1). \tag{6}$$

The following theorem establishes the relation between a solution of the 3-coloring problem and the minimum value of f in the constructed dNN with respect to a given graph G.

Theorem 2. *Let $G = (V, E)$ be a graph having n vertices over m edges. There exists a 3-coloring of V in G such that n_1, n_2, and n_3 number of vertices of G colored with the first, second, and third colors, respectively, with $n_1 \leq n_2 \leq n_3$ and $n_1 + n_2 + n_3 = n$, if and only if the minimum value of f is $-(\frac{5 \cdot n_3 + 3 \cdot n_2 + n_1}{2})$.*

Proof. Let there exist a 3-coloring of V in G, which colors n_i vertices of V with color i ($i \in \{1, 2, 3\}$). For each $v \in V$, set the value of θ_v as follows: If v is colored with color i, then set $\theta_v = i$. Consider the output f for an arbitrary edge $uv \in E$ represented in Fig. 5. As per the construction of the dNN, the edges denote the outputs of the preceding nodes in the network. Furthermore, the ith neurons in the first hidden layer are denoted by η_i^1 and the second hidden layer is denoted by η_i^2. Observe that, for $i \in \{1, 2, 3\}$, each node $v \in V$ colored with i contributes $\frac{1}{2} - i$ to the output $f(\theta)$. As per the definition of the 3-coloring problem, there does not exist an edge $uv \in E$ such that u and v are assigned with the same color. Therefore, the output of $\eta_{n+\ell}^2 = 0$. It proves that the minimum value of the function f is $-(\frac{5 \cdot n_3 + 3 \cdot n_2 + n_1}{2})$, if the graph is 3-colorable.

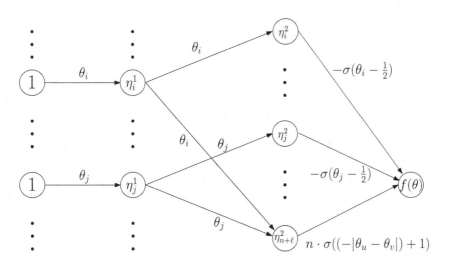

Fig. 5. Output with respect to two arbitrary nodes.

Conversely, assume that the minimum value of the output function f is $f(\theta) = -(\frac{5 \cdot n_3 + 3 \cdot n_2 + n_1}{2})$. We construct a 3-coloring assignment of V in G from f as follows: From the construction of the dNN, it is clear that, for each edge $uv \in E$, $(-|\theta_u - \theta_v|) \leq -1$. Otherwise, f does not achieve its minimum value. To prove this, assume that $(-|\theta_u - \theta_v|) > -1$. It follows that the neuron $\eta_{n+\ell}^2$ contributes $n \cdot ((-|\theta_u - \theta_v|) + 1) > 0$ to the output $f(\theta)$. This is a contradiction to the fact that f achieves its minimum value. So, it is clear that for any edge $uv \in E$ in G, the θ values of u and v are different and $(-|\theta_u - \theta_v|) \leq -1$. Each such vertex, which θ value is i ($i \in \{1, 2, 3\}$), contributes $\frac{1}{2} - i$ to $f(\theta)$ through η_i^2. Furthermore, it contributes a value of 0 to $f(\theta)$ through $\eta_{n+\ell}^2$. That means there are n_3 entries of value 3, n_2 entries of value 2, and n_1 entries of value 1 in θ. For each entry of θ with the value i ($i \in \{1, 2, 3\}$), take the corresponding vertex in G and assign color i. It is clear that the above assignment is a 3-coloring of V in G. □

Corollary 1. *The above-discussed dNN for the 3-coloring problem can be generalized for the k-coloring problem.*

Proof. Observe that the above-discussed dNN for the 3-coloring problem can handle the k-coloring problem by changing the trainable parameter θ. The values of all four layers are the same in the dNN except the θ value. The θ value was $\theta \in [1, 3]^n$ in the 3-coloring problem. However, for the k-coloring problem, we set the θ value $\theta \in [1, k]^n$. In this case, $f(\theta)$ can be represented in the same way as in the 3-coloring problem. However, the minimum value of the function is different. The construction and proof for the 3-coloring problem will follow for the k-coloring problem. □

6   Maximum Cardinality d-Distance Matching

In this section, we design a dNN for the maximum cardinality d-distance matching (MCDM) problem.

Let $G = (S, T, E)$ be a bipartite graph having n vertices over m edges. Let ℓ be the total number of pairs of edges present in G which share a common vertex $t_i \in T$. With respect to G, we construct a dNN f with trainable parameters $\theta \in [0, 1]^m$ for the MCDM problem as follows: The input to the dNN is all-ones vector e_m which does not depend upon any data. The output of the dNN is $f(e_m; \theta) = f(\theta) \in \mathbb{R}$. The dNN for the MCDM problem consists of four layers; one input layer, two hidden layers, and one output layer. The input layer e_m is connected with the first hidden layer through an elementwise product of the trainable parameters θ. The first hidden layer is connected to the second hidden layer by the binary matrix $W \in \{0, 1\}^{n \times (n+m)}$. Observe that the binary matrix is only dependent on G. At the second hidden layer, there exists a bias vector $b \in \{1, -\frac{1}{2}\}^{m+\ell}$. There is a fully connected weight matrix $w \in \{-1, n\}^{m+\ell}$ in the second hidden layer to the output layer. Note that all the parameters are defined as a function of G as follows:

$$f(e_m; \theta) = f(\theta) = w^T \cdot \sigma((W^T \cdot (e_m \odot \theta)) + b). \tag{7}$$

From an input graph $G = (S, T, E)$, the fixed parameters of f can be constructed as follows: In the binary matrix W, the first $n \times n$ submatrix represents the vertices S and T of G. Its weights are set equal to the identity matrix I_n. Furthermore, the remaining m columns of W represent edges of G and for each edge $t_i s_j$, the value of $t_i = s_j = 1$ in the column. For each edge $e_i \in E$ in G, the corresponding entry of m edges is $-\frac{1}{2}$ in the biased vector b. For each pair of edges, the corresponding value in the bias vector is set to -1. Finally, the value of -1 is assigned in the entries corresponding to the edges of G in vector w. For ℓ entries corresponding to the pair of edges in G, the value is set to n in w. So, the function in (7) can be rewritten as follows:

$$f(\theta) = - \sum_{e_i = t_k s_u \in E} (\sigma(\theta_i - \frac{1}{2}) + n \cdot \sum_{\substack{e_i = t_k s_u \in E \\ e_j = t_k s_v \in E}} \sigma((d - |u - v|) \cdot (\theta_i + \theta_j - 1)) \tag{8}$$

The following theorem establishes a relation between the minimum value of (8) and the size of the MCDM.

Theorem 3. *Given a graph $G = (S, T, E)$ and its corresponding dNN f. There exists an MCDM $\mathcal{M} \subseteq E$ in G of size k, if and only if the minimum value of f is $-\frac{k}{2}$.*

Proof. Let $\mathcal{M}$ be an MCDM of size k in G. For each $e_i = t_k s_u \in E$, set the value of θ_i as follows: If $e_i \in \mathcal{M}$, then set $\theta_i = 1$. Otherwise, set $\theta_i = 0$. Consider the output f represented in Fig. 6 for an arbitrary set of two edges $e_i, e_j \in \mathcal{M}$. As per the construction of the dNN, the pair of edges denote the outputs of the preceding nodes in the network. Furthermore, the ith neurons in the first hidden layer are denoted by η_i^1 and the second hidden layer is denoted by η_i^2. Observe

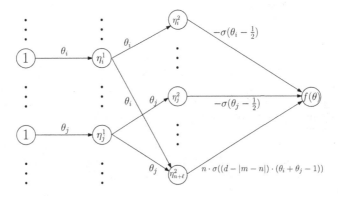

Fig. 6. Output with respect to two arbitrary edges.

that each edge e_i and e_j contribute $-\frac{1}{2}$ to the output $f(\theta)$. As per the definition of the MCDM, there does not exist a pair of edges $e_i = t_k s_u$ and $e_j = t_k s_v$ in $\mathcal{M}$ such that $|u - v| < d$. Therefore, the output of $\eta_{n+\ell}^2 = 0$. Moreover, the size of the MCDM in G is $|\mathcal{M}| = k$. It immediately proves that for an MCDM $\mathcal{M}$ of size k, $f(\theta) = -\frac{k}{2}$, that is the minimum value of the function.

Conversely, assume that the minimum value of the output function f is $f(\theta) = -\frac{k}{2}$. We construct an MCDM $\mathcal{M}$ of size k in G from f as follows: From the construction of the dNN, it is clear that, for each pair of edges $e_i = t_k s_u$ and $e_j = t_k s_v$ in $\mathcal{M}$ such that $|u - v| < d$, $(d - |u - v|) \cdot (\theta_i + \theta_j - 1) \le 0$. Otherwise, f does not achieve its minimum value. To prove this, assume that $(d - |u - v|) \cdot (\theta_i + \theta_j - 1) > 0$. It follows that the neuron $\eta_{n+\ell}^2$ contributes $n \cdot (d - |u - v|) \cdot (\theta_i + \theta_j - 1) > 0$ to the output $f(\theta)$. This is a contradiction to the fact that f achieves its minimum value. We can simply assign the value of θ_i or θ_j as zero and reduce the value of f further. So, it is clear that for any pair of edges $e_i = t_k s_u$ and $e_j = t_k s_v$ in $\mathcal{M}$ such that $|u - v| < d$, the θ value of at most one of the two edges can be 1. Each such edge, which θ value is one, contributes $-\frac{1}{2}$ to $f(\theta)$ through η_i^2. Furthermore, it contributes a value of 0 to $f(\theta)$ through $\eta_{n+\ell}^2$. That means there are k entries of value 1 in θ. For each entry of θ with value 1, take the corresponding edge in the MCDM $\mathcal{M}$. It is clear that $\mathcal{M}$ is an MCDM in G with size k. $\square$

7   Conclusion

In this paper, we have formulated novel differentiable representations for a variety of **NP-hard** problems. These representations open up opportunities for mixed discrete-continuous optimization solutions to combinatorial optimization problems. We proved the correctness of our formulations and leveraged the framework of dataless neural networks for their derivations. We leave as future work the application of backpropagation-based experiments using a variety of dataless neural networks. Our primary focus is to design a generalized dataless neural network that captures integer programming; a general purpose network will be more complex than the problem-specific models discussed in this paper.

References

1. Akiba, T., Iwata, Y.: Branch-and-reduce exponential/FPT algorithms in practice: a case study of vertex cover. Theoret. Comput. Sci. **609**, 211–225 (2016)
2. Alkhouri, I.R., Atia, G.K., Velasquez, A.: A differentiable approach to the maximum independent set problem using dataless neural networks. Neural Netw. **155**, 168–176 (2022)
3. Bengio, Y., Lodi, A., Prouvost, A.: Machine learning for combinatorial optimization: a methodological tour d'horizon. Eur. J. Oper. Res. **290**(2), 405–421 (2021)
4. Boppana, R., Halldórsson, M.M.: Approximating maximum independent sets by excluding subgraphs. BIT Numer. Math. **32**(2), 180–196 (1992)
5. Cygan, M., et al.: Parameterized Algorithms. Springer, Cham (2015). https://doi.org/10.1007/978-3-319-21275-3

6. Drori, I., et al.: Learning to solve combinatorial optimization problems on real-world graphs in linear time. In: 2020 19th IEEE International Conference on Machine Learning and Applications (ICMLA), pp. 19–24 (2020)
7. Festa, P.: A brief introduction to exact, approximation, and heuristic algorithms for solving hard combinatorial optimization problems. In: 2014 16th International Conference on Transparent Optical Networks (ICTON), pp. 1–20 (2014)
8. Flum, J., Grohe, M.: Parameterized Complexity Theory. TTCSAES, Springer, Heidelberg (2006). https://doi.org/10.1007/3-540-29953-X
9. Fomin, F.V., Kratsch, D.: Exact Exponential Algorithms. TTCSAES, Springer, Heidelberg (2010). https://doi.org/10.1007/978-3-642-16533-7
10. Gaspers, S.: Exponential Time Algorithms - Structures, Measures, and Bounds. VDM (2010)
11. Lamm, S., Sanders, P., Schulz, C., Strash, D., Werneck, R.F.: Finding near-optimal independent sets at scale. In: 2016 Proceedings of the Eighteenth Workshop on Algorithm Engineering and Experiments (ALENEX), pp. 138–150 (2016)
12. Li, Z., Chen, Q., Koltun, V.: Combinatorial optimization with graph convolutional networks and guided tree search. In: Advances in Neural Information Processing Systems, vol. 31 (2018)
13. Mazyavkina, N., Sviridov, S., Ivanov, S., Burnaev, E.: Reinforcement learning for combinatorial optimization: a survey. Comput. Oper. Res. **134**, 105400 (2021)
14. Niedermeier, R.: Invitation to Fixed-Parameter Algorithms. Oxford University Press (2006)
15. Segundo, P.S., Rodríguez-Losada, D., Jiménez, A.: An exact bit-parallel algorithm for the maximum clique problem. Comput. Oper. Res. **38**(2), 571–581 (2011)
16. Schuetz, M.J.A., Brubaker, J.K., Katzgraber, H.G.: Combinatorial optimization with physics-inspired graph neural networks. Nat. Mach. Intell. **4**(4), 367–377 (2022)
17. Wilder, B., Dilkina, B., Tambe, M.: Melding the data-decisions pipeline: Decision-focused learning for combinatorial optimization. In: Proceedings of the AAAI Conference on Artificial Intelligence, vol. 33, pp. 1658–1665 (2019)

When Advertising Meets Assortment Planning: Joint Advertising and Assortment Optimization Under Multinomial Logit Model

Chenhao Wang[1], Yao Wang[2(✉)], and Shaojie Tang[3]

[1] School of Data Science, The Chinese University of Hong Kong (CUHK-Shenzhen), Shenzhen, China
chenhaowang@link.cuhk.edu.cn
[2] Xi'an Jiaotong University, Xi'an, China
yao.s.wang@gmail.com
[3] The University of Texas at Dallas, Richardson, USA
shaojie.tang@utdallas.edu

Abstract. While the topic of assortment optimization has received a significant amount of attention, the relationship between advertising and its impact on this issue has not been well-explored. This paper aims to fill the gap in research by addressing the joint advertising and assortment optimization problem. We propose that advertising can influence product selection by increasing preference for certain products, and the extent of this effect is determined by the product-specific effectiveness of advertising and the resources allocated to advertising for that product. Our goal is to find an optimal solution, which comprises of a combination of advertising strategy and product assortment, that maximizes revenue, taking into account budget constraints on advertising. In this paper, we examine the characteristics of this problem and present efficient methods to solve it under various scenarios. Both the unconstraint and cardinality constraint settings are studied and the joint assortment, pricing, and advertising problem is also examined. We further extend our findings to account for consumer decision-making patterns.

Keywords: budget allocation · advertising effect · assortment optimization

1 Introduction

One of the major challenges faced by both online and offline retailers is the problem of assortment optimization, in which they choose a specific group of products to offer to customers such that their expected revenue can be maximized. The revenue generated by an assortment of products is usually determined by two factors: the revenue generated by selling each individual product, and the purchasing behavior of consumers. The latter is often captured by discrete choice models such as a multinomial logit (MNL) model [26] and a nested logit (NL) model [7]. Unlike previous studies that assume *fixed* choice models, we take into account the fact that customer purchasing behavior may be influenced by sophisticated selling practices such as advertising. Specifically, advertising is an important and effective strategy for establishing brand recognition and communicating the value of a product effectively to the public. Given the importance of advertising, determining how to allocate the promotional budget over products and time is a critical aspect of retailers' decision making [13, 19, 22], hence, it is important for a retailer to consider the impact of advertising on their product choices to increase revenue. To maximize this effect, the retailer should align their advertising and product recommendations.

In this paper, we propose and investigate a joint advertising and assortment optimization problem (JAAOP). We employ the MNL model to understand consumer purchasing behavior, in which every product, including the choice not to purchase, is assigned a random utility. Once presented with a variety of products, the consumer chooses the one with the highest utility. Our study differs from previous research on traditional assortment optimization by taking into account the influence of advertising. That is, rather than just selecting a group of products, we investigate the potential of combining advertising with traditional product selection to enhance the overall optimization. Specifically, we assume that the platform can increase the attractiveness (a.k.a. utility) of a product by advertising it, the effectiveness of which is represented by a product-specific response function and the amount of advertising efforts allocated to that product. With constraints on the advertising budget, our goal is to jointly determine which products to present to consumers and how to allocate the advertising budget among them in order to maximize expected revenue. In one extension of our work, we also examine the sequential choice behavior of consumers [15], a common feature on online shopping platforms such as Amazon and Taobao, where a large number of products are displayed to the consumer in stages. If the consumer does not select any products in a stage, they will move on to the next set of products. This requires the platform to not only select which products to display, but also their positions. We formulate this problem as a joint multi-stage advertising and assortment optimization problem.

1.1 Summary of Contributions

This section summarizes the major contributions of our work.

- We introduce the JAAOP in which the platform must concurrently select (1) an advertising strategy and (2) a set of products to present to consumers. By using the MNL model and assuming no constraint on the maximum number of products that can be displayed to a user, we can obtain an optimal revenue-ordered assortment and an efficient advertising strategy.
- When a constraint on the maximum number of products that can be displayed to a user is present, we analyze the problem under different response functions. If the response function is a log function, the optimal advertising

strategy is to allocate all the advertising budget to a single product. If the response function is a general concave function, we formulate our problem as a nonlinear continuous optimization problem and use McCormick inequalities to convert it into a convex optimization problem. We then develop an efficient algorithm to find the optimal strategy.
- In Appendix A and B, we study several extensions. In one extension of this study, we include the price of each product as a decision variable and consider the joint product assortment, pricing, and advertising optimization problem. We also extend our model to incorporate the multi-stage purchase behavior and investigate the structural properties of the problem. We develop a heuristic method that comes with a performance guarantee.
- We also conduct a series of experiments to evaluate the performance of our solutions and further confirm the value of advertising in Appendix C. Our proposed heuristic method is robust and outperforms other methods in different settings. Specifically, the results suggest that allocating the advertising budget uniformly or greedily leads to substantial revenue loss.

2   Literature Review

Our work is closely related to the assortment optimization problem in revenue management, which aims to select a subset of products to maximize the expected revenue. Various discrete choice models have been proposed to model consumer decision-making behavior, including the MNL model [26], the Nested Logit (NL) model, the d-level NL model and so on. Recently, several works have considered sequential choice behavior. For example, Flores et al. [12] investigated a two-stage MNL model, where the consumer sequentially browses two stages that are disjoint in terms of potential products and [17] extended to the multi-stage setting. Moreover, [15] developed a sequential MNL model, where the utility of the no-purchase option is fixed at the beginning instead of being resampled each time, and studied the assortment and pricing problem with impatient customers.

Another related problem is the advertising budget allocation problem. [2] proposed a model to allocate resources among multiple brands in a single period. [9] further considered advertising budget allocation across products and media channels. [11] considered the lagged effect of advertising and studied the dynamic marketing budget allocation problem for a multi-product, multi-country setting. [1] proposed a single-product spatiotemporal model that includes the spatial differences and sale dynamics.

Finally, our work belongs to the growing literature that aims to improve revenue through sophisticated selling practices beyond product selection. The approaches in this area include offering certain items only through lotteries [18] and making certain products unattractive to consumers [4]. [4] studied the refined assortment optimization problem for several regular choice models, including the MNL, latent-class MNL (LC-MNL), and random consideration set (RCS) models. While the authors in [4] focus on *reducing* the utilities of some products to improve revenue, our approach aims to increase revenue by *increasing* the utilities of some products. The main differences are: 1. [4] focus on strategically reducing the utilities of products, whereas our study centers on increasing the utilities of products. 2. [4] assume that changing the utility of a product has no cost, while our model takes into account the cost of increasing the utility of a product through advertising and considers budget constraints in the optimization problem. We also show that the platform has no incentive to decrease the utilities of any products in the MNL model under a cardinality constraint. A similar result was discovered independently by [4] for an unconstrained MNL model.

3 Preliminaries and Problem Formulation

3.1 MNL Model

We list the main notations in Table 1. Generally, the input of our problem is a set of n products $\mathcal{N} = \{1, 2, \cdots, n\}$. In the MNL model, each product $i \in \mathcal{N}$ has a utility $q_i + \epsilon_i$, where q_i is a constant that captures the initial utility of product i, and ϵ_i is a random variable that captures the error term. We assume that ϵ_i follows a Gumbel distribution with a location-scale parameter $(0, 1)$. Let $\mathbf{v}$ denote the preference vector of $\mathcal{N}$, where $v_i := e^{q_i}$ for each $i \in \mathcal{N}$. Given an assortment $S \subseteq \mathcal{N}$ and a preference vector $\mathbf{v}$, for each product $i \in \mathcal{N}$, a consumer will purchase product i with a probability of

$$\phi_i(S, \mathbf{v}) = \begin{cases} \frac{v_i}{1 + \sum_{j \in S} v_j} & \text{if } i \in S \\ 0 & \text{otherwise.} \end{cases} \tag{1}$$

The no-purchase probability is $\phi_0(S, \mathbf{v}) = \frac{1}{1 + \sum_{j \in S} v_j}$. Let $\mathbf{r}$ denote the revenue vector of $\mathcal{N}$, where for each product $i \in \mathcal{N}$, $r_i > 0$ represents the revenue from product i. Based on the above notations, the expected revenue $R(S, \mathbf{v})$ of the assortment S is given by

$$R(S, \mathbf{v}) = \sum_{i \in S} r_i \cdot \phi_i(S, \mathbf{v}) = \frac{\sum_{i \in S} r_i v_i}{1 + \sum_{i \in S} v_i}. \tag{2}$$

3.2 Joint Advertising and Assortment Optimization

We use a vector $\mathbf{x}$ to represent an advertising strategy where for each $i \in \mathcal{N}$, x_i represents the amount of advertising efforts allocated to i. Let $\mathbf{c}$ denote the *advertising effectiveness* of $\mathcal{N}$. We assume that the utility of each product $i \in \mathcal{N}$, increases by $f(c_i x_i)$ if it receives x_i advertising efforts from the platform, where

$f(\cdot)$ is called *response function* and c_i is the advertising effectiveness of product i. Intuitively, $\mathbf{c}$ and $f(\cdot)$ together determine the degree to which a product's preference weight is influenced by the advertising it receives from the platform. For a given preference vector $\mathbf{v}$, the expected revenue $R(S, \mathbf{v}, \mathbf{x})$ of an assortment S under an advertising strategy $\mathbf{x}$ is calculated as

$$R(S, \mathbf{v}, \mathbf{x}) = \frac{\sum_{i \in S} r_i e^{q_i + f(c_i x_i)}}{1 + \sum_{i \in S} e^{q_i + f(c_i x_i)}} = \frac{\sum_{i \in S} r_i v_i g(c_i x_i)}{1 + \sum_{i \in S} v_i g(c_i x_i)}, \tag{3}$$

where $g(\cdot) = e^{f(\cdot)}$. Hence, $R(S, \mathbf{v}) = R(S, \mathbf{v}, 0)$.

We next formally introduce the JAAOP.

Definition 1. *Let $\mathcal{X} = \{\mathbf{x} | \sum_{i=1}^{n} x_i \le B\}$ denote the set of all feasible advertising strategies subject to the advertising budget B. JAAOP aims to jointly find an assortment S of size at most K and a feasible advertising strategy $\mathbf{x} \in \mathcal{X}$ to maximize the expected revenue, that is,*

$$\max_{\mathbf{x} \in \mathcal{X}} \max_{S : |S| \le K} R(S, \mathbf{v}, \mathbf{x}). \tag{4}$$

Let S^* and $\mathbf{x}^*$ denote the optimal assortment and advertising strategy, respectively, subject to the advertising budget B and cardinality constraint K. In case of multiple optimal assortments, we select the one with the smallest number of items. For ease of presentation, let $S_{\mathbf{v}}$ denote the optimal assortment when $B = 0$, that is, $S_{\mathbf{v}} = \arg\max_{S : |S| \le K} R(S, \mathbf{v})$. In this paper, we make two assumptions about $g(\cdot)$.

Assumption 1. $g(\cdot)$ *is differentiable, concave, and monotonically increasing.*

We will now provide the reasoning behind this assumption. Several researchers have investigated the impact of advertising on customer utility, including [10,25], and [28]. These studies all assumed logarithmic response functions, which imply that market share is a concave function of advertising efforts, meaning that the benefit of incremental advertising decreases as advertising efforts increase. This property, also known as the law of diminishing returns, has been widely used in other works [3,9,21]. The assumption we made in our study, known as Assumption 1, captures this property effectively. For the market share of product i in assortment S, that is, $\frac{v_i g(c_i x_i)}{1 + \sum_{i \in S} v_i g(c_i x_i)}$, we can verify the concavity of the market share function = by observing the negativity of the second derivative. The second assumption states that the advertising effect is zero if a product receives zero amount of advertising efforts from the platform.

Assumption 2. $g(0) = 1$.

We present a useful lemma that states that there exists an optimal advertising strategy that always uses the entire advertising budget.

Lemma 1. *There exists an optimal advertising strategy $\mathbf{x}^*$ for problem (4) such that $\sum_{i \in S^*} x_i = B$.*

4   Unconstrained JAAOP

We start by examining a special case of the JAAOP, where $K = n$, meaning there is no limit on the assortment size.

In the absence of any size constraints and advertising budget, our problem becomes the standard unconstrained assortment optimization problem. As proven by [26], the optimal assortment in this scenario is a revenue-ordered assortment, i.e. all products generating revenue greater than a certain threshold are included. This threshold, as demonstrated in [24], is the expected optimal revenue.

Lemma 2 *[24, Theorem 3.2].   If $K = n$ and $B = 0$, there exists an optimal assortment $S_\mathbf{v}$ such that $S_\mathbf{v} = \{i \in \mathcal{N} | r_i > R(S_\mathbf{v}, \mathbf{v})\}$.*

This characteristic has been noted in other contexts as well, such as the joint pricing and assortment optimization problem [27] and the robust assortment optimization problem [24]. The optimal assortment, given a fixed advertising strategy, remains revenue-ordered. Thus, to find the best solution, we find the optimal advertising strategy for each possible revenue-ordered assortment, and choose the one with the highest expected revenue as the final result. The number of possible revenue-ordered assortments is at most n. The efficiency of this algorithm can be improved by taking into consideration the following observations.

Lemma 3. *There exists an optimal assortment S^* such that $S^* \subseteq S_\mathbf{v}$.*

Lemma 3 implies that to find the optimal advertising strategy, we must evaluate all the revenue-ordered assortments within $S_\mathbf{v}$ and determine the optimal advertising plan. Then, for any revenue-ordered assortment S, we find the optimal advertising strategy to obtain the complete solution, i.e.,

$$\max_{\mathbf{x} \geq 0} \quad \frac{\sum_{i \in S} r_i v_i g(c_i x_i)}{1 + \sum_{i \in S} v_i g(c_i x_i)} \tag{5}$$
$$\text{s.t.} \quad \sum_{i \in S} x_i = B.$$

With $u_i = v_i g(c_i x_i)$, (5) can be rewritten as:

$$\max_{\mathbf{u}} \quad \frac{A(\mathbf{u})}{B(\mathbf{u})} = \frac{\sum_{i \in S} r_i u_i}{1 + \sum_{i \in S} u_i} \tag{6}$$
$$\text{s.t.} \quad \mathbf{u} \in \mathcal{U},$$

where $\mathcal{U} = \{\mathbf{u} | \sum_{i=1}^{n} m_i(u_i) \leq B, u_i \geq v_i, i = 1, \dots, n\}$ and $m_i(\cdot) = g^{-1}(\frac{\cdot}{v_i})/c_i$. Here (6) is a single-ratio fractional programming (FP) problem. Before presenting our solution to (6), we show that $\mathcal{U}$ is a convex set.

Lemma 4. *The constraint set $\mathcal{U}$ in (6) is a convex set.*

This part describes our solution to (6) in detail. Lemma 4 indicates that (6) is a concave-convex FP problem. We apply the classical Dinkelbach transform [8] by iteratively solving the following parameterized problem:

$$\max_{\mathbf{u}} \quad h(y) = A(\mathbf{u}) - yB(\mathbf{u}) \tag{7}$$

$$\text{s.t.} \quad \mathbf{u} \in \mathcal{U}.$$

In particular, our algorithm starts with iteration $t = 0$ and $y_0 = \frac{A(\mathbf{v})}{B(\mathbf{v})}$, and in each subsequent iteration $t + 1$, we find $\mathbf{u}_{t+1}$ to maximize $h(y_t)$ by solving (7) and update $y_{t+1} = \frac{A(\mathbf{u}_{t+1})}{B(\mathbf{u}_{t+1})}$. This process iterates until the optimal solution of (7) is 0 and we output the corresponding maximizer $\mathbf{u}^F$. Equation (7) can be solved efficiently because $A(\mathbf{u})$ is a concave function and $B(\mathbf{u})$ is a convex function. This algorithm is guaranteed to converge to the optimal solution [8]. After solving (6) and obtaining $\mathbf{u}^F$, we transform (6) to an optimal advertising strategy such that for each $i \in S$, we set $x_i = m_i(u_i^F)$; that is, we allocate $m_i(u_i^F)$ efforts to i. A detailed description of our solution is presented in Algorithm 1.

Algorithm 1. Optimal Solution for Unconstrained JAAOP

Input: preference weight $\mathbf{v}$, revenue $\mathbf{r}$, advertising effectiveness $\mathbf{c}$, budget B
Output: optimal assortment S^*, advertising strategy $\mathbf{x}^*$
1: Solve the classic unconstrained assortment optimization problem, and obtain the optimal assortment $S_{\mathbf{v}}$ when $B = 0$
2: Solve (6) for each revenue-ordered assortment in $S_{\mathbf{v}}$, and return the best one as the final solution

---

5   Cardinality-Constrained JAAOP

We next study our problem under a cardinality constraint of $K > 0$. First, we examine a scenario where $g(\cdot)$ is a linear function, and then we delve into the general case where $g(\cdot)$ is a concave function.

5.1 $g(x)$ as Linear Function

We first study the scenario where $g(\cdot)$ is a linear function, expressed as $1 + ax$ for some $a \geq 0$. The next lemma demonstrates the existence of an optimal advertising strategy that allocates the entire budget to a single product. For each $i \in \mathcal{N}$, we define $\mathbf{x}_i$ as a vector in which the i-th element is B and all other elements are zero.

Lemma 5. *For any assortment S, the optimal solution for the following problem is achieved at $\mathbf{x}_i$ for some $i \in S$:*

$$\max_{\mathbf{x} \geq 0} \quad L(S, \mathbf{x}) = \frac{\sum_{i \in S} r_i v_i (1 + ac_i x_i)}{1 + \sum_{i \in S} v_i (1 + ac_i x_i)} \tag{8}$$

$$\text{s.t.} \quad \sum_{i \in S} x_i = B.$$

This lemma implies that to find the optimal advertising strategy, we need to consider at most n candidate advertising strategies: $\{\mathbf{x}_i | i \in \mathcal{N}\}$. Specifically, considering $\mathbf{x}_i$, we replace the original preference weight v_i of i using $v_i g(c_i B)$ and then solve the standard capacity-constrained assortment optimization problem to obtain an optimal assortment. Among the n returned solutions, we return the best one as the final solution. [23] showed that the standard cardinality-constrained assortment optimization problem for each $\mathbf{x}_i$ can be solved in $O(n^2)$ time. Thus, the overall time complexity of our solution is $n \times O(n^2) = O(n^3)$. Assume all products are indexed in non-increasing order of r_i. The next lemma shows that we can further narrow the search space and reduce the time complexity to $O(n^2 T)$, where $T = \max\{i | i \in S_{\mathbf{v}}\}$ represents the index of the product that has the smallest revenue in $S_{\mathbf{v}}$.

Lemma 6. *Assume all products are indexed in non-increasing order of r_i. Let $T = \max\{i | i \in S_{\mathbf{v}}\}$, there exists an optimal assortment S^* such that $S^* \subseteq \{1, 2, ..., T\}$.*

We present the detailed implementation of our algorithm in Algorithm 2.

Algorithm 2. Optimal Cardinality Constrained Solution for Log Response Function

---

Input: preference weight $\mathbf{v}$, revenue $\mathbf{r}$, cardinality constraint K, advertising effectiveness $\mathbf{c}$, budget B

Output: optimal assortment S^* and advertising strategy $\mathbf{x}^*$

1: Let $T = \max\{i | i \in S_{\mathbf{v}}\}$
2: **for** $i = 1, \ldots, T$ **do**
3: Compute an assortment S_i that maximizes $R(S, \mathbf{v}, \mathbf{x}_i)$
4: **end for**
5: Return the best $(S_j, \mathbf{x}_j)$.

5.2 $g(x)$ as a General Concave Function

We next discuss the general case. Before presenting our solution, we first construct an example to demonstrate that allocating the entire budget to a single product is not necessarily optimal.

Example 1. Consider three products with revenue $\mathbf{r} = (8, 7.5, 2.8)$, preference weight $\mathbf{v} = (1.2, 1, 1.7)$ and the effectiveness $\mathbf{c} = (0.9, 0.8, 1)$. Assume the cardinality constraint is $K = 2$ and the total advertising budget is $B = 10$. We consider a concave function $g(x) = \sqrt{x} + 1$. If we are restricted to allocating the entire budget to a single product, then the optimal advertising strategy is $(10, 0, 0)$, the optimal assortment is composed of the first two products, and the expected revenue of this solution is 6.75. However, the actual optimal advertising strategy is (approximately) $(8.285, 1.715, 0)$, the actual optimal assortment contains the first two products, and it achieves expected revenue of 6.812. The above example shows that the single-product advertising strategy is no longer optimal for a general concave response function.

We next present our solution. Let $u_i = v_i g(c_i x_i)$ and define $m_i(\cdot) = g^{-1}(\frac{\cdot}{v_i})/c_i$ for each $i \in \mathcal{N}$, we first transform (4) to an equivalent nonlinear mixed integer program (9) by replacing $\sum_{i=1}^{n} x_i = B$ using $\sum_{i=1}^{n} m_i(u_i) \leq B$,

$$\max_{\mathbf{u} \in \mathcal{U}} \quad \max_{\mathbf{t} \in \{0,1\}^n} \frac{\sum_{j=1}^{n} u_j r_j t_j}{1 + \sum_{j=1}^{n} u_j t_j} \tag{9}$$

$$\text{s.t.} \quad \sum_{i=1}^{n} t_i \leq K,$$

where $\mathcal{U} = \{\mathbf{u} \mid \sum_{i=1}^{n} m_i(u_i) \leq B, u_i \geq v_i, i = 1, \ldots, n\}$. We next present a useful lemma from [6].

Lemma 7 *(Theorem 1 [6]).* *The inner problem of (9) is equivalent to the following linear program*

$$\max_{\mathbf{w}, w_0} \quad \sum_{j=1}^{n} r_i w_i \tag{10}$$

$$\text{s.t.} \quad \sum_{i=1}^{n} w_i + w_0 = 1, \tag{11}$$

$$\sum_{i=1}^{n} \frac{w_i}{u_i} \leq K w_0, \tag{12}$$

$$0 \leq \frac{w_i}{u_i} \leq w_0 \quad \forall i \in \mathcal{N}. \tag{13}$$

Notice that (12) and (13) involve some nonlinear constraints if $\mathbf{u}$ is not fixed. Thus we introduce new variables $\ell_i = \frac{w_i}{u_i}$, $i \in \mathcal{N}$ and rewrite (9) as follows:

$$(\text{NO}) \quad \max_{\mathbf{u} \in \mathcal{U}} \max_{\mathbf{w}, \ell, w_0} \quad \sum_{j=1}^{n} r_i w_i \tag{14}$$

$$\text{s.t.} \quad \sum_{i=1}^{n} w_i + w_0 = 1, \tag{15}$$

$$\sum_{i=1}^{n} \ell_i \leq K w_0, \tag{16}$$

$$0 \leq \ell_i \leq w_0 \quad \forall i \in \mathcal{N}, \tag{17}$$

$$w_i = \ell_i u_i \quad \forall i \in \mathcal{N}. \tag{18}$$

We further use the classic McCormick inequalities ([20]) to relax the nonconvex constraints (18):

$$
\begin{aligned}
\text{(MC)} \quad & w_i \geq \ell_i v_i && \forall i \in \mathcal{N}, \\
& w_i \geq u_i + \ell_i v_i g(Bc_i) - v_i g(Bc_i) && \forall i \in \mathcal{N}, \\
& w_i \leq \ell_i v_i g(Bc_i) && \forall i \in \mathcal{N}, \\
& w_i \leq u_i + \ell_i v_i - v_i && \forall i \in \mathcal{N}.
\end{aligned}
$$

Through the above relaxation, we transform (NO) into a convex optimization problem that can be solved efficiently. After solving this relaxed problem and obtaining a solution $\mathbf{w}$, we can compute the final assortment S as follows: we first find the product for which the w_i is strictly larger than 0, that is $S^w = \{i | i \in \mathcal{N}, w_i \neq 0\}$. Then we sort the products in S^w by the value of w_i and choose the first K products. Notice that the advertising strategy that is obtained from solving the previous relaxed problem may not be optimal. One can solve (6) to find the optimal advertising strategy for S. Lastly, if the size of the input is large, we can reduce the problem size by selecting a smaller group of candidate products based on Lemma 6. A detailed description of our solution is listed in Algorithm 3.

Algorithm 3. Cardinality Constrained Solution for General Response Function

---

Input: preference weight $\mathbf{v}$, revenue $\mathbf{r}$, capacity constraint K, advertising effectiveness $\mathbf{c}$, budget B
Output: assortment S and advertising strategy $\mathbf{x}$
1: Let $T = \max\{i | i \in S_{\mathbf{v}}\}$
2: Solve the optimization problem (NO + MC) for the first T products to find an assortment S
3: Solve problem (6) for S to find the advertising strategy $\mathbf{x}$
4: Return $(S, \mathbf{x})$.

---

6   Conclusion

This paper considers the JAAOP problem under the MNL model, where the seller decides their advertising strategy for all products to improve the current revenue. We consider both the log and general concave response functions. If there are no capacity constraints, we show that the optimal assortment is still revenue-ordered. However, this result does not hold in the presence of a cardinality constraint. When the response function is a log function, we prove that the optimal advertising strategy is a single-product advertising strategy, thus the optimal solution could be found in polynomial time. For the general concave response function, we develop an efficient algorithm to find a near-optimal solution. We further consider the seller could adjust the price simultaneously, and show that such a problem can be efficiently solvable under unconstrained setting or be transformed as a mixed-integer nonlinear programming for the cardinality constraint setting. Additionally, as an extension, we study the multi-stage MNL choice model, in which the customer browses the assortments sequentially. Our results demonstrate that the seller has no incentive to decrease the utility of any product, even under the capacity constraint. Finally, we conduct extensive experiments to illustrate that the advertising strategy is more effective with small cardinality constraint and large no-purchase utility.

Appendix

A   Joint Assortment, Pricing, and Advertising Optimization

In this section, we study the case when the price of each product is also a decision variable. Formally, we assume that the preference weight of each product $i \in \mathcal{N}$ can be represented as $e^{q_i + f(c_i x_i) - p_i}$, whose value is jointly decided by i's initial utility q_i, i's price p_i, and the advertising efforts x_i received from the platform. Hence, the revenue r_i of each product $i \in \mathcal{N}$ is $r_i = p_i - d_i$, where d_i is the production cost of i. Based on the above notations, we can represent the expected revenue $R(S, \mathbf{p}, \mathbf{x})$ of an assortment S as

$$R(S, \mathbf{p}, \mathbf{x}) = \frac{\sum_{i \in S}(p_i - d_i)e^{q_i + f(c_i x_i) - p_i}}{1 + \sum_{i \in S} e^{q_i + f(c_i x_i) - p_i}} = \frac{\sum_{i \in S}(p_i - d_i)e^{q_i - p_i}g(c_i x_i)}{1 + \sum_{i \in S} e^{q_i - p_i}g(c_i x_i)}. \quad (A.1)$$

A.1   Unconstrained Case

If there is no cardinality constraint, our goal is to solve the following joint advertising, pricing, and assortment optimization problem:

$$\max_{\mathbf{p}, \mathbf{x}, S} \quad R(S, \mathbf{p}, \mathbf{x}) \tag{A.2}$$

$$\text{s.t.} \quad \sum_{j \in S} x_j \leq B.$$

Before describing our solution, we first present a useful lemma from [16].

Lemma A.1 *[16]. Given any assortment S, the optimal price for each product $i \in S$ is $p_i = W(\sum_{i \in S} e^{q_i - d_i - 1})g(c_i x_i) + d_i + 1$, where $W(\cdot)$ is the Lambert W function; that is, $W(\cdot)$ is the value of x that satisfies $xe^x = z$. Moreover, the revenue of the optimal solution is $W(\sum_{i \in S} e^{q_i - d_i - 1}g(c_i x_i))$.*

For any given advertising strategy $\mathbf{x}$ and assortment S, the optimal price and corresponding expected revenue are explicitly given by Lemma A.1. Because

$W(\cdot)$ is an increasing function, Lemma A.1 implies that the optimal assortment must include all products. Hence, we can transform (A.2) into

$$\max_{\mathbf{x}} \quad \sum_{i=1}^{n} e^{q_i - d_i - 1} g(c_i x_i) \tag{A.3}$$

$$\text{s.t.} \quad \sum_{i=1}^{n} x_i \leq B,$$

$$x_i \geq 0 \quad \forall i \in \mathcal{N}.$$

Denote $\alpha_i = e^{q_i - d_i - 1}$, and rewrite the above problem as

$$\max_{\mathbf{x}} \quad \sum_{i=1}^{n} \alpha_i g(c_i x_i) \tag{A.4}$$

$$\text{s.t.} \quad \sum_{i=1}^{n} x_i \leq B,$$

$$x_i \geq 0 \quad \forall i \in \mathcal{N}.$$

Because $g(\cdot)$ is a concave function (Assumption 1) and $\sum_{i=1}^{n} x_i \leq B$ is a linear constraint, (A.4) is a concave maximization problem with convex constraints. Hence, (A.4) is a convex minimization problem over a convex set, and the problem has efficient solutions [5].

A special case where $g(\cdot)$ is a linear function: If $g(\cdot)$ is a linear function, that is, $g(x) = ax + 1$ for some $a \geq 0$, then the optimal advertising strategy is to allocate the entire advertising budget to a single product.

Lemma A.2. *When $g(\cdot)$ is a linear function, the optimal solution to (A.4) is to allocate the entire advertising budget to the product with the largest $\alpha_i c_i$.*

A.2   Cardinality-Constrained Case

We next consider a case where the size of the assortment is at most $K \geq 0$. Lemma A.1 indicates that the optimal assortment contains the top K products that have the largest $e^{q_i + f(c_i x_i) - d_i - 1}$. We next show that if $e^{f(\cdot)}$ is a linear function, then we only need to consider two possible advertising strategies. Hence, this problem can be solved efficiently.

Lemma A.3. *Let $\alpha_i = e^{q_i - d_i - 1}$ and $g(x) = ax + 1$ for some $a \geq 0$. Assume all products are indexed in non-increasing order of α_i. Let $t_1 = \text{argmax}_{i \in \{1, \cdots, K\}} \{\alpha_i c_i\}$ and $t_2 = \text{argmax}_{j \in \{K+1, \ldots, n\}} \{\alpha_j (ac_j B + 1)\}$, then the optimal advertising strategy is $\mathbf{x}_{t_1}$ or $\mathbf{x}_{t_2}$.*

We next discuss a case with a general response function. For each product $i \in \mathcal{N}$, let $t_j = 1$ if a product j is offered in the assortment and let $t_j = 0$ otherwise. Our problem can be formulated as the following mixed-integer programming problem:

$$\max_{\mathbf{x},\mathbf{t}} \quad \sum_{i=1}^{n} \alpha_i t_i g(c_i x_i) \tag{A.5}$$

$$\text{s.t.} \quad \sum_{i=1}^{n} x_i \leq B,$$

$$\sum_{i=1}^{n} t_i \leq K,$$

$$x_i \geq 0 \quad \forall i \in \mathcal{N},$$

$$t_i \in \{0, 1\} \quad \forall i \in \mathcal{N}.$$

If all products have the same advertising effectiveness, that is, $c_i = c$, for all $i \in \mathcal{N}$, the optimal assortment is to select the top K products that have the largest α_i.

Lemma A.4. *Assume all products are indexed in non-increasing order of α_i and $c_i = c$ for all $i \in \mathcal{N}$. The optimal assortment is $S^* = \{1, \ldots, K\}$, and the optimal advertising strategy $\mathbf{x}^*$ satisfies $x_i^* \geq x_j^* \ \forall i \leq j$.*

To find the optimal advertising strategy under $S^* = \{1, \ldots, K\}$, we need to solve an optimization problem that is similar to (A.4). Because this problem is a concave maximization problem with convex constraints, it can be solved efficiently.

For a general case, where advertising effectiveness is heterogeneous, the objective function of (A.5) contains the bilinear terms $t_i g(c_i x_i)$. We linearize each of these terms by relaxation. Specifically, for each $t_i e^{f(c_i x_i)}$, we introduce a new continuous variable $w_i = t_i g(c_i x_i)$ and add the inequalities: $g(c_i x_i) - w_i \leq g(c_i B)(1 - t_i)$, $0 \leq w_i \leq g(c_i x_i)$, and $w_i \leq g(c_i B) t_i$. This leads to the following mixed-integer nonlinear programming problem:

$$\max_{\mathbf{x},\mathbf{t},\mathbf{w}} \quad \sum_{i=1}^{n} \alpha_i w_i \tag{A.6}$$

$$\text{s.t.} \quad \sum_{i=1}^{n} x_i \leq B,$$

$$\sum_{i=1}^{n} t_i \leq K,$$

$$g(c_i x_i) - w_i \leq g(c_i B)(1 - t_i) \quad \forall i \in \mathcal{N},$$

$$0 \leq w_i \leq g(c_i x_i) \quad \forall i \in \mathcal{N},$$

$$w_i \leq g(c_i B) t_i \quad \forall i \in \mathcal{N},$$

$$x_i \geq 0 \quad \forall i \in \mathcal{N},$$

$$t_i \in \{0, 1\} \quad \forall i \in \mathcal{N}.$$

B   Sequential Joint Advertising and Assortment Optimization

In this section, we extend our study to consider a sequential joint advertising and assortment problem. The model put forward by [15] examines the behavior of consumers who may visit multiple product assortments before making a purchase or leaving the store. The consumer is assumed to progress through a sequence of m stages, each featuring a different assortment ($\mathcal{S} = (S_1, \ldots, S_m)$). If the consumer chooses to buy a product in stage i, they will leave the store, but if they do not make a purchase, they will proceed to the next stage. If no product is selected after visiting all m assortments, the consumer exits the store without making a purchase. This choice model is referred to as the sequential multinomial logit (SMNL) choice model. For the purpose of simplicity, we assume that the consumer will continue visiting subsequent assortments if they do not make a purchase in the current stage. However, it should be noted that this assumption can be relaxed to include the factor of consumer patience.

We will now provide a detailed explanation of the SMNL model. Given a sequence of assortments $\mathcal{S}$, the consumer will purchase product i in stage k with a probability of

$$\phi_i^k(\mathcal{S}) = \frac{v_i}{(1 + \sum_{\ell=1}^{k-1} V(S_\ell))(1 + \sum_{\ell=1}^{k} V(S_\ell))}.$$

Let $V(S) = \sum_{i \in S} v_i$ and $W(S) = \sum_{i \in S} r_i v_i$. The expected revenue is represented as

$$R(\mathcal{S}) = \sum_{k=1}^{m} \frac{W(S_k)}{(1 + \sum_{\ell=1}^{k-1} V(S_\ell))(1 + \sum_{\ell=1}^{k} V(S_\ell))}.$$

Under the advertising strategy $\mathbf{x}$, the expected revenue increases to

$$R(\mathcal{S}, \mathbf{x}) = \sum_{k=1}^{m} \frac{W(S_k, \mathbf{x})}{(1 + \sum_{\ell=1}^{k-1} V(S_\ell, \mathbf{x}))(1 + \sum_{\ell=1}^{k} V(S_\ell, \mathbf{x}))}, \tag{B.1}$$

where $V(S, \mathbf{x}) = \sum_{i \in S} v_i g(c_i x_i)$ and $W(S, \mathbf{x}) = \sum_{i \in S} r_i v_i g(c_i x_i)$.

Based on the transformation in (5), the optimization problem can be written as

$$\max_{\mathbf{u}, \mathcal{S}} \quad \sum_{k=1}^{m} \frac{\sum_{i \in S_k} r_i u_i}{(1 + \sum_{l=1}^{k-1} \sum_{i \in S_l} u_i)(1 + \sum_{l=1}^{k} \sum_{i \in S_l} u_i)} \tag{B.2}$$
$$\text{s.t.} \quad \mathbf{u} \in \mathcal{U}.$$

We focus on the unconstrained setting. Given an arbitrary advertising strategy, [15] demonstrated that the optimal assortments are sequential revenue-ordered assortments. Specifically, there exists a set of decreasing thresholds $\{t_1^*, t_2^*, \ldots, t_{m+1}^*\}$, such that $S_k^* = \{i \in \mathcal{N} : t_{k+1}^* \leq r_i < t_k^*\}$ for $k \in \mathcal{M} = [1, 2, \ldots, m]$. The values of $\{t_i^*\}_{i=1}^{m+1}$ are given in the following lemma.

Lemma B.1 *[15, Theorem 3.1]. There exists an optimal solution $(S_1^*, \ldots, S_m^*)$ such that for $i \in S_k^*$, we have $t_{k+1}^* \leq r_i < t_k^*$. Let $R_k(S_1^*, \ldots, S_m^*) = \dfrac{W(S_k^*)}{(1+\sum_{\ell=1}^{k-1} V(S_\ell^*))(1+\sum_{\ell=1}^{k} V(S_\ell^*))}$. The value of t_k^* can be chosen as follows:*

$$t_1^* = +\infty, \quad t_k^* = \frac{R_{k-1}(S_1^*, \ldots, S_m^*) + R_k(S_1^*, \ldots, S_m^*)}{\frac{1}{1+\sum_{\ell=1}^{k-2} V(S_\ell^*)} - \frac{1}{1+\sum_{\ell=1}^{k} V(S_\ell^*)}} \quad \forall k \in \mathcal{M}\backslash\{1\}, \quad t_{m+1}^* = \frac{R_m(S_1^*, \ldots, S_m^*)}{\frac{1}{1+\sum_{\ell=1}^{m-1} V(S_\ell^*)}}.$$

Based on this lemma, we analyze the structure of the optimal assortments and the advertising strategy. We denote the optimal solution of B.2 as $\mathbf{u}^*$ and $\mathcal{S}^*$.

Lemma B.2. *For the optimization problem* (B.2), *we have* $\dfrac{\partial R(\mathcal{S}^*, \mathbf{u}^*)}{\partial u_i^*} \geq \dfrac{\partial R(\mathcal{S}^*, \mathbf{u}^*)}{\partial u_j^*} \geq 0$ *for all products $i, j \in \mathcal{N}$ and $i < j$.*

[4] showed that in the MNL choice model, the partial derivative $h_i^1 \geq 0$, indicating that the seller has no incentive to reduce the utilities of products in order to maximize their expected revenue. In Lemma B.2, we extend this result to the SMNL choice model. Moreover, due to the sequential revenue-ordered property stated in Lemma B.1 being maintained for any feasible set of products, this result remains valid even under capacity constraints, meaning that the seller has no incentive to decrease product utilities in the capacity-constrained scenario either. If the seller has the ability to enhance product utilities, the optimal advertising strategy would be to allocate the entire budget to the product that generates the highest revenue.

Lemma B.3. *Denote the optimal solution of the following optimization problem as $(\mathbf{x}^*, \mathcal{S}^*)$. $x_1^* = B$ and $x_i^* = 0$ for all $i \in \mathcal{N} \backslash \{1\}$.*

$$\max_{\mathbf{x}, \mathcal{S}} \sum_{k=1}^{m} \frac{\sum_{i \in S_k} r_i(v_i + x_i)}{(1 + \sum_{l=1}^{k-1} \sum_{i \in S_l}(v_i + x_i))(1 + \sum_{l=1}^{k} \sum_{i \in S_l}(v_i + x_i))} \tag{B.3}$$

$$\text{s.t.} \quad \sum_{i=1}^{n} x_i \leq B$$

In our setting, the allocation of budget x_i to product i increases its utility to $v_i e^{f(c_i x_i)}$, where f is the nonlinear response function. Due to the heterogeneous advertising effectiveness, utility and nonlinear response function, the optimal advertising strategy may be more complex than a single-product advertising strategy. Given a specific sequence of assortments, finding the optimal advertising strategy is equivalent to solving the following optimization problem.

$$\max_{\mathbf{u}} \sum_{k=1}^{m} \frac{\sum_{i \in S_k} r_i u_i}{(1 + \sum_{l=1}^{k-1} \sum_{i \in S_l} u_i)(1 + \sum_{l=1}^{k} \sum_{i \in S_l} u_i)} \tag{B.4}$$

$$\text{s.t.} \quad \mathbf{u} \in \mathcal{U}$$

where $\mathcal{U} = \{\mathbf{u} | \sum_{i=1}^{n} m_i(u_i) \leq B, u_i \geq v_i, i = 1, \ldots, n\}$ and $m_i(\cdot) = g^{-1}(\frac{\cdot}{v_i})/c_i$. When $m = 1$, this problem is a single-ratio FP problem, which can be solved efficiently. However, the sum-of-ratio problem is generally NP-complete [14]. Hence, even though the optimal assortments may be sequential revenue-ordered assortments, finding the optimal advertising strategy may not be straightforward. As a result, we propose a heuristic method as an alternative approach.

B.1   Heuristic Method

The design of our heuristic method (listed in Algorithm 4) is based on two key observations. Firstly, given an advertising strategy, the optimal sequence of assortments can be found efficiently in polynomial time. Secondly, given the set of products to be displayed, the single-stage optimal advertising strategy is computationally tractable. Specifically, Algorithm 4 iteratively updates the assortments and advertising strategy until the expected revenue cannot be improved any further.

Algorithm 4. Heuristic for Unconstrained Multi-stage JAAOP

Input: preference weight $\mathbf{v}$, revenue $\mathbf{r}$, advertising effectiveness $\mathbf{c}$, budget B
Output: approximate assortment $\mathcal{S}^*$, advertising strategy $\mathbf{x}^*$
1: $i = 0, rev^0 = 0$
2: Implement Algorithm 1 and obtain the advertising strategy $\mathbf{x}^0$
3: **repeat**
4:     $i = i + 1$
5:     Find the optimal sequence of assortments $\mathcal{S}^i$ and expected revenue rev^i based
       on the current advertising strategy $\mathbf{x}^{i-1}$
6:     Find the optimal advertising strategy for $S^i = \cup_{j=1}^{m} S_j$, denoted as $\mathbf{x}^i$
7: **until** $rev^i < rev^{i-1}$

By exploring the structure of the objective function in (B.2), we next show that our heuristic method achieves an approximation ratio of 50%.

Lemma B.4. *Let $(\mathcal{S}^*, \mathbf{x}^*), (\mathcal{S}^h, \mathbf{x}^h)$ be the optimal values of (B.2) and our heuristic method. We have $R(\mathcal{S}^h, \mathbf{x}^h) \geq \frac{1}{2} R(\mathcal{S}^*, \mathbf{x}^*)$.*

C   Numerical Study

In this section, we explore the effect of advertising on assortment optimization and validate the superiority of our algorithms compared with several heuristic methods on randomly generated instances and different response functions. The revenue of each product is drawn uniformly from the interval $[1, 10]$. For the preference weight v_i of product i, we first sample γ_i uniformly from the interval $[1, 10]$ and then assign $v_i = \gamma_i/\Delta$, where $\Delta = P_0 \sum_{i \in \mathcal{N}} \gamma_i/(1 - P_0)$. In this case, we guarantee the no-purchase probability when providing all products is exactly P_0. We consider three types of response functions: $g_1(x) = \sqrt{x} + 1$, $g_2(x) = \log(x+1) + 1$, $g_3(x) = 2 - e^{-x}$. For advertising effectiveness, we consider the following settings.

- Setting A: The advertising effectiveness c_i of each product $i \in \mathcal{N}$ is drawn uniformly from the interval $[0, 1]$.
- Setting B: The advertising effectiveness c_i of each product $i \in \mathcal{N}$ is drawn independently from a standard log-normal distribution and rescaled by a factor of $\frac{1}{2\sqrt{e}}$ to make sure the same mean as setting A. In this case, there is more dispersion in advertising effectiveness.

We choose the number of products from $\{50, 100, 200\}$, the cardinality constraint K from $\{5, 10, 20\}$, and the value of P_0 from $\{0.1, 0.3\}$. For the multi-stage problem, the stage m is chosen from $\{3, 5, 8\}$. For each setting, we randomly generate 10 instances and calculate the average percentage of improvement over the non-advertising strategy. Finally, we denote our heuristic algorithm as HA.

C.1   Compared Heuristics

For the cardinality-constrained single-stage problem, the main challenge lies in finding the optimal advertising strategy as the optimal assortment for a given advertising strategy can be found efficiently in polynomial time. In order to tackle this difficulty, we propose two practical advertising strategies.

- Uniform advertising (UA) strategy: for any assortment S, we have $x_i = B/|S|$ if $i \in S$.
- Revenue advertising (RA) strategy: for any assortment S, we have $x_i = B \cdot \frac{r_i}{\sum_{i \in S} r_i}$ if $i \in S$.

We start with the optimal assortment with no advertising strategy S^1. After allocating the budget according to the heuristic method, we recompute the optimal assortment S^2; if $S^1 \neq S^2$, then we reallocate the budget and compute the new assortment. This process continues until the assortment is unchanged with advertising (Table 2).

C.2   Performance Evaluation

Table 1 presents the average performance of three heuristic algorithms for the single-stage joint advertising and assortment problem, evaluated over 36 different parameter settings. Algorithm 3 demonstrates superior performance compared to the other heuristic algorithms, particularly when P_0 is large and the cardinality constraint is small. In most cases, the RA strategy performs slightly better than the UA strategy. The performance of each heuristic algorithm does not vary significantly with an increase in the dispersion of advertising effectiveness. When the set of products is less attractive and the cardinality constraint is small, advertising has a more significant impact, and the gap between our algorithm and the compared heuristic algorithms is even larger.

Table 1. Average Performance of Tested Heuristic Algorithm on Single Stage Problem

Parameters				$g_1(x)$			$g_2(x)$			$g_3(x)$		
Setting	n	K	P_0	HA	UA	RA	HA	UA	RA	HA	UA	RA
A	50.0	5.0	0.1	27.08	25.82	25.88	21.38	19.47	19.53	18.5	17.74	17.78
	50.0	10.0	0.1	14.45	13.27	13.34	10.56	8.49	8.55	9.56	8.19	8.24
	50.0	20.0	0.1	9.76	7.95	8.11	6.86	4.14	4.29	6.03	4.08	4.22
	50.0	5.0	0.3	57.26	53.02	53.26	43.78	38.05	38.29	36.98	34.1	34.18
	50.0	10.0	0.3	31.94	29.85	29.94	22.75	18.14	18.2	20.26	17.45	17.5
	50.0	20.0	0.3	19.48	17.28	17.57	12.88	8.69	8.94	11.65	8.57	8.81
	100.0	5.0	0.3	75.21	68.19	68.19	57.29	48.65	48.64	48.09	43.32	43.3
	100.0	10.0	0.3	44.16	42.09	42.14	30.21	26.13	26.17	27.55	25.0	25.03
	100.0	20.0	0.3	25.23	23.88	24.01	15.62	11.89	12.01	14.57	11.73	11.84
	100.0	5.0	0.1	41.13	39.28	39.32	32.01	29.57	29.59	27.79	26.66	26.66
	100.0	10.0	0.1	21.69	20.87	20.93	15.04	13.68	13.73	14.26	13.16	13.2
	100.0	20.0	0.1	11.2	10.72	10.8	7.48	5.63	5.7	6.94	5.56	5.62
	200.0	5.0	0.1	60.48	52.98	53.04	46.5	38.29	38.35	38.54	34.33	34.38
	200.0	10.0	0.1	32.24	30.17	30.22	22.58	19.1	19.14	20.42	18.3	18.34
	200.0	20.0	0.1	16.55	16.13	16.2	10.77	8.39	8.44	9.7	8.28	8.32
	200.0	5.0	0.3	90.46	77.76	77.78	67.21	54.17	54.2	57.44	47.98	48.03
	200.0	10.0	0.3	54.22	50.86	50.85	37.26	30.78	30.75	33.49	29.34	29.32
	200.0	20.0	0.3	32.7	31.11	31.19	19.69	15.3	15.37	18.44	15.07	15.13
B	50.0	5.0	0.1	28.7	25.83	25.96	22.67	19.26	19.38	18.77	16.83	16.84
	50.0	10.0	0.1	15.59	13.89	14.02	11.37	8.94	9.06	9.72	8.29	8.4
	50.0	20.0	0.1	9.3	7.42	7.6	6.35	3.6	3.74	5.51	3.51	3.64
	50.0	5.0	0.3	61.14	53.73	54.03	45.3	38.11	38.34	35.77	31.92	31.85
	50.0	10.0	0.3	32.77	29.65	29.83	22.76	17.71	17.85	18.64	16.4	16.44
	50.0	20.0	0.3	20.43	17.14	17.6	13.86	8.73	9.09	11.69	8.43	8.76
	100.0	5.0	0.1	43.47	36.75	36.84	33.61	26.12	26.2	25.99	22.66	22.66
	100.0	10.0	0.1	20.68	19.2	19.29	14.48	11.72	11.82	12.75	11.13	11.21
	100.0	20.0	0.1	11.96	10.65	10.75	8.4	5.64	5.7	7.3	5.47	5.52
	100.0	5.0	0.3	76.08	63.21	63.22	56.17	43.04	43.05	43.28	37.37	37.41
	100.0	10.0	0.3	43.16	38.01	38.12	31.0	22.11	22.22	25.18	20.68	20.8
	100.0	20.0	0.3	25.42	22.46	22.62	17.15	10.94	11.1	14.24	10.52	10.66
	200.0	5.0	0.1	67.19	52.12	52.16	52.04	37.22	37.27	38.03	32.2	32.23
	200.0	10.0	0.1	31.58	28.81	28.84	20.74	17.75	17.78	18.75	16.72	16.76
	200.0	20.0	0.1	17.15	15.66	15.72	11.61	8.28	8.33	9.91	7.92	7.98
	200.0	5.0	0.3	95.94	67.3	67.19	68.48	43.43	43.3	51.34	38.86	38.79
	200.0	10.0	0.3	56.04	47.21	47.27	37.58	27.28	27.35	31.49	25.82	25.86
	200.0	20.0	0.3	32.8	28.57	28.73	21.97	13.85	14.02	18.11	13.11	13.22

Table 2. Average Performance of Tested Heuristic Algorithm on Multiple Stage Problem

Parameters				$g_1(x)$			$g_2(x)$			$g_3(x)$		
Setting	n	m	P_0	HA	UA	RA	HA	UA	RA	HA	UA	RA
A	50.0	3.0	0.1	7.32	5.31	5.94	4.99	2.03	2.57	4.48	2.02	2.55
	50.0	5.0	0.1	6.83	4.85	5.58	4.61	1.75	2.34	4.14	1.74	2.33
	50.0	8.0	0.1	6.69	4.75	5.51	4.49	1.68	2.3	4.04	1.68	2.29
	50.0	3.0	0.3	14.48	10.83	12.21	8.88	3.81	4.86	8.14	3.8	4.84
	50.0	5.0	0.3	14.55	10.92	12.35	8.92	3.79	4.9	8.18	3.78	4.87
	50.0	8.0	0.3	14.48	10.96	12.37	8.86	3.81	4.89	8.14	3.8	4.87
	100.0	3.0	0.1	5.67	3.95	4.42	3.43	1.12	1.45	3.22	1.12	1.45
	100.0	5.0	0.1	5.42	3.65	4.25	3.29	0.98	1.37	3.07	0.98	1.36
	100.0	8.0	0.1	5.37	3.62	4.23	3.24	0.96	1.36	3.04	0.96	1.35
	100.0	3.0	0.3	11.32	8.17	9.31	6.09	2.12	2.83	5.73	2.11	2.82
	100.0	5.0	0.3	11.18	8.17	9.31	5.99	2.11	2.82	5.64	2.11	2.82
	100.0	8.0	0.3	11.17	8.2	9.33	5.98	2.12	2.82	5.63	2.12	2.82
	200.0	3.0	0.1	4.55	2.96	3.33	2.54	0.58	0.77	2.34	0.58	0.77
	200.0	5.0	0.1	4.1	2.56	3.02	2.24	0.46	0.66	2.07	0.46	0.65
	200.0	8.0	0.1	4.01	2.55	2.99	2.18	0.46	0.65	2.02	0.46	0.65
	200.0	3.0	0.3	8.27	5.72	6.55	3.87	1.0	1.35	3.66	1.0	1.35
	200.0	5.0	0.3	8.5	5.96	6.8	4.07	1.07	1.43	3.86	1.07	1.43
	200.0	8.0	0.3	6.87	5.19	6.08	4.7	0.92	1.45	3.51	0.97	1.05
B	50.0	3.0	0.1	9.03	5.65	6.3	6.4	2.32	3.0	5.0	2.23	2.84
	50.0	5.0	0.1	6.22	4.12	4.88	3.85	1.24	1.74	3.49	1.24	1.73
	50.0	8.0	0.1	6.07	4.2	4.91	3.97	1.37	1.89	3.61	1.36	1.87
	50.0	3.0	0.3	14.16	9.91	11.33	9.04	3.27	4.29	8.2	3.24	4.25
	50.0	5.0	0.3	14.6	10.14	11.71	9.17	3.2	4.36	8.08	3.19	4.33
	50.0	8.0	0.3	13.93	10.0	11.38	8.74	3.28	4.26	7.73	3.26	4.22
	100.0	3.0	0.1	6.36	4.05	4.54	4.21	1.16	1.53	3.58	1.16	1.52
	100.0	5.0	0.1	5.43	3.13	3.63	3.54	0.78	1.07	2.73	0.78	1.06
	100.0	8.0	0.1	4.88	3.25	3.75	2.97	0.82	1.11	2.73	0.82	1.11
	100.0	3.0	0.3	12.09	7.96	9.2	7.17	2.09	2.86	6.16	2.09	2.84
	100.0	5.0	0.3	10.48	7.28	8.2	6.18	1.91	2.46	5.34	1.9	2.44
	100.0	8.0	0.3	10.77	7.32	8.24	6.52	1.96	2.53	5.38	1.94	2.49
	200.0	3.0	0.1	4.02	2.7	3.01	2.16	0.53	0.67	1.99	0.53	0.67
	200.0	5.0	0.1	3.93	2.31	2.7	2.31	0.42	0.59	1.9	0.42	0.59
	200.0	8.0	0.1	4.54	2.53	2.99	2.99	0.5	0.72	2.39	0.5	0.72
	200.0	3.0	0.3	8.04	5.4	6.14	4.44	1.0	1.32	3.89	1.0	1.31
	200.0	5.0	0.3	8.33	5.54	6.33	4.61	1.02	1.37	4.11	1.02	1.37
	200.0	8.0	0.3	7.94	5.31	6.05	4.37	0.94	1.25	3.71	0.94	1.25

The multi-stage setting has a decreasing impact on advertising as the seller is given more stages. Even without a cardinality constraint, the revenue improvement can still be significant when the utility of the no-purchase option is relatively high. The improvement under the UA strategy can be less than 0.5%, while the improvement using the heuristic method is at least 2%. This shows the importance of advertising strategy on expected revenue (Fig. 2).

Finally, we evaluate the computational efficiency of our Algorithm 3. Table 3 shows its average running time for different parameters. Our algorithm has a low computational complexity as it only requires solving two linear programming problems to find the optimal assortment and a few convex optimization problems to find the corresponding advertising strategy. The results in Table 3 demonstrate that our algorithm has a running time of less than 2 s for all cases, making it highly efficient.

Table 3. Average Running Time of Algorithm 3

Parameters			$g_1(x)$		$g_2(x)$		$g_3(x)$	
n	K	P_0	A	B	A	B	A	B
50	5	0.1	0.076	0.065	1.499	1.533	1.131	1.1
50	10	0.1	0.074	0.075	1.543	1.452	1.454	1.445
50	20	0.1	0.072	0.129	1.342	0.74	1.651	1.472
50	5	0.3	0.111	0.136	0.625	0.562	0.986	0.778
50	10	0.3	0.131	0.122	0.57	0.57	0.677	0.671
50	20	0.3	0.127	0.154	0.764	0.721	0.96	0.87
100	5	0.1	0.09	0.142	1.042	1.129	1.046	0.916
100	10	0.1	0.128	0.115	0.83	0.775	1.28	0.768
100	20	0.1	0.165	0.131	0.749	0.818	1.305	1.105
100	5	0.3	0.167	0.193	0.615	0.744	0.935	0.739
100	10	0.3	0.196	0.197	0.749	0.742	0.751	0.744
100	20	0.3	0.163	0.174	0.826	0.827	0.932	0.851
200	5	0.1	0.219	0.166	0.764	1.06	1.593	1.474
200	10	0.1	0.169	0.226	0.895	0.838	1.179	1.033
200	20	0.1	0.165	0.166	1.015	1.136	0.807	0.948
200	5	0.3	0.368	0.289	0.962	1.086	1.039	0.986
200	10	0.3	0.289	0.238	1.054	1.106	0.967	1.019
200	20	0.3	0.215	0.235	1.084	1.059	1.141	1.125

C.3   Effect of Budget on Expected Revenue

In practicality, the seller must also decide on the advertising budget. Since the return per budget investment can reduce with increasing budget, this subsection examines the relationship between expected revenue and invested budget. The experiment has 100 products and the budget is varied from 0 to 50 while the revenue and advertising effectiveness are kept constant. 100 preference weights are sampled for each budget and response function. The results of the expected revenue for each of these settings are displayed in Fig. 1.

The expected revenue is shown to increase with an increase in advertising budget as illustrated by Fig. 1. For the first response function g_1, when the budget is adequate and $P_0 = 0.1$, the difference in revenue between the different cardinality constraints becomes small, as indicated by Fig. 1(a). Hence, when the seller has an adequate budget, limiting their focus to a small group of products does not result in a significant reduction in revenue. For the same response function, the trend of increasing expected revenue remains consistent across different values of P_0, with lower values leading to higher expected revenue. For the third response function, $g_3(x) = 2 - e^{-x}$, the increase in expected revenue becomes insignificant when more than 20 units of the budget are allocated to advertising.

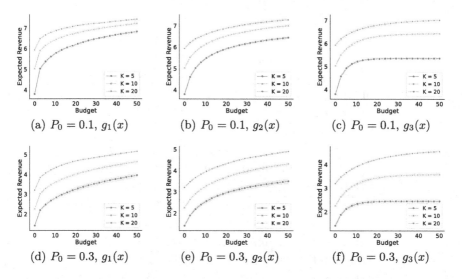

(a) $P_0 = 0.1$, $g_1(x)$ (b) $P_0 = 0.1$, $g_2(x)$ (c) $P_0 = 0.1$, $g_3(x)$

(d) $P_0 = 0.3$, $g_1(x)$ (e) $P_0 = 0.3$, $g_2(x)$ (f) $P_0 = 0.3$, $g_3(x)$

Fig. 1. The relationship between budget and expected revenue for 100 products in different settings

D   Omitted Proofs

Proof of Lemma 1: Consider an arbitrary advertising strategy $\mathbf{x}$ that satisfies $\sum_{i \in S^*} x_i = B_1 < B$, define $\Delta = B - B_1$ and $\tau = \arg\max_{i \in S^*} r_i$. To prove this lemma, we can increase the expected revenue by allocating the remaining budget Δ to the product τ. Let S^* denote the optimal assortment, and $\mathbf{x}'$ denote this new strategy. Because $\mathbf{x}'$ and $\mathbf{x}$ only differs in entry τ, we rewrite

$R(S^*, \mathbf{v}, \mathbf{x}')$ as $\frac{\beta + r_\tau v_\tau (g(c_\tau(x_\tau + \Delta)) - g(c_\tau x_\tau))}{\alpha + v_\tau(g(c_\tau(x_\tau + \Delta)) - g(c_\tau x_\tau))}$ and rewrite $R(S^*, \mathbf{v}, \mathbf{x})$ as $\frac{\beta}{\alpha}$, where $\beta = \sum_{i \in S^*} r_i v_i g(c_i x_i)$ and $\alpha = 1 + \sum_{i \in S^*} v_i g(c_i x_i)$. Notice that

$$R(S^*, \mathbf{v}, \mathbf{x}') - R(S^*, \mathbf{v}, \mathbf{x}) = \frac{(r_\tau \alpha - \beta) \cdot v_\tau(g(c_\tau(x_\tau + \Delta)) - g(c_\tau x_\tau))}{\alpha \cdot (\alpha + v_\tau(g(c_\tau(x_\tau + \Delta)) - g(c_\tau x_\tau)))}.$$

Because $\tau = \arg\max_{i \in S^*} r_i$, we have $r_\tau - r_i \geq 0, \forall i \in S^*$. Thus, $r_\tau + \sum_{i \in S^*}(r_\tau - r_i)(v_i g(c_i x_i)) > 0$, which is equivalent to $r_\tau(1 + \sum_{i \in S^*} v_i g(c_i x_i)) > \sum_{i \in S^*} r_i v_i g(c_i x_i)$. Hence, $r_\tau \alpha > \beta$. Because $g(\cdot)$ is an increasing function, we have $g(c_\tau(\Delta + x_\tau)) - g(c_\tau x_\tau) \geq 0$. Moreover, because both $r_\tau \alpha - \beta$ and $g(c_\tau(\Delta + x_\tau)) - g(c_\tau x_\tau)$ are non-negative, we obtain that $R(S^*, \mathbf{v}, \mathbf{x}') \geq R(S^*, \mathbf{v}, \mathbf{x})$. $\square$

Proof of Lemma 3: *Proof:* Because $(S^*, \mathbf{x}^*)$ is optimal solution, we have $R(S^*, \mathbf{v}, \mathbf{x}^*) \geq R(S_\mathbf{v}, \mathbf{v})$. According to Lemma 2, there exists an S^* such that $S^* = \{i \in \mathcal{N} | r_i > R(S^*, \mathbf{v}, \mathbf{x}^*)\}$. The following chain proves this lemma: $S^* = \{i \in \mathcal{N} | r_i > R(S^*, \mathbf{v}, \mathbf{x}^*)\} \subseteq \{i \in \mathcal{N} | r_i > R(S_\mathbf{v}, \mathbf{v})\} = S_\mathbf{v}$. $\square$

Proof of Lemma 4: *Proof:* Because $g(\cdot)$ is an increasing concave function, its inverse function $g^{-1}(x)$ is a convex function. Moreover, because $\frac{u}{v_i}$ is a linear function, its composition with $g^{-1}(\cdot)$ is also a convex function. Finally, because $\mathcal{U}_1 = \{\mathbf{u} | \sum_{i=1}^n m_i(u_i) \leq B\}$, which is the level set of $\sum_{i=1}^n m_i(u_i)$, is a convex set, its intersection with the convex set $\mathcal{U}_2 = \{\mathbf{u} | u_i \geq v_i, i = 1, \ldots, n\}$ is also a convex set. $\square$

Proof of Lemma 5: *Proof:* For a given assortment S, we can represent any feasible advertising strategy $\mathbf{y}$ that satisfies $\sum_{i \in S} y_i = B$ as a convex combination of $\mathbf{x}_i$; that is, $\mathbf{y} = \sum_{i \in S} \lambda_i \mathbf{x}_i$, where $\lambda_i = y_i / B$ and $\sum_{i \in S} \lambda_i = 1$. Assume $k = \arg\max_{j \in S} L(S, \mathbf{x}_j)$. We have $L(S, \mathbf{y}) \leq L(S, \mathbf{x}_k)$ based on the following observation:

$$L(S, \mathbf{y}) = L\left(S, \sum_{i \in S} \lambda_i \mathbf{x}_i\right) = \frac{\beta + aB \sum_{i \in S} \lambda_i r_i v_i c_i}{\alpha + aB \sum_{i \in S} \lambda_i v_i c_i},$$

where $\alpha = 1 + \sum_{i \in S} v_i$ and $\beta = \sum_{i \in S} r_i v_i$. By the definition of k, we have $L(S, \mathbf{x}_k) \geq L(S, \mathbf{x}_j), \forall j \in S$. Moreover, $L(S, \mathbf{x}_k) \geq L(S, \mathbf{x}_j)$ is equivalent to

$$\alpha c_k v_k r_k - \beta c_k v_k \geq \alpha c_j v_j r_j - \beta c_j v_j + aB c_k v_k (r_j - r_k) c_j v_j, \qquad (\text{D.1})$$

based on the following observation:

$$\begin{aligned} L(S, \mathbf{x}_k) - L(S, \mathbf{x}_j) &= \frac{\beta + aB r_k c_k v_k}{\alpha + aB c_k v_k} - \frac{\beta + aB r_j c_j v_j}{\alpha + aB c_j v_j} \\ &= \frac{(\beta + aB r_k c_k v_k)(\alpha + aB c_j v_j) - (\beta + aB r_j c_j v_j)(\alpha + aB c_k v_k)}{(\alpha + aB c_k v_k)(\alpha + aB c_j v_j)} \\ &= aB \cdot \frac{\alpha(c_k v_k r_k - c_j v_j r_j) + \beta(c_j v_j - c_k v_k) + aB c_k v_k (r_k - r_j) c_j v_j}{(\alpha + aB c_k v_k)(\alpha + aB c_j v_j)}. \end{aligned}$$

By multiplying λ_j by both sides of (D.1) for all $j \in S$ and summing up all inequalities, we have

$$\alpha c_k v_k r_k - \beta c_k v_k \geq \alpha \sum_{j \in S} \lambda_j c_j v_j r_j - \beta \sum_{j \in S} \lambda_j c_j v_j + aB c_k v_k \sum_{j \in S} \lambda_j (r_j - r_k) c_j v_j.$$

$$(\text{D.2})$$

Using a similar argument as the one used to prove the equivalence of $L(S, \mathbf{x}_k) \geq L(S, \mathbf{x}_j)$ and (D.1), we show that (D.2) is equivalent to $L(S, \mathbf{x}_k) \geq L(S, \sum_{i \in S} \lambda_i \mathbf{x}_i) = L(S, \mathbf{y})$. $\square$

Proof of Lemma 6: *Proof:* Let $Z = R(S_\mathbf{v}, \mathbf{v})$ denote the expected revenue of $S_\mathbf{v}$ when $B = 0$, we have $\sum_{i \in S_\mathbf{v}}(r_i - Z)v_i = Z$. If there exists a product $i \in S_\mathbf{v}$ such that $r_i < Z$, then removing this product from $S_\mathbf{v}$ would increase the expected revenue, which contradicts the assumption that $S_\mathbf{v}$ is the optimal assortment when $B = 0$. Thus, we have $S_\mathbf{v} \subseteq \{1, \ldots, T\}$, where $T = \max\{i | i \in S_\mathbf{v}\}$. Similarly, let $Z^* = R(S^*, \mathbf{v}, \mathbf{x}^*)$, we have $S^* \subseteq \{1, \ldots T^*\}$ where $T^* = \max_i\{i | r_i \geq Z^*\}$. Since $Z^* \geq Z$, we conclude that $r_{T^*} \geq r_T$, otherwise we have $Z^* \leq r_{T^*} < Z$ or $Z \leq r_{T^*} < r_T$. Thus we have $S^* \subseteq \{1, \ldots T^*\} \subseteq \{1, \ldots, T\}$. $\square$

Proof of Lemma A.2: *Proof:* When $g(x) = ax + 1$ for some $a \geq 0$, the objective function of (A.4) can be written as $\sum_{i=1}^{n} \alpha_i + a \sum_{i=1}^{n} \alpha_i c_i x_i$. Allocating the entire advertising budget to the product that has the largest $\alpha_i c_i$ maximizes $\sum_{i=1}^{n} \alpha_i + a \sum_{i=1}^{n} \alpha_i c_i x_i$. $\square$

Proof of Lemma A.3: *Proof:* Consider a fixed feasible assortment S. If $f(x) = \log(ax + 1)$ for some $a \geq 0$, then the objective function $R(S, \mathbf{p}, \mathbf{x})$ can be written as $\sum_{i \in S} \alpha_i + a \sum_{i \in S} \alpha_i c_i x_i$. It is easy to verify that the optimal advertising strategy for S must come from $\{\mathbf{x}_0, \mathbf{x}_1, \ldots, \mathbf{x}_n\}$, where $\mathbf{x}_0$ is an all-zero vector. Let $S(\mathbf{x})$ be the optimal assortment under the advertising strategy $\mathbf{x}$. Thus $S(\mathbf{x})$ contains the top K products that have the largest $\alpha_i(ac_i x_i + 1)$. Because $ac_i x_i \geq 0$ for all $i \in \{1, \ldots, n\}$, we have $S(\mathbf{x}_i) = S(\mathbf{x}_0) = \{1, \ldots, K\}$, and the expected revenue for $S(\mathbf{x}_i)$ is $W(\sum_{j=1}^{K} \alpha_j + a\alpha_i c_i B)$ for all $i \in \{1, ..., K\}$. When $j \in \{K + 1, ..., n\}$, there are two possible cases: $S(\mathbf{x}_j) \setminus S(\mathbf{x}_0) = \{\emptyset\}$ or $S(\mathbf{x}_j) \setminus S(\mathbf{x}_0) = \{j\}$.

Case 1 : When $S(\mathbf{x}_j) \setminus S(\mathbf{x}_0) = \{\emptyset\}$ for all $j \in \{K + 1, \ldots, n\}$, the optimal assortment is $\{1, \ldots, K\}$. Because the expected revenue for $S(\mathbf{x}_j)$ is $W(\sum_{j=1}^{K} \alpha_j)$ for all $j \in \{K + 1, \ldots, n\}$, which is the same as $S(\mathbf{x}_0)$, and $t_1 = \operatorname{argmax}_{i \in \{1, ..., K\}} W(\sum_{j=1}^{K} \alpha_j + a\alpha_i c_i B)$, the optimal advertising strategy is $\mathbf{x}_{t_1}$.

Case 2 : When $S(\mathbf{x}_j) \setminus S(\mathbf{x}_0) = \{j\}$ for some $j \in \{K + 1, ..., n\}$, the expected revenue for $S(\mathbf{x}_j)$ is $W(\sum_{i=1}^{K-1} \alpha_i + \alpha_j(ac_j B + 1))$. We denote this subset as S_c. Because $t_2 = \operatorname{argmax}_{j \in S_c} W(\sum_{i=1}^{K-1} \alpha_i + \alpha_j(ac_j B + 1))$, $\mathbf{x}_{t_2}$ is the best advertising strategy in $\{\mathbf{x}_{K+1}, \ldots, \mathbf{x}_n\}$. Moreover, because $\mathbf{x}_{t_1}$ is the best advertising strategy in $\{\mathbf{x}_0, \mathbf{x}_1, \ldots, \mathbf{x}_K\}$, the better strategy between $\mathbf{x}_{t_1}$ and $\mathbf{x}_{t_2}$ must be the optimal advertising strategy. $\square$

Proof of Lemma A.4: When $c_i = c$ for all $i \in \mathcal{N}$, the objective function of (A.5) can be simplified to $h(\mathbf{x}, S) = \sum_{i \in S} \alpha_i g(cx_i)$. To prove the first part of this lemma, we show that for any optimal solution $(S, \mathbf{x})$, we can construct a new solution $(S', \mathbf{x}')$, where $S' = \{1, \ldots, K\}$, which is no worse than $(S, \mathbf{x})$. Due to the monotonicity of the objective function, $|S| = K$ can be assumed. We construct such $\mathbf{x}'$ as follows: for each $i \in \{1, \ldots, K\}$, let $x'_i = x_{L(i)}$, where $L(i)$ represents the product that has the i-th largest α_i in S. Therefore $h(\mathbf{x}', S') - h(\mathbf{x}, S) =$

$\sum_{i \in S'}(\alpha_i - \alpha_{L(i)})g(cx_i) \geq 0$; the inequality exists because S' contains the top K products that have the largest α_i. Hence, $(\mathbf{x'}, S')$ is no worse than $(S, \mathbf{x})$.

We next prove that the optimal advertising strategy $\mathbf{x}^*$ satisfies $x_i^* \geq x_j^*$ $\forall i \leq j$ through contradiction. Assume there exist two products $i, j \in S^*$ such that $x_i^* < x_j^*$ and $i < j$. We can construct a new advertising strategy $\mathbf{x}$ such that $x_k = x_k^*$ for $k \notin \{i, j\}$, and $x_i = x_j^*, x_j = x_i^*$. The following chain proves that $h(\mathbf{x}, S^*) - h(\mathbf{x}^*, S^*) = (\alpha_i - \alpha_j) \cdot (g(cx_j^*) - g(cx_i^*))$:

$$h(\mathbf{x}, S^*) - h(\mathbf{x}^*, S^*) = \alpha_i g(cx_i) + \alpha_j g(cx_j) - \alpha_i g(cx_i^*) - \alpha_j g(cx_j^*)$$
$$= \alpha_i (g(cx_j^*) - g(cx_i^*)) + \alpha_j(g(cx_i^*) - g(cx_j^*))$$
$$= (\alpha_i - \alpha_j) \cdot (g(cx_j^*) - g(cx_i^*)).$$

Because $\alpha_i \geq \alpha_j$ and $x_i^* < x_j^*$, $\mathbf{x}$ is a better solution than $\mathbf{x}^*$ which contradicts to the assumption that $\mathbf{x}^*$ is the optimal solution. $\square$

Proof of Lemma B.2: *Proof:* For simplicity, let $h_i^k = \frac{\partial R(S^*, \mathbf{u}^*)}{\partial u_i^*}$ be the partial derivative for product i in assortment S_k^*, and denote $A_k = \sum_{l=1}^{k} V(S_l^*)$ and $B_k = \frac{W(S_k^*)}{V(S_k^*)}$. We have

$$h_i^k = \frac{(r_i - \frac{W(S_k^*)}{1+A_k})}{(1+A_{k-1})(1+A_k)} - \sum_{j=k+1}^{m} B_j \cdot \left(\frac{1}{(1+A_{j-1})^2} - \frac{1}{(1+A_j)^2} \right).$$

According to Lemma B.1, $S_k^* = \{i \in \mathcal{N} | t_{k+1}^* \leq r_i < t_k^*\}$. We first consider the products in the same stage, that is $i, j \in S_k^*$, and $i < j$. $h_i^k \geq h_j^k$ because $r_i \geq r_j$. Then, we consider the cases i and j in two different stages. The difference between h_i^k and h_j^{k+1} is

$$h_i^k - h_j^{k+1} = \frac{(r_i - \frac{W(S_k^*)}{1+A_k})}{(1+A_{k-1})(1+A_k)} - B_{k+1} \cdot \left(\frac{1}{(1+A_k)^2} - \frac{1}{(1+A_{k+1})^2} \right) - \frac{(r_j - \frac{W(S_{k+1}^*)}{1+A_{k+1}})}{(1+A_k)(1+A_{k+1})}$$
$$\geq \left[\frac{1}{(1+A_{k-1})(1+A_k)} - \frac{1}{(1+A_k)(1+A_{k+1})} \right] t_{k+1}^* - \frac{W(S_k^*)}{(1+A_{k-1})(1+A_k)^2}$$
$$- \frac{2(1+A_k)+V(S_{k+1}^*)}{(1+A_k)^2(1+A_{k+1})^2} \cdot W(S_{k+1}^*) + \frac{W(S_{k+1}^*)}{(1+A_k)(1+A_{k+1})^2}$$
$$= \frac{W(S_k^*)}{(1+A_{k-1})(1+A_k)^2} + \frac{W(S_{k+1}^*)}{(1+A_k)^2(1+A_{k+1})} - \frac{W(S_k^*)}{(1+A_{k-1})(1+A_k)^2}$$
$$- \frac{2(1+A_k)+V(S_{k+1}^*)}{(1+A_k)^2(1+A_{k+1})^2} \cdot W(S_{k+1}^*) + \frac{W(S_{k+1}^*)}{(1+A_k)(1+A_{k+1})^2}$$
$$= \frac{2+A_k+A_{k+1}-2(1+A_k)+V(S_{k+1}^*)}{(1+A_k)^2(1+A_{k+1})^2} \cdot W(S_{k+1}^*)$$
$$= 0.$$

The first inequality uses the fact that $r_i \geq t_{k+1}^* \geq r_j$. Lastly, for the product in the last assortment S_m, we have $h_i^m = \frac{r_i - \frac{W(S_m^*)}{1+A_m}}{(1+A_{m-1})(1+A_m)}$. In this case, $r_i \geq t_{m+1}^* = \frac{W(S_m^*)}{1+A_m}$, which means $h_i^m \geq 0$. $\square$

Proof of Lemma B.3: *Proof:* Let $Q(\mathbf{x}) = R(\mathcal{S}^*, \mathbf{x})$. Based on the analysis in Lemma B.2, $\frac{\partial Q(\mathbf{x})}{\partial x_i} \geq \frac{\partial Q(\mathbf{x})}{\partial x_j} \geq 0$ for all $i < j$. For any $\mathbf{x}$ satisfying the budget constraint, through the mean value theorem, we have

$$
\begin{aligned}
Q(\mathbf{x}) - Q(\mathbf{x}^*) &= \nabla Q(\mathbf{x} + (1-c)\mathbf{x}^*)^T \cdot (\mathbf{x} - \mathbf{x}^*) \\
&= Q(\mathbf{x}^c)^T \cdot (\mathbf{x} - \mathbf{x}^*) \\
&= \sum_{i=2}^{n} \frac{\partial Q(\mathbf{x}^c)}{\partial x_i} x_i^c + \frac{\partial Q(\mathbf{x}^c)}{\partial x_1}(x_1^c - B) \\
&\leq \frac{\partial Q(\mathbf{x}^c)}{\partial x_1} \sum_{i=2}^{n} x_i^c + \frac{\partial Q(\mathbf{x}^c)}{\partial x_1}(x_1^c - B) \\
&= \frac{\partial Q(\mathbf{x}^c)}{\partial x_1}(\sum_{i=1}^{n} x_i^c - B) \\
&\leq 0.
\end{aligned}
$$

Here, $c \in (0,1)$, and we denote $\mathbf{x} + (1-c)\mathbf{x}^*$ as $\mathbf{x}^c$. The first inequality exists because $\frac{\partial Q(\mathbf{x}^c)}{\partial x_i} \leq \frac{\partial Q(\mathbf{x}^c)}{\partial x_1}$, and the last inequality is due to the budget constraint. $\square$

Proof of Lemma B.4: *Proof:* Let $T_k^* = \cup_{i=1}^{k} S_i^*$. We have

$$
\begin{aligned}
R(\mathcal{S}^*, \mathbf{x}^*) &= \sum_{k=1}^{m} \cdot \frac{W(T_k^*, \mathbf{x}^*) - W(T_{k-1}^*, \mathbf{x}^*)}{(1 + V(T_{k-1}^*, \mathbf{x}^*))(1 + V(T_k^*, \mathbf{x}^*))} \\
&= \sum_{k=1}^{m-1} \frac{W(T_k^*, \mathbf{x}^*)}{1 + V(T_k^*, \mathbf{x}^*)} \left\{ \frac{1}{1 + V(T_{k-1}^*, \mathbf{x}^*)} - \frac{1}{1 + V(T_{k+1}^*, \mathbf{x}^*)} \right\} + \frac{W(T_m^*, \mathbf{x}^*)}{(1 + V(T_{m-1}^*, \mathbf{x}^*))(1 + V(T_m^*, \mathbf{x}^*))} \\
&\leq \max_{S, \mathbf{x}} \frac{W(S, \mathbf{x})}{1 + V(S, \mathbf{x})} \left[\sum_{k=1}^{m-1} \left\{ \frac{1}{1 + V(T_{k-1}^*, \mathbf{x}^*)} - \frac{1}{1 + V(T_{k+1}^*, \mathbf{x}^*)} \right\} + \frac{1}{1 + V(T_{m-1}^*, \mathbf{x}^*)} \right] \\
&= \max_{S, \mathbf{x}} \frac{W(S, \mathbf{x})}{1 + V(S, \mathbf{x})} \left[1 + \frac{1}{1 + V(T_1^*, \mathbf{x}^*)} - \frac{1}{1 + V(T_m^*, \mathbf{x}^*)} \right] \\
&\leq 2 \max_{S, \mathbf{x}} \frac{W(S, \mathbf{x})}{1 + V(S, \mathbf{x})} \\
&\leq 2R(\mathcal{S}^h, \mathbf{x}^h).
\end{aligned}
$$

The last inequality holds because our heuristic method starts with the optimal solution of the single-stage problem and iteratively improves upon it. $\square$

References

1. Aravindakshan, A., Peters, K., Naik, P.A.: Spatiotemporal allocation of advertising budgets. J. Mark. Res. **49**(1), 1–14 (2012)
2. Basu, A.K., Batra, R.: ADSPLIT: a multi-brand advertising budget allocation model. J. Advert. **17**(2), 44–51 (1988)
3. Beltran-Royo, C., Zhang, H., Blanco, L., Almagro, J.: Multistage multiproduct advertising budgeting. Eur. J. Oper. Res. **225**(1), 179–188 (2013)
4. Berbeglia, G., Flores, A., Gallego, G.: The refined assortment optimization problem. arXiv preprint arXiv:2102.03043 (2021)

5. Bertsekas, D.: Nonlinear Programming, vol. 4. Athena Scientific (2016)
6. Davis, J., Gallego, G., Topaloglu, H.: Assortment planning under the multinomial logit model with totally unimodular constraint structures (2013, work in progress)
7. Davis, J.M., Gallego, G., Topaloglu, H.: Assortment optimization under variants of the nested logit model. Oper. Res. **62**(2), 250–273 (2014)
8. Dinkelbach, W.: On nonlinear fractional programming. Manage. Sci. **13**(7), 492–498 (1967)
9. Doyle, P., Saunders, J.: Multiproduct advertising budgeting. Mark. Sci. **9**(2), 97–113 (1990)
10. Dubé, J.P., Hitsch, G.J., Manchanda, P.: An empirical model of advertising dynamics. Quant. Mark. Econ. **3**(2), 107–144 (2005)
11. Fischer, M., Albers, S., Wagner, N., Frie, M.: Practice prize winner-dynamic marketing budget allocation across countries, products, and marketing activities. Mark. Sci. **30**(4), 568–585 (2011)
12. Flores, A., Berbeglia, G., Van Hentenryck, P.: Assortment optimization under the sequential multinomial logit model. Eur. J. Oper. Res. **273**(3), 1052–1064 (2019)
13. Freimer, M., Horsky, D.: Periodic advertising pulsing in a competitive market. Mark. Sci. **31**(4), 637–648 (2012)
14. Freund, R.W., Jarre, F.: Solving the sum-of-ratios problem by an interior-point method. J. Global Optim. **19**(1), 83–102 (2001)
15. Gao, P., et al.: Assortment optimization and pricing under the multinomial logit model with impatient customers: sequential recommendation and selection. Oper. Res. **69**(5), 1509–1532 (2021)
16. Hopp, W.J., Xu, X.: Product line selection and pricing with modularity in design. Manuf. Serv. Oper. Manage. **7**(3), 172–187 (2005)
17. Liu, N., Ma, Y., Topaloglu, H.: Assortment optimization under the multinomial logit model with sequential offerings. INFORMS J. Comput. **32**(3), 835–853 (2020)
18. Ma, W.: When is assortment optimization optimal? Manage. Sci. **69**, 2088–2105 (2022)
19. Mahajan, V., Muller, E.: Advertising pulsing policies for generating awareness for new products. Mark. Sci. **5**(2), 89–106 (1986)
20. McCormick, G.P.: Computability of global solutions to factorable nonconvex programs: part i - convex underestimating problems. Math. Program. **10**(1), 147–175 (1976)
21. Mesak, H.I.: An aggregate advertising pulsing model with wearout effects. Mark. Sci. **11**(3), 310–326 (1992)
22. Park, S., Hahn, M.: Pulsing in a discrete model of advertising competition. J. Mark. Res. **28**(4), 397–405 (1991)
23. Rusmevichientong, P., Shen, Z.J.M., Shmoys, D.B.: Dynamic assortment optimization with a multinomial logit choice model and capacity constraint. Oper. Res. **58**(6), 1666–1680 (2010)
24. Rusmevichientong, P., Topaloglu, H.: Robust assortment optimization in revenue management under the multinomial logit choice model. Oper. Res. **60**(4), 865–882 (2012)
25. Sriram, S., Kalwani, M.U.: Optimal advertising and promotion budgets in dynamic markets with brand equity as a mediating variable. Manage. Sci. **53**(1), 46–60 (2007)

26. Talluri, K., Van Ryzin, G.: Revenue management under a general discrete choice model of consumer behavior. Manage. Sci. **50**(1), 15–33 (2004)
27. Wang, R.: Capacitated assortment and price optimization under the multinomial logit model. Oper. Res. Lett. **40**(6), 492–497 (2012)
28. Yang, C., Guo, L., Zhou, S.X.: Customer satisfaction, advertising competition, and platform performance. Prod. Oper. Manag. **31**(4), 1576–1594 (2022)

Twin-Treewidth: A Single-Exponential Logic-Based Approach

Maurício Pires$^{(\boxtimes)}$ [iD], Uéverton S. Souza [iD], and Bruno Lopes [iD]

Institute of Computing, Fluminense Federal University, Niterói, Brazil
mspires@id.uff.br, {ueverton,bruno}@ic.uff.br

Abstract. An equivalence class in a set is a subset of elements considered equivalent according to some criterion. This concept is applied to different graph parameters, such as neighborhood diversity, twin-cover, twin-width, and modular width. In this work, we introduce a new parameter in graphs called twin-treewidth, which explores the equivalence classes of twins. This parameter generalizes treewidth and neighborhood diversity, two of the most studied parameters in parameterized complexity. We demonstrate the usefulness of this parameter by proposing a simple exponential-time generic procedure to solve problems that can be expressed in a fragment of a variant of Second-Order Monadic Logic.

Keywords: Parameterized complexity · Dynamic programming · Twin vertices · Treewidth · Model checking

1 Introduction

Parameterized algorithms are an effective alternative to deal with NP-hard problems. These algorithms aim to explore particular characteristics of the inputs to obtain polynomial algorithms once the parameter value is fixed, i.e., algorithms whose runtime is $f(\kappa) \cdot n^c$ where n is the input instance size, c is a constant independent of n and $f(\kappa)$ is any computable function in the parameter. Problems that admit algorithms of this type are considered fixed-parameter tractable (FPT). We will highlight treewidth and neighborhood diversity among all the well-known parameters in the literature.

Many NP-hard graph problems are efficiently solvable when restricted to the class of the trees. In that context, the graph parameter treewidth measures the tree-likeliness of general graphs. The measure is obtained by decomposing the input graph in non-disjoint subgraphs connected in a tree-like manner. This

This research has received funding from Rio de Janeiro Research Support Foundation (FAPERJ) under grant agreements E-26/201.344/2021 and SEI-260003/001674/2021, Coordenação de Aperfeiçoamento de Pessoal de Nível Superior (CAPES), National Council for Scientific and Technological Development (CNPq) under grant agreement 309832/2020-9.

W. Wu and J. Guo (Eds.): COCOA 2023, LNCS 14462, pp. 43–55, 2024.
https://doi.org/10.1007/978-3-031-49614-1_3

organization allows the development of dynamic programming algorithms [7], which explore the subgraphs in a bottom-up computation in the tree structure. The class of bounded treewidth graphs includes relevant graph families such as cactus graphs, outerplanar graphs, series-parallel graphs, Halin graphs, and Apollonian networks [4].

The neighborhood diversity introduced by Lampis [13] is an alternative to strengthen the vertex cover of a graph as a parameter. Typically, dense graphs (which have a quadratic number of edges concerning their number of vertices) have an extensive vertex coverage, which makes it challenging to use them as a parameter. On the other hand, some graph problems dispense with the exhaustive computation in classes of vertices considered equivalent. In this way, by limiting the possibilities of types of equivalence classes, Lampis reinforces the idea of vertex coverage as a structural parameter.

The most prominent work on treewidth is Courcelle's Theorem [6], which generalizes the work of Thatcher and Wright [15] about trees and Monadic Second-Order Logic (MSO). In the context of graphs, consider the Monadic-Second Order Logic with quantification over vertices and edges subsets (MSO_2) as the main logic framework to express graph problems. Courcelle's work establishes that given an MSO_2 formula φ and an input graph G with bounded treewidth exists a procedure that decides if the formula is satisfied in the graph in FPT-time, i.e., in $f(|\varphi|, twd(G)) \cdot \mathcal{O}(n)$ time where $|\varphi|$ is the formula's size and $twd(G)$ is the treewidth of G. Its result has a subjacent procedure to compute the solution of the model-checking problem derived from the graph problem, turning it into an *algorithmic meta theorem* [10–13]. This theorem is a relevant tool for designing parameterized algorithms due to its dual nature. While providing an efficient classification tool, they also provide a generic procedure to solve different problems.

Although considered efficient in the parameterized complexity, FPT algorithms may suffer performance losses due to the type of parameter function obtained. The procedure underlying Courcelle's theorem implies a parameter function as an exponential tower. The height of this tower depends directly on the number of alternations of the quantifiers used in the formula that describes the problem. In this context, Lampis obtains an algorithm whose parameter function is an exponential fixed-sized tower when considering neighborhood diversity. However, Lampis's result only concerns properties in graphs expressed in MSO_1, the MSO fragment that deals only with the quantification of vertices and vertex sets.

Since 1990, Courcelle's ideas evolved into many works exploring different logics and parameters searching for meta results. Exploring treewidth, Pilipczuk [14] proposed a modal logic to deal with neighborhood property-based problems, the Existential Counting Modal Logic (ECML). This work resulted in a meta-theorem establishing a single-exponential FPT-tractability for ECML-expressible problems concerning treewidth, even involving connectivity requirements. In 2019, Knop et al. [16] presented new meta-theorems about treewidth but using extensions of Monadic Second-Order Logic (MSO) with global and local cardinality constraints (CardMSO and MSO-LCC) and optimizing the fair objective function (fairMSO). Besides treewidth, this last work also studied neighborhood diversity.

When the input graph has large cliques, a challenge presents itself as the size of the maximum clique is a lower bound on the treewidth of a graph. On the other hand, neighborhood diversity equivalence classes can sometimes be an alternative to deal with this problem. This work introduces the notion of twin-treewidth, a new graph parameter extending treewidth and neighborhood diversity. The idea behind this parameter is to reduce possible redundancies in verifying solutions for problems in graphs with classes of true/false twins. As we will see later, our parameter provides an algorithmic meta theorem for some problems representable in a fragment of ECML. The procedure obtained for verifying the desired property in the input graph preserves, with appropriate differences, the simple exponential time achieved by Michal Pilipczuk in his work.

2 Preliminaries

In this work, we utilize the standard notations commonly used in the fields of graph theory, parameterized complexity, and logic. For undefined notations, we recommend the following references to the readers: Bondy et al. [5], Downey and Fellows [8,9], and Cygan et al. [7], Van Dallen [17], and Biggs et al. [3].

Graph Theory and Parameterized Complexity. Let $G = \langle V, E \rangle$ be a graph. Two vertices u and v are *true twin* if their closed neighborhoods are equal, i.e., $N[u] = N[v]$. The vertices will be false twins if the open neighborhoods are equal., i.e., $N(v) = N(u)$. As a consequence of the definition of twin, if a set $C \in V(G)$ is such that its vertices are pairwise true twins, then C is a clique. However, if two by two vertices are false twins, then C is an independent set.

The treewidth is a parameter that quantifies the similarity between a graph and a tree. This notion is one of the results of the work of Bertele and Brioschi [2]. They proposed to organize some subgraphs of the original graph in a tree manner, obeying some restrictions, as we see in 1. Once we have an association between tree nodes and subgraphs represented by its vertex subset, the *width of the tree decomposition*, denoted by $width(\mathcal{T})$, is the maximum cardinality of these subsets, i.e., $width(\mathcal{T}) = \max\{|B_t| \mid t \in T\}$. Then, the *treewidth* of a graph G, denoted by $twd(G)$, is the minimum width among all the possible tree decompositions for this graph. Although the tree decomposition is not an invariant of a graph, the treewidth is.

Definition 1 (Tree decomposition). *Let G be a graph. A tree decomposition of G is a pair $\mathcal{T} = \langle T, \{B_t\}_{\forall t \in V(T)} \rangle$ where T is a tree and $\{B_t\}_{t \in V(T)}$ is the set of bags of T. A bag B_t is a vertex subset of the graph G assigned to a tree node t, i.e., $\forall t \in V(T) : B_t \subseteq V(G)$. Every bag represents a subgraph of G, and the assignment must hold the following properties:*

1. *Every vertice belongs to at least one bag, which implies $\bigcup_{t \in V(T)} B_t = V(G)$*
2. *Every edge in graph G belongs to a bag, i.e., $\forall uv \in E(G)$ such that $u, v \in V(G)$, there's a node $t \in T$ such that $\{u, v\} \subseteq B_t$*

3. For every vertex v, the set of tree nodes whose bags contain v induce a subtree of T, i.e., $\forall v \in V(G)$, the set $T_v = \{t \in T | v \in B_t\}$ induces a subtree

Usually, tree decompositions have no roots. Nevertheless, it could be convenient to consider variations where we have rooted trees. One of these variations is the *extended nice tree decomposition* [7]. Besides the traditional restrictions of the tree decompositions, extended nice tree decompositions classify the tree nodes. Then, every node in the tree could be one of the following types:

Leaf is a node t without children, and $B_t = \emptyset$.

Introduce vertex v node is a node t with child t', such that $B_t = B_{t'} \cup \{v\}$, for some $v \in V(G)$.

Introduce edge uv node is a node t with child t' such that $B_t = B_{t'}$, and the edge uv appears only from t, given $uv \in E(G)$.

Forget vertex v node is a node t with child t', such that $B_t = B_{t'} - \{v\}$.

Join node is a node t with children t' and t'', such that $B_t = B_{t'} = B_{t''}$.

The idea of neighborhood diversity comes as a way to eliminate redundancies on the entrance graph. Redundant vertices are said to belong to the same type, formally described in Definition 2. Therefore, the premise of the parameter is to group vertices that share a similar neighborhood, as shown in Definition 3. In other words, the neighborhood diversity of a graph is the number of vertices that this graph will have after the contraction of the vertices with the same type.

Definition 2 (Type). *Let G be a graph. Two vertices u and v are of the same type if $N(u)\backslash\{v\} = N(v)\backslash\{u\}$.*

Definition 3 (Neighborhood diversity). *Let G be a graph. The neighborhood diversity of G is w, denoted by $nd(G) = w$, if G has an unique minimal partition $\{V_1, V_2, \cdots, V_w\}$ of its vertices which every set V_i, $1 \leq i \leq w$, is of vertices of the same type. Moreover, the partition can be computed in linear time.*

Existential Counting Modal Logic. *Counting Modal Logic* (CML) is a modal logic tailored to deal with properties in the neighborhood of a vertex. In addition to propositional operators (conjunction, disjunction, negation, implication) and vertex and edge sets predicates, CML has modalities of the form $\Diamond^S$, where S is a finitely recognizable set. The meaning of formulas $\Diamond^S \varphi$ is "the number of neighbors of a vertice v in which φ is satisfied belongs to S". That is the counting part of the logic. As a short syntax, we write $\Box^S \varphi$ to denote $\neg\Diamond^S\neg\varphi$. When $S = \mathbb{N}^+$, one omits the set S and $\Diamond$ means "at least one neighbor" and $\Box$ means "all the neighbors."

In [14], Pilipczuk introduces *Existential Counting Modal Logic* (ECML) for graphs. First, consider the following elements of such a logic: an input graph G; a vector $\overline{X}$ where each entry $X[i]$ of $\overline{X}$ will be mapped to a vertex subset of the input graph G; a vector $\overline{Y}$ where each entry $Y[i]$ of $\overline{Y}$ will be mapped to an edge subset of the input graph G; a vector $\overline{FX}$ where each entry represent a set of highlighted vertices of G (such as terminal vertices of Steiner instances);

a vector $\overline{FY}$ where each entry represent a set of highlighted edges of G; and a vector of non-negative integers $\overline{k}$ used in constraints on the cardinalities of sets X and Y.

An Existential Counting Modal Logic formula has the following form:

$$\varphi \;=\; \exists_{\overline{X}} \exists_{\overline{Y}} \left[\; \phi \;\wedge\; \forall_v \; (G, \overline{FX}, \overline{FY}, \overline{X}, \overline{Y}, v) \models \psi \;\right]$$

where ϕ is an arbitrary quantifier-free arithmetic formula over the parameters, cardinalities of sets of vertices and edges of G and cardinalities of fixed and quantified sets; and ψ is a CML formula evaluated on the graph G supplied with all the fixed and quantified sets.

Informally, ECML formulas express "there are sets of vertices $X_1, X_2, \ldots,$ $X_{|\overline{X}|}$ and sets of edges $Y_1, Y_2, \ldots, Y_{|\overline{Y}|}$ such that the sets satisfy the arithmetic constraints on ϕ (such as $X_i \leq 5$) and for every vertex v: given the problem instance and fixed sets $X_1, X_2, \ldots, X_{|\overline{X}|}, Y_1, Y_2, \ldots, Y_{|\overline{Y}|}$, we have that the formula ψ is satisfied in v.

Let us state some ECML-expressible problems. The vertex cover problem consists of verifying if a set V of at most k vertices of a graph G exists such that it covers all its edges, i.e., every edge of G has an endpoint in V. Since every edge must have an endpoint in the cover, all its neighbors do if a vertex v is not in V. Thus, vertex cover is an example of an ECML-expressible problem. We represent it as

$$\exists_{X \subseteq V(G)} [|X| \leq k \wedge \forall_v (G, X, v \models \neg X \Rightarrow \Box X)]$$

A graph G is hamiltonian if a cycle containing all the vertices of G exists. For this type of problem, connectivity requirements are crucial. Thus, one must ensure that an edge set induces precisely one connected component and that all vertices of G have only two edges that are incident to it. Let $cc(X)$ and $cc(Y)$ be the number of connected components induced by the vertex set X, and the edge set Y. The problem of determining if exists a Hamiltonian cycle in G is denoted by

$$\exists_{Y \subseteq E(G)} [|cc(Y)| = 1 \wedge |Y| = |V(G)| \wedge \forall_v (G, Y, v \models \Diamond^{\{2\}} Y]$$

Since tree decompositions explore neighborhoods to perform computations, Pilipczuk proposes a dynamic programming approach to model check ECML-expressible properties. His strategy involves storing information about vertex and neighborhoods to ensure a formula is satisfied. Predicates in CML are tests of membership for vertex and edge sets. Thus, dynamic programming benefits the structure of extended nice tree decompositions, where vertex and edge introduction is gradual. It permits updating information about a vertex once its neighborhood is built. His algorithm runs over $|\mathcal{J}|^{2twd(G)} \mathcal{O}(n)$ where $|\mathcal{J}|$ is the stored data's size. More details about the syntax and the semantics of CML and ECML could be consulted in the work of Pilipczuk [14].

ECML$_1$. The *Existential Counting Modal Logic with quantification over vertex sets* (ECML$_1$) stands for ECML dropped the edges predicates, and the highlighted edges sets. Therefore ECML$_1$-expressible problems are those that are expressed depending exclusively on vertex sets. For example, one can note vertex cover, dominating set, Steiner tree, odd cycle transversal, and some of its variants are ECML$_1$ problems. However, problems like cycle cover, Hamiltonian cycle, matching, longest path, max-cut, and spanning tree are not expressible in this ECML fragment because they demand edge verification.

3   Twin-Treewidth

An equivalence relation is a relation that is transitive, reflexive, and symmetric. Thus, the equality of open/closed neighborhoods, which defines the concept of false/true twins, is a type of equivalence relation [13]. It is interesting to note that this relation preserves an underlying structure in the graph. Vertices belonging to the same class are neighbors of the same vertices that do not belong to its class, i.e., if $u, v \in S$, such that $S \subseteq V(G)$, then $N(v) \backslash S = N(u) \backslash S$. In doing so, we prevent the original neighborhoods of the graph from being altered when contracting the twin classes. Let G be an input graph. The twin class graph W_G is obtained by contracting the equivalence classes of false or true twins. In W_G, two vertices are adjacent if and only if the classes they denote are adjacent, i.e., all vertices in one class are adjacent to all vertices in the other. Refer to Fig. 1 for an example of graph contraction into a twin-class graph. A graph has bounded twin-treewidth, denoted by *ttw*, if and only if its twin-class graph has bounded treewidth. Thus, we have $ttw(G) = twd(W_G)$.

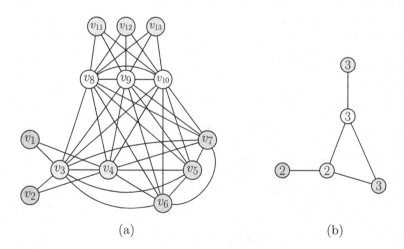

(a)   (b)

Fig. 1. Example of (a) a graph with a max-clique size of 8 and (b) its twin-class graph

Although twin-treewidth generalizes treewidth, there are limitations. The first is the loss of the ability to handle problems involving edges and sets of edges. Considering the combinatorial aspects between the elements of one twin class and those of another, obtaining a result that would be simply exponential in the parameter would be compromised. It is because we do not impose a prior limit on the size of twin cliques or independent sets. While introducing such a limitation might provide some control over the algorithm's complexity, it would lose the intended generality.

It is necessary to emphasize that the dynamic programming developed considers the decomposition of the twin-class graph. Thus, each processed vertex denotes a clique or an independent set. Therefore, we must consider all the necessary information to validate that the formula representing the problem is satisfied in every vertex of the original graph.

3.1   Evaluation Information Function

Pilipczuk's work considered three types of information as crucial for verifying the truth of the CML formula in a vertex of the graph: (1) The vertex's membership in highlighted and quantified sets; (2) The number of neighbors satisfying formulas in the scope of a modality; (3) The satisfaction of subformulas headed by a modality in teh current vertex. Remember that the modality denotes the occurrence of some property in the neighbors of the current vertex. Given that the only predicates in the logic are membership tests to sets, these three pieces of data and a flag f indicating the type of twin class are sufficient to perform the verification in a single vertex.

Although the size of twin classes is not limited a priori, the number of possible memberships for each vertex in the class is. Consider a case where three sets are quantified. There are eight possible membership configurations, ranging from not belonging to any of the three sets to belonging to all three simultaneously. Thus, for q quantified sets, there are 2^q possible membership configurations. We will refer to this component as *grouping* η, where each of the 2^q entries represents a membership configuration, and the value of that entry informs us of how many vertices have that configuration.

Given that twin classes do not have a predefined size limit, we will restrict the studied problems to those in which the expected number of behaviors is limited for each class. Consider a vertex domination problem. When analyzing this problem in a clique, we observe that the number of expected behaviors is limited: either we dominate the clique by choosing one of its vertices, or the clique is dominated by a vertex outside. An analogous behavior occurs in the independent set case. Dominating an independent set is possible by selecting either one set neighbor or all the set members.

The grouping allows for the verification of formulas at modal depth 0, i.e., outside the scope of a modality. In the case of formulas within the scope of a modality, it will be necessary to analyze the neighborhood of each vertex in the original graph. However, since all vertices in a class have the same neighborhood, vertices belonging to the same sets and the same class will yield the same result regarding the validity of the property (formula) to be verified. Therefore, information regarding the satisfiability of subformulas within the scope of a modality must be computed for each possible configuration of pertinence.

Consider that the formula representing the problem has subformulas $\varphi_i = \lozenge^{S_i}\varphi'_i$, $1 \leq i \leq l$. For each component of the grouping, we create binary strings of length l. These strings will encode the satisfaction of the subformulas φ_i. This component, which we will call the *evaluator* μ, gathers 2^q binary strings of length l. The evaluator's data is computed considering the data collected in the third component of data: *perception*.

The perception is the component that keeps track of how many neighbors of each vertex satisfy the subformulas φ'_i. This component consists of two parts. The *external perception* α gathers the count of vertices outside the twin class that satisfies the subformulas. From the perspective of the vertices within a class, those outside the class always have the same configuration. On the other hand, internally, this is not the case. Therefore, the *internal perception* β should be divided into 2^q entries, where each entry counts the number of neighbors within the class that satisfy the subformulas φ'_i. Thus, the external perception has only one entry of length l, while the internal perception has 2^q entries of length l. The values used in this counting depend on the finitely recognizable set S_i that indexes the modality of the subformula φ_i. In the case of the false twin class of vertices, all the entries of the internal perception must be zero.

Thus, the information evaluation function regarding the twin-treewidth takes the form $\iota : V(G_t) \rightarrow \{0,1\} \times \mathcal{P}^l \times (\mathcal{P}^l)^{2^q} \times (\{0,1\}^l)^{2^q} \times \mathcal{N}^{2^q}$, where $\mathcal{P}$ and $\mathcal{N}$ are monoids to which the finitely recognizable sets S_i and the set of expected behaviors for the given problem have been mapped, respectively.

3.2   Dynamic Programming

The following dynamic programming algorithm assists in verifying the existence of a solution for the problem in the input graph. This algorithm considers two important elements. The first one is a vector τ of non-negative integers of size q. Each entry in this vector is associated with the cardinality of one of the q sets quantified by the formula. Thus, when filling the dynamic programming table, we must consider the cardinality constraints imposed by the problem. Cells in which at least one of the entries in τ does not satisfy the cardinality constraints should be filled with false. The second element is the information evaluation function ι. Like any dynamic programming algorithm based on treewidth, the filling of a cell in the table depends on the type of node associated with the cell.

Leaf Node. Since leaves in the extended nice tree decompositions have no associated information, the sets of the partial solution and the evaluation information function are empty. That is the only valid combination of entries of A when node t is a leaf, which implies the relation defined in Eq. 1. All operations performed at node type leaf are constant time. The processing of leaf nodes consumes time $|\iota|^{ttw(G)}n^q$.

$$A_x(\bar{\tau}, \iota) = \begin{cases} true & \text{iff} \quad \bar{\tau} = \bar{0} \wedge \iota = \emptyset \\ false & \text{otherwise} \end{cases} \tag{1}$$

Introducing Vertex Node. The vertex introduction nodes represent the insertion of a true twin clique of a false twin independent set in constructing the solution. The class is inserted in isolation. Therefore, the external perception should indicate that no external neighbors satisfy the subformulas. On the other hand, the internal perceptions should be filled according to the grouping and the type of the flag f. If $f = 0$, then we are inserting a class of false twins. Otherwise, the inserted class is of true twin. Additionally, the entries of the evaluator should be consistent with the information present in the internal perception, as the information in the evaluator depends on the addition of the external perception to each entry of the internal perception. If the inserted vertex denotes an independent set, besides the external perception, the internal perceptions must also be empty. Equation 2 gives us the recurrence relation for a node x when x is a node introducing the vertex v.

$$A_x(\bar{\tau}, \iota) = \begin{cases} false & \text{if} \quad card(\bar{\tau}) \\ false & \text{if} \quad \sum_{i=1}^{2^q} \eta_{v,i} \neq |C| \\ false & \text{if} \quad \exists i \in \{1, 2, \cdots, q\} \; : \; \bar{\tau}_i < set(\eta_v)_i \\ false & \text{if} \quad \alpha_v \neq e_{\mathcal{H}} \\ false & \text{if} \quad \neg\, check(v) \\ false & \text{if} \quad v \notin \iota \\ false & \text{if} \quad \beta_v \neq e_{\mathcal{H}} \; \wedge \; f = 0 \\ A_y(\bar{\tau} - set(\eta_v), \iota \backslash \{v\}) & \text{otherwise} \end{cases} \tag{2}$$

The conditions presented in Eq. 2 show types of entries of $A(\bar{\tau}, \iota)$ that are not valid. The function $card(\bar{\tau})$ checks if the cardinalities of the sets satisfy the restrictions expressed by the subformulas involving the relational operators. This function takes constant time since the formula is fixed. The second condition checks the consistency of the grouping. The function $set(\eta_v)$ returns the values added to the cardinalities of each set when v is included in the partial solution. Each coordinate of the grouping denotes a subset of quantified sets that will receive some vertices from the class. Since we defined the structure of grouping, this function takes constant time. The third condition ensures that the difference between the cardinalities of the set and the contribution of including the class C, encoded in vertex v, in the partial solution is greater than or equal to zero. The fourth condition guarantees that the external perception is empty since the class C is currently isolated. The function $check(v)$ verifies the consistency between the internal perception, evaluator, and grouping. This verification also takes constant time. Therefore, the processing of introducing a vertex in the partial solution takes time $|\iota|^{ttw(G)} n^q$.

Introducing Edge Node. The introduction of an edge in the twin-class graph defines a neighborhood between the vertices of one class and those of another. Thus, this type of node denotes the change in the external perception of the classes (vertices) that are becoming adjacent. Note that changes in the external perception impact the information the evaluator presents. Equation 3 gives the recurrence relation for x when introducing edge uv.

$$A_x(\bar{\tau}, \iota) = \begin{cases} false & if \quad card(\bar{\tau}) \\ false & if \quad v \notin \iota \lor u \notin \iota \\ \bigvee_{\delta \in \Delta} A_y(\bar{\tau}, \delta) & otherwise \end{cases} \qquad (3)$$

The set Δ denotes the set of configurations of ι such that if in a node x we have $\iota(v) = (\alpha_v, \beta_v, \mu_v, \eta_v)$ and $\iota(u) = (\alpha_u, \beta_u, \mu_u, \eta_u)$, then in its child y we must have $\delta(v) = (\alpha_v - satn(u), \beta_v, \mu'_v, \eta_v)$ and $\delta(u) = (\alpha_u - satn(v), \beta_u, \mu'_u, \eta_u)$. The function $satn(u)$ returns a vector with the numbers of vertices in the class denoted by u that satisfy all subformulas φ'_i, $1 \le i \le l$. The components μ'_u and μ'_v denote consistent configurations of the evaluators given the new configurations of the external perception of u and v. This consistency check requires constant time, as the formula is fixed. The processing of edge introduction nodes consumes time $|\iota|^{ttw(G)} n^q$.

Forgetting Vertex Node. In the context of a tree decomposition, forgetting a vertex means that its processing has been completed. All neighbors of this vertex have appeared in the partial solution, and all necessary information is available. It implies the completion of verifying the desired property for this vertex. Since the property must be satisfied in all vertices, we need to check the configurations of A that imply the satisfaction of the property. Equation 4 provides the recurrence relation for the case when x is a forget node.

$$A_x(\bar{\tau}, \iota) = \begin{cases} false & if \quad card(\bar{\tau}) \\ \bigvee_{(\alpha, \beta, \mu, \eta) \in \mathcal{D}} A_y(\bar{\tau}, \iota \cup \{(v, (\alpha, \beta, \mu, \eta))\}) & otherwise \end{cases} \qquad (4)$$

The set $\mathcal{D}$ is the set of valid configurations for ι, i.e., those configurations for which the evaluator and grouping imply the satisfaction of the property for every group (type of vertex membership in the grouping). Determining whether a configuration of $A(\bar{\tau}, \iota)$ is valid or not takes constant time since the desired property only involves membership tests for the current vertex (grouping) and its neighbors (external and internal perception). Thus, processing forget nodes takes time $|\iota|^{ttw(G)} n^q$.

Remember that the root of the tree decomposition has an empty bag. Therefore, the roots are always forget nodes. In particular, the result of formula satisfiability for the entire graph is obtained at the root.

Join Node. Joining two different branches of the tree decomposition means composing information computed in two separate fragments of the neighborhood of the vertices in the node's bag. To count the number of solutions in join nodes, it's necessary to combinate all possible configurations which lead to a solution. The merge of partial solutions of different branches implies updating the quantified sets' cardinalities and external perception and evaluator of the vertices in the bag. Equation 5 gives the recurrence relation for a join node x with children y and z.

$$A_x(\overline{\tau}, \iota) = \begin{cases} false & if \quad card(\overline{\tau}) \\ false & if \quad \exists v \in X_x : \sum_{i=1}^{2^q} \eta_{v,i} \neq |C_v| \\ \displaystyle\bigvee_{\overline{\tau}' + \overline{\tau}'' = \overline{\tau} + \gamma(B_x)} \bigvee_{\iota' + \iota'' = \iota} A_y(\overline{\tau}', \iota') A_z(\overline{\tau}'', \iota'') & otherwise \end{cases} \tag{5}$$

The factor $\gamma(B_x)$ deals with the redundancy of the operation $\overline{\tau}' + \overline{\tau}''$, which is related to the cardinality of the bag B_x. The expression $\iota' + \iota'' = \iota$ implies the combination of the stored information for each vertex in B_x. Since the grouping and internal perception must be the same in x and its children y and z, the external perceptions are added, eliminating some redundancy, and the evaluator for each group must be consistent with the values of its perceptions. All these operations take constant time. Due to the combinatorial nature of composing solutions, the recurrence relation for join nodes takes time $|\iota|^{2ttw(G)} n^q$.

Tree decompositions have $\mathcal{O}(n)$ nodes [1]. Thus, the final complexity depends on the computation that demands more operations. Join nodes are expensive because of the combinatorics performed in the node children. So, our dynamic programming performs in time $(2^{l2^q} |\mathcal{P}| |\mathcal{N}|^{2^q+1})^{ttw(G)} \mathcal{O}(n)$. Since the formula is fixed, l, q, $|\mathcal{P}|$ and $|\mathcal{N}|$ are constants, implying our algorithm is single-exponential in the parameter $ttw(G)$. We remark that the key property to preserve the single-exponential dependence on the parameter is that since the ECML$_1$ has constant size, the equivalence classes will be mapped to monoids having constant size.

Therefore, the following theorem holds.

Theorem 1. *Given a graph G and an ECML$_1$ formula ϕ, one can decide if G satisfy ϕ in $2^{\mathcal{O}(ttw)} \cdot n^{\mathcal{O}(1)}$ time, where ttw is the twin-treewidth of G.*

4   Final Remarks

With twin-treewidth, we gain a stronger parameter than treewidth and neighborhood diversity, which is FPT-computable. We provide a well-known strategy for implementing algorithms through dynamic programming with simple exponential time in the parameter. On the other hand, we lose the ability to guarantee, using our framework, the parameterized tractability for problems that involve edge verification, such as graph coloring, Hamiltonian cycle, cycle cover, and

MaxCut. However, this study points to a promising result as it surpasses the class of graphs with bounded treewidth using a strategy based on elements of simple computation. Furthermore, it is important to highlight that, although we found our strategy on the results obtained by Pilipczuk [14], we still need to study how to address the connectivity requirements present in the $ECML_1$ fragment. Since the foundations of our strategy are the equivalence classes and the dynamic programming provided is independent of the applied equivalence relation, it is natural to ask what results can be obtained by replacing the equivalence relation for false/true twins for a more general equivalence relation.

References

1. Althaus, E., Ziegler, S.: Optimal tree decompositions revisited: a simpler linear-time FPT algorithm. In: Gentile, C., Stecca, G., Ventura, P. (eds.) Graphs and Combinatorial Optimization: from Theory to Applications. ASS, vol. 5, pp. 67–78. Springer, Cham (2021). https://doi.org/10.1007/978-3-030-63072-0_6
2. Bertele, U., Brioschi, F.: On non-serial dynamic programming. J. Comb. Theor. Ser. A **14**(2), 137–148 (1973)
3. Biggs, N.L., et al.: Discrete Mathematics. Oxford University Press (2002)
4. Bodlaender, H.L.: A partial k-arboretum of graphs with bounded treewidth. Theoret. Comput. Sci. **209**(1–2), 1–45 (1998)
5. Bondy, J.A., Murty, U.S.R., et al.: Graph Theory with Applications, vol. 290. Macmillan, London (1976)
6. Courcelle, B.: The monadic second-order logic of graphs iii: tree-decompositions, minors and complexity issues. RAIRO-Theoret. Inf. Appl. **26**(3), 257–286 (1992)
7. Cygan, M., et al.: Parameterized Algorithms, vol. 5. Springer, Cham (2015). https://doi.org/10.1007/978-3-319-21275-3
8. Downey, R.G., Fellows, M.R.: Fundamentals of Parameterized Complexity, vol. 4. Springer, London (2013). https://doi.org/10.1007/978-1-4471-5559-1
9. Downey, R.G., Fellows, M.R.: Parameterized Complexity. Springer, Heidelberg (2012)
10. Fomin, F.V., Golovach, P.A., Stamoulis, G., Thilikos, D.M.: An algorithmic meta-theorem for graph modification to planarity and fol. arXiv preprint arXiv:2106.03425 (2021)
11. Grohe, M., Kreutzer, S.: Methods for algorithmic meta theorems. Model Theoret. Meth. Finite Comb. **558**, 181–206 (2011)
12. Kreutzer, S.: Algorithmic meta-theorems. In: Grohe, M., Niedermeier, R. (eds.) IWPEC 2008. LNCS, vol. 5018, pp. 10–12. Springer, Heidelberg (2008). https://doi.org/10.1007/978-3-540-79723-4_3
13. Lampis, M.: Algorithmic meta-theorems for restrictions of treewidth. Algorithmica **64**(1), 19–37 (2012)
14. Pilipczuk, M.: Problems parameterized by treewidth tractable in single exponential time: a logical approach. In: Murlak, F., Sankowski, P. (eds.) MFCS 2011. LNCS, vol. 6907, pp. 520–531. Springer, Heidelberg (2011). https://doi.org/10.1007/978-3-642-22993-0_47
15. Thatcher, J.W., Wright, J.B.: Generalized finite automata theory with an application to a decision problem of second-order logic. Math. Syst. Theor. **2**(1), 57–81 (1968)

16. Knop, D., Koutecký, M., Masařík, T., Toufar, T.: Simplified algorithmic metatheorems beyond MSO: treewidth and neighborhood diversity. In: Bodlaender, H.L., Woeginger, G.J. (eds.) WG 2017. LNCS, vol. 10520, pp. 344–357. Springer, Cham (2017). https://doi.org/10.1007/978-3-319-68705-6_26
17. Van Dalen, D.: Logic and Structure, vol. 3. Springer, Heidelberg (1994). https://doi.org/10.1007/978-3-662-02962-6

Highway Preferential Attachment Models for Geographic Routing

Ofek Gila$^{(\boxtimes)}$ (ID), Evrim Ozel (ID), and Michael Goodrich (ID)

University of California, Irvine, CA 92617, USA
{ogila,eozel,goodrich}@uci.edu

Abstract. In the 1960 s, the world-renowned social psychologist Stanley Milgram conducted experiments that showed that not only do there exist "short chains" of acquaintances between any two arbitrary people, but that these arbitrary strangers are able to find these short chains. This phenomenon, known as the *small-world phenomenon*, is explained in part by any model that has a low diameter, such as the Barabási and Albert's *preferential attachment* model, but these models do not display the same efficient routing that Milgram's experiments showed. In the year 2000, Kleinberg proposed a model with an efficient $\mathcal{O}(\log^2 n)$ greedy routing algorithm. In 2004, Martel and Nguyen showed that Kleinberg's analysis was tight, while also showing that Kleinberg's model had an expected diameter of only $\Theta(\log n)$—a much smaller value than the greedy routing algorithm's path lengths. In 2022, Goodrich and Ozel proposed the *neighborhood preferential attachment* model (NPA), combining elements from Barabási and Albert's model with Kleinberg's model, and experimentally showed that the resulting model outperformed Kleinberg's greedy routing performance on U.S. road networks. While they displayed impressive empirical results, they did not provide any theoretical analysis of their model. In this paper, we first provide a theoretical analysis of a generalization of Kleinberg's original model and show that it can achieve expected $\mathcal{O}(\log n)$ routing, a much better result than Kleinberg's model. We then propose a new model, *windowed NPA*, that is similar to the neighborhood preferential attachment model but has provable theoretical guarantees w.h.p. We show that this model is able to achieve $\mathcal{O}(\log^{1+\epsilon} n)$ greedy routing for any $\epsilon > 0$.

Keywords: small worlds · social networks · random graphs

1 Introduction

Stanley Milgram, a social psychologist, popularized the concept of the ***small-world phenomenon*** through two groundbreaking experiments in the 1960 s [13,16]. In these experiments, Milgram determined that the median number of hops from a random volunteer in Nebraska and Boston to a stockbroker in Boston was six, thereby giving rise to the expression "six degrees of separation".

W. Wu and J. Guo (Eds.): COCOA 2023, LNCS 14462, pp. 56–80, 2024.
https://doi.org/10.1007/978-3-031-49614-1_4

A common and well-studied method for modeling real-world social networks is the **preferential attachment** model, popularized by Barabási and Albert in 1999 [1]. In this model, nodes are added to the graph one at a time, and each node is connected to m other nodes with probability proportional to their degree. Put simply, in this model, nodes with a greater degree are more likely to obtain an even greater degree, in what is commonly referred to as a "rich-get-richer" process. Such a process leads to power law degree distributions, meaning that the number of nodes with degree k is proportional to $k^{-\alpha}$ for some constant $\alpha > 1$. In 2009, Dommers, Hofstad, and Hooghiemstra showed that the diameter of the preferential attachment model is $\Omega(\log n)$ when the power law exponent $\alpha > 3$, and $\Omega(\log \log n)$ when $\alpha \in (2,3)$ [6]. While such preferential attachment models indeed display small diameters, therefore explaining how these short paths *exist*, they do not explain how these paths are *found*. In other words, individual nodes in these models, using only local information, cannot find short paths to other nodes, unlike in Milgram's experiments.

In 2003 Dodds, Muhamad, and Watts conducted an experiment similar to Milgram's using email, with more than 60,000 volunteers and 18 targets in 13 countries. This experiment determined that the average number of hops was around five if the target was in the same country and seven if the target was in a different country, largely in line with Milgram's results. Interestingly, this experiment asked participants the reasons for picking their next particular acquaintance, finding that, especially during the early stages of routing, geographical proximity was the dominant factor [5]. This result suggests that realistic models aiming to explain the small-world phenomenon should incorporate geographical information.

1.1 Kleinberg's Model

In 2000, Jon Kleinberg proposed a famous model that, while not incorporating true geographical information, does consider a notion of geographic distance by placing nodes on an $n \times n$ grid. Kleinberg's model connects nodes using two types of connections—**local connections**, in which nodes are connected to all neighbors within a fixed lattice distance, and **long-range connections**, in which nodes are connected to random nodes in the graph. Importantly, these long-range connections are chosen with distance in mind, namely that closer nodes are picked more often as long-range connections than farther nodes. Specifically, each node u picks long-range connection v with probability proportional to $d(u,v)^{-s}$, where $d(u,v)$ is the lattice distance between u and v and s is the clustering exponent. This model mimics how individuals in a social network are more likely to know people who are geographically closer to them, but also have a small probability of knowing people who are farther away. Kleinberg showed that, for $s = 2$, a greedy routing algorithm can find paths of length $\mathcal{O}(\log^2 n)$ with high probability (w.h.p.), and that this is optimal for any s[1] [9]. In 2004, Martel and Nguyen proved tight bounds of expected $\Theta(\log^2 n)$ hops for greedy

---

[1] for 2-d grids.

routing, and of expected diameter of $\Theta(\log n)$—highlighting the large discrepancy between the two [12]. We are not aware of any other work that achieves an asymptotically better expected number of greedy routing hops using a constant average node degree and using only a constant average amount of local information per node.

1.2 The Neighborhood Preferential Attachment Model

In 2022, Goodrich and Ozel proposed a new model that combines the preferential attachment model with Kleinberg's model, which they call the ***neighborhood preferential attachment*** model [8]. In this model, as in the Barabási-Albert model, nodes are added to the graph one at a time, but instead of connecting to nodes solely based on their degree as in the preferential attachment models, they also take into account the distance between the nodes, as in Kleinberg's model. Specifically, each node u picks a node v with probability proportional to $\deg(v)/d(u,v)^s$, where $\deg(v)$ is the current degree of vertex v. Furthermore, Goodrich and Ozel expanded all three models (Barabási-Albert, Kleinberg, and their own) to work with underlying distances defined by a road network rather than a grid. In their work, they conducted rigorous experiments on U.S.A. road networks and showed that their model is able to outperform both the constituent models in terms of average greedy routing hops between randomly chosen pairs of nodes. In their paper, they describe how road networks serve as good proxies for social networks since the density of road infrastructure is correlated with population density. Their model was, at the time, the only randomized model to not only capture a proxy for the position of nodes in a social network, but also the power law distribution of node degrees that is widely common social networks. These two facts allowed this model to be the first randomized model able to reproduce results from Stanley Milgram's original small-worlds social experiment using a small average degree (only of around 30). However, importantly, they did not prove any theoretical bounds on their model. Our paper can be seen as a theoretical complement to their work, as we prove high probability bounds on the average greedy routing path length of a grid version of a very similar model, showing that it is far better than the $\Theta(\log^2 n)$ bound of Kleinberg's model.

1.3 Our Results

As stated before, our main goal for this paper was to provide theoretical results for the work of Goodrich and Ozel, or more generally, for preferential attachment variations of Kleinberg's model. In this paper, we propose three new models, each combining aspects of both Kleinberg's model and the preferential attachment model. We prove that, for grid networks, each of our networks are able to asymptotically outperform Kleinberg's original model in terms of average greedy routing path length, while using only a constant average amount of local information per node and while maintaining an expected constant average node degree.

We note that greedy routing can be improved by relaxing either of these two constraints. For example, if we allow nodes in the Kleinberg model to have access to more local information, we can improve greedy routing to $\mathcal{O}(\log^{3/2} n)$. Similarly, if we allow nodes to have a higher, $\mathcal{O}(\log n)$, average degree, then we can improve greedy routing to $\mathcal{O}(\log n)$ hops [12]. The latter of these two relaxations reveals that greedy routing can be greatly improved by getting to—and staying on—high degree nodes. With this in mind, we consider a node *highway*—a set of interconnected nodes that each have higher than average degrees. Our first two models introduce a parameter k that controls both the size of the highway and the degree of nodes on the highway. Specifically, the degree of nodes on the highway is proportional to k while the number of nodes on the highway is inversely proportional to k, such that the average degree of the entire graph is constant.

Our first model, the ***Kleinberg highway model*** (KH), works by embedding a Kleinberg grid within an $n \times n$ grid, such that there are n^2/k nodes on the highway. Each of the nodes on the highway grid only chooses long-range connections to other nodes on the highway grid, while local connections are still made to all neighbors within a fixed lattice distance as in the original Kleinberg model. Our second model, the ***randomized highway model*** (RH), is similar to the first, but instead of embedding a perfect Kleinberg grid inside the original graph, nodes are chosen uniformly at random to be on the highway grid. More specifically, each node has probability $1/k$ to become a highway node, leading to an expected $\Theta(n^2/k)$ highway nodes w.h.p. Both of these generalizations reduce to the original Kleinberg model when $k = 1$, that is when every node is a highway node, and adds a constant number of long-range connections per node. Importantly, both models reach a global minimum of $\mathcal{O}(\log n)$ hops when $k = \Theta(\log n)$, a much better result than Kleinberg's (see Fig. 1).

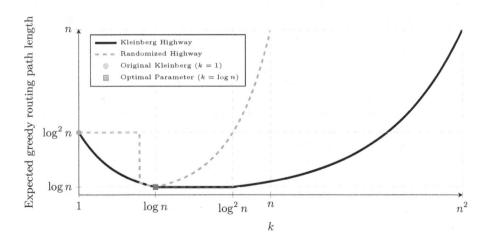

Fig. 1. The average greedy routing path length of the Kleinberg highway model for different values of parameter k.

Our final model is the **_windowed neighborhood preferential attachment model_** (windowed NPA), which like Goodrich and Ozel's neighborhood preferential attachment model (NPA), is based on both Kleinberg's model and the preferential attachment model. There are two main differences between the models. First, in the NPA model, the power law degree distribution naturally arises from the rich-get-richer selection property when adding new edges. In contrast, in our model, the power law degree distribution is strictly enforced, with each node picking a popularity k with probability $\propto k^{-\alpha}$. Each node node then adds a number of long-range connections proportional to its popularity. In order to maintain a constant average degree, the power law exponent α must be greater than 2, so $\alpha \geq 2 + \epsilon$ for any $\epsilon > 0$. The second main difference is that instead of there existing a probability of any two nodes being connected, in the windowed NPA model, nodes are only connected to other nodes within a constant factor of their popularity. The idea being that a residential street is more likely to connect to an alley, another residential street, or an arterial road, than it is to connect directly to a highway. This constant factor is controlled by a parameter A, and any node u with popularity k_u can only have long-range connections to nodes with popularity k_v such that $k_v \in [k_u/A, k_u \cdot A]$. We prove that for any arbitrarily small $\epsilon > 0$, the average greedy routing path length of the windowed NPA model is $\mathcal{O}(\log^{1+\epsilon} n)$ w.h.p.[2] While this result only holds for grid networks, we provide experimental results of our new model on both grid and road networks, showing that the windowed NPA model is able to outperform Kleinberg's model on both types of networks.

2   Preliminaries

As stated before, for the theoretical analysis, we will be using an $n \times n$ grid, such that the total number of nodes $|V| = n^2$. For simplicity, we will assume that our grid has wrap-around edges, as is common when analyzing grid networks [12], although our results can be extended to non-wrap-around grids. Let $d(u,v)$ be defined as the lattice distance between two nodes u and v in the grid, i.e. $d(u,v) = \min(\delta_x, n - \delta_x) + \min(\delta_y, n - \delta_y)$, where δ_x and δ_y are the absolute differences in the x and y coordinates of u and v, respectively. Let $B_d(u)$ denote the set of nodes within lattice distance d from u. All three models have the notion of **local connections** and **long-range connections**. Without loss of generality, we will only consider the case where we only add immediately adjacent local connections, that is, each node is only connected to the four nodes directly above, below, to the left, and to the right of it. Equivalently, we can say that each node is connected to all other nodes in $B_1(u)$, as in the case when $p = 1$ in Kleinberg's original model. In this paper, when we refer to a node's degree $\deg(u)$, we will be referring to the number of outgoing long-range connections from u.

---

[2] We proved this for a slightly modified greedy routing algorithm.

3   Kleinberg Highway

As stated before, Kleinberg's model is defined on a graph $\mathcal{G}$ comprising of an $n \times n$ grid where each node u adds local connections to all nodes in $B_P(u)$ (all nodes within lattice distance P of u), and Q long-range connections to other nodes. The probability of adding a long-range connection to node v is proportional to $d(u, v)^{-r}$. In our model, we will set P to 1 w.l.g., and we will set $r = 2$, as this is the value that Kleinberg showed was optimal for 2-dimensional grids, and Goodrich and Ozel hypothesized could be optimal for road networks [8,9]. Furthermore, in our model, we will define a subgraph $\mathcal{G}_H$, known as the **highway**, which for this model is an $n_H \times n_H$ evenly spaced grid in $\mathcal{G}$. We introduce a new parameter k in the range of $1 \leq k \leq n^2$, where $1/k$ of the nodes are designated as **highway nodes**, meaning that n_H is equal to $n/\sqrt{k}$ (which for simplicity we assume is a whole number). Now, we introduce two forms of local connections, the first connects all nodes in the entire graph $\mathcal{G}$ to their neighbors, and the second connects all nodes in the highway subgraph $\mathcal{G}_H$ to their highway neighbors. Finally, and importantly, only highway nodes are able to add long-range connections, and these long-range connections are *directed* edges added only *to other highway nodes* (see Fig. 2). Since there are fewer highway nodes, we are able to add proportionally more long-range connections per node to maintain the same constant average degree Q. In particular, each highway node is able to add $Q \times k$ long-range connections, where Q, as in the original Kleinberg model, represents the average highway degree. Put simply, $\mathcal{G}_H$ is a Kleinberg graph with Kleinberg parameters: $n = n_H = n/\sqrt{k}$, $p = 1$, $q = Q \times k$, $r = 2$. We call the entire graph $\mathcal{G}$ the **Kleinberg highway** model.

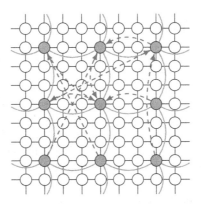

Fig. 2. An example of the Kleinberg highway model with $n = 9$, $k = 9$, and $Q = 1/9$. The solid black and curved solid blue lines represent local connections for the entire grid and for the highway grid, respectively. The value of Q was picked such that each highway node has only one long-range connection (represented by the dashed light green directed lines) to make the graph less cluttered. If Q were 1, each highway node would have 9 long-range connections.

3.1   Results

Our results depend on whether or not the structure of the highway is known to the vertices. Due to the structured nature of the highway, we will assume that its layout is known to all vertices (a constant amount of information), such that nodes know the location of the closest highway node to them. We will include both results for completeness, and both have the same optimum value and result, but our standard definition of our model will include this natural assumption.

We split our decentralized algorithm to route from s to t into three steps:

1. We use local connections in $\mathcal{G}$ to route from s to the closest highway node.
2. We traverse the highway ($\mathcal{G}_H$) using standard Kleinberg routing towards t.
3. Finally, we use the local connections in $\mathcal{G}$ to route to t.

A straightforward proof, included for completeness in Sect. 7.2, produces the following result:

Theorem 1. *The expected decentralized routing time in a* **Kleinberg highway** *network is* $\mathcal{O}(\sqrt{k} + \log^2(n)/k + \log n)$ *for* $1 \leq k \leq n^2$ *when each node knows the positioning of the highway grid, and* $\mathcal{O}(k + \log^2(n)/k)$ *otherwise.*

Reassuringly, both results are consistent with the original Kleinberg model when k is constant, with the expected routing time being $\mathcal{O}(\log^2 n)$. Our key observation, however, is that the expected routing time reaches a global minimum when $\Theta(\log n) \leq k \leq \Theta(\log^2 n)$ when the positioning of the highway is known, or just when $k \in \Theta(\log n)$ in general, in which case the expected routing time becomes $\mathcal{O}(\log n)$, as shown visually in Fig. 1. This is a major improvement over the original Kleinberg model.

4   Randomized Highway

The key difference between this model and the Kleinberg highway model is that in this model highway nodes are distributed randomly through the entire graph $\mathcal{G}$ instead of the unrealistic expectation that they are distributed perfectly uniformly. As in the previous model, nodes are laid out in an $n \times n$ grid with wraparound, where each node is connected to its 4 directly adjacent neighbors. Each node independently becomes a highway node with probability $1/k$ for $1 \leq k \leq n^2/\log n$ such that there are an expected $\Theta(n^2/k)$ highway nodes total w.h.p., and each highway node adds $Q \times k$ long-distance connections to other highway nodes such there is an expected average of Q long-distance connections per node w.h.p.[3]. As before, each highway node only considers other highway nodes as candidates for long-distance connections, and the probability that highway node u picks highway node v as a long-distance connection is proportional to $d(u,v)^{-2}$. An important difference, however, is that there is no clear notion of local connections between highway nodes in this graph, which will affect the decentralized greedy routing results. See Fig. 3.

---

[3] This holds for $k \in o(n^2/\log n)$ when $k \in \Theta(n^2/\log n)$, the density is at most αQ w.h.p. for a large enough constant α.

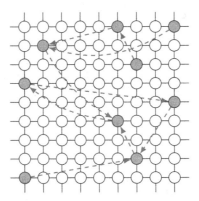

Fig. 3. An example of the randomized highway model with $n = 9$, $k = 9$, and $Q = 1/9$. The solid black and curved solid blue lines represent local connections for the entire grid. In this model, there are no local connections for the highway subgraph. The value of Q was picked such that each highway node has only one long-range connection (represented by the dashed light green directed lines) to make the graph less cluttered. If Q were 1, each highway node would have 9 long-range connections.

4.1   Results

As before, we split our decentralized routing algorithm into three steps: reaching a highway node from s, traversing the highway, and reaching t from the highway. While traversing the highway, we will only take local connections that improve our distance to t by at least $4\sqrt{k}$, for reasons that will be clear from the proof of Lemma 3. We will show that the expected time to reach a highway node from s is $\mathcal{O}(k + \log n)$ w.h.p., the expected time to traverse the highway is $\mathcal{O}(\log^2 n)$ w.h.p. for $k \in o(\log n)$ or $\mathcal{O}(\log n)$ w.h.p. for $k \in \Omega(\log n)$, and the expected time to reach t from the highway is $\mathcal{O}(k + \log n)$ w.h.p. From these results, we will obtain:

Theorem 2. *For $k \in o\left(\frac{\log n}{\log\log\log n}\right)$, the expected decentralized greedy routing path length is $\mathcal{O}(\log^2 n)$ w.h.p., while for $\Theta\left(\frac{\log n}{\log\log\log n}\right) \le k < \Theta(\log n)$, the expected decentralized greedy routing path length is $\mathcal{O}(\log^2(n)/k)$ w.h.p., and finally for $\Theta(\log n) \le k \le \Theta(n)$, the expected decentralized greedy routing path length is $\mathcal{O}(k)$. Finally, for $k \in \Omega(n)$, the expected decentralized greedy routing path length is $\mathcal{O}(n)$.*

Note that importantly, the results of Theorem 2 are worse than the results of Theorem 1 for values of k between $\Theta(1)$ to $o\left(\frac{\log n}{\log\log\log n}\right)$, and for values of k greater than $\Theta(n)$. This can be attributed to two facts, the first being that the location of the closest highway node to s is not known, and the second being that there is no notion of local connections between the highway nodes.

4.2 Greedy Routing Sketch

Proving the expected decentralized greedy routing path length results for the randomized highway model in Theorem 2 follows similar steps to the proof for the Kleinberg highway model in Theorem 1. We include a sketch below, leaving the complete proofs for the appendix in Sect. 7.3.

We start by proving a lower bound on the probability that a long-range connection exists between two arbitrary highway nodes. In order to do this, we need to find a high probability upper bound on the normalization constant z for any arbitrary highway node.

Lemma 1. *The normalization constant z for any arbitrary highway node is at most $25 \log \log \log n + \frac{41}{9} \frac{\log n}{k \log \log n} + 26 \frac{\log n}{k}$ for $n > 5$ w.h.p. (for at most $\mathcal{O}(\log^2 n)$ invocations).*

This result gives us a normalization constant that is in $\mathcal{O}(\log(n)/k)$ for $k \in o\left(\frac{\log n}{\log \log \log n}\right)$, and in $\mathcal{O}(\log \log \log n)$ for $k \in \Omega\left(\frac{\log n}{\log \log \log n}\right)$. Note that this bound is worse for large values of k than the bound we obtained for the Kleinberg highway model in Lemma 7. We can, however, improve this bound, but without the same high probability guarantees:

Lemma 2. *The normalization constant z for any arbitrary highway node is at most $10 + 37 \frac{\log n}{k}$ for $n > 2$ with probability at least $1/2$. From now on, we will refer to this tighter bound as z'.*

This improved bound gives us a normalization constant that is in $\mathcal{O}(\log(n)/k)$ for $k \in o(\log n)$ in $\mathcal{O}(1)$ for $k \in \Omega(\log n)$, a result in line with the Kleinberg highway model. We want to be able to use this improved bound when calculating the probability of halving the distance to the destination.

Lemma 3. *Using the improved normalization constant bound z' incurs at most a constant factor to the probability of halving the distance to the destination while routing w.h.p.*

Now we can use these improved normalization constant bounds to find the probability of halving our distance. Suppose we are in phase j where $\log(c(k + \log n)) \leq j \leq \log n$ (for some constant c we will discuss later), and the current message holder u is a highway node. Let us find the probability that we have a long-range contact that is in a better phase. First, we find the number of highway nodes in a better phase than us, i.e., within the ball of radius 2^j around t $(B_{2^j}(t))$.

Lemma 4. *There are at least $2^{2j-2}/k$ highway nodes in a ball of radius 2^j for $\log(c(k + \log n)) \leq j \leq \log n$ with high probability (with probability at least $1 - n^{-0.18c^2}$).*

Each of these nodes has lattice distance less than 2^{j+2}, allowing us to bound the probability of them being a specific long-range contact of u. Then, we can obtain an identical result (in asymptotic notation) to the result in Lemma 8:

Lemma 5. *In the randomized highway model, the probability that a node u has a long-range connection to a node v that halves its distance to the destination is proportional to at most $k/\log n$ for $k \in \mathcal{O}(\log n)$ and is constant for $k \in \Omega(\log n)$.*

Once we reach phase $j = \log(c(k + \log n))$, we are at distance $\mathcal{O}(k + \log n)$ from the destination, reaching it in $\mathcal{O}(k + \log n)$ local hops. As stated up until now, we would be able to perform greedy routing with results equivalent to those of Theorem 1 assuming no knowledge about the positioning of the highway nodes ($\mathcal{O}(k + \log^2(n)/k)$ routing). However, we have not yet addressed the elephant in the room: the fact that there is no notion of local contacts between highway nodes. In simple terms, while routing, if there are no long-range contacts that improve your distance, you must leave the highway. And when you leave the highway, it may take a while to get back onto it. We will show that this is not a problem for large values of k, i.e. values of $k \in \Omega(\log n)$, but for smaller values of k the bound will be worse than before, becoming $\mathcal{O}(\log^2 n)$ expected routing instead of $\mathcal{O}(\log^2(n)/k)$ (note that we do not prove that the bound is tight). In Sect. 7.5 we propose a variant which trivially achieves the improved $\mathcal{O}(\log^2(n)/k)$ expected routing for small values of k. We consider this variant slightly less elegant, and since it maintains the same optimal results, we do not consider it further.

5   Windowed Neighborhood Preferential Attachment

Our previous models have a binary distinction between highway nodes and normal nodes, represented by a fixed value of k. We now describe a new model with a continuous transition, where each node picks its own value of k, such that the distribution of the values of k, and consequently the degree distribution, exhibits a power law. Each node independently picks their probability k from a distribution $\Pr(k) \propto 1/k^{2+\epsilon}$ for $\epsilon > 0$. Each node u then adds $\epsilon Q \times k$ long-range connections, but only to nodes within a given range, or "window", of popularity. Specifically, let the window of popularity for a given node u with popularity k_u be popularities in the range $[k_u/A, Ak_u]$.

5.1   Results

While at first glance this model may seem irreconcilable from the previous models, consider referring to all nodes with popularity $\log n \leq k \leq A \log n$ as the "highway". We expect to have $\mathcal{O}(1/\log^{1+\epsilon} n447)$ highway nodes. Ignoring all long-range connections that do not connect two highway nodes, we find an instance of the randomized highway model embedded within the windowed NPA model, albeit with a small (but nevertheless constant) value of Q. With these key observations, we are able to prove:

Theorem 3. *The windowed NPA model has a decentralized greedy algorithm that routes in $\mathcal{O}(\log^{1+\epsilon}(n))$ hops w.h.p.*

The complete proof for this theorem can be found in Sect. 7.6. Furthermore, experimental results confirming that this model greedily routes significantly better than Kleinberg's can be found in Sect. 7.1.

5.2   Efficient Construction

The neighborhood preferential attachment model of Goodrich and Ozel [8] takes $\mathcal{O}(|V|^2)$ time to construct and there is no more efficient construction currently known. The windowed NPA model can similarly be constructed sequentially in $\mathcal{O}(|V|^2)$ time. However, due to how each node picks their connections independently, this model is embarrassingly parallel, and can be constructed in $\mathcal{O}(|V|)$ time with $|V|$ processors, without any communication between processors.

6   Future Work

It would be interesting to be able to prove whether our bounds are tight for our models. Specifically, whether the bounds for the randomized highway model can be improved to be more in line with the Kleinberg highway results. While the diameter of models with constant degree is at least $\Omega(\log n)$, there is no such lower bound when dealing with constant *average* degree. It would be interesting to either bridge the gap or show that a true gap exists between the lower bound on the diameter of our networks, $\Omega(\log n / \log \log n)$, and the upper bound on greedy routing, $\mathcal{O}(\log n)$. Also, it would be interesting to prove whether it is possible to achieve a greedy routing time of $\log n + \sqrt{k}$ for larger values of k if each node knows the location of the nearest highway node (a constant amount of additional information). This result would improve the expected running time of the windowed NPA model to just $\mathcal{O}(\log n)$ for $0 < \epsilon \leq 1$. Finally, our analysis for the randomized highway model depends on the network having a mostly even spread of nodes. Experimentally, both our model and the original NPA model perform worse on Alaska, a highly unevenly spread out state. It would be interesting to generalize our results if some form of density condition is met.

7   Appendix

7.1   Experimental Analysis

Goodrich and Ozel's paper on the neighborhood preferential model [8] was able to show that a hybrid model combining elements from Kleinberg's model with preferential attachment is able to outperform both individual models for decentralized greedy routing on road networks by showing many experimental results. In the previous sections, we provided some theoretical justification for their results, by proving asymptotically better greedy routing times for a similar model. In this section, we complete our comparisons by reproducing their key experimental results with our new model. Our experimental framework is nearly identical to theirs, except that we implement directed versions of each algorithm, i.e. where each long-range connection is directed (local connections are by definition always undirected). This allows us to run experiments much more efficiently—we sample between 30,000 to 200,000 source/target pairs for each data point, as compared to their 1,000 pairs—but results in all algorithms having a worse performance. For our experiments we picked $\epsilon = 0.5$ and $A = 1.01$. It is possible that other parameters would yield better results.

Key Results. Our main key result is that our windowed NPA model outperforms Kleinberg's model for road networks by a factor of 2, as shown in Fig. 4. This result is directly in line with Goodrich and Ozel's experimental results with their similar model [8]. It is worth mentioning that our directed version of the model is worse than the undirected version from Goodrich and Ozel's paper by roughly a factor of 2.

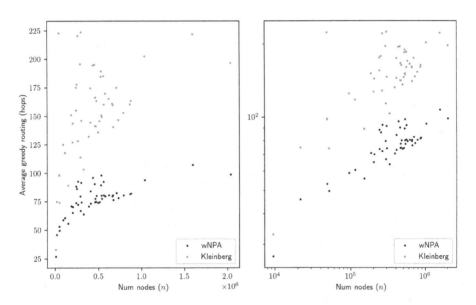

Fig. 4. Comparison of greedy routing times for Kleinberg's model and the windowed NPA model when $Q = 1, \epsilon = 0.5, A = 1.01$. The right plot is in log scale.

Similarly, we show that by increasing the degree density to 32 we can achieve a result of less than 20 degrees of separation, which again is roughly twice the results from Goodrich and Ozel's paper (see Fig. 5), which we attribute primarily to the directed implementation of the models for our experiments.

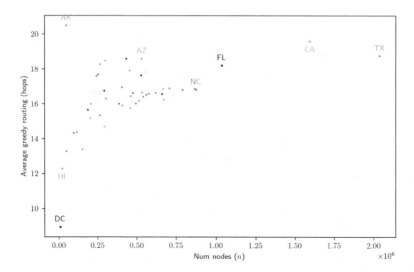

Fig. 5. The greedy routing times for the windowed NPA model on the 50 US states when $Q = 32$, $\epsilon = 0.5$, and $A = 1.01$.

7.2   Kleinberg Highway Proofs

In this section, we prove Theorem 1 by proving upper bounds on each of the three steps of the greedy routing algorithm: routing from s to the highway using local connections, within the highway towards t using standard Kleinberg routing, and finally from the highway to t again using local connections.

Lemma 6. *It is possible to route from any node $s \in \mathcal{G}$ to a highway node $h \in \mathcal{G}_H$ in at most $\sqrt{k}$ hops, if the location of h is known, or in at most $k - 1$ hops, if the location of h is not known.*

Proof. Without loss of generality, let's assume highway nodes are located wherever $\mod(x, \sqrt{k}) = 0$ and $\mod(y, \sqrt{k}) = 0$. Then, the maximum distance in the x dimension to a highway node is $\delta_x = \min(\mod(x, \sqrt{k}), \sqrt{k} - \mod(x, \sqrt{k})) = \left\lfloor \frac{\sqrt{k}}{2} \right\rfloor$, and an equivalent result holds for δ_y. Therefore, the maximum lattice distance to a highway node is the sum of both, or at most $2 \left\lfloor \frac{\sqrt{k}}{2} \right\rfloor \leq \sqrt{k}$. If the location of h is known, then we can route to it directly taking a number of hops equal to the lattice distance to h. If the location of h is not known, we can visit every node in a $\sqrt{k} \times \sqrt{k}$ square, guaranteeing that we will encounter a highway node h, in $k - 1$ hops.

After we reach the highway subgraph $\mathcal{G}_H$, we can use the standard Kleinberg routing algorithm towards t. As in Kleinberg's original analysis, we first prove a lower bound on the probability that a long-range connection exists between two arbitrary highway nodes.

Lemma 7. *The normalization constant z for $\mathcal{G}_H$ is upper bounded by $z \leq 4\ln(6n_H) \leq 4\ln(6n)$. As such, the probability of any two highway nodes u and v being connected is at least $[4\ln(6n)d_H(u,v)^2]^{-1}$, where $d_H(u,v)$ is the lattice distance between u and v in $\mathcal{G}_H$.*

Proof. This result follows directly from Kleinberg's original analysis on the highway subgraph $\mathcal{G}_H$.

In Kleinberg's analysis, the probability that a node u has a long-range connection to a node v that halves its distance to the destination is proportional to $[\log n]^{-1}$, when a node has a constant number of long-range connections Q. In our case, each highway node has $Q \times k$ long-range connections, where k does not need to be constant. This gives us improved distance-halving probabilities:

Lemma 8. *In the Kleinberg highway model, the probability that a node u has a long-range connection to a node v that halves its distance to the destination is proportional to at most $k/\log n$ for $k \in \mathcal{O}(\log n)$ and is constant for $k \in \Omega(\log n)$.*

Proof. Following Kleinberg's analysis, the probability that a single long-range connection from u halves its distance to the destination is still proportional to $[\log n]^{-1}$. Therefore, the probability that a single long-range connection does *not* halve its distance to the destination is proportional to $1 - [\log n]^{-1}$. The probability that all Qk long-range connections do not halve the distance is therefore proportional to $\left(1 - [\log n]^{-1}\right)^{Qk} = \left[\left(1 - [\log n]^{-1}\right)^{\log n}\right]^{\frac{Qk}{\log n}} \leq e^{-\frac{Qk}{\log n}}$. Finally, the probability that any one of the Qk succeed in halving the distance is therefore proportional to $1 - e^{-\frac{Qk}{\log n}}$. When $k \in \omega(\log n)$, the exponential term tends towards zero, and the probability tends towards one. For smaller values of k, a Taylor expansion of $e^{-\frac{Qk}{\log n}}$ shows that this probability is proportional to at least $1 - \left[1 - \frac{Qk}{\log n} + \mathcal{O}\left(\left[\frac{Qk}{\log n}\right]^2\right)\right] = \frac{Qk}{\log n} - \mathcal{O}\left(\left[\frac{Qk}{\log n}\right]^2\right)$. When $k \in o(\log n)$, the lower order terms become asymptotically negligible, and we are left with a probability proportional to $\frac{Qk}{\log n} = \mathcal{O}(k/\log n)$. When $k = \Theta(\log n)$, we are left with a constant dependent on Q.

Importantly, this result reproduces Kleinberg's original result when k is constant, since we are left with a probability proportional to $1/\log n$. Finally, we can prove the main result of this section:

Proof (of Theorem 1). It is possible to describe the greedy routing path in terms of at most $\log n$ phases, where a node u in phase j if it is at a lattice distance between 2^j and 2^{j+1} from the destination t. It is easy to see that halving the distance to the destination results in reducing what phase a node is in by one. The expected amount of hops spent in each phase is therefore $1/\Pr(\text{distance halving}) = \mathcal{O}(\log(n)/k)$. Note that importantly, when no long-range connections halve the distance, we take local connections on the *highway graph* towards t, as in the original Kleinberg model. Since there are at most $\log n$

phases, we expect to spend at most $\mathcal{O}(\log n(\log(n)/k+1))$ hops on the highway[4]. Finally, the final highway node is known to be at most $\sqrt{k}$ hops away from the destination t. The theorem follows from these results along with the results from Lemma 6.

7.3 Randomized Highway Proofs

We now present proofs of theorems and lemmas discussed in Sect. 4.2.

The Nested Lattice Construction. For our proofs, similarly to the Kleinberg highway model, we will conceptually subdivide the highway into a lattice of balls of various sizes (see Fig. 6 for an example nested lattice structure), and show upper and lower bounds on the number of highway nodes within each ball with varying degrees of probability bounds. Specifically we will prove:

Lemma 9. *Results from the nested lattice structure:*

1. *All balls of radius $3\sqrt{k\log n}$, centered around any of the n^2 nodes, contain at least $9\log n$ highway nodes with high probability in n.*
2. *All balls of radius $3\sqrt{k\log n}$, centered around any of the n^2 nodes, contain fewer than $41\log n$ highway nodes with high probability in n.*
3. *$\mathcal{O}(\log^2 n)$ balls of radius $3\sqrt{k\log\log n}$, centered around any $\mathcal{O}(\log^2 n)$ nodes, contain fewer than $41\log\log n$ highway nodes with high probability in $\log n$.*
4. *Any arbitrary ball of radius $2\sqrt{k}$ has at most 18 highway nodes with probability at least $1/2$. This result is not a high probability bound, and is only independent for balls centered around nodes with lattice distance greater than $4\sqrt{k}$ between them.*

Proof. Consider balls of radius $a\sqrt{k\log n}$ for some constant a. There are at least $2a^2 k\log n$-many nodes within each ball of radius $a\sqrt{k\log n}$. The probability that any node is a highway node is $1/k$, so the expected number of highway nodes within each ball is $\mu \geq 2a^2\log n$. We can lower bound the number of highway nodes within each ball by using a Chernoff bound. Letting X be the number of highway nodes within each ball, we have:

$$\Pr(X \leq (1-\delta)\mu) \leq e^{-\frac{\delta^2\mu}{2}} = e^{-a^2\delta^2\log n} = n^{-\frac{a^2\delta^2}{\ln 2}}$$

By union bound, the probability this fails for a ball centered at any of the n^2 vertices is at most $n^{2-\frac{a^2\delta^2}{\ln 2}}$. Setting $\delta = 1/2$ and $a = 3$, we obtain that all balls with radius $3\sqrt{k\log n}$ have at least $9\log n$ highway nodes with probability at least $1 - n^{-1.24}$, which is w.h.p. For an upper bound, we first note that there

---

[4] Some minor details regarding the final $\log\log n$ phases are omitted for brevity.

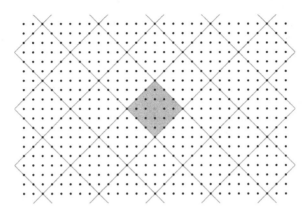

Fig. 6. The nested lattice construction showing balls of radius 3, centered around an orange node. The central ball is depicted in solid light green, while the 8 adjacent balls are shown in dashed yellow.

are fewer than $3a^2 k \log n$-many nodes within each ball of radius $a\sqrt{k \log n}$ for radii of at least 3. Using another Chernoff bound:

$$\Pr(X \geq (1+\delta)\mu) \leq e^{-\frac{\delta^2 \mu}{2+\delta}} = e^{-\frac{2a^2 \delta^2 \log n}{2+\delta}} = n^{-\frac{3a^2 \delta^2}{\ln 2(2+\delta)}}$$

By setting $\delta = 1/2$ and $a = 3$, we obtain that all balls with radius $3\sqrt{k \log n}$ have fewer than $41 \log n$ highway nodes w.h.p. (with probability at least $1 - n^{-1.89}$). We can obtain similar bounds for smaller balls, although with worse probabilities. For example, for balls of radius $a\sqrt{k \log \log n}$, we expect $\mu < 3a^2 \log \log n$ highway nodes for radii of at least 3. Using another Chernoff bound with $\delta = 1/2$ and $a = 3$, we obtain that any given ball with radius $3\sqrt{k \log \log n}$ has more than $41 \log \log n$ highway nodes with probability less than $\log^{-3.89} n$. Assuming we will only invoke this bound at most $\mathcal{O}(\log^2 n)$ times, the probability that any of the invocations fail is negligible (at most $\mathcal{O}(\log^{-1.89} n)$). Finally, we consider balls of radius only $2\sqrt{k}$, which have at most 18 highway nodes with probability at least $1/2$.

Finding the Normalization Constant. The probability that highway node u picks highway node v as a long-range connection is $d(u,v)^{-2} / \left[\sum_{w \neq u} d(u,w)^{-2} \right]$, where each w in the summation is a highway node. In order to lower bound this probability, we must upper bound the denominator, known as the ***normalization constant*** z.

Proof (of Lemma 1). Let's consider a lattice of balls centered around an arbitrary highway node u. Let's define a notion of "ball distance" b to measure the distance between two balls in this ball lattice. Let $\mathcal{B}_b(u)$ be the set of all balls at ball distance b from a ball centered at u. There is 1 ball at ball distance 0 ($|\mathcal{B}_0(u)| = 1$), 8 balls at ball distance 1, and in general at most $8b$ balls at distance b for

$b > 0$ (see Fig. 6). The minimum distance between u to a node in another ball at distance b is $2b - 1$ times the ball radius for $b > 0$. Let's consider a lattice of balls with radius $3\sqrt{k \log n}$. From Lemma 9.2 we know that there are at most $41 \log n$ highway nodes within this ball w.h.p. Let's also find the normalization constant in two parts, first due to highway nodes in $b > 0$ ($z_{>0}$), and then due to highway nodes within the same ball (z_0).

Note that any two balls are separated by ball distance at most $2n$/twice the ball radius, or $\frac{n}{3\sqrt{k \log n}}$.

$$z_{>0} \leq \sum_{b=1}^{\frac{n}{3\sqrt{k \log n}}} \frac{(\text{max \# highway nodes in } \mathcal{B}_b(u))}{(\text{min distance to node in } \mathcal{B}_b(u))^2}$$

$$\leq \sum_{b=1}^{\frac{n}{3\sqrt{k \log n}}} \frac{41 \log n \times 8b}{(2b-1)^2 \times 9k \log n} < \frac{37}{k} \sum_{b=1}^{\frac{n}{3\sqrt{k \log n}}} \frac{b}{(2b-1)^2}$$

$$\leq \frac{37}{k} \sum_{b=1}^{\frac{n}{3\sqrt{k \log n}}} \frac{1}{b} = \frac{37}{k} \mathcal{H}\left(\frac{n}{3\sqrt{k \log n}}\right)$$

$$\leq \frac{37}{k} \mathcal{H}\left(\frac{n}{3\sqrt{\log n}}\right) < 26 \frac{\log n}{k} \text{ for } n > 2$$

Now that we showed the contribution of highway nodes in different balls from u, let's bound the contribution due to highway nodes within the same ball. We are only interested in the normalization constant for nodes that we visit along the highway, which we will show is at most $\mathcal{O}(\log^2 n)$ nodes. Knowing this, we can use the improved bound for balls of radius $3\sqrt{k \log \log n}$, which from Lemma 9.3 we know contain fewer than $41 \log \log n$ highway nodes w.h.p. Let's consider the worst case where they are all bunched up around u. Let's denote their contribution $z_{0,\text{inner}}$.

$$z_{0,\text{inner}} \leq \sum_{j=1}^{\lceil\sqrt{41 \log \log n}\rceil} \frac{4j}{j^2} < 4\mathcal{H}\left(\sqrt{41 \log \log n} + 1\right)$$

$$< 25 \log \log \log n \text{ for } n > 5$$

Recall that we can still have up to $41 \log n$ highway nodes in in the same (large) ball as u. Let's assume they are all as close as possible, meaning that they are all at the edge of the inner ball. Let's denote their contribution $z_{0,\text{outer}}$.

$$z_{0,\text{outer}} < \frac{41 \log n}{(3\sqrt{k \log \log n})^2} = \frac{41}{9} \frac{\log n}{k \log \log n}$$

Combining these results, we obtain:

$$z < 25 \log \log \log n + \frac{41}{9} \frac{\log n}{k \log \log n} + 26 \frac{\log n}{k} \text{ for } n > 5$$

w.h.p., for at most $\mathcal{O}(\log^2 n)$ invocations.

We provide a tighter bound for the normalization constant, z', in a similar fashion:

Proof (of Lemma 2). Recall from Lemma 9.4 that balls of radius $2\sqrt{k}$ have at most 18 highway nodes with probability at least $1/2$. When this occurs, $z_{0,\text{inner}}$ can be improved to:

$$z_{0,\text{inner}} < \sum_{j=1}^{5} \frac{4j}{j^2} = 4\mathcal{H}(5) < 10$$

Meanwhile, $z_{0,\text{outer}}$ changes to:

$$z_{0,\text{outer}} < \frac{41 \log n}{(2\sqrt{k})^2} = \frac{41}{4}\frac{\log n}{k}$$

Overall, with probability at least $1/2$, we obtain the improved bounds on the normalization constant:

$$z' < 10 + 37\frac{\log n}{k} \text{ for } n > 2$$

Probability of Distance Halving. As explained before, the first step is to show that we can use the improved bounds on the normalization constant by incurring only an increase in a constant factor to the probability of halving the distance:

Proof (of Lemma 3). The probability of the improved normalization constant bound z' applying is at least $1/2$, and this probability is independent for any nodes a distance of at least $4\sqrt{k}$ apart (see Lemma 9.4). For values of $k \in o\left(\frac{\log n}{\log \log \log n}\right)$, the improved normalization constant bound is already only a constant factor better. For values of $k \in \Omega\left(\frac{\log n}{\log \log \log n}\right)$ we will show that we can always visit highway nodes that are at least $4\sqrt{k}$ apart, so that we have independence. All our routing algorithms expect to take $\mathcal{O}(\log n)$ hops on the highway, or $a \log n$ hops for some constant a. We expect at least $\frac{1}{2}a \log n$ of the highway nodes visited to have the improved bounds apply. By Chernoff bound, we visit at least $\frac{1}{4}a \log n$ highway nodes with the improved bounds w.h.p. (with probability at least $1 - n^{-\frac{a}{16 \ln 2}}$). Since a can be picked arbitrarily large, then with high probability we will visit $\mathcal{O}(\log n)$-many nodes with the improved bounds along our path, which is the same as our original expectation of how many nodes we will visit, meaning our results are the same up to a constant hidden by the asymptotic notation. Note that a similar reasoning works for smaller values of k as well.

Next, we need to prove a lower bound on how many nodes are in a better phase than us w.h.p.:

Proof (of Lemma 4). Kleinberg showed that there are more than 2^{2j-1} nodes within lattice distance 2^j of t [9], for $\log \log n \leq j < \log n$. Within this range, we expect there to be at least $2^{2j-1}/k$ highway nodes. Since we are only considering the case where $j \geq \log(c(k + \log n))$, we can use this to create a Chernoff bound (with $\delta = 1/2$). Letting X be the number of highway nodes:

$$\Pr(X \leq \mu/2) \leq e^{-\frac{\mu}{8}} = e^{-\frac{2^{2j-1}}{8k}} \leq e^{-\frac{2^2 \log(c(k+\log n))}{16k}}$$

$$= e^{-\frac{[c(k+\log n)]^2}{16k}} < e^{-\frac{c^2(2k\log n)}{16k}} = n^{-\frac{c^2}{8\ln 2}}$$

$$< n^{-0.18c^2}$$

In summary, since we picked $\delta = 1/2$, we expect at least $2^{2j-2}/k$ highway nodes, to be within lattice distance 2^j of t w.h.p. (with probability at least $1 - n^{-0.18c^2}$).

Finally, we use these results to prove the main lemma of this section, the probability of halving the distance:

Proof (of Lemma 5). From our previous results, we know we can use the improved bounds for the normalization constant, $z' = 10 + 37\frac{\log n}{k}$, with at most a constant factor increase in the probability of halving the distance. Furthermore, we know that there exist at least $2^{2j-2}/k$ highway nodes in better phases w.h.p. Since they are in phase j or better, they are each within lattice distance $< 2^{j+1} + 2^j < 2^{j+2}$ from u. Using this, and letting v be an arbitrary long-range connection of u, we obtain:

$$\Pr(v \in B_{2^j}(u)) > [64kz']^{-1} > [64k \times 37(1 + \log(n)/k)]^{-1}$$

The probability of v not being in a better phase is similarly $1 - \Pr(v \in B_{2^j}(u))$. Recalling that each highway node has Qk independently chosen random long-range connections, the probability of none of them being connected to a better phase is therefore $(1 - \Pr(v \in B_{2^j}(u)))^{Qk} \leq e^{-Qk\Pr(v\in B_{2^j}(u))}$. The probability of any one of them being connected is therefore:

$$\Pr(\exists v \in B_{2^j}(u)) \geq 1 - e^{-Qk\Pr(v\in B_{2^j}(u))} > 1 - e^{-\frac{Qk}{2368(k+\log n)}}$$

When $k \in o(\log n)$, the $\log n$ term in the denominator dominates, and we obtain similar asymptotic results to Lemma 8. When $k \in \Omega(\log n)$, the k term in the denominator dominates, cancelling out the k term in the numerator, and leaving us with a constant term dependent on Q. It is worth noting that the constant factors in this analysis are very loose, and also considerably decrease for larger values of n. In any case, we obtain that the probability of halving the distance is at least in $\mathcal{O}(k/\log n)$ for $k \in o(\log n)$, and at least $f(Q) = \mathcal{O}(1)$ for $k \in \Omega(\log n)$.

7.4 Removing Local Contact Dependence

In this section, we complete the proof of Theorem 2 by removing the dependence on local connections. The results of the theorem directly follow.

If we do find a long-range connection that takes us to the next phase, we can just take it, but what do we do when there aren't any? To continue the Kleinberg analogy, we would just keep taking local connections to keep re-rolling the dice, and as long as we never traverse any space twice and never traverse any space that is within $4\sqrt{k}$ of previous spaces (because of Lemma 9.4), we can assume each step taken is independent of other steps. The obvious problem here is that there is no notion of "local connections" in this randomly selected highway. We could either greedily take local connections in the entire graph until we happen to reach a highway node again (in expected $\mathcal{O}(k)$ time), or we can simply pick any long-range connection that takes us closer to the destination by at least $4\sqrt{k}$. For values of $k \in o(\log n)$, we will use the first method (greedily taking local connections), and for values of $k \in \Omega(\log n)$, we will use the second.

Values of $k \in o\left(\frac{\log n}{\log\log\log n}\right)$. For these smaller values of k, from Lemma 5, we expect to take $\mathcal{O}(\log(n)/k)$ hops on highway nodes to reach the next phase, and since there are at most $\log n$ total phases, we expect to visit at most $\mathcal{O}(\log^2(n)/k)$ highway nodes throughout the entire routing process w.h.p. In the worst case, whenever we can't halve the distance, we never have any closer long-range connections, so we would need to greedily move along local contacts towards t until reaching another highway node. Recalling that each node has probability $1/k$ of being a highway node, and that we expect to visit a highway node every k independent hops. In order to avoid visiting highway nodes within $4\sqrt{k}$ of each other, we can first walk $4\sqrt{k}$ hops before checking for highway nodes, which we will expect to find after $4\sqrt{k} + k \in \mathcal{O}(k)$ hops. Over the entire duration of the routing, we expect to spend $\mathcal{O}(\log^2(n)/k \times k) = \mathcal{O}(\log^2 n)$ hops using local connections to reach highway nodes w.h.p.

Values of $k \in \Omega\left(\frac{\log n}{\log\log\log n}\right)$ For these larger values of k, we will prove that we can find a long-range connection to an arbitrary highway node u in phase $\log(c(k+\log n)) \le j < \log n$ that is at least $4\sqrt{k}$ closer to the destination t, w.h.p. Recall that long-range connections are always only between highway nodes, so taking them will always keep us on the highway. To find the probability of one of these connections existing, we consider a ball of radius $d - 4\sqrt{k}$ centered on the destination t ($B_{d-4\sqrt{k}}(t)$), where d is the distance from u to t ($d = d(u,t)$). Let's lower bound the probability of an arbitrary long-range connection of u going into this ball. We can assume w.l.o.g. that u shares either an x or a y coordinate with t (see Lemma 13). As before, let's consider the nested lattice construct, where this time u sits at the edge of one such ball. There are exactly $2b-1$ balls closer to t than u is at ball distance b, for $1 \le b \le \frac{2d-2}{6\sqrt{k}\log n}$. In order to enforce the condition that we improve the distance by at least $4\sqrt{k}$, we can dismiss the outer layer of balls, leaving us with $2b - 3$ balls for $2 \le b \le \frac{d-1}{3\sqrt{k}\log n} - 1$. The maximum distance from u to any node in one of these balls is $2b \times 3\sqrt{k}\log n$.

From Lemma 9.1, we know that each ball of radius $3\sqrt{k \log n}$ has at least $9 \log n$ highway nodes w.h.p. This lower bound must apply w.h.p. for any highway node along our path, so we must use the looser normalization constant bound, z. We can now lower bound the probability that v is in one of these closer balls:

$$
\Pr(v \in B_{d-4\sqrt{k}}) \geq \sum_{b=2}^{\frac{d-1}{3\sqrt{k \log n}}-1} \frac{(\text{min \# dist } b \text{ highway nodes})}{z(\text{max dist to node at dist } b)^2}
$$

$$
\geq \sum_{b=2}^{\frac{d-1}{3\sqrt{k \log n}}-1} \frac{(2b-3) \times 9 \log n}{z(2b \times 3\sqrt{k \log n})^2}
$$

$$
= \frac{2}{9kz} \sum_{b=2}^{\frac{d-1}{3\sqrt{k \log n}}-1} \frac{2b-3}{b^2}
$$

$$
> \frac{2}{9kz} \left[\ln\left(\frac{d-1}{3\sqrt{k \log n}} - 1 \right) \right]
$$

$$
> \frac{\ln\left(\frac{d}{3\sqrt{k \log n}} \right)}{9kz}
$$

Note that this result holds for $d \geq c(k + \log n)$ for large enough constant c.

This result holds for a single long-range connection of u. The probability that none of u's long-range connections are closer is:

$$
\Pr(\text{none closer}) < \left[1 - \frac{\ln\left(\frac{d}{3\sqrt{k \log n}} \right)}{9kz} \right]^{Qk}
$$

$$
= \left(\left[1 - \frac{\ln\left(\frac{d}{3\sqrt{k \log n}} \right)}{9kz} \right]^{kz} \right)^{\frac{Q}{z}}
$$

$$
< e^{-\frac{Q}{9z} \ln\left(\frac{d}{3\sqrt{k \log n}} \right)}
$$

$$
< e^{-\frac{Q \ln d}{9z}} = d^{-\frac{Q}{9z}}
$$

again, holding for large enough constant c.

With this probability established, let's try seeing how many hops we can take before we hit a dead end. Let's do this in two parts. First, let's see if we can get to within a distance of $(a \log n)^{bz}$ from t for some constants a and b. Since the probability of hitting a dead end only increases as we get closer, the probability of hitting a dead end while in this range is always going to be $< (a \log n)^{-\frac{bQ}{9}}$. This gives us an expected number of hops of $\Omega\left((a \log n)^{\frac{bQ}{9}} \right)$ w.h.p. When setting b large enough, we can get this to be $\Omega(\log^2 n)$, which is more than the maximum number of steps we expect to spend in routing.

In the second part, we are within distance $(a \log n)^{bz} \geq d \geq c(k + \log n)$ of t. From Lemma 1, we know that our normalization constant z is at most $\mathcal{O}(\log \log \log n)$ for $k \in \Omega\left(\frac{\log n}{\log \log \log n}\right)$ w.h.p., so $z < w \log \log \log n$ for some constant w. This gives us probability of hitting a dead end of less than $(c(k + \log n))^{-\frac{bQ}{9w \log \log \log n}}$. Setting constant c large enough, we can expect to take at least $\Omega\left(\log n^{\frac{Q}{9w \log \log \log n}}\right)$ hops on the highway within this range before hitting a dead end w.h.p. Let's call this our "allowance". While this is less than the maximum number of steps we expect to spend while routing, we only have at most $bz \log(a \log n)$ phases left in this second part, while we spend at most $\mathcal{O}(\log \log \log n)$ highway hops per phase. Putting this together, we expect to take at most $f(\log \log \log n)^2 \log \log n$ hops in this second part of the routing for some large enough constant f. Let's determine if our allowance is enough to get us to t, by considering the ratio r between our allowance and the number of remaining highway hops:

$$r = \lim_{n \to \infty} \frac{\log n^{\frac{Q}{9w \log \log \log n}}}{f(\log \log \log n)^2 \log \log n}$$

$$\log r = \lim_{n \to \infty} \frac{Q \log \log n}{9w \log \log \log n} - \log(f(\log \log \log n)^2 \log \log n)$$

$$= \lim_{n \to \infty} \frac{\log \log n}{\log \log \log n} - \log((\log \log n)^3) = \infty$$

Since $\log r$ tends towards infinity, r tends towards infinity, meaning that for a large enough constant c, our allowance is enough to get us to t w.h.p. for arbitrarily large n. Combining these results, we can conclude that we can reach a highway node within distance $c(k + \log n)$ of t w.h.p. while only taking long-range connections that improve our distance by at least $4\sqrt{k}$, thus eliminating the need for local connections.

7.5 Randomized Highway Variant

If it is desired to improve the greedy decentralized routing time of the randomized highway model for smaller values of k to be inline with the Kleinberg highway model, it is possible to reintroduce local connections within the highway nodes, despite the fact that nodes are picked arbitrarily. One straightforward way to do so is to add a local connection between each highway node to an arbitrary highway node in each of the 8 adjacent balls of radius $3\sqrt{k \log n}$ (see Fig. 6). From Lemma 9.1 we know that at least one highway node will exist in each of those balls w.h.p. At least one of these adjacent highway nodes will be at least $3\sqrt{k \log n}$ closer to the destination. With this variant, the routing time for smaller values of k is improved to $\log^2(n)/k$, while only increasing the average degree by a constant, inline with the randomized highway model. However, this model is not as clean as the original, and still maintains the same optimal parameter k of $\Theta(\log n)$ with the same result of $\Theta(\log n)$ hops, so we will not consider it further.

7.6   Windowed NPA Proofs

In this section, we prove that the windowed NPA model maintains a constant average degree while having a greedy, decentralized routing algorithm taking at most $\mathcal{O}(\log^{1+\epsilon} n)$ hops w.h.p. Specifically, we will define the routing algorithm as follows: define the subgraph made of nodes with popularity $\log n \leq k \leq A \log n$ as the highway, ignoring any long-range connections that do not connect two "highway" nodes. We expect to have $\mathcal{O}(1/\log^{1+\epsilon} n)$ highway nodes. Using the results from the previous section, we are able to route in $\mathcal{O}(\log^{1+\epsilon} n)$ hops w.h.p.

First, we prove the expected constant average degree:

Lemma 10. *The average node degree in the windowed NPA model is Q.*

Proof.

$$\int_{k=1}^{\infty} \epsilon Q k / k^{2+\epsilon} dk = \epsilon Q \int_{k=1}^{\infty} 1/k^{1+\epsilon} dk = \epsilon Q \times 1/\epsilon = Q$$

Where the normalization constant to pick k is:

$$\int_{k=1}^{\infty} 1/k^{2+\epsilon} dk = \frac{1}{1+\epsilon}$$

Next, we show that there are an expected $\mathcal{O}(1/\log^{1+\epsilon} n)$ highway.

Lemma 11. *There are $\Theta(\log^{1+\epsilon} n)$ highway nodes w.h.p.*

Proof. Now, let's find the probability that a node has popularity between $\log n$ and $A \log n$:

$$\Pr(\log n \leq k \leq A \log n) = \int_{k=\log n}^{A \log n} \Pr(k) dk$$

$$= \int_{k=\log n}^{A \log n} 1/k^{2+\epsilon} dk$$

$$= \frac{(A^{1+\epsilon} - 1)\ln^{1+\epsilon}(2)}{(1+\epsilon)A^{1+\epsilon}} \frac{1}{\log^{1+\epsilon} n}$$

Since A and ϵ are predetermined constants, the probability that a node has a popularity in this range is $\propto \log^{-(1+\epsilon)}(n)$.

Importantly, each node within this range of popularities considers all other points within this range of popularities as long-distance node candidates with equal likelihoods, a requirement important for the analysis of the randomized highway model. Next we must prove:

Lemma 12. *Each highway node expects to connect a constant fraction of its connections to other highway nodes, where the constant is at least $[1 + A^{1+\epsilon}]^{-1}$.*

Proof. The case where there is the least probability of overlap is when $k = \log n$. Let's consider v, an arbitrary long-range connection of node u, where $k_u = \log n$. The probability that v is part of the highway is:

$$\Pr(v \in \text{highway}) = \frac{\int_{k=\log n}^{A \log n} k^{-2-\epsilon} dk}{\int_{k=\log n/A}^{A \log n} k^{-2-\epsilon} dk} = [1 + A^{1+\epsilon}]^{-1}$$

This is enough to set up an instance of the randomized highway model. An (N, P, Q, ϵ, A) instance of the windowed NPA model corresponds with an $(N' = N, P' = P, Q' = \epsilon Q[1 + A^{1+\epsilon}]^{-1}, k' = \log^{1+\epsilon} n)$ instance with a few minor modifications. The highway graph, instead of consisting of nodes with degrees k, consists of nodes with degrees $\log n \le k \le A \log n$.

A little nuance applies since while $k = \log^{1+\epsilon} n$, each of the nodes has fewer connections, only $\mathcal{O}(\log n)$. However, the constant probability of halving the distance analysis still holds, and this algorithm achieves $\mathcal{O}(\log^{1+\epsilon} n)$ expected total greedy-routing steps. This concludes the proof for Theorem 3.

7.7   Miscellaneous Proofs

Lemma 13. *Let $S_d(w)$ denote the set of vertices at lattice distance d away from any vertex w. Let u be any vertex, and let v be any vertex such that $v \in S_d(u)$, and let $B = B_d(u)$. Then $|S_j(v) \cap B|$ is $\Theta(j)$ for all $1 \le j \le 2d$.*

Proof. Consider the ratio $R_{j,v} = \frac{|S_j(v) \cap B|}{|S_j(v)|}$ at each $1 \le j \le 2d$. It is clear that no matter where v is located in $S_j(u)$, $R_{j,v}$ always grows smaller as j increases. The value of j that minimizes $R_{j,v}$ for a particular $v \in S_d(u)$ is then $2d$, and we can achieve $\min_v(R_{v,2d})$ when v is a non-corner vertex in $S_d(u)$, in which case $R_{v,2d} = \frac{d}{8d} = 1/8$. Therefore at every $1 \le j \le 2d$, we have that $\frac{1}{8} \le \frac{|S_j(v) \cap B|}{4j}$, and therefore $|S_j(v) \cap B| \ge j/2$. Since we already have that $|S_j(v) \cap B| \le |S_j(v)| \le 4j$, the lemma follows.

References

1. Barabási, A.L., Albert, R.: Emergence of scaling in random networks. Science **286**(5439), 509–512 (1999). https://doi.org/10.1126/science.286.5439.509
2. Berger, N., Borgs, C., Chayes, J.T., D'Souza, R.M., Kleinberg, R.D.: Competition-induced preferential attachment. In: Díaz, J., Karhumäki, J., Lepistö, A., Sannella, D. (eds.) Automata, Languages and Programming: 31st International Colloquium, ICALP 2004, Turku, Finland, July 12–16, 2004. Proceedings. Lecture Notes in Computer Science, vol. 3142, pp. 208–221. Springer (2004). https://doi.org/10.1007/978-3-540-27836-8_20
3. Bollobás, B., Riordan, O.M.: Mathematical results on scale-free random graphs. In: Bornholdt, S., Schuster, H.G. (eds.) Handbook of Graphs and Networks: From the Genome to the Internet, chap. 1, pp. 1–34. Wiley (2002). https://doi.org/10.1002/3527602755.ch1

4. Borgs, C., Chayes, J.T., Daskalakis, C., Roch, S.: First to market is not everything: an analysis of preferential attachment with fitness. In: Johnson, D.S., Feige, U. (eds.) Proceedings of the 39th Annual ACM Symposium on Theory of Computing, San Diego, California, USA, June 11–13, 2007, pp. 135–144. ACM (2007). https://doi.org/10.1145/1250790.1250812

5. Dodds, P.S., Muhamad, R., Watts, D.J.: An experimental study of search in global social networks. Science **301**(5634), 827–829 (2003). https://doi.org/10.1126/science.1081058, https://www.science.org/doi/abs/10.1126/science.1081058

6. Dommers, S., van der Hofstad, R., Hooghiemstra, G.: Diameters in preferential attachment models. J. Stat. Phys. **139**(1), 72–107 (2010). https://doi.org/10.1007/s10955-010-9921-z

7. Flaxman, A.D., Frieze, A.M., Vera, J.: A geometric preferential attachment model of networks. Internet Math. **3**(2), 187–205 (2007). https://doi.org/10.1080/15427951.2006.10129124

8. Goodrich, M.T., Ozel, E.: Modeling the small-world phenomenon with road networks. In: Renz, M., Sarwat, M. (eds.) Proceedings of the 30th International Conference on Advances in Geographic Information Systems, SIGSPATIAL 2022, Seattle, Washington, November 1–4, 2022, pp. 46:1–46:10. ACM (2022). https://doi.org/10.1145/3557915.3560981

9. Kleinberg, J.M.: The small-world phenomenon: an algorithmic perspective. In: Yao, F.F., Luks, E.M. (eds.) Proceedings of the Thirty-Second Annual ACM Symposium on Theory of Computing, May 21–23, 2000, Portland, OR, USA, pp. 163–170. ACM (2000). https://doi.org/10.1145/335305.335325

10. Kumar, R., Liben-Nowell, D., Tomkins, A.: Navigating low-dimensional and hierarchical population networks. In: Azar, Y., Erlebach, T. (eds.) Algorithms - ESA 2006, 14th Annual European Symposium, Zurich, Switzerland, September 11–13, 2006, Proceedings. Lecture Notes in Computer Science, vol. 4168, pp. 480–491. Springer (2006). https://doi.org/10.1007/11841036_44

11. Liben-Nowell, D., Novak, J., Kumar, R., Raghavan, P., Tomkins, A.: Geographic routing in social networks. Proc. Natl. Acad. Sci. U.S.A. **102**(33), 11623–11628 (2005). https://doi.org/10.1073/pnas.0503018102

12. Martel, C.U., Nguyen, V.: Analyzing Kleinberg's (and other) small-world models. In: Chaudhuri, S., Kutten, S. (eds.) Proceedings of the Twenty-Third Annual ACM Symposium on Principles of Distributed Computing, PODC 2004, St. John's, Newfoundland, Canada, July 25–28, 2004, pp. 179–188. ACM (2004). https://doi.org/10.1145/1011767.1011794

13. Milgram, S.: The small world problem. Psychol. Today **1**(1), 61–67 (1967)

14. Mitzenmacher, M.: A brief history of generative models for power law and lognormal distributions. Internet Math. **1**(2), 226–251 (2004)

15. Slivkins, A.: Distance estimation and object location via rings of neighbors. In: Aguilera, M.K., Aspnes, J. (eds.) Proceedings of the Twenty-Fourth Annual ACM Symposium on Principles of Distributed Computing, PODC 2005, Las Vegas, NV, USA, July 17–20, 2005, pp. 41–50. ACM (2005). https://doi.org/10.1145/1073814.1073823

16. Travers, J., Milgram, S.: An experimental study of the small world problem. Sociometry **32**(4), 425–443 (1969)

Complexity and Approximation

Restricted Holant Dichotomy on Domains 3 and 4

Yin Liu$^{(\boxtimes)}$, Austen Z. Fan, and Jin-Yi Cai

University of Wisconsin-Madison, Madison, WI 53715, USA
yinliuchr@gmail.com, {afan,jyc}@cs.wisc.edu

Abstract. Holant$^*(f)$ denotes a class of counting problems specified by a constraint function f. We prove complexity dichotomy theorems for Holant$^*(f)$ in two settings: (1) f is any symmetric arity-3 real-valued function on input of domain size 3. (2) f is any symmetric arity-3 $\{0,1\}$-valued function on input of domain size 4.

Keywords: Holant problem · Dichotomy · Higher domain

1 Introduction and Background

Counting problems arise in many branches in computer science, machine learning and statistical physics. Holant problems encompass a broad class of counting problems [1,2,7,8,10,12,16,17,19,22–24]. For symmetric constraint functions (a.k.a. signatures) this is also equivalent to edge-coloring models [20,21]. These problems extend counting constraint satisfaction problems. Freedman, Lovász and Schrijver proved that some prototypical Holant problems, such as counting perfect matchings, cannot be expressed as vertex-coloring models known as graph homomorphisms [14,18]. The complexity classification program of counting problems is to classify the computational complexity of these problems.

Formally, a Holant problem on domain D is defined on a graph $G = (V, E)$ where edges are variables and vertices are constraint functions. Given a set of constraint functions $\mathcal{F}$ defined on D, a *signature grid* $\Omega = (G, \pi)$ assigns to each vertex $v \in V$ an $f_v \in \mathcal{F}$. The aim is to compute the following partition function

$$\text{Holant}_\Omega = \sum_{\sigma:E \to D} \prod_{v \in V} f_v\left(\sigma|_{E(v)}\right).$$

The computational problem is denoted by Holant($\mathcal{F}$). E.g., on the Boolean domain, it is over all $\{0,1\}$-edge assignments. On domain size 3, it is over all $\{R,G,B\}$-edge assignments, signifying three colors Red, Green and Blue. On domain size 4, it is over all $\{R,G,B,W\}$-edge assignments. As an example, on the Boolean domain, if every vertex has the EXACT-ONE function (which evaluates to 1 if exactly one incident edge is 1, and evaluates to 0 otherwise), then the partition function gives the number of perfect matchings. As another example, on domain size k, if every vertex has the ALL-DISTINCT function, then the partition function gives the number of valid k-edge colorings.

W. Wu and J. Guo (Eds.): COCOA 2023, LNCS 14462, pp. 83–96, 2024.
https://doi.org/10.1007/978-3-031-49614-1_5

A *symmetric* signature is a function that is invariant under any permutation of its variables. The value of such a signature depends only on the numbers of each color assigned to its input variables. The number of variables is its arity; unary, binary, ternary signatures have arities 1, 2, 3. We denote a symmetric ternary signature g on domain size 3 by a "triangle" consisting of 10 numbers:

$$
\begin{array}{ccccccc}
& & & g_{3,0,0} & & & \\
& & g_{2,1,0} & & g_{2,0,1} & & \\
& g_{1,2,0} & & g_{1,1,1} & & g_{1,0,2} & \\
g_{0,3,0} & & g_{0,2,1} & & g_{0,1,2} & & g_{0,0,3}
\end{array}
$$

where $g_{i,j,k}$ is the value on inputs having i Red, j Green and k Blue. Similarly, we denote a symmetric ternary signature g on domain size 4 by a "tetrahedron":

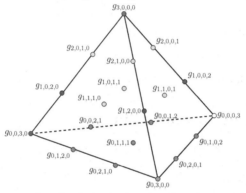

While much progress has been made for the classification of counting CSP [3] [4,5,13], and some progress for Holant problems [6,11,15], classifying Holant problems on higher domains is particularly challenging. One of the few existing work on a higher domain is [11], in which a dichotomy for Holant*(f) is proved where f is a ternary complex symmetric function on domain size 3 and the * means all unary functions are available. (Note that Holant problems with signatures of arity ≤ 2 are all P-time tractable; the interesting case where both tractable and #P-hardness occur starts with ternary signatures.)

In this work, we attempt to extend this to Holant problems on domain size 4. Our effort only met with partial success. We are able to prove a complexity dichotomy for Holant*(f) for any $\{0,1\}$-valued symmetric ternary constraint function f (see Theorem 4).

Our technique is to try to reduce a domain 4 problem to a domain 3 problem, and then analyze the situation using the existing domain 3 dichotomy [11]. To do so, we will need to be able to construct (or interpolate) a suitable constraint function that allows us to effectively restrict the problem to a domain 3 problem, where the new domain elements are superpositions of old domain elements under a holographic transformation. This turns out to be a nontrivial task. And one reason that we cannot extend to a more general domain 4 dichotomy is that for some real-valued signatures, it is impossible to construct such a constraint function. On the other hand, for $\{0,1\}$-valued domain 4 signatures, we are able to succeed in this plan (using several different constructions).

To carry out this plan, we use the domain 3 dichotomy [11] extensively. This motivates us to examine the domain 3 dichotomy more closely, when the constraint function f is real-valued. Since the domain 3 dichotomy [11] applies to all complex-valued functions it certainly also applies to real-valued functions[1]. However, we found out that applying the dichotomy for complex-valued functions directly is very cumbersome, so much so that the attempt to use it for our exploration in domain 4 grinds to a halt.

So, we return to domain size 3, and found that there is a cleaner form of the dichotomy of Holant$^*(f)$ on domain size 3 for real-valued f. This turns out to be a non-trivial adaptation, as we prove that certain tractable forms in the complex case *cannot* occur for real-valued signatures (see Theorem 2). In the proof of this real Holant* dichotomy, orthogonal holographic transformations are heavily used.

Armed with this more effective form of the domain 3 dichotomy, we return to Holant problems on domain 4, and prove a dichotomy for Holant$^*(f)$ where f is $\{0,1\}$-valued (see Theorem 4). We use several strategies that are more generally applicable and are worth mentioning (see Sect. 3.1). For general real-valued ternary symmetric signatures on domain 4, we prove Theorem 3. It gives some broad classes of constraint functions that define P-time tractable Holant problems. We conjecture that this is actually a complexity dichotomy.

Some Preliminaries. We can picture a signature as a vertex with several dangling edges as its input variables. Connecting a unary signature u to another signature f of arity $r \geq 1$ creates a signature of arity $r-1$. If f is symmetric then this does not depend on which variable (dangling edge of f) u is connected to, and the resulting signature is denoted by $\langle f, u \rangle$. In particular, if f is also a unary, then $\langle f, u \rangle$ is a scalar value, equal to the dot product of their signature entries. One should note that, for complex unary signatures, this dot product (without conjugation) is not the usual inner product and it is possible that $\langle u, u \rangle = 0$ for $u \neq 0$. We call a vector u *isotropic* if $\langle u, u \rangle = 0$ (including $u = 0$).

A signature is identified with a tensor, listing its values lexicographically as a truth table. We use Holant $(\mathcal{R} \mid \mathcal{G})$ to denote bipartite Holant problems on bipartite graphs $H = (U, V, E)$, where each signature on a vertex in U or V is from $\mathcal{R}$ or $\mathcal{G}$, respectively. Suppose T is an invertible matrix of the same size as that of the domain. We say that there is a *holographic transformation* from Holant $(\mathcal{R} \mid \mathcal{G})$ to Holant $(\mathcal{R}' \mid \mathcal{G}')$ by T, if $\mathcal{R}' = \mathcal{R}T^{-1}$ and $\mathcal{G}' = T\mathcal{G}$, where $\mathcal{R}T^{-1} = \{f(T^{-1})^{\otimes r(f)} \mid f \in \mathcal{R}\}$, $T\mathcal{G} = \{T^{\otimes r(f)} f \mid f \in \mathcal{G}\}$ and $r(f)$ is the arity of f. Here each signature is written as a column/row vector in lexicographical order as a truth-table. We also write Tf for $T^{\otimes r(f)} f$ when the arity $r(f)$ is clear.

---

[1] There is a slight issue that for a real-valued f, Holant$^*(f)$ naturally refers to having free real-valued unary functions, while the existing Holant* dichotomy for complex-valued f assumes all complex-valued unary functions are available for free. In Lemma 9 we address this technical difficulty.

Theorem 1 (Valiant's Holant Theorem [23]**).** *Suppose there is a holographic transformation from* Holant $(\mathcal{R} \mid \mathcal{G})$ *to* Holant $(\mathcal{R}' \mid \mathcal{G}')$, *then* Holant $(\mathcal{R} \mid \mathcal{G}) \equiv_T$ Holant $(\mathcal{R}' \mid \mathcal{G}')$, *where* $\equiv_T$ *means equivalence up to a P-time reduction.*

Therefore, if there is a holographic transformation from Holant $(\mathcal{G} \mid \mathcal{R})$ to Holant $(\mathcal{G}' \mid \mathcal{R}')$, then one problem is in P iff the other one is, and similarly one problem is #P-hard iff the other one is. For any general graph, we can make it bipartite by adding an additional vertex on each edge (thus forming the vertex-edge incidence graph), and assigning those new vertices the binary EQUALITY signature ($=_2$) of the corresponding domain size. Note that for the binary EQUALITY, if T is orthogonal, then it is unchanged under the holographic transformation by T. Hence, Holant($\mathcal{F}$) $\equiv_T$ Holant ($=_2 \mid \mathcal{F}$) $\equiv_T$ Holant($T\mathcal{F}$) for any orthogonal matrix T.

2 A Real Dichotomy for Holant*(f) on Domain 3

In this section we prove a complexity dichotomy of Holant*(f) for any *real-valued* symmetric ternary function f over the domain $\{R, G, B\}$. We investigate the three tractable forms of Theorems 3.1 and 3.2 in [11] when f is real-valued. It turns out that they take more special forms, and one tractable case for complex-valued f does not occur for real-valued f. However, complex tensors are still needed to express one of the two tractable forms. Lemmas 2, 3 and 4 address each of the three tractable families.

Lemma 1. *For all $\beta \in \mathbb{C}^3$, if $\langle \beta, \beta \rangle = 0$ then there exists a 3-by-3 real orthogonal matrix T, such that $T\beta = c(1, i, 0)^T$ where $c \in \mathbb{R}$.*

Proof. Write $\beta = \gamma + \delta i$, $\gamma, \delta \in \mathbb{R}^3$. Then $0 = \langle \beta, \beta \rangle = \langle \gamma, \gamma \rangle - \langle \delta, \delta \rangle + 2\langle \gamma, \delta \rangle i$. Considering its real and imaginary parts separately, $\|\gamma\| = \|\delta\|$ and $\gamma \perp \delta$. Then there exists a real orthogonal T, such that $T\gamma = ce_1$ and $T\delta = ce_2$, where $c = \|\gamma\| \in \mathbb{R}$. It follows that $T\beta = T(\gamma + \delta i) = c(1, i, 0)^T$.

Lemma 2. *If there exist $\alpha, \beta, \gamma \in \mathbb{C}^3$ such that $f = \alpha^{\otimes 3} + \beta^{\otimes 3} + \gamma^{\otimes 3}$, $\langle \alpha, \beta \rangle = \langle \beta, \gamma \rangle = \langle \gamma, \alpha \rangle = 0$, and f is real-valued, then there exist $\alpha', \beta', \gamma' \in \mathbb{R}^3$, s.t., $f = \alpha'^{\otimes 3} + \beta'^{\otimes 3} + \gamma'^{\otimes 3}$, $\langle \alpha', \beta' \rangle = \langle \beta', \gamma' \rangle = \langle \gamma', \alpha' \rangle = 0$. Thus, there is a real orthogonal transformation T, such that $Tf = ae_1^{\otimes 3} + be_2^{\otimes 3} + ce_3^{\otimes 3}$, for some $a, b, c \in \mathbb{R}$.*

Proof. Let $M_i = \langle f, e_i \rangle = \alpha_i \alpha^{\otimes 2} + \beta_i \beta^{\otimes 2} + \gamma_i \gamma^{\otimes 2}$, $i = 1, 2, 3$, then M_i is a real symmetric matrix.

If there is any v among $\{\alpha, \beta, \gamma\}$ that is non-isotropic, i.e., $\langle v, v \rangle \neq 0$, then by symmetry, assume it is α. Then $\alpha \neq 0$. At least one of $\alpha_i \neq 0$. Say $\alpha_1 \neq 0$.

We have $M_i \alpha = \lambda_i \alpha$, where $\lambda_i = \alpha_i \langle \alpha, \alpha \rangle$. So λ_i is an eigenvalue of a real symmetric matrix M_i, therefore it is real ($i = 1, 2, 3$). As $\langle \alpha, \alpha \rangle \neq 0$ and $\alpha_1 \neq 0$, $\frac{\alpha_i}{\alpha_1} = \frac{\lambda_i}{\lambda_1}$ is real and is well defined ($i = 1, 2, 3$). We can then write $\alpha = \mu u$, where $\mu \in \mathbb{C}, u \in \mathbb{R}^3$ and $\|u\| = 1$. As $\alpha \neq 0$, we have $\mu \neq 0$ and $\langle u, \beta \rangle = \langle u, \gamma \rangle = 0$.

Thus $f = \mu^3 u^{\otimes 3} + \beta^{\otimes 3} + \gamma^{\otimes 3}$. We have $\langle f, u \rangle = \mu^3 u^{\otimes 2}$. Since f and u are both real, and $u \neq 0$, we have $\mu^3 \in \mathbb{R}$. Thus, $\mu^3 = t^3$ for some real $t \in \mathbb{R}$. It follows that $\alpha^{\otimes 3} = t^3 u^{\otimes 3} = (tu)^{\otimes 3}$.

Then we replace α with tu. Similarly, if β or γ is a non-isotropic vector, we can replace it with a real vector, without changing f. Thus we get a new form $f = \alpha^{\otimes 3} + \beta^{\otimes 3} + \gamma^{\otimes 3}$ where α, β, γ are either real or isotropic (and since the zero vector is real we may further assume α, β, γ are either real or nonzero isotropic.)

If they are all real, then we can use a real orthogonal matrix T to transform f, i.e., $Tf = ae_1^{\otimes 3} + be_2^{\otimes 3} + ce_3^{\otimes 3}$, for some $a, b, c \in \mathbb{R}$. Then we are done.

Now suppose there is at least one nonzero isotropic vector among $\{\alpha, \beta, \gamma\}$. W.o.l.o.g., we can assume γ is nonzero and isotropic. By Lemma 1, there exists a real orthogonal T, s.t., $T\gamma = r(1, i, 0)^T, r \in \mathbb{R} \setminus \{0\}$. Then since $\langle T\alpha, T\gamma \rangle = \langle T\beta, T\gamma \rangle = 0$, $T\alpha, T\beta$ each must have the form $(c, ci, d)^T$. If, in addition, α is also isotropic, then $T\alpha$ must have the form $(c, ci, 0)^T$, which is a multiple of $T\gamma$. As γ is nonzero, α is hence a multiple of γ. Then α can be absorbed into γ and form a new isotropic vector γ', and $\langle \gamma', \beta \rangle = 0$, $f = \beta^{\otimes 3} + \gamma'^{\otimes 3}$. We have the same argument for β. So w.o.l.o.g., we can write $f = \alpha^{\otimes 3} + \beta^{\otimes 3} + \gamma^{\otimes 3}$, and there is *at most one*, and therefore, *exactly one*, nonzero isotropic vector among $\{\alpha, \beta, \gamma\}$, and the others are real vectors and could be zero. We have $f = \gamma^{\otimes 3} + R$, where R is some real-valued tensor. Hence $\gamma^{\otimes 3}$ is real. Since $\gamma \neq 0$, we have $\gamma^{\otimes 3} \neq 0$. But then, $\langle\langle\langle \gamma^{\otimes 3}, \gamma \rangle, \gamma \rangle, \gamma \rangle = \langle \gamma^{\otimes 3}, \gamma^{\otimes 3} \rangle \neq 0$. In particular, $0 \neq \langle \gamma^{\otimes 3}, \gamma \rangle = \langle \gamma, \gamma \rangle \gamma^{\otimes 2} = 0$. This is a contradiction.

Lemma 3. *If there exist $\alpha, \beta_1, \beta_2 \in \mathbb{C}^3$ such that $f = \alpha^{\otimes 3} + \beta_1^{\otimes 3} + \beta_2^{\otimes 3}, \langle \alpha, \beta_i \rangle = \langle \beta_i, \beta_i \rangle = 0, i = 1, 2$, and f is real-valued, then there is a real orthogonal transformation T, such that $cTf = \epsilon(\beta_0^{\otimes 3} + \overline{\beta_0}^{\otimes 3}) + \lambda e_3^{\otimes 3}$, where $\beta_0 = (1, i, 0)^T, \epsilon \in \{0, 1\}$ and $c, \lambda \in \mathbb{R}, c \neq 0$. Thus, there exist $\alpha \in \mathbb{R}^3$ and $\beta \in \mathbb{C}^3$, s.t., $f = \alpha^{\otimes 3} + \beta^{\otimes 3} + \overline{\beta}^{\otimes 3}, \langle \alpha, \beta \rangle = \langle \beta, \beta \rangle = 0$.*

Proof. First we assume $\beta_1 = \beta_2 = 0$. As f is real, by a similar argument as in the proof of Lemma 2, we know that there exists an $\alpha' \in \mathbb{R}^3, f = \alpha^{\otimes 3} = \alpha'^{\otimes 3}$. Thus there is a real orthogonal T, $T\alpha' = (0, 0, d)^T, d \in \mathbb{R}$, so $Tf = (T\alpha')^{\otimes 3} = d^3 e_3^{\otimes 3}$.

Now without loss of generality, we can assume $\beta_1 \neq 0$.

Suppose $\langle \alpha, \alpha \rangle = 0$. By Lemma 1, there exists a real orthogonal transformation T, such that $T\beta_1 = c(1, i, 0)^T$ for some $c \neq 0$. Since $\langle T\alpha, T\beta_1 \rangle = 0$, and that α is also isotropic, $T\alpha$ must have the form $(a, ai, 0)^T$, thus a multiple of $T\beta_1$. So, α is a multiple of β_1. Then α can be absorbed into β_1 and form a new isotropic vector $\mu\beta_1$ for some μ. Then we replace β_1 with $\mu\beta_1$ and $f = \beta_1^{\otimes 3} + \beta_2^{\otimes 3}$.

Else, $\langle \alpha, \alpha \rangle \neq 0$, then $\alpha \neq 0$. Let $M_i = \langle f, e_i \rangle, i = 1, 2, 3$, then M_i is a real symmetric matrix. $M_i \alpha = \alpha_i \langle \alpha, \alpha \rangle \alpha$. By the same argument as in the proof of Lemma 2, there exists an $\alpha' \in \mathbb{R}^3$, s.t., $\alpha^{\otimes 3} = \alpha'^{\otimes 3}$.

Hence in both cases, we get a new form $f = \alpha^{\otimes 3} + \beta_1^{\otimes 3} + \beta_2^{\otimes 3}$, where $\alpha \in \mathbb{R}^3$ is real (possibly 0), $\langle \alpha, \beta_i \rangle = \langle \beta_i, \beta_i \rangle = 0, i = 1, 2$. Since α is real, there is some real orthogonal T, $T\alpha = te_3 = (0, 0, t)^T, t \in \mathbb{R}$. In this new form, if $\beta_1 = \beta_2 = 0$, we are done by the same argument. Thus we may assume $\beta_1 \neq 0$ for the new form as well.

Case 1: $\alpha \neq 0$, i.e., $t \neq 0$. Since $\langle T\beta_i, T\alpha \rangle = 0, i = 1, 2$ and β_i is isotropic, we have $T\beta_1 = u(1, \pm i, 0)^T$, $T\beta_2 = v(1, \pm i, 0)^T, u \neq 0$. If β_2 is a multiple of β_1, it can be absorbed into β_1 and form a new isotropic vector $s(1, \pm i, 0)^T$ for some s. Then we get $Tf = (te_3)^{\otimes 3} + s^3((1, \pm i, 0)^T)^{\otimes 3}$. As Tf is real and t is real, we get $s = 0$ and we are done. Else, β_1, β_2 are independent, i.e., $Tf = (te_3)^{\otimes 3} + u^3((1, \pm i, 0)^T)^{\otimes 3} + v^3((1, \mp i, 0)^T)^{\otimes 3}$ where $uv \neq 0$. As Tf is real, it follows that $u^3 + v^3 \in \mathbb{R}$ and $u^3 - v^3 = 0$, and thus $u^3 = v^3 \in \mathbb{R}$. So $\frac{1}{u^3} Tf = \beta_0^{\otimes 3} + \overline{\beta_0}^{\otimes 3} + \frac{t^3}{u^3} e_3^{\otimes 3}$, where $\beta_0 = (1, i, 0)^T$.

Case 2: $\alpha = 0$, i.e., $t = 0$. We have $f = \beta_1^{\otimes 3} + \beta_2^{\otimes 3}$, $\langle \beta_i, \beta_i \rangle = 0, i = 1, 2$, and $\beta_1 \neq 0$. By Lemma 1, there exists a real orthogonal T, $T\beta_1 = u(1, i, 0)^T, u \in \mathbb{R} \setminus \{0\}$. If $\beta_2 = 0$, then $Tf = T\beta_1^{\otimes 3} = u^3((1, i, 0)^T)^{\otimes 3}$, where the LHS is real but the RHS is not, which is a contradiction. So we have $\beta_2 \neq 0$. Let $\beta_i' = T\beta_i$, and we have $Tf = \beta_1'^{\otimes 3} + \beta_2'^{\otimes 3}$. Let $M_i = \langle Tf, e_i \rangle = \beta_{1i}' \beta_1'^{\otimes 2} + \beta_{2i}' \beta_2'^{\otimes 2}, i = 1, 2, 3$. Both $\beta_1', \beta_2' \neq 0$. Then,

$$\begin{cases} M_i \beta_1' = \lambda_{2i} \beta_2' \\ M_i \beta_2' = \lambda_{1i} \beta_1', \end{cases} \quad \text{where} \quad \begin{cases} \lambda_{1i} = \beta_{1i}' \langle \beta_1', \beta_2' \rangle \\ \lambda_{2i} = \beta_{2i}' \langle \beta_1', \beta_2' \rangle. \end{cases}$$

Applying M_i twice, we get $M_i^2 \beta_1' = \lambda_{2i} \lambda_{1i} \beta_1'$. Since M_i is real symmetric, so is M_i^2. As $\beta_1' \neq 0$, $\lambda_{2i} \lambda_{1i}$ is an eigenvalue of M_i^2, therefore real, i.e., $\beta_{1i}' \beta_{2i}' \langle \beta_1', \beta_2' \rangle^2$ is real, $i = 1, 2, 3$. Recall $\beta_1' = u(1, i, 0)^T$, and now let $\beta_2' = (x, y, z)^T \in \mathbb{C}^3$. Let $\tau = x + yi$, and $\mu = \langle \beta_1', \beta_2' \rangle = u\tau$. Then we have $\begin{cases} \beta_{11}' \beta_{21}' \mu^2 = ux \cdot u^2 \tau^2 = u^3 \tau^2 x \\ \beta_{12}' \beta_{22}' \mu^2 = uiy \cdot u^2 \tau^2 = u^3 \tau^2 yi, \end{cases}$
both of which are real. Since $u \in \mathbb{R} \setminus \{0\}$, we know $\tau^2 x, \tau^2 yi \in \mathbb{R}$.

If $\tau = 0$, then $y = xi$ and hence $z = 0$ (because β_2' is isotropic). It follows that β_2' can be absorbed into β_1'. We can then rewrite $Tf = \lambda((1, i, 0)^T)^{\otimes 3}$ for some λ. As Tf is real, we know $\lambda = 0$ and we are done.

Now we can assume $\tau \neq 0$. Since both $\tau^2 x, \tau^2 yi \in \mathbb{R}$, adding them we know τ^3 is real. Then for some $k \in \{0, 1, 2\}$, $\omega^k \tau \in \mathbb{R}$, where $\omega^3 = 1$. Replacing β_2' by $\omega^k \beta_2'$, which satisfies $(\omega^k \beta_2')^{\otimes 3} = \beta_2'^{\otimes 3}$, we may assume $\tau \in \mathbb{R} \setminus \{0\}$.

We have $y = (x - \tau)i$. Since $\tau^2 x \in \mathbb{R}$, we know $x \in \mathbb{R}$. Because $Tf = \beta_1'^{\otimes 3} + \beta_2'^{\otimes 3}$ is real, we know $\begin{cases} (Tf)_{2,1,0} = u^3 i + x^2 y = (u^3 + x^2(x - \tau))i \in \mathbb{R} \\ (Tf)_{0,3,0} = y^3 - u^3 i = -(u^3 + (x - \tau)^3)i \in \mathbb{R}. \end{cases}$
From the fact that $x, \tau \in \mathbb{R}$ and the above relations, we know $x - \tau = -u$ (which gives $y = -ui$), and $x = \pm u$. As β_2' is isotropic, it follows that $z = 0$ and hence $\beta_2' = \pm u(1, \mp i, 0)^T$. If it is $+u(1, -i, 0)^T$, then $\frac{1}{u^3} Tf = \beta_0^{\otimes 3} + \overline{\beta_0}^{\otimes 3}$. If it is $-u(1, +i, 0)^T$, then $f = 0$.

Lemma 4. *If f is a real-valued signature such that there exist $\beta, \gamma \in \mathbb{C}^3$, $f = f_\beta + \beta^{\otimes 2} \otimes \gamma + \beta \otimes \gamma \otimes \beta + \gamma \otimes \beta^{\otimes 2}$, where $\beta \neq 0$, $\langle \beta, \beta \rangle = 0$, and f_β is a complex ternary signature satisfying $\langle f_\beta, \beta \rangle = 0$, then there exists a real orthogonal transformation T such that $Tf = \lambda e_3^{\otimes 3}, \lambda \in \mathbb{R}$. It implies that there is an $\alpha \in \mathbb{R}^3$, $f = \alpha^{\otimes 3}$.*

Proof. Let $\beta = (\beta_1, \beta_2, \beta_3)^T$, we have $\langle f, \beta \rangle = \langle \gamma, \beta \rangle \beta^{\otimes 2}$. Let $M_i = \langle f, e_i \rangle$, for $i = 1, 2, 3$, so M_i is real symmetric. Then $M_i \beta = \langle \langle f, e_i \rangle, \beta \rangle = \langle \langle f, \beta \rangle, e_i \rangle = \beta_i \langle \gamma, \beta \rangle \beta$, for $i = 1, 2, 3$. As $\beta \neq 0$, we know $\beta_i \langle \gamma, \beta \rangle$ is a real eigenvalue of M_i.

Suppose $\langle \gamma, \beta \rangle \neq 0$. Since $\beta \neq 0$, we can, without loss of generality, assume $\beta_1 \neq 0$. Then $\frac{\beta_i}{\beta_1} = \frac{\beta_i \langle \gamma, \beta \rangle}{\beta_1 \langle \gamma, \beta \rangle}$ is real and well defined ($i = 1, 2, 3$). We can then write $\beta = \lambda u$, where $\lambda \in \mathbb{C}, u \in \mathbb{R}^3$ and $\lambda \neq 0, u \neq 0$. In particular, $0 \neq \lambda^2 \langle u, u \rangle = \langle \beta, \beta \rangle = 0$. This is a contradiction.

So we have $\langle \gamma, \beta \rangle = 0$. Then, $\langle f, \beta \rangle = 0$. From Lemma 1, we know there exists some real orthogonal matrix T such that $T\beta = t\beta_0$ where $t \in \mathbb{R}, \beta_0 = (1, i, 0)^T$. Since $\beta \neq 0$, we have $t \neq 0$. Then $0 = \langle f, \beta \rangle = \langle T\beta, Tf \rangle$. So Tf has the form

$$
\begin{array}{cccc}
 & & a & \\
 & ai & b & \\
 -a & bi & c & \\
-ai & -b & ci & d
\end{array}
$$

Since Tf is real, we get $a = b = c = 0$. Thus, $Tf = de_3^{\otimes 3}$ for some $d \in \mathbb{R}$.

Theorem 2. *Let f be a real-valued symmetric ternary function over domain $\{R, G, B\}$. Then Holant*$^*(f)$ is #P-hard unless the function f in expressible as one of the following two forms, in which case the problem is in FP.*

1. *$f = \alpha^{\otimes 3} + \beta^{\otimes 3} + \gamma^{\otimes 3}$ where $\alpha, \beta, \gamma \in \mathbb{R}^3$ and $\langle \alpha, \beta \rangle = \langle \beta, \gamma \rangle = \langle \gamma, \alpha \rangle = 0$.*
2. *$f = \alpha^{\otimes 3} + \beta^{\otimes 3} + \overline{\beta}^{\otimes 3}$ where $\alpha \in \mathbb{R}^3$, $\langle \alpha, \beta \rangle = \langle \beta, \beta \rangle = 0$.*

This is equivalent to the existence of a real orthogonal transformation T, s.t.,

1. *$Tf = ae_1^{\otimes 3} + be_2^{\otimes 3} + ce_3^{\otimes 3}$ for some $a, b, c \in \mathbb{R}$.*
2. *$cTf = \epsilon(\beta_0^{\otimes 3} + \overline{\beta_0}^{\otimes 3}) + \lambda e_3^{\otimes 3}$ where $\beta_0 = (1, i, 0)^T$, $\epsilon \in \{0, 1\}$, and for some $c, \lambda \in \mathbb{R}$ and $c \neq 0$.*

Proof. This follows from Theorems 3.1 and 3.2 in [11] and Lemmas 2, 3, and 4.

Theorem 2 is the adapted real dichotomy for Holant$^*(f)$ with any real-valued signature f of arity 3 on domain 3. We showed that in the real case, we can take a real orthogonal transformation to the corresponding canonical forms. Also the third tractable case in the scenario of complex dichotomy does not exist anymore in the real case, which will simplify and expedite the analysis of further exploration of real dichotomies on domain 4.

3   Holant$^*(f)$ Dichotomy for $\{0, 1\}$-Valued f on Domain 4

In this section we give a dichotomy for a single $\{0, 1\}$-valued arity-3 symmetric signature f which is defined on a domain of size 4. We will prove this dichotomy theorem using a dichotomy for Holant$^*(f)$ on domain 2 [9], and the Holant$^*(f)$ dichotomy for real-valued signatures f on domain 3 from Sect. 2.

Theorem 3. *Let f be a real symmetric ternary function defined on a domain of size 4. If there is a real orthogonal transformation T such that Tf has one of the following forms, then* $\mathrm{Holant}^*(f)$ *is P-time computable, where* $\beta_0 = (1, i, 0, 0)^T$ *and* $\beta_1 = (0, 0, 1, i)^T$.

1. *For some $a, b, c, d \in \mathbb{R}$, $Tf = ae_1^{\otimes 3} + be_2^{\otimes 3} + ce_3^{\otimes 3} + de_4^{\otimes 3}$.*
2. *For some $c, \lambda_1, \lambda_2 \in \mathbb{R}$, and $c \neq 0$, $cTf = \beta_0^{\otimes 3} + \overline{\beta_0}^{\otimes 3} + \lambda_1 e_3^{\otimes 3} + \lambda_2 e_4^{\otimes 3}$.*
3. *For some $\lambda_1, \lambda_2 \in \mathbb{R}$, $Tf = \lambda_1(\beta_0^{\otimes 3} + \overline{\beta_0}^{\otimes 3}) + \lambda_2(\beta_1^{\otimes 3} + \overline{\beta_1}^{\otimes 3})$.*

The proof is by a holographic transformation. The details will be given in the full version. Notice that Theorem 3 does not claim to be a dichotomy, although we believe that the listed tractability in fact is a complete list.

Theorem 4. *Let f be a $\{0, 1\}$-valued symmetric ternary function defined on a domain of size 4. If f is not among the P-time computable cases in Theorem 3, then the problem* $\mathrm{Holant}^*(f)$ *is #P-hard. Moreover, for $\{0, 1\}$-valued f, only cases 1 and 2 in Theorem 3 occur.*

We remark that for $\{-1, 1\}$-valued symmetric ternary functions, the third tractable case of Theorem 3 does occur. The following is an example:

$$Q = \begin{bmatrix} \frac{1+\sqrt{2}}{2\sqrt{\sqrt{2}+2}} & -\frac{1}{2\sqrt{\sqrt{2}+2}} & \frac{-\sqrt{2}-1}{2\sqrt{\sqrt{2}+2}} & \frac{1}{2\sqrt{\sqrt{2}+2}} \\ \frac{-1+\sqrt{2}}{2\sqrt{2-\sqrt{2}}} & \frac{1}{2\sqrt{2-\sqrt{2}}} & \frac{-1+\sqrt{2}}{2\sqrt{2-\sqrt{2}}} & \frac{1}{2\sqrt{2-\sqrt{2}}} \\ \frac{-3+2\sqrt{2}}{2\sqrt{10-7\sqrt{2}}} & \frac{1-\sqrt{2}}{2\sqrt{10-7\sqrt{2}}} & \frac{3-2\sqrt{2}}{2\sqrt{10-7\sqrt{2}}} & \frac{-1+\sqrt{2}}{2\sqrt{10-7\sqrt{2}}} \\ \frac{-3-2\sqrt{2}}{2\sqrt{7\sqrt{2}+10}} & \frac{1+\sqrt{2}}{2\sqrt{7\sqrt{2}+10}} & \frac{-3-2\sqrt{2}}{2\sqrt{7\sqrt{2}+10}} & \frac{1+\sqrt{2}}{2\sqrt{7\sqrt{2}+10}} \end{bmatrix}$$

On the left is a signature g. On the right is an orthogonal matrix Q. In fact, under the transformation, $Qg = \sqrt{2-\sqrt{2}}(\beta_0^{\otimes 3} + \overline{\beta_0}^{\otimes 3}) - \sqrt{2+\sqrt{2}}(\beta_1^{\otimes 3} + \overline{\beta_1}^{\otimes 3})$, which is one example of the third tractable case of Theorem 3.

There are only a finite (albeit a large) number of $\{0, 1\}$-valued symmetric ternary signatures on domain 4. We will prove Theorem 4 by going through all signatures using five general strategies. When one signature could not be identified as #P-hard by any of the five strategies in Sect. 3.1, it is shown that it actually satisfies the first or second tractable conditions in Theorem 3.

3.1 Strategies

There are five different strategies we use to identify #P-hard signatures.

1. Use gadgets to form a binary symmetric signature which when written as a matrix M has rank 2. Then apply an orthogonal holographic transformation T, which transforms M to the form $\begin{bmatrix} 0 & 0 & 0 & 0 \\ 0 & 0 & 0 & 0 \\ 0 & 0 & \lambda & 0 \\ 0 & 0 & 0 & \mu \end{bmatrix}$, $\lambda\mu \neq 0$. We then interpolate $\mathrm{diag}(0, 0, 1, 1)$, a Boolean equality on the last two (new) domain elements by Lemma 6. Finally apply the Boolean domain dichotomy.

2. Similarly, we form a binary symmetric signature which when written as a matrix has rank 3. Then apply an orthogonal transformation T and get $\begin{bmatrix} 0 & 0 & 0 & 0 \\ 0 & \lambda_1 & 0 & 0 \\ 0 & 0 & \lambda_2 & 0 \\ 0 & 0 & 0 & \lambda_3 \end{bmatrix}$, $\lambda_1\lambda_2\lambda_3 \neq 0$, and interpolate $\mathrm{diag}(0,1,1,1)$ by Lemma 5. Finally, apply Theorem 2 to Tf on the last three (new) domain elements.

3. Find a nonzero unary signature $u \in \mathbb{R}^4$, such that $\langle f, u \rangle = 0$. Then define an orthogonal matrix T with (normalized) u as the first row. T transforms f to a signature supported on a lower domain (all 0's except the bottom face of the signature tetrahedron). Then apply the corresponding dichotomy.

4. Find some nonzero unary signature $u \in \mathbb{R}^4$, and nonzero $c \in \mathbb{R}$, such that $\langle f, u \rangle = cu \cdot u^T$. Define an orthogonal T using (normalized) u to be the first row. T will transform f to be domain separated (where the first new domain element R' is separated from the rest $\{G', B', W'\}$, i.e., Tf evaluates to 0, when R' is among its input, *except* possibly on (R', R', R')). Then apply the domain 3 dichotomy Theorem 2.

5. Use gadgets to construct a symmetric binary signature M which when written as a matrix has rank 4, and its 4 eigenvalues $\lambda_1, \lambda_2, \lambda_3, \lambda_4$ satisfy some condition under which we can interpolate (by Lemma 7 and Lemma 8) either $\mathrm{diag}(0,0,1,1)$ or $\mathrm{diag}(0,1,1,1)$ from $\mathrm{diag}(\lambda_1, \lambda_2, \lambda_3, \lambda_4)$. Form the orthogonal matrix Q such that $QMQ^T = \mathrm{diag}(\lambda_1, \lambda_2, \lambda_3, \lambda_4)$. Finally apply the corresponding lower domain dichotomy to the corresponding part of Qf.

We now show several examples using some of the strategies above.

Consider the tetrahedron on the left; we call this signature g. (Ignore the triangle on the right for now.)

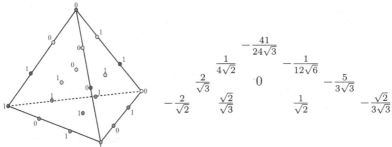

We use strategy 5. Connecting a unary $e_4 = (0,0,0,1)^T$ to g produces a binary symmetric function $M = \begin{bmatrix} 1 & 1 & 0 & 1 \\ 1 & 0 & 1 & 1 \\ 0 & 1 & 1 & 1 \\ 1 & 1 & 1 & 0 \end{bmatrix}$ whose eigenvalues are $\{3, 1, -1, -1\}$. We then use an orthogonal matrix $Q = \frac{1}{2\sqrt{3}}\begin{bmatrix} \sqrt{3} & \sqrt{3} & \sqrt{3} & \sqrt{3} \\ -1 & -1 & -1 & 3 \\ -\sqrt{6} & 0 & \sqrt{6} & 0 \\ \sqrt{2} & -2\sqrt{2} & \sqrt{2} & 0 \end{bmatrix}$ to diagonalize M, i.e., $QMQ^T = \mathrm{diag}(3, -1, 1, -1)$. It follows that we are able to interpolate $\mathrm{diag}(0,1,1,1)$, an equality on the new domain subset $\{G', B', W'\}$ (by Lemma 7). We build a new signature grid where we add a binary vertex on each edge between two Qg's, and assign $\mathrm{diag}(0,1,1,1)$ on all the new degree 2 vertices. Restricting Qg on $\{G', B', W'\}$ gives the domain 3 function depicted as the triangle on the right in the figure above. The binary $\text{diag}(0,1,1,1)$ restricts all edges in the new signature grid to be assigned a color only from $\{G', B', W'\}$ (no R') in order the evaluation of any product term in the partition function to be nonzero. It follows that we have a problem on domain size 3, which is defined by the ternary signature on domain 3 shown on the right as a "triangle". We can apply Theorem 2, and find that it is #P-hard. Therefore, the problem $\text{Holant}^*(g)$ is #P-hard.

Let's show a tractable example:

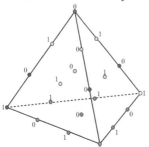 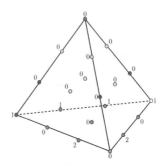

On the left is g, a ternary signature on domain size 4. There is a unary $u = (1,-1,0,0)^T$ such that $\langle u, g \rangle = 0$. Hence, we construct an orthogonal $Q = \begin{bmatrix} \frac{1}{\sqrt{2}} & -\frac{1}{\sqrt{2}} & 0 & 0 \\ \frac{1}{\sqrt{2}} & \frac{1}{\sqrt{2}} & 0 & 0 \\ 0 & 0 & 1 & 0 \\ 0 & 0 & 0 & 1 \end{bmatrix}$ (where its first row is the normalized u), such that under the transformation, g becomes Qg (on the right) and $Qg = \alpha^{\otimes 3} + \beta^{\otimes 3}$ where $\alpha = (0, -2^{\frac{1}{6}}, 2^{-\frac{1}{3}}, 2^{-\frac{1}{3}})^T$ and $\beta = (0, 2^{\frac{1}{6}}, 2^{-\frac{1}{3}}, 2^{-\frac{1}{3}})^T$. Here, after we use strategy number 3 to transform the problem to one on domain 3, we find it is tractable.

For all $\{0,1\}$-valued ternary signatures on domain size 4, we went through them using one of the above 5 strategies. It is found that we either can prove it is #P-hard, or when it fails to do so, it is tractable by being in one of the tractable forms in Theorem 4.

Tricks in Applying the Domain 3 Dichotomy. To apply the domain 3 dichotomy, there are also several tricks that can help simplify the calculation.

1. When checking whether a real ternary domain 3 signature satisfies the first tractable form $f = \alpha^{\otimes 3} + \beta^{\otimes 3} + \gamma^{\otimes 3}$, we can use different unary signatures u to connect to it and get different binary functions. Written as a matrix form $M = \langle f, u \rangle$, it's obviously symmetric. Write its eigen-decomposition as $M = Q\Lambda Q^T$, where Q is an orthogonal matrix $Q = [q_1, q_2, q_3]$ where q_i $(i = 1, 2, 3)$ are column vectors, and $\Lambda = \text{diag}(\lambda_1, \lambda_2, \lambda_3)$. If $\lambda_1, \lambda_2, \lambda_3$ are all distinct (at most one of them can be 0), and if f falls into the first tractable case, then the set $\{\pm q_1, \pm q_2, \pm q_3\}$ is independent of u.

Proof. If f satisfies the first tractable form $f = \alpha^{\otimes 3} + \beta^{\otimes 3} + \gamma^{\otimes 3}$, then $M = \langle u, f \rangle = \langle u, \alpha \rangle \alpha\alpha^T + \langle u, \beta \rangle \beta\beta^T + \langle u, \gamma \rangle \gamma\gamma^T$. Then $M\alpha = \langle u, \alpha \rangle \langle \alpha, \alpha \rangle \alpha$, so α is an eigenvector of M (if nonzero). Similarly, nonzero β, γ are both eigenvectors of M. As rank $M \geq 2$, at most one of α, β, γ can be 0. So at least two of them are scalar multiples of q_i $(i = 1, 2, 3)$. So, at least two of $\pm q_i$ $(i = 1, 2, 3)$ do not depend on the choice of u, and the third is uniquely determined by those two up to a $\pm$ factor.

2. When checking whether a real ternary domain 3 signature is in the first tractability case, we can search for whether there exists a unary u such that $\langle u, f \rangle = \langle u, u \rangle uu^T$. If a nonzero f does not have such a nonzero unary, it cannot be in the first tractable case since α, β, γ satisfy this relation.
3. To check whether a real ternary domain 3 signature is in the second tractability case, assume it has the form $f = \alpha^{\otimes 3} + \beta^{\otimes 3} + \overline{\beta}^{\otimes 3}$. If $\beta = 0$, then it also falls into the first tractability category. Now we assume $\beta \neq 0$. Then we have a nonzero β such that $\langle \beta, f \rangle = \langle \beta, \overline{\beta} \rangle \overline{\beta} \, \overline{\beta}^T$ and $\langle \beta, \beta \rangle = 0$. So if there does not exist such a nonzero isotropic β, it is not in the second tractable case.

3.2   Interpolate Restricted Equalities

Suppose for some unary u, the 4-by-4 matrix $A = \langle u, f \rangle$ has rank 3. Because A is a real symmetric matrix, we can construct an orthogonal matrix T so that

$$TAT^T = \begin{bmatrix} 0 & 0 & 0 & 0 \\ 0 & \lambda_1 & 0 & 0 \\ 0 & 0 & \lambda_2 & 0 \\ 0 & 0 & 0 & \lambda_3 \end{bmatrix} \tag{1}$$

where $\lambda_1 \lambda_2 \lambda_3 \neq 0$. We denote by $=_{G,B,W}$ the binary function in (1) if all $\lambda_i = 1$.

Lemma 5. *Let $H : \{R, G, B, W\}^2 \to \mathbb{R}$ be a rank 3 binary function of the form (1). Then for any $\mathcal{F}$ containing H, we have*

$$\text{Holant}(\mathcal{F} \cup \{=_{G,B,W}\}) \leq_T \text{Holant}(\mathcal{F}).$$

Similarly, we denote by $=_{B,W}$ the binary function in (1) if $\lambda_1 = 0$ and $\lambda_2 = \lambda_3 = 1$. We have

Lemma 6. *Let $H : \{R, G, B, W\}^2 \to \mathbb{R}$ be a rank 2 binary function of the form $\begin{bmatrix} 0 & 0 & 0 & 0 \\ 0 & 0 & 0 & 0 \\ 0 & 0 & \lambda & 0 \\ 0 & 0 & 0 & \mu \end{bmatrix}$ where $\lambda\mu \neq 0$. Then for any signature set $\mathcal{F}$ containing H, we have*

$$\text{Holant}(\mathcal{F} \cup \{=_{B,W}\}) \leq_T \text{Holant}(\mathcal{F}).$$

Lemma 5 and 6 enable us to construct instances on a lower domain which can help establish #P-hardness. In the next two lemmas we interpolate a lower domain equality directly from a rank 4 real symmetric matrix.

Lemma 7. *Let $H : \{R, G, B, W\}^2 \to \mathbb{R}$ be a rank 4 symmetric binary function. Let Q be the orthogonal matrix such that $QHQ^T = \begin{bmatrix} \lambda_1 & 0 & 0 & 0 \\ 0 & \lambda_2 & 0 & 0 \\ 0 & 0 & \lambda_3 & 0 \\ 0 & 0 & 0 & \lambda_4 \end{bmatrix}$ where $\lambda_1 \lambda_2 \lambda_3 \lambda_4 \neq 0$. If the four eigenvalues $\lambda_1, \lambda_2, \lambda_3, \lambda_4$ further satisfy the condition: for all $s, a, b, c \in \mathbb{Z}$, $s > 0$, if $a + b + c = s$ then $\lambda_1^s \neq \lambda_2^a \lambda_3^b \lambda_4^c$. Then for any $\mathcal{F}$ containing H, we have*

$$\text{Holant}(Q\mathcal{F} \cup \{=_{G,B,W}\}) \leq_T \text{Holant}(\mathcal{F}).$$

Similarly, we have Lemma 8.

Lemma 8. *Let* $H : \{R, G, B, W\}^2 \to \mathbb{R}$ *be a rank 4 symmetric binary function. Let* Q *be the orthogonal matrix such that* $QHQ^T = \begin{bmatrix} \lambda_1 & 0 & 0 & 0 \\ 0 & \lambda_2 & 0 & 0 \\ 0 & 0 & \lambda_3 & 0 \\ 0 & 0 & 0 & \lambda_4 \end{bmatrix}$ *where* $\lambda_1 \lambda_2 \lambda_3 \lambda_4 \neq 0$. *If the four eigenvalues* $\lambda_1, \lambda_2, \lambda_3, \lambda_4$ *further satisfy the condition: for all* $s, t, a, b \in \mathbb{Z}$, $s, t \geq 0$, $s + t > 0$, *if* $a + b = s + t$ *then* $\lambda_1^s \lambda_2^t \neq \lambda_3^a \lambda_4^b$. *Then for any* $\mathcal{F}$ *containing* H, *we have*

$$\text{Holant}(Q\mathcal{F} \cup \{=_{B,W}\}) \leq_T \text{Holant}(\mathcal{F}).$$

Let $\mathcal{U}$ be the set of all complex unaries and f be a possibly complex signature, both over domain 3. Then by definition, $\text{Holant}^*(f) = \text{Holant}(f \cup \mathcal{U})$. When f is a real-valued signature, we temporarily define $\text{Holant}^{r*}(f)$ as the Holant problem where only all the **real** unaries are available. In Lemma 9, we prove that in fact the complexity of the problem $\text{Holant}^{r*}(f)$ is the same as $\text{Holant}^*(f)$. (Hence, after Lemma 9, this new notation $\text{Holant}^{r*}(f)$ will be seen as unimportant, as far as the complexity of the problem is concerned.) The same lemma also holds for any domain k; the proof of Lemma 9 is easily adapted.

Lemma 9. *For any real signature* f *over domain 3, we have*

$$\text{Holant}^*(f) \leq_T \text{Holant}^{r*}(f).$$

Proof. By Theorem 3.1 in [11], the problem $\text{Holant}^*(f)$ is either polynomial-time solvable, or it is #P-hard. If it is polynomial-time solvable, then the reduction trivially holds. Otherwise, it is #P-hard. This means that there exist finitely many unaries $u_1, u_2, \ldots, u_k$, possibly complex-valued, such that the problem $\text{Holant}(\{f, u_1, \ldots, u_k\})$ is #P-hard. We prove that $\text{Holant}(\{f, u\}) \leq_T \text{Holant}^{r*}(f)$, for any complex-valued u, and then use induction, as k is a constant.

Now consider an arbitrary instance I of $\text{Holant}(\{f, u\})$ and suppose the unary function u appears m times in I. We write $u = [x, y, z]$ where $x, y, z \in \mathbb{C}$. We may assume $u \neq 0$, and by symmetry we may assume $z \neq 0$. We stratify the edge assignments according to the number of Red, Green and Blue assigned to the u's. Specifically, let ρ_{ij} denote the sum of products of the evaluations of all f's, where the sum is over all assignments with exactly i many times Red, j many times Green, and $(m - i - j)$ many times Blue are assigned to the u's. Then the Holant value on the instance I can be written as $\sum_{i=0}^{m} \sum_{j=0}^{m-i} \rho_{ij} x^i y^j z^{m-i-j} = z^m \sum_{i=0}^{m} \sum_{j=0}^{m-i} \rho_{ij} (\frac{x}{z})^i (\frac{y}{z})^j$. Observe that once we know all the values of ρ_{ij}, we can compute the Holant value of I in P-time. We now interpolate the values of ρ_{ij}.

We construct the instances I_k, $1 \leq k \leq \binom{m+2}{2}$, for the problem $\text{Holant}^{r*}(f)$. For any such k, let I_k be the same signature grid as I except we replace every appearance of u by the real unary $[3^k, 2^k, 1]$. Then, the Holant value for I_k

equals to $\sum_{i=0}^{m} \sum_{j=0}^{m-i} \rho_{ij} 3^{ki} 2^{kj}$. Therefore, we can write a non-degenerate Vandermonde system, where the matrix has entries $3^{ki} 2^{kj}$ with the columns indexed by lexicographical order of the tuples (i, j) and the rows indexed by k, and the unknown variables are ρ_{ij}. We can then solve for ρ_{ij} and therefore compute the Holant value of I in polynomial time. This finishes the proof of the lemma.

References

1. Backens, M.: A full dichotomy for Holant^c, inspired by quantum computation. SIAM J. Comput. **50**(6), 1739–1799 (2021). https://doi.org/10.1137/20M1311557
2. Backens, M., Goldberg, L.A.: Holant clones and the approximability of conservative Holant problems. ACM Trans. Algorithms **16**(2), 23:1-23:55 (2020). https://doi.org/10.1145/3381425
3. Bulatov, A.A.: A dichotomy theorem for constraint satisfaction problems on a 3-element set. JACM **53**(1), 66–120 (2006). https://doi.org/10.1145/1120582.1120584
4. Cai, J.Y., Chen, X.: Complexity of counting CSP with complex weights. JACM **64**(3), 19:1-19:39 (2017). https://doi.org/10.1145/2822891
5. Cai, J.Y., Chen, X., Lu, P.: Nonnegative weighted# CSP: an effective complexity dichotomy. SIAM J. Comput. **45**(6), 2177–2198 (2016). https://doi.org/10.1137/15M1032314
6. Cai, J., Guo, H., Williams, T.: The complexity of counting edge colorings and a dichotomy for some higher domain Holant problems. In: 55th IEEE Annual Symposium on Foundations of Computer Science, FOCS 2014, Philadelphia, PA, USA, October 18–21, 2014, pp. 601–610. IEEE Computer Society (2014). https://doi.org/10.1109/FOCS.2014.70
7. Cai, J.Y., Guo, H., Williams, T.: A complete dichotomy rises from the capture of vanishing signatures. SIAM J. Comput. **45**(5), 1671–1728 (2016). https://doi.org/10.1137/15M1049798
8. Cai, J.Y., Lu, P.: Holographic algorithms: from art to science. J. Comput. Syst. Sci. **77**(1), 41–61 (2011). https://doi.org/10.1016/j.jcss.2010.06.005
9. Cai, J.Y., Lu, P., Xia, M.: Holographic algorithms by Fibonacci gates and Holographic reductions for hardness. In: 2008 49th Annual IEEE Symposium on Foundations of Computer Science, pp. 644–653 (2008). https://doi.org/10.1109/FOCS.2008.34
10. Cai, J.Y., Lu, P., Xia, M.: Holant problems and counting CSP. In: Proceedings of the Forty-First Annual ACM Symposium on Theory of Computing, pp. 715–724 (2009). https://doi.org/10.1145/1536414.1536511
11. Cai, J.Y., Lu, P., Xia, M.: Dichotomy for Holant* problems with domain size 3. In: Proceedings of the Twenty-Fourth Annual ACM-SIAM Symposium on Discrete Algorithms, SODA 2013, New Orleans, Louisiana, USA, January 6–8, 2013, pp. 1278–1295. SIAM (2013). https://doi.org/10.1137/1.9781611973105.93
12. Cai, J.Y., Lu, P., Xia, M.: Holographic algorithms with matchgates capture precisely tractable planar #CSP. SIAM J. Comput. **46**(3), 853–889 (2017). https://doi.org/10.1137/16M1073984
13. Dyer, M., Richerby, D.: An effective dichotomy for the counting constraint satisfaction problem. SIAM J. Comput. **42**(3), 1245–1274 (2013). https://doi.org/10.1137/100811258

14. Freedman, M., Lovász, L., Schrijver, A.: Reflection positivity, rank connectivity, and homomorphism of graphs. J. Amer. Math. Soc. **20**(1), 37–51 (2007)
15. Fu, Z., Yang, F., Yin, M.: On blockwise symmetric matchgate signatures and higher domain #CSP. Inf. Comput. **264**, 1–11 (2019). https://doi.org/10.1016/j.ic.2018. 09.012
16. Guo, H., Huang, S., Lu, P., Xia, M.: The complexity of weighted Boolean #CSP modulo k. In: 28th International Symposium on Theoretical Aspects of Computer Science, STACS, pp. 249–260 (2011). https://doi.org/10.4230/LIPIcs.STACS.2011. 249
17. Guo, H., Lu, P., Valiant, L.G.: The complexity of symmetric Boolean parity Holant problems. SIAM J. Comput. **42**(1), 324–356 (2013). https://doi.org/10. 1137/100815530
18. Hell, P., Nešetřil, J.: Graphs and Homomorphisms, Oxford Lecture Series in Mathematics and its Applications, vol. 28. Oxford University Press (2004)
19. Kowalczyk, M., Cai, J.Y.: Holant problems for 3-regular graphs with complex edge functions. Theory Comput. Syst. **59**(1), 133–158 (2016). https://doi.org/10.1007/ s00224-016-9671-7
20. Szegedy, B.: Edge coloring models and reflection positivity. J. Am. Math. Soc. **20**(4), 969–988 (2007)
21. Szegedy, B.: Edge Coloring Models as Singular Vertex Coloring Models, pp. 327–336. Bolyai Society Mathematical Studies. Springer, Berlin Heidelberg pp (2010)
22. Valiant, L.G.: Accidental algorithms. In: 47th Annual IEEE Symposium on Foundations of Computer Science, FOCS, pp. 509–517. IEEE (2006). https://doi.org/ 10.1109/FOCS.2006.7
23. Valiant, L.G.: Holographic algorithms. SIAM J. Comput. **37**(5), 1565–1594 (2008). https://doi.org/10.1137/070682575
24. Xia, M.: Holographic reduction: a domain changed application and its partial converse theorems. Int. J. Softw. Informatics **5**(4), 567–577 (2011). https:// www.ijsi.org/ch/reader/view_abstract.aspx?file_no=i109   https://www.ijsi.org/ ch/reader/view_abstract.aspx?file_no=i109 https://www.ijsi.org/ch/reader/view_ abstract.aspx?file_no=i109

Earliest Deadline First Is a 2-Approximation for DARP with Time Windows

Barbara M. Anthony[1]📷, Christine Chung[2]📷, Ananya Das[3(✉)]📷, and David Yuen[4]📷

[1] Southwestern University, Georgetown, TX 78626, USA
anthonyb@southwestern.edu
[2] Connecticut College, New London, CT 06320, USA
cchung@conncoll.edu
[3] Middlebury College, Middlebury, VT 05753, USA
adas@middlebury.edu
[4] Kapolei, HI 96707, USA
yuen888@hawaii.edu

Abstract. Dial-a-Ride problems (DARP) require determining a schedule to efficiently serve transportation requests in various scenarios. We consider a variant of offline DARP in a uniform metric space where requests have release times and deadlines, and are all of equal duration and value. The goal is for a single unit-speed, unit-capacity server to serve as many requests as possible by an overall time limit, and this problem is NP-hard. We show that a natural greedy algorithm, Earliest Deadline First, is a 2-approximation, and this is tight.

Keywords: Approximation algorithms · Deadline scheduling · Dial-a-Ride problems

1 Introduction

The widely studied Dial-a-Ride Problem (DARP) requires scheduling one or more servers to complete a sequence of rides, each consisting of a pickup location (or *source*) and delivery location (or *destination*). Common optimality criteria include minimizing makespan (i.e., the time the server has completed the last request), minimizing the average flow time (i.e., the difference in a request's completion and release times), or maximizing the number of served requests within a specified time limit. In many variants *preemption* is not allowed, so if the server begins to serve a request, it must do so until completion. Applications include the transport of people and goods, including package delivery services, ambulances, ride-hailing services, and paratransit services. For an overview of DARP and its many variants, please refer to the surveys [13,18,23].

In this work we study offline DARP on the uniform metric space with a single unit-capacity server, where each request has a source, destination, release time, and deadline. Requests can be served only between their release time and

© The Author(s), under exclusive license to Springer Nature Switzerland AG 2024
W. Wu and J. Guo (Eds.): COCOA 2023, LNCS 14462, pp. 97–110, 2024.
https://doi.org/10.1007/978-3-031-49614-1_6

their deadline and may not be preempted. The server also has a specified time limit T after which no more requests can be served, and the goal is to maximize the number of requests served within T. This variant may be useful for settings where several equal-length requests must be completed within a deadline and each one has a specified time window for completion.

For the remainder of this paper, we refer to this request-maximizing, **Time-bounded DARP** variant, where each request has a specified **Time Window** (its release time to its deadline) as TDARPTW. TDARPTW is NP-hard because a special case of the problem without time windows, TDARP, where each request can be considered to have release time 0 and deadline T, has already been shown to be NP-hard in [1]. At any time t, given the current server location, we refer to a request as *servable* if it has not already been served, and can feasibly be served by its deadline. In this work, we show that the algorithm, which we refer to as EARLIEST DEADLINE FIRST (EDF), that continuously serves the servable request with the earliest deadline, has approximation ratio at most 2 for TDARPTW.

A common objective studied in previous DARP work is to minimize the time needed to serve all requests, as inspired by the original objective of the Traveling Salesperson Problem, since DARP is a generalization of TSP [8,17]. We note that this more classical DARP objective can be reduced to the time-bounded variant we study here, which has the objective of maximizing the number of requests served by time T. A solution for our time-bounded request-maximizing TDARP can be invoked a polynomial number of times to find the minimum time-bound where all requests can be served. Hence TDARP is at least as hard as its corresponding more classical DARP variants.

There has been limited prior work on approximation algorithms for TDARP in the offline setting. The work in [5] gives a $O(\log n)$-approximation for the closely related Vehicle Routing with Time Windows problem where n nodes in a metric space have release times and deadlines and the goal is to visit as many nodes as possible within their time windows. The work of [1] presented a 3/2-approximation algorithm for the uniform metric version of the problem without the constraint of time windows, or equivalently, where the time window for each request was assumed to be $[0, T]$. The work of [2] showed that the Segmented Best Path algorithm of [11,12] is a 4-approximation for TDARP on non-uniform metric spaces. That same work also shows that a greedy algorithm that repeatedly serves the fastest set of k remaining requests has an approximation ratio of $2 + \lceil \lambda \rceil / k$, where λ denotes the aspect ratio of the metric space.

In much of the voluminous and wide-ranging DARP literature, which is most often generated by applied researchers in operations and management, the term "time windows" often refers to slightly different notions than in our model, where we simply have a release time and a deadline per request. DARP in settings with time windows has been extensively studied with various parameters [16,19,21, 24], and the particular definition of time windows in our model has also been studied in works such as [26–28], which unlike all the empirical research in DARP, lies instead in the domain of competitive analysis of online DARP problems. In this domain, there have been a number of notable recent developments, albeit with different objectives [4,6,7].

ODAPRTW, investigated by [27], is the same as our TDARPTW problem, except the server has no overarching time limit, and their problem is online, so the requests are not known a priori, but instead arrive over time. They also assume the time windows are of uniform length for all requests, and their goal is to maximize the number of requests that meet their deadlines. They give an algorithm with competitive ratio $(2-\Delta)/(2\Delta)$, where Δ denotes the diameter of the metric space. In [26] they consider time windows of non-uniform length, as we do, but in the online setting. They find that a greedy-by-deadline algorithm is 3-competitive in the uniform metric.

DARP problems also generalize scheduling problems, as they require requests to be served with the additional constraint of a metric space in which the server must move to reach each request it wishes to serve. Scheduling requests with a given deadline for each request is part of the setting in many classical problems studied in scheduling theory literature, and the *Deadline Scheduling Problem*, for example, is well-known to be (strongly) NP-hard [20].

The EDF scheduling algorithm is a simple greedy approach that is well-known to be effective in many different contexts—offline, online, and in real-time systems—for various scheduling objectives where jobs to be scheduled have deadlines (for a sample of such settings, see [9,10,14,15,20,25]). Due to space constraints, we omit a more detailed discussion of the scheduling literature here. To our knowledge, our work is the first to present an approximation guarantee for greedy EDF scheduling of DARP requests with release times and deadlines.

2   Formalizing the Problem and Algorithm

The input for an instance of our problem, TDARPTW, consists of a uniform metric space, an origin, a set of requests S, and an overall time limit T. The origin o, a point in the metric space, indicates where the server is at time 0; a request may start at o but need not. Each request is an ordered tuple (s, d, a, b) consisting of the source s, destination d, release time a, and deadline b. We focus on the offline variant, where all information is known at time 0.

We assume without loss of generality that all time values (e.g., release times, deadlines, T) are integers. We do not allow preemption so if the server begins serving a request, it will serve it to completion. For any request i, let a_i and b_i denote the release time and deadline of i, respectively, with $0 \le a_i < b_i$. We refer to $[a_i, b_i]$ as the *time window* for request i. The goal for TDARPTW is to maximize the number of requests that a single unit-capacity server can serve within their time windows by the time limit T.

For an algorithm ALG and a TDARPTW instance I, ALG(I) denotes the schedule created by ALG on I, i.e., the action prescribed by ALG for the server at each time unit, and OPT(I) denotes an optimal schedule on I. |ALG(I)| and |OPT(I)| denote the number requests served by ALG and OPT, respectively, on instance I. When the instance is clear from the context, (I) may be dropped.

We define a request to be "servable" at a given time if it can be reached and served within its time window. We also define "drive" below, similarly to [1].

In the uniform metric space, all drives (empty or service) each take one time unit. We emphasize that because the schedule is part of the definition of a drive, serving the same request in $\text{EDF}(I)$ and $\text{OPT}(I)$, even during the same time unit, is considered two distinct drives.

Definition 1. *For a request i and a time t, let $h_i(t)$ be an indicator variable where $h_i(t) = 0$ if the server is at the source of request i at time t and $h_i(t) = 1$ otherwise. We characterize request i as **servable** at time t if request i has not already been served by time t and $a_i \leq t + h_i(t)$ and $b_i \geq t + h_i(t) + 1$.*

Definition 2. *A **drive** refers to one movement of the server from one point in the metric space to a second (not necessarily distinct) point at a specific time in a given schedule (e.g. EDF or OPT). A drive that serves a request is a **service drive**. A drive that does not serve a request but simply re-positions the server or allows it to remain at the current location for one time unit is an **empty drive**.*

2.1 The Earliest Deadline First (EDF) Algorithm

One might be inclined to think it is trivial to find an algorithm that serves a request every other time unit, giving a 2-approximation. However, due to the constraints of the release times and deadlines, such schedules may not always exist or be straightforward to find.

The focus of this paper is on an algorithm inspired by the well-established earliest deadline first greedy algorithms from the scheduling literature (see, e.g., [20,25]). Informally, we greedily consider all servable requests, serving one with the soonest deadline, breaking ties arbitrarily, as long as time remains. Our EDF algorithm is formally described in Algorithm 1. Note that EDF does not simply sort by earliest deadline, as a request is servable only if it has been released. Furthermore, EDF will first consider requests that can be served in the next two time units rather than only those in the next time unit, as only those requests that begin where the server is currently located can be served in the next time unit. Requiring the server to prioritize requests at its current location could result in serving requests with later deadlines while requests with earlier deadlines are available. This is not necessarily beneficial as we show in Theorem 4.

3 Our Results

We first lower bound the performance of EDF by showing that there exist instances of TDARPTW such that $|\text{EDF}| = |\text{OPT}|/2$.

Theorem 1. *There exist arbitrarily large instances of TDARPTW such that $|\text{OPT}| = 2 \cdot |\text{EDF}|$.*

Proof. Consider an instance which consists of T requests, each with release time 0 and deadline T, where the destination of one request is the source of the next request, and $T/2$ requests, each with release time 1 and deadline $T-1$, where all

---

Algorithm 1. The Earliest Deadline First (EDF) algorithm

1: Input: Metric space of points, set S of requests, time limit T, origin o.
2: **while** $t \leq T - 2$ **do**
3:     **if** there are no servable requests (recall definition of servable) **then**
4:         $t = t + 1$
5:     **else**
6:         Choose a request, r, with the earliest deadline among all servable requests.
7:         Let s and d denote the source and destination, respectively, of r.
8:         **if** the server is at s **then**
9:             Serve r during time $[t, t+1]$. Set $t = t + 1$.
10:        **else**
11:            During $[t, t+1]$, move the server to s and during $[t+1, t+2]$ serve r.
12:            Set $t = t + 2$.
13:        **end if**
14:    **end if**
15: **end while**
16: **if** $t = T - 1$ and there is a servable request **then**
17:    Serve any such servable request during $[t, T]$.
18: **end if**

---

the sources and destinations are distinct points in the metric space. No requests have their source at the origin.

OPT can serve all $T - 1$ of the requests with release time 0 and deadline T; no better solution is possible in time T as there is no request beginning at the origin. Thus, $|\text{OPT}| = T - 1$. EDF chooses a request with deadline $T - 1$ over any request with deadline T. Accordingly, EDF serves all $(T-1)/2$ requests with deadline $T - 1$ and must do an empty drive between every pair of requests, giving $|\text{EDF}| = (T-1)/2 = |\text{OPT}|/2$. $\qquad\square$

For the upper bound, we will show by induction that $|\text{OPT}(I)| \leq 2|\text{EDF}(I)| + 1$ for any input instance I. The most involved case of the proof is when $\text{EDF}(I)$ has as least two consecutive empty drives, so we define terminology for that scenario and a given instance I, enabling us to derive various facts that we can then use in the overall proof. Key to our work is considering what requests are or could have been servable before the first time EDF makes two consecutive empty drives, since without two consecutive empty drives, EDF would be serving at least one request every other time unit, immediately giving us the ratio of 2. Accordingly, let the earliest such occurrence of two consecutive empty drives by $\text{EDF}(I)$ be during the interval $[\tau, \tau + 2]$, for some time $\tau \leq T - 2$. Let w be the number of requests $\text{EDF}(I)$ has served by time τ.

We now focus on requests that are served before τ, and in particular, requests whose deadlines are after τ, since our induction will be based on a smaller instance created by removing such requests. Let E^* be the set of empty drives that $\text{OPT}(I)$ makes during $[0, \tau]$. Let R^* be the set of requests with deadline $\geq \tau + 2$ that $\text{OPT}(I)$ serves during $[0, \tau + 2]$. We define a set R to be analogous to R^* except served by $\text{EDF}(I)$ instead of by $\text{OPT}(I)$. Formally, let R be the set of requests with deadline $\geq \tau + 2$ that EDF serves during $[0, \tau + 2]$. Note that the deadline requirement for R^* requests ensures that $R^* \subseteq R$, since if a request in R^* was not served by EDF during $[0, \tau]$, it could have then been served during $[\tau, \tau + 2]$ in which EDF makes two empty drives. The size of the set $R - R^*$ is important in the inductive proof of Theorem 2.

We further partition the requests in R into two sets depending upon whether or not they were served in EDF(I) after an empty drive. Formally, define disjoint sets X and Y with $X \cup Y = R$ as follows: $Y \subseteq R$ are requests served in EDF(I) after time interval $[0, 1]$ that do not immediately follow another request in EDF(I); they must follow an empty drive in EDF(I). $X \subseteq R$ are requests which are either served at time 0 or immediately follow another request in EDF(I), which could have been in X but could likewise have been in Y or need not have been in R at all if its deadline was less than $\tau + 2$.

Let the set Q be all requests served in EDF(I) by time τ that were not preceded in EDF(I) by an empty drive. Thus, all requests in Q are either served at time 0 or immediately follow another request served by EDF. Q is a superset of X, as unlike X, the requests in Q are not constrained to have a deadline $\geq \tau + 2$.

Since τ is the first time at which two consecutive empty drives occur in EDF's schedule, every empty drive by EDF in $[0, \tau]$ is followed by a request served. We can thus add the number of requests served by EDF by time τ, or w, to the number of empty drives by EDF prior to τ, or $w - |Q|$, to get

$$\tau = 2w - |Q| \tag{1}$$

Understanding how EDF differs from OPT allows us to bound EDF's performance. As such, we construct an alternating sequence of drives that we call a *trace* that allows us to catalog EDF's non-optimal choices, as well as what OPT was doing when EDF served a different request, "tracing" through the two parallel schedules to the roots of the discrepancies between the two schedules. See Fig. 1 for an example.

Definition 3. *An **alternating trace**, or simply a **trace**, is a sequence of distinct drives that alternates between drives of OPT and drives of EDF. A trace is constructed using the rules below, beginning with the initialization in Rule 1 and then iterating between Rule 2 and Rule 3 as applicable until a termination condition (Rule 4-Rule 6) is reached. A **maximal alternating trace** is a trace that is not a proper subsequence of any other trace.*

Rule 1 Begin the trace with a drive in EDF from R or a drive in OPT from R^*.

Rule 2 If the drive most recently added to the trace is a request served by OPT, it is immediately followed in the trace by the EDF drive serving the same request. (Note that we know EDF has indeed served the same request at some point due to Lemma 1 below.)

Rule 3 If the drive most recently added to the trace is a drive EDF did at some time t and is preceded in EDF by an empty drive, then the next drive in the trace is the drive OPT did at time t.

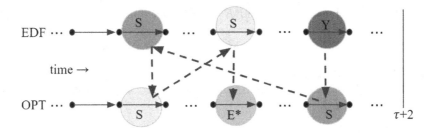

Fig. 1. An alternating trace. This figure depicts the schedule of drives in EDF in parallel with that of OPT, with time increasing to the right. Each drive (empty or service) is represented by a directed edge to the right between two points. The trace itself is represented by the dashed edges leading from a drive in one schedule to a drive in the other. The trace shown here starts with a drive in EDF serving a request from $Y \subseteq R$, and ends with an empty drive of OPT (in E^*). Drives that are the same shade serve the same request.

Rule 4 If the drive most recently added to the trace is one that EDF did at time t that was not preceded in EDF by an empty drive at time $t - 1$, then the trace terminates after said drive.

Rule 5 An empty drive in OPT ends the trace.

Rule 6 If the preceding rules ever result in an attempt to add a drive that is already part of the trace, the trace instead terminates (e.g., if there is a request served by both EDF and OPT at the same time, and applying Rule 2 would cause the same EDF drive to be added to the trace a second time). This rule prevents the same drive within a schedule from being repeated in the trace.

Maximal alternating traces are disjoint from one another due to the deterministic nature of their construction. We use the following lemma for Rule 2, above, as well as for our main result in Theorem 2.

Lemma 1. *In a maximal trace, every request served by a drive of* OPT *must also be served by a drive of* EDF, *and these drives never occur past* $\tau + 2$.

Proof Idea. Suppose there is a trace with a drive of OPT before time $\tau + 2$ serving a request $q \in R^*$. Then EDF could have served q at time $\tau + 1$, but did not because EDF has two empty moves during $[\tau, \tau + 2]$. Therefore EDF must have already served q before time τ, implying $q \in R$. The difficult case is when $q \notin R^*$, having a deadline before $\tau + 2$, and hence also $q \notin R$. For this case, we strengthen the inductive hypothesis, and show any request satisfying certain time window constraints (which any request served by OPT must also satisfy) will be served by EDF. Due to space limitations, the proof is deferred to the full version ([3]). □

To aid in categorizing maximal traces, in Lemma 2 we state their possible termination types. Note that every request in R and R^* appears exactly once in some maximal trace. To relate |EDF| to |OPT| we endeavor to find a relation between the sizes of $R - R^*$, Q, and E^*. An increase in $|R - R^*|$ can decrease what EDF serves after time τ, an increase in $|Q|$ increases what EDF serves before time τ, and an increase in $|E^*|$ decreases what OPT serves before time τ.

Lemma 2. *Other than traces that consist of a drive from* OPT *and a drive from* EDF *that serve the same request in the same time unit[2], any maximal alternating trace ends in Q or E^*.*

Proof. By Lemma 1 and Rule 2, a maximal trace cannot end at a request served by OPT except in the case of Rule 6. Thus the only way of ending a trace is if EDF serves a request in Q (Rule 4), or if OPT does an empty drive (Rule 5). □

Let C denote the collection of all maximal alternating traces. Due to Lemma 2, we may partition C into six categories C_0 through C_5 as described below. (Also see Fig. 2.) For the remainder of this work, the term "trace" always refers to maximal traces.

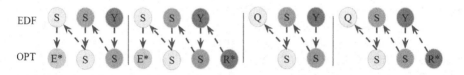

Fig. 2. Small examples depicting four general categories of traces (from left to right): C_1, C_2, C_3, and C_4. Recall that $Y \subseteq R$. To avoid crossing lines, these are samples of traces that proceed from later in the schedules to earlier, but as shown in Fig. 1, the upward edges may, more generally, also point forward in time rather than backward.

Trace categories C_0 and C_5 are not shown in Fig. 2, as they are special cases where the trace includes only a single request.

C_0: traces consisting solely of requests from X (terminated immediately due to Rule 4)

C_1: traces beginning with a request from $Y \subseteq R$ and ending with an empty drive from E^* (terminated due to Rule 5); one such trace was also seen in Fig. 1.

C_2: traces beginning with a request from R^* and ending with an empty drive from E^* (terminated due to Rule 5)

C_3: traces beginning with a request from $Y \subseteq R$ and ending with a request from Q (terminated due to Rule 4)

C_4: traces beginning with a request from R^* and ending with a request from Q (terminated due to Rule 4)

C_5: traces consisting only of two drives: the same request served at the same time in both R^* and Y (terminated due to Rule 6)

Definition 4. *For any set of requests P, we define P_i to be the number of requests from P that occur in a trace from category C_i.*

---

[2] In this special case, represented by category C_5, the trace ends in R^* or Y.

We derive the following facts about the size of R^* by considering trace categories C_0 to C_5. From Lemma 1, we know that $R_i^* \leq R_i = X_i + Y_i$ for any trace category $i = 1 \ldots 5$, and we also know from Rule 2 that every request of X_i or Y_i was preceded in the trace by a request of R_i^*, so $R_i^* = X_i + Y_i$, unless the trace started with an additional drive from EDF. So for trace categories that start with an initial EDF request (from Y, in the case of C_1 or C_3, or from X, in the case of C_0) we must subtract $|C_i|$ in computing the size of R^*.

Fact 0: $R_0^* = X_0 + Y_0 - |C_0|$. (Here, $Y_0 = 0$ and $X_0 = |C_0|$, so $R_0^* = 0$.)
Fact 1: $R_1^* = X_1 + Y_1 - |C_1|$. (Here, $X_1 = 0$.)
Fact 2: $R_2^* = X_2 + Y_2$. (Here, $X_2 = 0$.)
Fact 3: $R_3^* = X_3 + Y_3 - |C_3|$.
Fact 4: $R_4^* = X_4 + Y_4$.
Fact 5: $R_5^* = X_5 + Y_5$. (Here, $X_5 = 0$.)

We are now ready to prove the main theorem of this work.

Theorem 2. $|\text{OPT}(I)| \leq 2|\text{EDF}(I)| + 1$ *for any input instance I of TDARPTW.*

Proof. We proceed by induction on $|\text{EDF}(I)| + T$. Base step: $|\text{EDF}(I)| + T = 1$. In this case $T = 1$ and $|\text{EDF}(I)| = 0$. Since EDF cannot serve any requests, then neither can OPT. So $|\text{OPT}(I)| = 0$, and the claim is satisfied.

Inductive step: For the inductive step, we thus consider $|\text{EDF}(I)| + T \geq 2$ and suppose the theorem holds for any instance I' where $|\text{EDF}(I')| + T < |\text{EDF}(I)| + T$.

We consider two cases depending upon the size of T compared to $|\text{EDF}(I)|$.
Case 1: $T \leq 2|\text{EDF}(I)| + 1$. In this straightforward case, because of the time limit, OPT can serve at most $2|\text{EDF}(I)| + 1$ requests, so the claim holds.
Case 2: $T \geq 2|\text{EDF}(I)| + 2$. Since $T - |\text{EDF}(I)| \geq |\text{EDF}(I)| + 2$, EDF(I) must have $|\text{EDF}(I)| + 2$ or more empty drives, and thus at least two consecutive empty drives. As defined above, the earliest such occurrence is during the interval $[\tau, \tau + 2]$.

Summing the six labeled facts above and noting that $|R| = |X| + |Y| = X_0 + X_3 + X_4 + Y_1 + Y_2 + \ldots + Y_5$, we have

$$|R| - |R^*| = |C_1| + |C_3| + |C_0|. \tag{2}$$

We now define a sub-instance I' of the original instance I as follows. We remove all requests that EDF served by time τ, and we also remove any requests with deadline at most $\tau + 2$. The origin is at the point in the metric space where EDF(I) (EDF's schedule on the original instance) places the server at time $\tau + 1$; denote said point as m. The new time limit is $T - (\tau + 1)$, but to avoid confusion, we will view the time in I' as running from time $\tau + 1$ to time T.

Observe that by construction there are no servable requests in this sub-instance I' at time $\tau + 1$: any request with release time $\tau + 1$ or earlier and deadline greater than $\tau + 2$ would have already been served by EDF because otherwise EDF(I) would not have had two consecutive empty drives at time τ.

Let w^* denote the number of requests served by OPT(I) in $[0, \tau + 2]$. To give a lower bound on OPT's performance on the sub-instance (OPT(I')), we consider the subschedule of $\mathrm{OPT}(I)$ starting at time $\tau + 2$ continuing on $\mathrm{OPT}(I)$'s path to time T. We let P denote this subschedule concatenated with a single drive from m to the front. The requests served by following P on I' number at least $|\mathrm{OPT}(I)| - w^* - (|R| - |R^*|)$: this accounts for all of the requests that $\mathrm{OPT}(I)$ serves in time $[\tau + 2, T]$ (the $|\mathrm{OPT}(I)| - w^*$) as well as those served by $\mathrm{EDF}(I)$ by time $\tau + 2$ with deadline $\tau + 2$ or later (collectively $|R|$), except for those $\mathrm{OPT}(I)$ had served by time $\tau + 2$ (which total $|R^*|$). Hence,

$$|\mathrm{OPT}(I)| - w^* - (|R| - |R^*|) \leq |\mathrm{OPT}(I')|. \tag{3}$$

By induction on the sub-instance I', since no request can be served in the first time unit, we have $|\mathrm{OPT}(I')| \leq 2|\mathrm{EDF}(I')|$. Observe that $\mathrm{EDF}(I')$ has the same schedule as the suffix of $\mathrm{EDF}(I)$ starting from $\tau + 1$. Thus, since $\mathrm{EDF}(I)$ serves w requests by time τ, $\mathrm{EDF}(I')$ serves precisely $|\mathrm{EDF}(I)| - w$ requests. Thus,

$$|\mathrm{OPT}(I)| \leq w^* + (|R| - |R^*|) + 2|\mathrm{EDF}(I)| - 2w. \tag{4}$$

To bound w^*, we let z be a binary indicator variable denoting whether or not $\mathrm{OPT}(I)$ serves a request at time $[\tau + 1, \tau + 2]$, with value 1 if it does. Thus, $w^* \leq (\tau + 1) + z - |E^*|$. Combined with Eqs. 2 and 4, we get

$$|\mathrm{OPT}(I)| \leq \tau + 1 + z - |E^*| + (|C_1| + |C_3| + |C_0|) + 2|\mathrm{EDF}(I)| - 2w. \tag{5}$$

We now consider how $|E^*|$ relates to other quantities, defining E' informally as the empty drives of OPT not appearing in any trace. Formally, let $E' = E^* - (E_1^* \cup E_2^*)$, since empty drives of OPT appear in traces of category C_1 or C_2 only. Using the observations that $E_1^* = |C_1|$ and $E_2^* = |C_2|$, we have

$$|E^*| = |E'| + |C_1| + |C_2|. \tag{6}$$

Substituting Eqs. 1 and 6 into Eq. 5 gives:

$$|\mathrm{OPT}(I)| \leq -|Q| + 1 + z - |E'| - |C_2| + |C_3| + |C_0| + 2|\mathrm{EDF}(I)|. \tag{7}$$

Because each C_0, C_3 or C_4 trace has an element from Q and such requests are all distinct, $|Q| \geq |C_0| + |C_3| + |C_4|$. Combining that with Eq. 7 gives

$$|\mathrm{OPT}(I)| \leq 1 + z - |E'| - |C_2| - |C_4| + 2|\mathrm{EDF}(I)|. \tag{8}$$

Observe that if $z = 1$, the OPT request at time $[\tau + 1, \tau + 2]$ begins a C_2 or C_4 trace, ensuring $|C_4| + |C_2| \geq z$, whether z is 1 or 0. Thus,

$$|\mathrm{OPT}(I)| \leq 1 - |E'| + 2|\mathrm{EDF}(I)|, \tag{9}$$

allowing us to conclude that $|\mathrm{OPT}(I)| \leq 2|\mathrm{EDF}(I)| + 1$. □

If $\mathrm{OPT}(I)$ did not serve a request in the first time unit we can remove the additive 1 term in the upper bound, getting a stronger formulation. Note that this case will always occur if there is no request whose source is the origin.

Theorem 3. *For any instance I of TDARPTW on which $\text{OPT}(I)$ does not serve a request during the time $[0,1]$, $|\text{OPT}(I)| \leq 2|\text{EDF}(I)|$.*

Proof. In the straightforward case where $T \leq 2|\text{EDF}(I)| + 1$, if OPT does not serve a request in the first time unit, then OPT can serve at most $2|\text{EDF}(I)|$, satisfying the claim.

For the other case, where $T \geq 2|\text{EDF}(I)| + 2$, we begin with Eq. 9 from Theorem 2, $|\text{OPT}(I)| \leq 1 - |E'| + 2|\text{EDF}(I)|$, which holds for any instance of the problem, and now consider how it can be refined when $\text{OPT}(I)$ does not serve a request during $[0,1]$. We consider whether $\text{EDF}(I)$ served a request during $[0,1]$. If $\text{EDF}(I)$ also did not serve a request during $[0,1]$ then $\text{OPT}(I)$'s empty drive at $[0,1]$ does not appear in any trace (since according to the rules of traces, empty drives of EDF do not belong in any trace), but is in E', so $|E'| \geq 1$, and thus $|\text{OPT}(I)| \leq 2|\text{EDF}(I)|$, satisfying the claim.

If $\text{EDF}(I)$ does serve a request, call it r_1, during $[0,1]$ we now show by induction on a smaller instance that the same conclusion is reached. This smaller instance $\mathcal{I}$ differs from the original input I in that its origin $\bar{o}$ is the destination of r_1, request r_1 is removed from the set of requests, and the time limit is $T - 1$. Note that $|\text{EDF}(\mathcal{I})| = |\text{EDF}(I)| - 1$.

Let P denote the path that proceeds from $\bar{o}$ to where $\text{OPT}(I)$ is at time 2 (call that location f) and then continues on $\text{OPT}(I)$'s path for the remaining $T - 2$ units of time. We will use P to serve as a proxy for OPT to bound the value of $\text{OPT}(\mathcal{I})$. Let z be an indicator variable denoting whether or not there is a servable request at time 2 from $\bar{o}$ to f. We lower bound the number of requests P served by considering that while P largely aligns with $\text{OPT}(I)$, if OPT served r_1, P will not because r_1 is no longer present in $\mathcal{I}$ and P may now make a different move during time $[1,2]$. Accordingly, $|P| \geq |\text{OPT}(I)| - 2 + z$, or $|\text{OPT}(I)| \leq |P| + 2 - z$. By Theorem 2, $|P| \leq |\text{OPT}(\mathcal{I})| \leq 2(|\text{EDF}(I)| - 1) + 1$.

When $z = 1$, $|\text{OPT}(I)| \leq 2(|\text{EDF}(I)| - 1) + 1 + 2 - z = 2|\text{EDF}(I)|$. In the case $z = 0$, if $|P| = 2(|\text{EDF}(I)| - 1) + 1$, then P is an optimal path on $\mathcal{I}$. As such, the inductive hypothesis, $|\text{OPT}(\mathcal{I})| \leq 2(|\text{EDF}(\mathcal{I})|)$, guarantees $|P| \leq 2(|\text{EDF}(I)| - 1)$. Thus, when $z = 0$, $|\text{OPT}(I)| \leq |P| + 2 - z \leq 2(|\text{EDF}(I)| - 1) + 2 - z = 2|\text{EDF}(I)|$.

Hence, the proof is complete. □

From Theorems 1 and 3, we obtain the following corollary.

Corollary 1. *When no requests have their source at the origin, the approximation ratio of the EDF algorithm for TDARPTW is 2 (and this is tight).*

One may wonder if a better performance can be achieved if the server does not make an empty drive to serve an earlier-deadline request when there is a request available at its current location. Informally, the opportunistic EDF algorithm, denoted EDFO, determines the set of servable requests that begin where the server is located. If that set is nonempty, it chooses a request arbitrarily among those in the set with the earliest deadline; if that set is empty, it chooses a request as in EDF. Whenever t is incremented, it repeats this procedure. We show that the same upper bound holds for EDFO as for EDF. Moreover, this EDFO algorithm need not be better than EDF. The full version ([3]) provides instances for which $|\text{OPT}| = |\text{EDF}|$ but $|\text{OPT}|/|\text{EDFO}|$ is arbitrarily close to 2.

Theorem 4. EDFO *is a 2-approximation for TDARPTW (and this is tight).*

Proof. Apply the proof of Theorem 2, replacing EDF with EDFO, except for the distinctions we now note. Few changes are needed since when EDF serves a request that is preceded by an empty drive, it chooses the earliest deadline among all servable requests. As such, all claims about the traces and almost all equations still hold. However, we need to consider the situation in which OPT serves a request in the first time slot, $[0, 1]$. In such a case, though EDF may not have served a request during $[0, 1]$, EDFO must; call this request $\bar{r}$.

Reviewing the definitions of various sets of requests, $\bar{r} \in Q$ as it is served by EDFO prior to time τ and not preceded by an empty drive. Accordingly, we have to adjust the proof of Theorem 2 from right after Eq. 7 where we first consider the size of Q. Because $\bar{r}$ is not counted in C_0, C_3 or C_4 but is part of Q, we can state that for EDFO we have $|Q| \geq |C_0| + |C_3| + |C_4| + 1$. Combining this inequality with Eq. 7 gives

$$|\text{OPT}(I)| \leq -|C_4| + z - |E'| - |C_2| + 2|\text{EDFO}(I)|.$$

Identically to the proof of Theorem 2, if $z = 1$, the OPT request at time $[\tau + 1, \tau + 2]$ begins a C_2 or C_4 trace, ensuring $|C_4| + |C_2| \geq z$, and hence that $|\text{OPT}(I)| \leq 2|\text{EDFO}(I)|$.

The matching lower bound is deferred to the full version [3] due to space limitations.   □

4   Concluding Remarks

Preliminary experimental results indicate that while EDF may serve as few as half of the requests that OPT serves, that is rare; EDF often performs close to optimally and determines a schedule quickly [22].

A natural extension to TDARPTW would be to consider the problem in the online setting, where requests are not known until their release times. The work of [26] studies ODARPTW, which is the online form of TDARPTW, but without the overall time limit T. They show that an online algorithm that schedules requests greedily by "waiting time," which is a proxy for the deadline of the request, is 3-competitive. For the online form of TDARPTW, preliminary investigations provide instances guaranteeing that the online form of EDF has a competitive ratio no better than 3; our conjecture is that the online form of EDF is in fact 3-competitive for the online form of TDARPTW, consistent with [26]. However, if requests were known one time unit prior to their release, perhaps akin to how users call-ahead for ride services, the ratio of 2 from the offline setting holds. This is because our offline EDF algorithm uses information about a request only one time unit before it serves the request.

References

1. Anthony, B.M., et al.: Maximizing the number of rides served for dial-a-ride. In: 19th Symposium on Algorithmic Approaches for Transportation Modelling, Optimization, and Systems, vol. 75, pp. 11:1–11:15 (2019)
2. Anthony, B.M., Christman, A.D., Chung, C., Yuen, D.: Serving rides of equal importance for time-limited dial-a-ride. In: Mathematical Optimization Theory and Operations Research, pp. 35–50 (2021)
3. Anthony, B.M., Chung, C., Das, A., Yuen, D.: Earliest deadline first is a 2-approximation for DARP with time windows (2023). http://cs.conncoll.edu/cchung/research/publications/cocoa2023full.pdf
4. Baligács, J., Disser, Y., Mosis, N., Weckbecker, D.: An improved algorithm for open online dial-a-ride. In: Chalermsook, P., Laekhanukit, B. (eds.) Approximation and Online Algorithms. WAOA 2022. Lecture Notes in Computer Science, vol. 13538, pp. 154–171. Springer International Publishing, Cham (2022). https://doi.org/10.1007/978-3-031-18367-6_8
5. Bansal, N., Blum, A., Chawla, S., Meyerson, A.: Approximation algorithms for deadline-TSP and vehicle routing with time-windows. In: 36th Annual ACM Symposium on Theory of Computing, pp. 166–174 (2004)
6. Birx, A., Disser, Y.: Tight analysis of the Smartstart algorithm for online dial-a-ride on the line. SIAM J. Discret. Math. **34**(2), 1409–1443 (2020)
7. Birx, A., Disser, Y., Schewior, K.: Improved bounds for open online dial-a-ride on the line. Algorithmica **85**(5), 1372–1414 (2023)
8. Charikar, M., Raghavachari, B.: The finite capacity dial-a-ride problem. In: 39th Annual Symposium on Foundations of Computer Science, pp. 458–467 (1998)
9. Chen, S., He, T., Wong, H.Y.S., Lee, K.W., Tong, L.: Secondary job scheduling in the cloud with deadlines. In: IEEE International Symposium on Parallel and Distributed Processing Workshops and PhD Forum, pp. 1009–1016 (2011)
10. Chetto, H., Chetto, M.: Some results of the earliest deadline scheduling algorithm. IEEE Trans. Software Eng. **15**(10), 1261–1269 (1989)
11. Christman, A., Chung, C., Jaczko, N., Milan, M., Vasilchenko, A., Westvold, S.: Revenue maximization in online dial-a-ride. In: 17th Workshop on Algorithmic Approaches for Transportation Modelling, Optimization, and Systems, vol. 59, pp. 1:1–1:15. Dagstuhl, Germany (2017)
12. Christman, A.D., et al.: Improved bounds for revenue maximization in time-limited online dial-a-ride. Oper. Res. Forum **2**(3), 1–38 (2021). https://doi.org/10.1007/s43069-021-00076-x
13. Cordeau, J.F., Laporte, G.: The dial-a-ride problem: models and algorithms. Ann. Oper. Res. **153**(1), 29–46 (2007)
14. Dertouzos, M.L.: Control robotics: the procedural control of physical processes. In: Proceedings IFIP Congress, 1974 (1974)
15. Dertouzos, M.L., Mok, A.K.: Multiprocessor online scheduling of hard-real-time tasks. IEEE Trans. Software Eng. **15**(12), 1497–1506 (1989)
16. Desrosiers, J., Dumas, Y., Soumis, F.: A dynamic programming solution of the large-scale single-vehicle dial-a-ride problem with time windows. Am. J. Math. Manag. Sci. **6**(3–4), 301–325 (1986)
17. Frederickson, G.N., Hecht, M.S., Kim, C.E.: Approximation algorithms for some routing problems. In: 17th Annual Symposium on Foundations of Computer Science, pp. 216–227 (1976)

18. Ho, S.C., Szeto, W., Kuo, Y.H., Leung, J.M., Petering, M., Tou, T.W.: A survey of dial-a-ride problems: literature review and recent developments. Transp. Res. Part B: Methodol. **111**, 395–421 (2018)
19. Jaw, J.J., Odoni, A.R., Psaraftis, H.N., Wilson, N.H.: A heuristic algorithm for the multi-vehicle advance request dial-a-ride problem with time windows. Transp. Res. Part B: Methodol. **20**(3), 243–257 (1986)
20. Leung, J.Y.: Handbook of Scheduling: Algorithms, Models, and Performance Analysis. CRC Press (2004)
21. Madsen, O.B., Ravn, H.F., Rygaard, J.M.: A heuristic algorithm for a dial-a-ride problem with time windows, multiple capacities, and multiple objectives. Ann. Oper. Res. **60**, 193–208 (1995)
22. Medina, A., Anthony, B.M.: Evaluating an earliest deadline first algorithm for a dial-a-ride problem [poster]. Tapia Celebration of Diversity in Computing (2023)
23. Molenbruch, Y., Braekers, K., Caris, A.: Typology and literature review for dial-a-ride problems. Ann. Oper. Res. **3**, 295–325 (2017). https://doi.org/10.1007/s10479-017-2525-0
24. Solomon, M.M., Desrosiers, J.: Survey paper-time window constrained routing and scheduling problems. Transp. Sci. **22**(1), 1–13 (1988)
25. Stankovic, J.A., Spuri, M., Ramamritham, K., Buttazzo, G.C.: Fundamentals of EDF Scheduling. In: Deadline Scheduling for Real-Time Systems. The Springer International Series in Engineering and Computer Science, vol. 460, pp. 27–65. Springer, US, Boston, MA (1998). https://doi.org/10.1007/978-1-4615-5535-3_3
26. Yi, F., Song, Y., Xin, C., Walter, I., Kai, K.W.: Online dial-a-ride problem with unequal-length time-windows. In: 2009 International Conference on Management and Service Science, pp. 1–5 (2009)
27. Yi, F., Tian, L.: On the online dial-a-ride problem with time-windows. In: Megiddo, N., Xu, Y., Zhu, B. (eds.) AAIM 2005. LNCS, vol. 3521, pp. 85–94. Springer, Heidelberg (2005). https://doi.org/10.1007/11496199_11
28. Yi, F., Xu, Y., Xin, C.: Online dial-a-ride problem with time-windows under a restricted information model. In: Algorithmic Aspects in Information and Management, pp. 22–31 (2006)

Improved Approximation for Broadcasting in k-Path Graphs

Hovhannes A. Harutyunyan and Narek Hovhannisyan[✉]

Department of Computer Science and Software Engineering, Concordia University,
Montreal H3G 1M8, Canada
haruty@cs.concordia.ca, narek.hovhannisyan@mail.concordia.ca

Abstract. Broadcasting is an information dissemination primitive where a message is passed from one node (called originator) to all other nodes in the network. With the increasing interest in interconnection networks, an extensive amount of research was dedicated to broadcasting. Two main research goals of this area are finding inexpensive network structures that maintain efficient broadcasting and finding the broadcast time for well-known and widely used network topologies. In the scope of this paper, we will mainly focus on determining the broadcast time and the optimal broadcasting scheme for graphs. Determination of the broadcast time of a node x in an arbitrary network G is known to be NP-hard. Polynomial time solutions are known only for a few classes of networks. There also exist various heuristic and approximation algorithms for different network topologies. In this paper, we will consider networks that can be represented as k-path graphs. We will present a polynomial time 2-approximation algorithm for the broadcast time problem in k-path graphs.

Keywords: Interconnection networks · Information dissemination · Broadcasting · Approximation algorithms

1 Introduction

Broadcasting is one of the most important information dissemination processes in an interconnected network. Over the last four decades, a large amount of research work has been published concerning broadcasting in networks under different models [12,18,21]. These models can have different numbers of originators, numbers of receivers at each time unit, distances of each call, numbers of destinations, and other characteristics of the network such as the knowledge of the neighborhood available to each node. In the context of this paper, we are going to focus on the classical model of broadcasting. The network is modeled as an undirected connected graph $G = (V, E)$, where $V(G)$ and $E(G)$ denote the vertex set and the edge set of G, respectively. The classical model follows the below-mentioned basic assumptions.

© The Author(s), under exclusive license to Springer Nature Switzerland AG 2024
W. Wu and J. Guo (Eds.): COCOA 2023, LNCS 14462, pp. 111–122, 2024.
https://doi.org/10.1007/978-3-031-49614-1_7

1. The broadcasting process is split into discrete time units.
2. The only vertex that has the message at the first time unit is called *originator*.
3. In each time unit, an informed vertex (*sender*) can *call* at most one of its uninformed neighbors (*receiver*).
4. During each unit, all calls are performed in parallel.
5. The process halts as soon as all the vertices in the graph are informed.

We can represent each call in this process as an ordered pair of two vertices (u, v), where u is the sender and v is the receiver. The *broadcast scheme* is the order of calls made by each vertex during a broadcasting process and can be represented as a sequence $(C_1, C_2, ..., C_t)$, where C_i is the set of calls performed in time unit i. An informed vertex v is *idle* in time unit t if v does not make any call in time t. A broadcast scheme is called *busy* if any informed vertex sends a message to one of its uninformed neighbors during each round. These schedules guarantee that as long as there remains an uninformed neighbor, vertices are never idle.

Given that every vertex, other than the originator, can be informed by exactly one vertex, the broadcast scheme forms a directed spanning tree (*broadcast tree*) rooted at the originator. We are also free to omit the direction of each call in the broadcast tree.

Definition 1. *The **broadcast time** of a vertex v in a given graph G is the minimum number of time units required to broadcast in G if v is the originator and is denoted by $b(v, G)$. The broadcast time of a given graph G, is the maximum broadcast time from any originator in G, formally $b(G) = max_{v \in V(G)}\{b(v, G)\}$.*

A broadcast scheme for an originator v that uses $b(v, G)$ time units is called an optimal broadcast scheme. Obviously, by the assumption (3), the number of informed vertices after each time unit can at most be doubled. Meaning, in general, the number of informed vertices after time unit i is upper bounded by 2^i. Therefore, it is easy to see that $b(v, G) \geq \lceil \log n \rceil$, where n is the number of vertices in G, which implies that $b(G) \geq \lceil \log n \rceil$.

The general *broadcast time* decision problem is formally defined as follows. Given a graph $G = (V, E)$ with a specified set of vertices $V_0 \subseteq V$ and a positive integer k, is there a sequence $V_0, E_1, V_1, E_2, V_2, ..., E_k, V_k$ where $V_i \subseteq V$, $E_i \subseteq E(1 \leq i \leq k)$, for every $(u, v) \in E_i, u \in V_{i-1}, v \in V_i, v \notin V_{i-1}$, and $V_k = V$. Here k is the total broadcast time, V_i is the set of informed vertices at round i, and E_i is the set of edges used at round i. It is obvious that when $| V_0 |= 1$ then this problem becomes our broadcast problem of determining $b(v, G)$ for an arbitrary vertex v in an arbitrary graph G.

Generally, the broadcast time decision problem in an arbitrary graph is NP-complete [11,25]. Moreover, the minimum broadcast time problem was proved to be NP-complete even for some restricted graph families, such as 3-regular planar graphs [23]. The study of the parameterized complexity of the broadcast time problem was initiated in [9]. There is a very limited number of graph families, for which an exact algorithm with polynomial time complexity is known for the broadcast time problem. Exact linear time algorithms are available for the broadcast time problem in trees [24,25], in connected graphs with only one cycle (unicyclic graphs) [17,18], in necklace graphs (chain of rings) [14], in k-restricted cactus graphs [7], in fully connected trees [13], and in Harary-like graphs [3,4]. For a more detailed introduction to broadcasting, we refer the reader to [10,16,19,20].

In this paper, we discuss the broadcast time problem in k-path graphs, which are a subfamily of 2-connected series-parallel graphs. We present a polynomial-time 2-approximation algorithm for the broadcast time problem in k-path graphs, which improves the currently known best approximation ratio of $(4 - \epsilon)$ [2]. Moreover, the proposed algorithm has $\mathcal{O}(k)$ time complexity, which is better than $\mathcal{O}(|V| + k \log k)$ complexity of the one in [2].

The rest of this paper is organized as follows. In Section 2, we discuss some previous results on broadcasting in k-path graphs. Further, in Section 3, we present our approximation algorithm for the broadcast time problem in general k-path graphs when the originator is a junction vertex. In Section 4, we briefly present our approximation algorithm for broadcasting in k-path graphs when the originator is an internal vertex. Finally, we will conclude the paper in Section 5.

2 k-Path Graphs

In this section, we discuss a simple subfamily of 2-connected series-parallel graphs that are usually referred to as *k-path graphs* or *melon graphs*. A k-path graph $G = (P_1, P_2, ..., P_k)$ is obtained from a pair of vertices u and v, by adding $k \geq 2$ internally vertex-disjoint paths $P_1, P_2, ..., P_k$ between u and v. Vertices u and v are called junctions of G (Fig. 1).

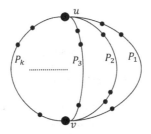

Fig. 1. Example of a k-path graph.

We assume that paths are indexed in a non-increasing order of their lengths. Formally, $l_1 \geq l_2 \geq ... \geq l_k$, where l_i is the number of vertices on path P_i (excluding u and v) for all $1 \leq i \leq k$. Note that since we consider all input graphs to be simple, only l_k can be 0.

Since k-path graphs are one of the simplest graphs that contain intersecting cycles, they are very interesting to research in terms of the broadcast time problem.

The broadcast time problem was researched for general series-parallel graphs as well as for subclasses of series-parallel graphs [21,22]. In [22], the authors prove that there is a $\mathcal{O}(\log n/\log\log n)$-approximation algorithm for the minimum broadcast time problem in graphs with bounded *treewidth* (for an introduction to treewidth we refer the reader to [5,6]). As series-parallel graphs have a treewidth of 2, the result also applies to series-parallel graphs.

Currently, the best-known results for the broadcast time problem in k-path graphs were introduced in [2]. The authors introduce a $(4 - \epsilon)$-approximation algorithm for the broadcast time problem in general k-path graphs. Additionally, for some particular subclasses of k-path graphs, the authors give better approximations or optimal algorithms. The complete set of results introduced in [2] is presented in Table 1.

The main algorithm, introduced in [2], first counts the number of uninformed vertices on each path in each time unit. Then, based on the relation between these numbers decides on the order of calls placed by the junction vertices. In this paper, we design a simple algorithm that achieves a better approximation ratio without counting the remaining lengths of the paths.

Table 1. Summary of known results for k-path graphs [2]

Case	Algorithm	Result
General k-path	S_{path}	$(4 - \epsilon)$-approximation
$l_j \geq l_{j+1} + 2$ and $k \leq l_k + 1$	S_{path}	optimal
$l_j = l_{j+1}$ and $k \leq l_k + 1$	S_{path}	optimal
$l_j = l_{j+1} + 1$ and $k \leq l_k + 1$	S_{path}	$(\frac{4}{3} - \epsilon)$-approximation
$l_j = l_{j+1} + 1$, $k \leq l_k + 1$ and u is the originator	A_{path}	$(\frac{7}{6} - \epsilon)$-approximation

In the same paper, the authors also presented several lower bounds and other auxiliary results for the broadcast time problem.

Given a k-path graph G_k and junction vertex u authors proved the following.

Lemma 1 ([2]). *There exists a minimum broadcast scheme from the originator u in G_k in which the shortest path P_k is informed in the first time unit.*

Similarly, given an internal vertex w authors proved the following.

Lemma 2 ([2]). *There exists a minimum broadcast scheme from the originator w in G_k in which w first sends the message along a shorter path towards a junction vertex.*

3   Broadcasting from a Junction Vertex

Let $G = (P_1, P_2, ..., P_k)$ be a k-path graph with junction vertices u and v.

It is easy to see that in any broadcast scheme where a junction vertex is the originator, an internal (non-junction) vertex w has at most one uninformed neighbor after being informed. Thus, to describe a busy broadcast scheme, it is sufficient to specify the order of calls placed by u and v. Algorithm 1 describes the behavior of u and v in our approximation algorithm, assuming, without loss of generality ($w.l.o.g.$), that u is the originator.

Algorithm 1. Broadcasting from a junction vertex

    Input A k-path graph $G = (P_1, P_2, ..., P_K)$, junction vertices u and v, and an originator u

    Output A broadcast scheme with time $b_{Alg}(u, G)$ for the graph G and the originator u

1: **procedure** BROADCASTINGFROMJUNCTION
2:    In the first time, unit u passes the message along the path P_k
3:    **for** $2 \leq i \leq k$ **do**
4:        **if** P_{i-1} contains an uninformed vertex **then**
5:            u passes the message along the path P_{i-1} in time unit i
6:        **end if**
7:    **end for**
8:    v gets the message in time unit $l_k + 1$
9:    **for** $2 \leq j \leq k$ **do**
10:       **if** P_{k-j+1} contains an uninformed vertex **then**
11:          v passes the message along the path P_{k-j+1} in time unit $l_k + j$
12:       **end if**
13:    **end for**
14: **end procedure**

3.1   Complexity Analysis

Note that all basic steps of Algorithm 1 require constant time. Moreover, the order of calls placed by each vertex is predefined and does not require any computation. Meaning that the time required to inform each path can be computed in constant time. Hence, Algorithm 1 returns a broadcast scheme in $\mathcal{O}(1)$ time and can be used to calculate the corresponding broadcast time in $\mathcal{O}(k)$ time.

3.2   The Approximation Ratio

Theorem 1. *Algorithm 1 is a polynomial-time 2-approximation algorithm for general k-path graphs when the originator is a junction vertex. Moreover, when $l_k \geq 1$, it guarantees $(2 - \epsilon)$-approximation for some $0 < \epsilon \leq 1$.*

Proof. Let P_i be a path that contained an uninformed vertex before the last time unit of the broadcast scheme described in Algorithm 1. In order to calculate $b_{Alg}(u, G)$, we will consider three cases: vertices in P_i were informed by both u and v, P_i was informed only by u, and P_i was informed only by v.

Case 1: Only u (but not v) placed a call towards the path P_i.

According to our algorithm, u informs path P_i in time unit $i + 1$. Then, it will require $l_i - 1$ time units for the path P_i to get fully informed. Hence,

$$b_{Alg}(u, G) = l_i + i \tag{1}$$

In [2], the authors showed that for any $1 \leq i \leq k - 1$, the following lower bound for the broadcast time holds.

$$b(u, G) \geq \left\lceil \frac{l_k + l_i + i + 1}{2} \right\rceil \geq \frac{l_k + l_i + i + 1}{2} \tag{2}$$

The approximation ratio directly follows from Eqs. 1 and 2.

$$\frac{b_{Alg}(u, G)}{b(u, G)} \leq \frac{l_i + i}{\frac{l_k + l_i + i + 1}{2}} < 2 \tag{3}$$

Case 2: Only v (but not u) placed a call towards the path P_i.

According to our algorithm, u informs path P_i in time unit $i + 1$. Since, in this case, path P_i was informed solely by v, then we can claim that it was fully informed in time unit $i + 1$ as the latest. Hence,

$$b(u, G) \leq i + 1 \leq k + 1 \tag{4}$$

In [2], the authors also proved the following lower bound.

$$b(u, G) \geq \left\lceil \frac{l_k + k + 1}{2} \right\rceil \geq \frac{l_k + k + 1}{2} \tag{5}$$

Again, the approximation ratio directly follows from Eqs. 4 and 5.

$$\frac{b_{Alg}(u, G)}{b(u, G)} \leq \frac{k + 1}{\frac{l_k + k + 1}{2}} \leq 2 \tag{6}$$

Case 3: Both u and v placed a call towards the path P_i.

According to our algorithm, u informs path P_i in time unit $i + 1$, and v informs path P_i no later than time unit $l_k + k - i + 1$. To calculate $b_{Alg}(u, G)$, we need to consider two subcases.

Subcase 3.a: $i + 1 \leq l_k + k - i + 1$.

In this case, a single vertex on path P_i is informed in each time unit between $i+1$ and $l_k + k - i + 1$. Afterward, up to 2 vertices can be informed. Hence,

$$
\begin{aligned}
b(u, G) &= l_k + k - i + \left\lceil \frac{l_i - (l_k + k - i - i)}{2} \right\rceil \\
&= \left\lceil \frac{2l_k + 2k - 2i + l_i - l_k - k + 2i)}{2} \right\rceil \\
&= \left\lceil \frac{l_k + k + l_i}{2} \right\rceil \leq \frac{l_k + k + l_i + 1}{2}
\end{aligned} \tag{7}
$$

Subcase 3.b: $i + 1 > l_k + k - i + 1$.
Similar to the previous case,

$$
\begin{aligned}
b(u, G) &= i + \left\lceil \frac{l_i - (i - (l_k + k - i))}{2} \right\rceil \\
&= i + \left\lceil \frac{l_i - i + l_k + k - i)}{2} \right\rceil \\
&= \left\lceil \frac{2i + l_i - 2i + l_k + k)}{2} \right\rceil \\
&= \left\lceil \frac{l_i + l_k + k}{2} \right\rceil \leq \frac{l_k + k + l_i + 1}{2}
\end{aligned} \tag{8}
$$

Hence, in both subcases, we have the same upper bound for $b_{Alg}(u, G)$.
The following lower bound for the broadcast time is easy to see from Eqs. 2 and 5.

$$
\begin{aligned}
b(u, G) = \frac{b(u, G)}{2} + \frac{b(u, G)}{2} &\geq \frac{l_k + k + 1}{4} + \frac{l_k + l_i + i + 1}{4} \\
&> \frac{l_k + k + 1}{4} + \frac{l_i}{4} \\
&= \frac{l_k + k + l_i + 1}{4}
\end{aligned} \tag{9}
$$

From Eqs. 7, 8 and 9,

$$
\frac{b_{Alg}(u, G)}{b(u, G)} < \frac{\frac{l_k + k + l_i + 1}{2}}{\frac{l_k + k + l_i + 1}{4}} = 2 \tag{10}
$$

Thus, we showed that $\frac{b_{Alg}(u, G)}{b(u, G)} \leq 2$ in all possible cases. Moreover, we can see from Eqs. 3 and 6 that the equality is only possible when $l_k = 0$.

4   Broadcasting from an Internal Vertex

Let $G = (P_1, P_2, ..., P_K)$ be a k-path graph with junction vertices u and v. Let w be a vertex on path P_m, where $1 \leq m \leq k$.

It is easy to see that in any broadcast scheme where w is the originator, any internal (non-junction) vertex other than w has at most one uninformed neighbor after being informed. Thus, in order to describe a busy broadcast scheme, it is sufficient to specify the order of calls placed by u, v, and w.

Let d be the length of the path $\overline{wu}$, and hence, $l_m + 1 - d$ be the length of the path $\overline{wv}$. W.l.o.g assume $d \leq l_m + 1 - d$. Also let $\tau(m) = l_m + 2 - 2d$. Note that $d \geq 1$ and $\tau(m) \geq 1$.

In the proposed algorithm, w passes the message along the shorter path towards u in the first time unit. Then, in the second time unit, w informs the vertex on path P_m towards v. Clearly, u gets informed in time unit d. Moreover, there will be $l_m + 1 - d - (d - 1) = \tau(m)$ uninformed vertices on path P_m. After that, if $\tau(m) \geq l_k + 1$, then u informs its neighbor on path P_k. Otherwise, u follows the main broadcasting scheme. We let t_v denote the time unit when v is informed. Algorithm 2 describes the behavior of u, v, and w in our approximation algorithm.

Algorithm 2. Broadcasting from an internal vertex

Input A k-path graph $G = (P_1, P_2, ..., P_K)$, junction vertices u and v, and an originator w on path P_m

Output A broadcast scheme with time $b_{Alg}(w, G)$ for the graph G and the originator w

1: **procedure** BROADCASTINGFROMINTERNAL
2:     In the first time unit, w passes the message along the path $\overline{wu}$
3:     In the second time unit, w passes the message along the path $\overline{wv}$
4:     **for** $1 \leq i \leq k$ **do**
5:         **if** u has an uninformed neighbor on path P_i **then**
6:             u passes the message along the path P_i in time unit $d + i$
7:         **end if**
8:     **end for**
9:     v gets the message in time unit $l_k + 1$
10:    **for** $1 \leq j \leq k$ **do**
11:        **if** v has an uninformed neighbor on path P_j **then**
12:            v passes the message along the path P_j in time unit $t_v + k - j$
13:        **end if**
14:    **end for**
15: **end procedure**

4.1   Complexity Analysis

Algorithm 2 has a complexity similar to the one of Algorithm 1. Hence, Algorithm 2 returns a broadcast scheme in $\mathcal{O}(1)$ time and can be used to calculate the corresponding broadcast time in $\mathcal{O}(k)$ time.

4.2 The Approximation Ratio

Theorem 2. *Algorithm 2 is a polynomial-time 2-approximation algorithm for general k-path graphs when the originator is an internal vertex and $\tau(m) < l_k+1$. Moreover, when $l_k \geq 1$, it guarantees $(2 - \epsilon)$-approximation for some $0 < \epsilon \leq 1$.*

Proof. Since $\tau(m) < l_k+2$, then vertex v will be informed via the direct path from w, and hence, $t_v = d + \tau(m)$. We will use a technique similar to the one used for proving Theorem 1. Let P_i be a path that contained an uninformed vertex before the last time unit of the broadcast scheme described in Algorithm 2.

First, let us consider the case when $i = m$. According to our broadcast scheme, path P_m will be fully informed in time unit $1 + l_m + 1 - d - 1 = l_m + 1 - d \leq l_m$. Moreover, any optimal broadcast scheme will fully inform path P_m no earlier than $b(w, G) \geq \lceil \frac{l_m + l_k}{2} \rceil \geq \frac{l_m + l_k}{2}$. The approximation ratio follows.

$$\frac{b_{Alg}(w, G)}{b(w, G)} \leq \frac{l_m}{\frac{l_m + l_k}{2}} = 2 \tag{11}$$

Next, we prove the theorem for any $i \neq m$.

Case 1: Only u (but not v) placed a call towards the path P_i.

Path P_i requires $l_i - 1$ time units to be fully informed, after time unit $d + i$ when the first call to it happens. Hence,

$$b_{Alg}(w, G) = d + i + l_i - 1 \tag{12}$$

In [2], the authors showed that for any $1 \leq i \leq k - 1$, $i \neq m$, the following lower bound for the broadcast time holds.

$$b(w, G) \geq d + \left\lceil \frac{l_i + \tau(m) + i - 1}{2} \right\rceil \geq d + \frac{l_i + \tau(m) + i - 1}{2} \tag{13}$$

The approximation ratio directly follows from Eqs. 12 and 13.

$$\frac{b_{Alg}(w, G)}{b(w, G)} \leq \frac{d + l_i + i - 1}{d + \frac{l_i + \tau(m) + i - 1}{2}} < 2 \tag{14}$$

Case 2: Only v (but not u) placed a call towards the path P_i.

According to our algorithm, u informs path P_i in time unit $d + i$. Since, in this case, path P_i was informed solely by v, then we can claim that it was fully informed in time unit $i + 1$ as the latest. Hence,

$$b(w, G) \leq d + i \leq d + k \tag{15}$$

In [2], the authors also proved the following lower bound.

$$b(w, G) \geq d + \left\lceil \frac{k + \tau(m) - 1}{2} \right\rceil \geq d + \frac{k + \tau(m) - 1}{2} \tag{16}$$

Again, the approximation ratio directly follows from Eqs. 4 and 5.

$$\frac{b_{Alg}(w,G)}{b(w,G)} \leq \frac{d+k}{d + \frac{k+\tau(m)-1}{2}} < 2 \tag{17}$$

Case 3: Both u and v placed a call towards the path P_i.

According to our algorithm, u informs path P_i in time unit $d+i$, and v informs path P_i in time unit $d+\tau(m)+k-i$.

Subcase 3.a: $d+i \leq d+\tau(m)+k-i$.

When this is the case, in every time unit between $d+i$ and $d+\tau(m)+k-i$ a single vertex on path P_i is informed. After $d+\tau(m)+k-i$ up to 2 vertices can be informed in each time unit. Hence,

$$\begin{aligned} b(w,G) &= d+\tau(m)+k-i-1+ \left\lceil \frac{l_i - (d+\tau(m)+k-i-d-i)}{2} \right\rceil \\ &= \left\lceil \frac{2d+2\tau(m)+2k-2i-2+l_i-\tau(m)-k+2i)}{2} \right\rceil \\ &= \left\lceil \frac{2d+\tau(m)+l_i+k-2}{2} \right\rceil \leq \frac{2d+\tau(m)+l_i+k-1}{2} \end{aligned} \tag{18}$$

Subcase 3.b: $d+i > d+\tau(m)+k-i$.

Similar to the previous case,

$$\begin{aligned} b(w,G) &= d+i-1+ \left\lceil \frac{l_i - (d+i-(d+\tau(m)+k-i))}{2} \right\rceil \\ &= d+i-1+ \left\lceil \frac{l_i - i+\tau(m)+k-i}{2} \right\rceil \\ &= \left\lceil \frac{2d+2i-2+l_i-i+\tau(m)+k-i)}{2} \right\rceil \\ &= \left\lceil \frac{2d+\tau(m)+l_i+k-2}{2} \right\rceil \leq \frac{2d+\tau(m)+l_i+k-1}{2} \end{aligned} \tag{19}$$

Hence, in both subcases, we have the same upper bound for $b_{Alg}(u,G)$.

The following lower bound for the broadcast time is easy to see from Eqs. 13 and 16.

$$\begin{aligned} b(w,G) &= \frac{b(w,G)}{2} + \frac{b(w,G)}{2} \geq \frac{2d+l_i+\tau(m)+i-1}{4} + \frac{2d+k+\tau(m)-1}{4} \\ &> \frac{2d+l_i+\tau(m)-1}{4} + \frac{k}{4} = \frac{2d+l_i+k+\tau(m)-1}{4} \end{aligned} \tag{20}$$

From Eqs. 18, 19 and 20,

$$\frac{b_{Alg}(w,G)}{b(w,G)} < \frac{\frac{2d+\tau(m)+l_i+k-1}{2}}{\frac{2d+\tau(m)+l_i+k-1}{4}} = 2 \tag{21}$$

Thus, we showed that $\frac{b_{Alg}(w,G)}{b(w,G)} \leq 2$ in all possible cases. Moreover, we can see from Eqs. 11 that the equality is only possible when $l_k = 0$.

Theorem 3. *Algorithm 2 is a polynomial-time 2-approximation algorithm for general k-path graphs when the originator is an internal vertex and $\tau(m) \geq l_k+1$. Moreover, when $l_k \geq 1$, it guarantees $(2 - \epsilon)$-approximation for some $0 < \epsilon \leq 1$.*

The proof of Theorem 3 uses a technique similar to the one used for Theorem 2 and is omitted due to space limitations.

5 Conclusion and Future Work

In this paper, we devised a polynomial-time broadcasting algorithm for arbitrary k-path graphs that guarantees an approximation ratio of 2. Moreover, we showed that the introduced algorithm is a $(2 - \epsilon)$-approximation algorithm for k-path graphs where the junction vertices are not adjacent. Our algorithm improves the best previously known approximation algorithm, introduced in [2], by a factor of 2. So far, there is no known inapproximability result for the broadcast time problem on k-path graphs, nor is it known to be NP-hard. Hence, in the future, we plan on either improving the known approximation ratio or introducing an inapproximability lower bound for this problem. Potentially, a polynomial-time exact algorithm for the broadcast time problem in k-path graphs can still be devised. Another direction for future research is applying the techniques used in this paper to other graph families that share some similarities with k-path graphs. For instance, it may be possible to apply the algorithm to flower graphs (also called k-cycle graphs [1]). Currently, the best known approximation algorithm for broadcasting in flower graphs was introduced in [8].

References

1. Bhabak, P., Harutyunyan, H.A.: Constant approximation for broadcasting in k-cycle graph. In: Ganguly, S., Krishnamurti, R. (eds.) CALDAM 2015. LNCS, vol. 8959, pp. 21–32. Springer, Cham (2015). https://doi.org/10.1007/978-3-319-14974-5_3
2. Bhabak, P., Harutyunyan, H.A.: Approximation algorithm for the broadcast time in k-path graph. J. Interconnection Netw. **19**(04), 1950006 (2019)
3. Bhabak, P., Harutyunyan, H.A., Kropf, P.: Efficient broadcasting algorithm in Harary-like networks. In: 46th International Conference on Parallel Processing Workshops (ICPPW 2017), pp. 162–170. IEEE (2017)
4. Bhabak, P., Harutyunyan, H.A., Tanna, S.: Broadcasting in Harary-like graphs. In: 17th International Conference on Computational Science and Engineering (CSE 2014), pp. 1269–1276. IEEE (2014)
5. Bodlaender, H.L.: A partial k-arboretum of graphs with bounded treewidth. Theoret. Comput. Sci. **209**(1–2), 1–45 (1998)
6. Bodlaender, H.L., Koster, A.M.: Combinatorial optimization on graphs of bounded treewidth. Comput. J. **51**(3), 255–269 (2008)

7. Čevnik, M., Žerovnik, J.: Broadcasting on cactus graphs. J. Comb. Optim. **33**, 292–316 (2017)
8. Ehresmann, A.L.: Approximation Algorithms for Broadcasting in Flower Graphs. Master's thesis, Concordia University (2021)
9. Fomin, F.V., Fraigniaud, P., Golovach, P.A.: Parameterized complexity of broadcasting in graphs. In: International Workshop on Graph-Theoretic Concepts in Computer Science (WG 2023) (2023)
10. Fraigniaud, P., Lazard, E.: Methods and problems of communication in usual networks. Discret. Appl. Math. **53**(1–3), 79–133 (1994)
11. Garey, M.R., Johnson, D.S.: Computers and Intractability, vol. 174. freeman San Francisco (1979)
12. Gholami, S., Harutyunyan, H.A.: Broadcast graphs with nodes of limited memory. In: Pacheco, D., Teixeira, A.S., Barbosa, H., Menezes, R., Mangioni, G. (eds.) Complex Networks XIII. Springer Proceedings in Complexity. Springer, Cham (2022). https://doi.org/10.1007/978-3-031-17658-6_3
13. Gholami, S., Harutyunyan, H.A., Maraachlian, E.: Optimal broadcasting in fully connected trees. J. Interconnection Netw. **23**(01), 2150037 (2023)
14. Harutyunyan, H.A., Hovhannisyan, N., Maraachlian, E.: Broadcasting in chains of rings. In: Submitted to International Workshop on Combinatorial Algorithms (IWOCA 2023) (2023)
15. Harutyunyan, H.A., Li, Z.: A simple construction of broadcast graphs. In: Du, D.-Z., Duan, Z., Tian, C. (eds.) COCOON 2019. LNCS, vol. 11653, pp. 240–253. Springer, Cham (2019). https://doi.org/10.1007/978-3-030-26176-4_20
16. Harutyunyan, H.A., Liestman, A.L., Peters, J.G., D., R.: Broadcasting and Gossiping. In: Handbook of Graph Theory, pp. 1477–1494. Chapman and Hall (2013)
17. Harutyunyan, H., Maraachlian, E.: Linear algorithm for broadcasting in unicyclic graphs. In: Lin, G. (ed.) COCOON 2007. LNCS, vol. 4598, pp. 372–382. Springer, Heidelberg (2007). https://doi.org/10.1007/978-3-540-73545-8_37
18. Harutyunyan, H.A., Maraachlian, E.: On broadcasting in unicyclic graphs. J. Comb. Optim. **16**(3), 307–322 (2008)
19. Hedetniemi, S.M., Hedetniemi, S.T., Liestman, A.L.: A survey of gossiping and broadcasting in communication networks. Networks **18**(4), 319–349 (1988)
20. Hromkovič, J., Klasing, R., Monien, B., Peine, R.: Dissemination of information in interconnection networks (Broadcasting & Gossiping). In: Du, DZ., Hsu, D.F. (eds.) Combinatorial Network Theory. Applied Optimization, vol. 1. Springer, Boston, MA (1996). https://doi.org/10.1007/978-1-4757-2491-2_5
21. Kortsarz, G., Peleg, D.: Approximation algorithms for minimum-time broadcast. SIAM J. Discret. Math. **8**(3), 401–427 (1995)
22. Marathe, M.V., Ravi, R., Sundaram, R., Ravi, S.S., Rosenkrantz, D.J., Hunt, H.B., III.: Bicriteria network design problems. J. Algorithms **28**(1), 142–171 (1998)
23. Middendorf, M.: Minimum broadcast time is NP-complete for 3-regular planar graphs and deadline 2. Inf. Process. Lett. **46**(6), 281–287 (1993)
24. Proskurowski, A.: Minimum broadcast trees. IEEE Trans. Comput. **30**(05), 363–366 (1981)
25. Slater, P.J., Cockayne, E.J., Hedetniemi, S.T.: Information dissemination in trees. SIAM J. Comput. **10**(4), 692–701 (1981)

The Fine-Grained Complexity
of Approximately Counting Proper
Connected Colorings (Extended Abstract)

Robert D. Barish$^{(\boxtimes)}$ and Tetsuo Shibuya

Division of Medical Data Informatics, Human Genome Center,
Institute of Medical Science, University of Tokyo, 4-6-1 Shirokanedai, Minato-ku,
Tokyo 108-8639, Japan
`rbarish@ims.u-tokyo.ac.jp, tshibuya@hgc.jp`

Abstract. A k-proper connected 2-coloring for a graph is an edge biparti-
tition which ensures the existence of at least k vertex disjoint simple
alternating paths (i.e., paths where no two adjacent edges belong to
the same partition) between all pairs of vertices. In this work, for every
$k \in \mathbb{N}_{>0}$, we show that exactly counting such colorings is $\#P$-hard under
many-one counting reductions, as well as $\#P$-complete under many-one
counting reductions for $k = 1$. Furthermore, for every $k \in \mathbb{N}_{>0}$, we rule
out the existence of a $2^{o\left(\frac{n}{k^2}\right)}$ time algorithm for finding a k-proper connected 2-coloring of an order n graph under the ETH, or for exactly
counting such colorings assuming the moderated Counting Exponential
Time Hypothesis (#ETH) of (Dell et al.; *ACM Trans. Algorithms* **10**(4);
2014). Finally, assuming the Exponential Time Hypothesis (ETH), and
as a consequence of a recent result of (Dell & Lapinskas; *ACM Trans.
Comput. Theory* **13**(2); 2021), for every $k \in \mathbb{N}_{>0}$ and every $\epsilon > 0$, we are
able to rule out the existence of a $2^{o\left(\frac{n}{k^2}\right)}/\epsilon^2$ time algorithm for approximating the number of k-proper connected 2-colorings of an order n graph
within a multiplicative factor of $1 + \epsilon$.

Keywords: proper connected coloring · edge coloring · counting
complexity · approximate counting · Exponential Time Hypothesis
(ETH) · Counting Exponential Time Hypothesis (#ETH) · NP · $\#P$

1 Introduction

The notion of an *alternating path* or *alternating trail* in a graph with an edge
partition, where no two successive edges along the path or trail belong to the
same partition, appears to trace its origins to the factor enumeration scheme in

This work was supported by JSPS Kakenhi grants {20K21827, 20H05967, 21H04871},
and JST CREST Grant JPMJCR1402JST.

the original 19th century proof argument of Petersen's 2-factor theorem [35,36] (i.e., that every $2k$-regular simple graph can be partitioned into k edge disjoint 2-factors). Since this time, alternating paths and trails have appeared in numerous pure and applied graph theoretic contexts (see, e.g., [4] for a review). This has ranged from Tutte's extensions [39,40] of Petersen's work to characterize the existence of f-factors for every $f \in \mathbb{N}_{>0}$, to Edmond's [17] $\mathcal{O}\left(|V|^2 \cdot |E|\right)$ time "blossom algorithm" for finding maximum matchings in general graphs, and Micali & Vazirani's [34] $\mathcal{O}\left(\sqrt{|V|} \cdot |E|\right)$ time improvement of the "blossom algorithm".

In the style of Menger's theorem [33], there have also been extensive efforts to develop a theory of connectivity for edge colored graphs based on alternating paths, cycles, and trails. While it is not possible to exhaustively survey this literature – see [4,5,12,32,38,42] for a review – we can highlight the somewhat surprising fact that it was shown to be NP-complete to find a pair of vertex disjoint alternating paths connecting a pair of vertices in a graph (Manoussakis [32] credits a private communication with R. Häggkvist for the proof of this fact). This finding was later extended by Abouelaoualim et al. [2] to show that, for every $k \in \mathbb{N}_{>1}$, it is NP-complete to decide the existence of k vertex disjoint or edge disjoint paths or trails connecting a given pair of vertices in an edge colored graph. On the other hand, finding a single alternating path between a pair of vertices in a w-edge colored graph was shown to be linear time solvable for every $w \in \mathbb{N}_{>0}$ [5,38].

In this work, we concern ourselves with Borozan et al.'s [9] and Andrews et al.'s [3] more recent variation (developed independently) on the aforementioned Menger-type theorems for alternating paths. Here, in lieu of finding alternating paths between a fixed pair of vertices in an edge colored graph, and akin to Chartrand et al.'s [10,11] rainbow connectivity problem, the objective is instead to find an edge coloring of a simple graph guaranteeing the existence of some number of vertex disjoint alternating paths (i.e., properly colored paths or *proper paths*) between all pairs of vertices. More specifically, we can ask for a *k-proper connected w-coloring* of a graph [3,9], which corresponds to a partition of the graph's edge set into at most w distinct color classes under the constraint that there must be at least k vertex disjoint proper paths between all pairs of vertices.

We now remark that, despite impressive efforts on characterizing the existence of k-proper connected w-colorings for special classes of graphs [3,9,16,18, 20,22,29,31] (e.g., bipartite graphs [9,18,22] and Erdős-Rényi random graphs [20]), and despite applications to a realistic variant of the frequency assignment problem (see Sect. 2), very little is known concerning the complexity of finding k-proper connected w-colorings. In particular, beyond a circa 2020 proof of Huang & Li [23,24][1] that deciding the existence of a 1-proper connected 2-coloring is

---

[1] We became aware of Huang & Li's result [23,24] only after completing an earlier draft of the proof for Theorem 1 of the current work, which, despite being a counting complexity result, yields an independent proof that deciding the existence of a 1-proper connected 2-coloring is NP-complete. We additionally remark that the proof strategy of Huang & Li [23,24] requires the construction of a complete graph on

NP-complete, the complexity of finding k-proper connected w-colorings remains open for any ordered pair (k, w) where $k \in \mathbb{N}_{>0}$ and $w \in \mathbb{N}_{>1}$. Concerning counting complexity, we remark that there exist some $\#P$-completeness results for counting k-proper connected 2-colorings as part of a study on counting special types of bipartite graph orientations [7]. However, these hardness results are transitively via reduction from evaluating the Tutte polynomial at the point $T_G(0, 2)$, and thus, do little to inform us as to the complexity of finding or approximately counting such colorings (e.g., approximability of the Tutte polynomial at this point remains uncharacterized [19]).

To fill this gap, we carry out a fine-grained complexity analysis of both finding and counting k-proper connected $(w = 2)$-colorings of graphs. In particular, we sketch a proof – note that the current work is an extended abstract due to space constraints – that counting k-proper connected 2-colorings is $\#P$-complete under right-bit-shift reductions in the case where $k = 1$ (Theorem 1) and $\#P$-hard under many-one counting reductions for every $k \in \mathbb{N}_{>0}$ (Corollary 1). We furthermore show that, for every $k \in \mathbb{N}_{>0}$, no $2^{o\left(\frac{n}{k^2}\right)}$ time algorithm can exist for either finding or counting k-proper connected 2-colorings of an order n graph under the ETH or the #ETH, respectively (Corollary 2). Finally, for any $k \in \mathbb{N}_{>0}$ and any input error parameter $\epsilon > 0$, we note a recent finding of Dell & Lapinskas [14] that allows us to rule out the existence of a $2^{o\left(\frac{n}{k^2}\right)}/\epsilon^2$ time ϵ-approximation algorithm for the number of k-proper connected 2-colorings of an order n graph (Corollary 3).

2   Applications of k-Proper Connected w-Colorings to the Frequency Assignment Problem

As explicitly noted by Li & Magnant [29], k-proper connected w-colorings have direct application in the context of a variation on the well-known frequency assignment problem [1, 21, 41]. To elaborate, in the typical frequency assignment problem [1, 21, 41] one asks for a proper coloring of a geometric intersection graph of disks corresponding to an assignment of sparse and expensive frequency bands (colors) to radio stations or other emitters (vertices representing disks) with overlapping emission profiles (encoded as edges). However, we can instead ask for an edge coloring of such a graph under the fairly realistic constraint that the incoming and outgoing signals at any particular station have a sufficient frequency separation (e.g., $\approx 50 - -100\, kHz$ [41]) to avoid interference [29]. Here, if we have w available frequency bands (colors), and have a redundancy criterion of requiring k vertex disjoint paths between all pairs of vertices satisfying this non-interference criterion, we arrive at exactly the problem of finding a k-proper connected w-coloring.

$2n + m + 1$ vertices, where n and m correspond to the number of variables and clauses, respectively, in an input NAE-3-SAT instance. Accordingly, this only allows for the exclusion of a $2^{o(\sqrt{n})}$ time algorithm for finding a 1-proper connected 2-coloring under the ETH.

If we now consider the frequency assignment problem in the context of designing a communication network to survive natural and man-made disasters – e.g., in the manner of the fault-tolerant adaptable packet switching models realized in the form of the late 1960s ARPANET project [6,37] – an important question arises as to how "robust" the network is, as well as our ability to randomly sample suitable assignments of frequencies to base stations. We remark that our ability to answer such questions fundamentally relates to our ability to exactly or approximately count witnesses for the frequency assignment problem (see, e.g., [27,28] concerning the direct relationship between counting and uniform sampling), or in the current context, k-proper connected w-colorings.

3   Preliminaries

3.1   Graph Theoretic Notions and Terminology

With regards to basic graph theoretic terminology, we will generally follow Diestel [15], or where appropriate, Bondy & Murty [8]. All graphs in this work should be assumed to be simple (i.e., loop and multiedge-free), and unless otherwise specified, undirected. All references to paths in graphs should be understood as references to *simple paths* which do not revisit vertices or edges.

Concerning some less standard terminology, when we refer to a w-edge coloring of a graph, we mean an edge coloring of the graph using a palette of at most w colors. When we *identify* a vertex v_a with a vertex v_b, we delete v_a and v_b, then construct a new vertex v_c adjacent to every vertex formerly adjacent to either v_a or v_b. When we identify a vertex v_a with a graph S (assumed to be disjoint from the graph containing v_a), we delete v_a, then create an instance of the graph S where every vertex in the graph is made adjacent to all vertices formerly adjacent to v_a. In addition, letting V and E be the vertex and edge sets for a graph G, recall that an *automorphism* is an isomorphism of a graph onto itself, on in other words, a bijection $f : V \to V$ where $v_i \leftrightarrow v_j \in E$ if and only if $f(v_i) \leftrightarrow f(v_j) \in E$. If G is edge colored in this context, we say that f is an *edge-color-preserving* automorphism if it additionally ensures that $v_i \leftrightarrow v_j$ and $f(v_i) \leftrightarrow f(v_j)$ have the same coloration. Finally, when we refer to a path *ingressing* (resp. *egressing*) from a subgraph (e.g., a gadget), we mean that there exists an ordered pair of vertices (v_a, v_b) along some orientation of the path, where v_a occurs prior to v_b, where v_a (resp. v_b) is disjoint from the subgraph or gadget, and where v_b (resp. v_a) belongs to the subgraph or gadget.

3.2   Counting Complexity

The class $\#P$ consists of all integer counting problems of the form $f : \Sigma^* \to \mathbb{N}_0$, wherein the objective is to count the number of witnesses for an instance of an NP language. To reduce a problem $f \in \#P$ to a problem $h \in \#P$ via a many-one counting reduction, we require two polynomial time computable functions $R_1 : \Sigma^* \to \Sigma^*$ and $R_2 : \mathbb{N} \to \mathbb{N}$ such that $f(x) = R_2(h(R_1(x)))$. If R_2 is the identity function, then we may refer to the many-one counting reduction as a *parsimonious reduction*. Additionally, if R_2 corresponds to integer division by a power of 2 (i.e., a "right-bit-shift" operation in binary), then we may refer to the many-one counting reduction as a *right-bit-shift reduction* (see, e.g., [30]).

3.3   Approximate Counting

Let ϕ be an arbitrary instance of some counting problem $\#X$, and let f be an oracle for $\#X$ (i.e., a function which returns the exact answer to the counting problem). In this context, an ϵ-approximation algorithm for $\#X$ is a function $\widehat{f}$ which satisfies the constraint that $(1 - \epsilon) f (\phi) \leq \widehat{f} (\phi) \leq (1 + \epsilon) f (\phi)$ for some input *error parameter $\epsilon > 0$*.

3.4   Exponential Time Hypothesis (ETH) and Counting Exponential Time Hypothesis (#ETH)

The *Exponential Time Hypothesis* (ETH) of Impagliazzo & Paturi [25] is defined as follows:

Definition 1. *Exponential Time Hypothesis (ETH) [25]. Letting n and m be the number of variables and clauses for a k-SAT instance, where we assume $k \geq 3$, and letting $s_k = inf\{\delta \; : \; k\text{-}SAT \text{ can be solved in } 2^{(\delta \cdot n)} \cdot poly (m) \text{ time}\}$, we have that $s_k > 0$.*

In this context, letting $\#k$-SAT be the $\#P$ problem of counting the number of solutions for an instance of k-SAT, the *Counting Exponential Time Hypothesis* (#ETH) of Dell et al. [13] can be defined as the following moderated conjecture:

Definition 2. *Counting Exponential Time Hypothesis (#ETH) [13]. Letting n and m be the number of variables and clauses for a $\#k$-SAT instance, where we assume $k \geq 3$, and letting $s_k = inf\{\delta \; : \; \#k\text{-}SAT \text{ can be solved in } 2^{(\delta \cdot n)} \cdot poly (m)$ time}, we have that $s_k > 0$.*

3.5   Variants of Not-All-Equal SAT

For $i, j \in \mathbb{N}_{>0}$, let $x_{(i,j)}$ be a Boolean variable, let $w_{(i,j)}$ (resp. $\neg w_{(i,j)}$) be a positive (resp. negative) literal corresponding to $x_{(i,j)}$, and let $w^*_{(i,j)}$ be a literal corresponding to $x_{(i,j)}$ that must later be specified to be positive or negative. As an example, the 3-SAT formula $\phi = \left(w_{(1,1)} \vee \neg w_{(1,2)} \vee w_{(2,1)}\right) \wedge \left(\neg w_{(1,1)} \vee w_{(1,2)} \vee \neg w_{(3,4)}\right)$ has Boolean variables $\{x_{(1,1)}, x_{(1,2)}, x_{(2,1)}, x_{(3,4)}\}$, literals $\{w_{(1,1)}, \neg w_{(1,1)}, w_{(1,2)}, \neg w_{(1,2)}, w_{(2,1)}, \neg w_{(3,4)}\}$, and can also be written as $\left(w^*_{(1,1)} \vee w^*_{(1,2)} \vee w^*_{(2,1)}\right) \wedge \left(w^*_{(1,1)} \vee w^*_{(1,2)} \vee w^*_{(3,4)}\right)$.

Now let f_{NAE} be a function which accepts any combination of positive and negative literals, then returns "True" if and only if not all of the input literals uniformly evaluate to "True" or uniformly evaluate to "False". Here, we can define

Not-All-Equal-k-SAT (NAE-k-SAT) as the problem of deciding the satisfiability of $f_{NAE}\left(w^*_{(1,1)}, w^*_{(1,2)}, \ldots, w^*_{(1,k)}\right) \wedge f_{NAE}\left(w^*_{(2,1)}, w^*_{(2,2)}, \ldots, w^*_{(2,k)}\right) \wedge \ldots$, where each positive or negative literal $w^*_{(i,j)}$ may be equivalent to (or distinct from) any other positive or negative literal. In this context, we can also define Monotone NAE-k-SAT (Mon-NAE-k-SAT) as a variant of NAE-k-SAT where all literals are strictly positive.

4 Exactly and Approximately Counting Proper Connected Colorings

In this section, after first proving a pair of helper lemmas (Lemma 1 and Lemma 2), we will proceed to sketch a proof of our main Theorem 1 concerning the #P-completeness of counting 1-proper connected 2-colorings under right-bit-shift reductions. We will subsequently establish the corollaries of this theorem detailed at the end of Sect. 1 (i.e., Corollary 1 through 3).

Lemma 1. *Counting satisfying assignments for arbitrary Monotone Not-All-Equal 3-SAT formula, #Mon-NAE-3-SAT, is #P-complete under right-bit-shift reductions.*

Proof. Observe first that there is a right-bit-shift reduction from #3-SAT to #NAE-4-SAT. Specifically, following the notation given in Sect. 3.5 and letting y be a positive literal, observe that any 3-SAT formula of the form $\phi = \left(w^*_{(1,1)} \vee w^*_{(1,2)} \vee w^*_{(1,3)}\right) \wedge \ldots \wedge \left(w^*_{(n,1)} \vee w^*_{(n,2)} \vee w^*_{(n,3)}\right)$ can be parsimoniously reduced to a 3-SAT formula of the form $\phi' = \left(w^*_{(1,1)} \vee w^*_{(1,2)} \vee w^*_{(1,3)}\right) \wedge \ldots \wedge \left(w^*_{(n,1)} \vee w^*_{(n,2)} \vee w^*_{(n,3)}\right) \wedge (y \vee y \vee y)$. Observe further that, letting z be another positive literal, the 3-SAT formula ϕ' can be reduced via a right-bit-shift reduction to an NAE-4-SAT formula $\phi'' = f_{NAE}\left(w^*_{(1,1)}, w^*_{(1,2)}, w^*_{(1,3)}, z\right) \wedge \ldots \wedge f_{NAE}\left(w^*_{(n,1)}, w^*_{(n,2)}, w^*_{(n,3)}, z\right) \wedge f_{NAE}(y, y, y, z)$, such that there are exactly two satisfying assignments of ϕ'' per satisfying assignment of ϕ. This is a consequence of our requiring that $y \neq z$, and our being able to everywhere reverse the assignment of "True" and "False" values to variables to obtain another satisfying assignment for ϕ''. As #NAE-4-SAT is straightforwardly in #P, we therefore have that #NAE-4-SAT is #P-complete under right-bit-shift reductions.

Next, letting $y_{(i,j)}$ (resp. $\neg y_{(i,j)}$) be a positive (resp. negative) literal for any $i, j \in \mathbb{N}_{>0}$, observe that there is a parsimonious reduction from #NAE-4-SAT to #NAE-3-SAT, where we decompose each NAE-4-SAT clause of the form $f_{NAE}\left(w^*_{(i,1)}, w^*_{(i,2)}, w^*_{(i,3)}, w^*_{(i,4)}\right)$ into the following five clause NAE-3-SAT expression: $f_{NAE}\left(w^*_{(i,1)}, w^*_{(i,2)}, y_{(i,1)}\right) \wedge f_{NAE}\left(w^*_{(i,1)}, \neg\left(w^*_{(i,3)}\right), y_{(i,1)}\right) \wedge f_{NAE}\left(\neg\left(w^*_{(i,1)}\right), w^*_{(i,4)}, \neg y_{(i,1)}\right) \wedge f_{NAE}\left(w^*_{(i,3)}, w^*_{(i,4)}, y_{(i,2)}\right) \wedge f_{NAE}\left(y_{(i,1)}, y_{(i,1)}, y_{(i,2)}\right)$.

Finally, observe that we can parsimoniously reduce #NAE-3-SAT to #Mon-NAE-3-SAT by replacing any negative literal $\neg p$ (e.g., where $p = w_{(...)}$ or $p = y_{(...)}$) with a new positive literal q, then adding the NAE-3-SAT clause $f_{NAE}(p, q, q)$ to require $q = \neg p$.

Putting everything together, as #Mon-NAE-3-SAT is trivially in #P, and as we doubled the number of witnesses in reducing #3-SAT to #NAE-4-SAT prior to parsimoniously reducing #NAE-4-SAT to #Mon-NAE-3-SAT, #Mon-NAE-3-SAT is accordingly #P-complete under right-bit-shift reductions.   □

Lemma 2. *A w-edge coloring for a graph with n vertices and m edges can be verified as a 1-proper connected w-coloring in $\mathcal{O}\left(n^2 \cdot m\right)$ time for every $w \in \mathbb{N}_{>0}$.*

Proof. Let G be a graph with n vertices, m edges, and an edge coloring using at most $w \in \mathbb{N}_{>0}$ colors. If $w = 1$, observe that we simply need to verify in $\mathcal{O}(m)$ time if G is a clique. Otherwise, for $w = 2$, we can observe a result of Bang-Jensen & Gutin [5], or for $w \geq 2$, a more general result of Szeider [38], that a proper path between any pair of vertices (if one exists) can be found in $\mathcal{O}(m)$ time. Accordingly, to check if a given edge coloring is a 1-proper connected ($w \geq 2$)-coloring, it suffices to use this procedure to check for the existence of a proper path between all pairs of vertices in $\mathcal{O}\left(n^2 \cdot m\right)$ time.   □

Theorem 1. *Counting 1-proper connected 2-colorings is #P-complete under right-bit-shift reductions.*

Proof sketch. Let ϕ_{MonNAE} be an arbitrary Mon-NAE-3-SAT formula with n variables and m clauses of the form $\phi_{MonNAE} = C_1 \wedge C_2 \wedge \ldots \wedge C_m$. We will proceed by giving a right-bit-shift reduction from the #Mon-NAE-3-SAT problem of counting witnesses for ϕ_{MonNAE} – which we can recall is #P-complete under right-bit-shift reductions by Lemma 1 – to the problem of counting 1-proper connected 2-colorings of a graph.

Our reduction involves three basic steps. In (Step 1.1), for each of the m clauses of ϕ_{MonNAE}, we construct an instance of the "NAE-3-SAT Clause Gadget" shown in Fig. 1(a) (see also Fig. 1(b) for a simplified abstraction). In (Step 1.2), for each of the n variables of ϕ_{MonNAE}, we construct an instance of the "NAE-3-SAT Variable Gadget" shown in Fig. 2(a) (see also Fig. 2(b) for a simplified abstraction) where the number of output edges, z, is specified to be equal to the number of times the given variable occurs in ϕ_{MonNAE}.

Finally, in (Step 1.3), for each literal $w_{(i,j)}$ corresponding to some variable x_j of ϕ_{MonNAE}, and belonging to a clause C_i of ϕ_{MonNAE}, let γ_i be the "NAE-3-SAT Clause Gadget" constructed for C_i in (Step 1.1) and let ζ_j be the "NAE-3-SAT Variable Gadget" constructed for x_j in (Step 1.2). Here, for some pair of vertices $\{v_{(out,r,1)}, v_{(out,r,2)}\}$ in ζ_j, where the selected value of r is unique for each literal corresponding to the variable x_j, we delete the vertex $v_{(out,r,2)}$ and identify the vertex $v_{(out,r,1)}$ with any degree 1 vertex in the set $\{v_{(in,1,1)}, v_{(in,1,2)}, v_{(in,1,3)}\}$ of input vertices for γ_i.

Let us now make the following assumption: (Assumption 1.1) the edge coloration schemes for the "NAE-3-SAT Clause Gadget", shown in Fig. 1(c.1)

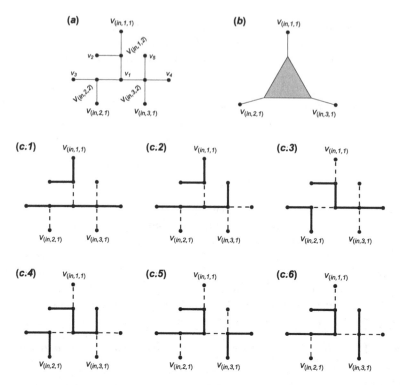

Fig. 1. Illustration and details of the "NAE-3-SAT Clause Gadget"; **(a)** the "NAE-3-SAT Clause Gadget" with vertex labels and edges left uncolored, having 3 input edges: $\{v_{(in,1,1)} \leftrightarrow v_{(in,1,2)}, v_{(in,2,1)} \leftrightarrow v_{(in,2,2)}, v_{(in,3,1)} \leftrightarrow v_{(in,3,2)}\}$; **(b)** gadget-level abstraction of the "NAE-3-SAT Clause Gadget" having 3 input edges; **(c.1)** through **(c.6)** show all possible manners of coloring the edges of the "NAE-3-SAT Clause Gadget" using a palette of 2 distinct colors (indicated using dashed and solid thick lines for edges), up to color inversion (though not, in this case, edge-color-preserving automorphism), such that there may exist a proper path between any two vertices in a given reduction construct.

through 1(c.6), and the "NAE-3-SAT Variable Gadget", shown in Fig. 2(a), are the only possible edge coloration schemes, up to edge-color-preserving automorphism and color inversion, that allow the reduction construct to be 1-proper connected using a palette of 2 distinct colors.

Under (Assumption 1.1), it is straightforward to understand how the reduction works. Specifically, let c_1 (solid thick coloration in Fig. 1(c.1) through Fig. 1(c.6)) and c_2 (dashed coloration in Fig. 1(c.1) through Fig. 1(c.6)) be the two distinct edge colorations, and arbitrarily specify that c_1 corresponds to "True" and c_2 corresponds to "False". Observe that the "NAE-3-SAT Clause Gadget" will allow the reduction construct to be 1-proper connected if and only if one of its input edges – which will correspond to the edges $\{v_{(in,1,1)} \leftrightarrow v_{(in,1,2)}, v_{(in,2,1)} \leftrightarrow v_{(in,2,2)}, v_{(in,3,1)} \leftrightarrow v_{(in,3,2)}\}$, respectively – has color c_1

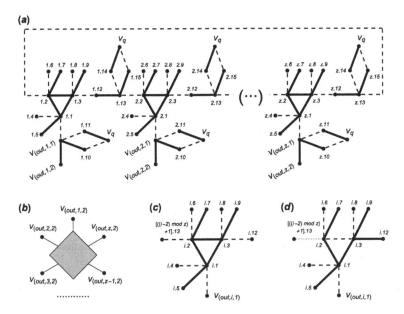

Fig. 2. Illustration and details of the "NAE-3-SAT Variable Gadget"; **(a)** the "NAE-3-SAT Variable Gadget" with vertex labels, having z output edges – $v_{(out,1,1)} \leftrightarrow v_{(out,1,2)}, v_{(out,2,1)} \leftrightarrow v_{(out,2,2)}, \ldots, v_{(out,z,1)} \leftrightarrow v_{(out,z,2)}$ – where we show the only possible manner of coloring the edges of the gadget using a palette of 2 distinct colors (indicated using dashed and solid thick lines for edges), up to color inversion and edge-color-preserving automorphism, such that there may exist a proper path between any two vertices in a given reduction construct, and where all vertices labeled v_q correspond to the same vertex; **(b)** gadget-level abstraction the "NAE-3-SAT Variable Gadget" with z output edges; **(c)** subgraph of the "NAE-3-SAT Variable Gadget" induced by the vertices in the set $\{i.1, i.2, i.3, i.4, i.5, i.6, i.7, i.8, i.9, i.12, (((i-2) \mod z) + 1).13\}$, where $1 \le i \le z$, with the edge coloration scheme shown in (a); **(d)** subgraph of the "NAE-3-SAT Variable Gadget" showing a "pathological" edge coloring scheme, forced (up to edge-color-preserving automorphism and color inversion) only by a failure to monochromatically color the edges of the C_3 cycle induced by the vertex set $\{i.1, i.2, i.3\}$, where we have that no proper path can connect the vertices $i.6$ and $i.9$ regardless of the coloration of the (thin dotted) edge $(((i-2) \mod z) + 1).13 \leftrightarrow i.2$.

(encoding the input "True") and one of the edges it is adjacent to has color c_2 (encoding the input "False"), and furthermore, that there are two possible edge coloration schemes for each permutation of inputs. For example, the edge coloration scheme shown in Fig. 1(c.1) and Fig. 1(c.2) are the only two edge coloring schemes for the "NAE-3-SAT Clause Gadget" that accept the input $\{v_{(in,1,1)} \leftrightarrow v_{(in,1,2)}, v_{(in,2,1)} \leftrightarrow v_{(in,2,2)}, v_{(in,3,1)} \leftrightarrow v_{(in,3,2)}\} = \{c_1, c_2, c_2\}$, and their color inverses are likewise the only two edge coloring schemes that accept the input $\{v_{(in,1,1)} \leftrightarrow v_{(in,1,2)}, v_{(in,2,1)} \leftrightarrow v_{(in,2,2)}, v_{(in,3,1)} \leftrightarrow v_{(in,3,2)}\} = \{c_2, c_1, c_1\}$.

Observe further that each of the outputs of the "NAE-3-SAT Variable Gadget" – encoded in the color of the edge $v_{(out,i,1)} \leftrightarrow v_{(out,i,2)}$ for each $1 \leq i \leq z$ – will have the same coloration. Additionally, observe that we may invert the colorations for each of the $3z$ pairs of edges connecting degree 1 vertices to a common vertex, and likewise invert the colors of the set of edges in each of the $2z$ length 4 loops containing the vertex v_q to obtain an edge-color-preserving automorphism of the "NAE-3-SAT Variable Gadget". We therefore have that, up to color inversion of all edges in the graph constructed for the reduction, there are exactly 2^{5z} edge coloring schemes for each "NAE-3-SAT Variable Gadget" encoding a variable set to either "True" or "False".

We now make an additional assumption (Assumption 1.2), that provided (Assumption 1.1) is correct, for any truth assignment to the variables of the ϕ_{MonNAE} formula encoded in the aforementioned reduction construct, there will be a 1-proper connected 2-coloring of the reduction construct if and only if the truth assignment is a witness for ϕ_{MonNAE}. With this assumption, we can now observe the following lemma:

Lemma 3. *Assuming the correctness of (Assumption 1.1) and (Assumption 1.2), our reduction from #Mon-NAE-3-SAT to counting 1-proper connected 2-colorings of graphs will multiply the number of witnesses for an input Mon-NAE-3-SAT formula with m clauses by a factor of exactly $2^{(16m+1)}$, implying that counting 1-proper connected 2-colorings of graphs is #P-hard under right-bit-shift reductions.*

Proof. Let ϕ_{MonNAE} be the original Mon-NAE-3-SAT formula with m clauses in the reduction of an instance of #Mon-NAE-3-SAT to an instance of counting 1-proper connected 2-colorings. For a given satisfying truth assignment for the variables in ϕ_{MonNAE}, recall that, up to inversion of all colors in the graph constructed for the reduction, there will be exactly 2^{5z} edge coloring schemes for each "NAE-3-SAT Variable Gadget" (where z corresponds to the number of gadget outputs) permitting the reduction construct to have a 1-proper connected 2-coloring. Furthermore, recall that, for a given satisfying truth assignment for the variables in ϕ_{MonNAE}, and for a specific coloration of the edges in each "NAE-3-SAT Variable Gadget", there will be exactly two permitted colorations of each "NAE-3-SAT Clause Gadget". Finally, recall that, assuming the correctness of (Assumption 1.1) and (Assumption 1.2), any 1-proper connected 2-coloring of the reduction construct must correspond to a witness for ϕ_{MonNAE}.

Putting everything together and accounting for color inversions, we have that there will be exactly $2^m \cdot 2^{(5 \times 3m)} \cdot 2 = 2^{(16m+1)}$ instances of 1-proper connected 2-colorings for the reduction construct per satisfying assignment of ϕ_{MonNAE}. This implies that our reduction from #Mon-NAE-3-SAT to counting 1-proper connected 2-colorings is a right-bit-shift reduction, and by Lemma 1, that the latter problem is accordingly #P-hard under right-bit-shift reductions.   □

To complete the current proof, it now suffices to establish the correctness of (Assumption 1.1) and (Assumption 1.2). Concerning (Assumption 1.2), we will in particular need to show that, for every edge coloration of the reduction construct corresponding to a witness for the encoded Mon-NAE-3-SAT formula, there will exist a proper path in a reduction construct connecting a pair of vertices v_a and v_b in each of the following cases: (case 1.1) $v_a = v_q$ and v_b is internal to a "NAE-3-SAT Variable Gadget"; (case 1.2) $v_a = v_q$ and v_b is internal to a "NAE-3-SAT Clause Gadget"; (case 1.3) v_a and v_b are both internal to the same "NAE-3-SAT Variable Gadget"; (case 1.4) v_a and v_b are internal to distinct instances of the "NAE-3-SAT Variable Gadget"; (case 1.5) v_a is internal to a "NAE-3-SAT Variable Gadget" and v_b is internal to a "NAE-3-SAT Clause Gadget"; (case 1.6) v_a and v_b are both internal to the same "NAE-3-SAT Clause Gadget"; (case 1.7) v_a and v_b are internal to distinct instances of the "NAE-3-SAT Clause Gadget".

Here, as proofs of correctness for (Assumption 1.1) and (Assumption 1.2) ultimately required a long and complex case analysis, due to space constraints we are forced to omit the details in this extended abstract.

Putting everything together, we can appeal to Lemma 3 to determine that counting 1-proper connected 2-colorings of graphs is #P-hard under right-bit-shift reductions. It remains to observe that the problem at hand is also in #P as a straightforward consequence of Lemma 2, yielding the current theorem. □

Corollary 1. *Counting k-proper connected 2-colorings is #P-hard under many-one counting reductions for every $k \in \mathbb{N}_{>0}$.*

Proof. Recall that a *orientation* of an undirected graph G is a directed graph generated by assigning a direction to each of the edges of G, and that a *unilateral orientation* of G is a directed graph where, for every pair of vertices v_a or v_b, there exists at least one directed path from v_a to v_b or from v_b to v_a. Here, we can also introduce the notion of a *k-unilateral orientation* [7], wherein we require $k \in \mathbb{N}_{>0}$ vertex disjoint directed paths connecting any pair of vertices v_a and v_b, and where each such path is permitted to originate at v_a and end at v_b or vice versa. Now observe that, in the specific context of a bipartite graph, 1-unilateral connected orientations and 1-proper connected 2-colorings are in bijection according to what Bang-Jensen & Gutin [4] refer to as the "BB correspondence" – simply color each edge in accordance with the partite set membership of the vertex that it is oriented towards, and notice that any directed path becomes a proper path. From this we can immediately infer that k-unilateral connected orientations and k-proper connected 2-colorings are also in bijection.

To proceed, we will slightly modify a construction given in "Theorem 3.6" of [7], which allows one to transform any bipartite graph H with vertex set V_H – and two non-adjacent degree 1 vertices attached to a pair of adjacent degree 3 vertices – in $\mathcal{O}\left(|V_H| \cdot f(k)\right)$ time into a graph M with $z\left(k, |V_H|\right)$ instances of k-unilateral connected orientations (equiv. k-proper connected 2-colorings) per unilateral orientation (equiv. 1-proper connected 2-coloring) of H, where $f(k) \in \mathcal{O}\left(k^2\right)$ is a polynomial time computable function of k and $z\left(k, |V_H|\right)$ is a polynomial time computable function of k and $|V_H|$.

Here, let G be an arbitrary simple undirected graph of order ≥ 3, with vertex set V_G, constructed from an input Mon-NAE-3-SAT formula via the reduction given in Theorem 1, and let S_G be any spanning tree of G (which can be computed in $\mathcal{O}\left(|V_G|\right)$ time). Next, exchange the input bipartite graph H for S_G in the "Theorem 3.6" reduction from [7] – treating any pair of non-adjacent degree 1 vertices as the degree 1 vertices attached to adjacent degree 3 vertices in the original "Theorem 3.6" reduction – and call the graph resulting from this reduction $\mathcal{W}$. Finally, create a graph $\mathcal{W}'$ by adding the edges in G that do not occur in S_G. While these changes may impact the "BB correspondence", it can be observed that the exact relationship between the number of 1-proper connected 2-colorings of H and k-proper connected 2-colorings of M from the original "Theorem 3.6" reduction will still hold for 1-proper connected 2-colorings of G and k-proper connected 2-colorings of $\mathcal{W}'$ in our modified construction. Accordingly, $\mathcal{W}'$ will have exactly $z\left(k, |V_G|\right)$ instances of k-proper connected 2-colorings per 1-proper connected 2-coloring of G, yielding the corollary at hand.   □

Corollary 2. *For every $k \in \mathbb{N}_{>0}$, it holds that no $2^{o\left(\frac{n}{k^2}\right)}$ time algorithm can exist for either finding a k-proper connected 2-coloring assuming the ETH, or for counting such colorings assuming the #ETH.*

Proof. Observe that Theorem 1 gives an $\mathcal{O}\left(n \cdot m\right)$ time many-one counting reduction, transitively from an arbitrary instance of #3-SAT with n variables and m clauses, to the problem of counting 1-proper connected 2-colorings. Here, as the sparsification lemma of Impagliazzo et al. [26] implies that no $2^{o(n)}$ time algorithm can exist for sparse (i.e., $m \in \Theta\left(n\right)$) instances of 3-SAT or #3-SAT under the ETH and #ETH, respectively, we have that no $2^{o(n)}$ time algorithm can exist for determining the existence of or counting 1-proper connected 2-colorings under the ETH or #ETH, respectively. Next, observe that Corollary 1 gives a many-one counting reduction, transitively from an arbitrary instance of #3-SAT with n variables and m clauses, to the problem of counting k-proper connected 2-colorings for every $k \in \mathbb{N}_{>0}$, with a $\mathcal{O}\left(k^2\right)$ overhead on the time complexity of the Theorem 1 reduction. Thus, for every $k \in \mathbb{N}_{>0}$, we have that no $2^{o\left(\frac{n}{k^2}\right)}$ time algorithm can exist for determining the existence of or counting k-proper connected 2-colorings under the ETH or #ETH, respectively.   □

Corollary 3. *Assuming the ETH, for every $k \in \mathbb{N}_{>0}$ and any input error parameter $\epsilon > 0$, no $2^{o\left(\frac{n}{k^2}\right)}/\epsilon^2$ time ϵ-approximation algorithm exists for number of k-proper connected 2-colorings of an order n graph.*

Proof. This result follows from Corollary 2 and a finding of Dell & Lapinskas [14] that, for any input error parameter $\epsilon > 0$, under the ETH no $2^{o(n)}/\epsilon^2$ time ϵ-approximation algorithm can exist for an n variable instance of #3-SAT.   □

5   Concluding Remarks

We strongly believe that our results can be extended to rule out at least the existence of a $2^{o\left(\frac{n}{k^2}\right)}$ time algorithm for finding or counting k-proper connected w-colorings under the ETH and #ETH, respectively, for each ordered pair of parameters (k, w), where $k \in \mathbb{N}_{>0}$ and $w \in \mathbb{N}_{>1}$. We leave this problem open.

References

1. Aardal, K.I., van Hoesel, S.P.M., Koster, A.M.C.A., Mannino, C., Sassano, A.: Models and solution techniques for frequency assignment problems. Ann. Oper. Res. **153**(1), 79–129 (2007)
2. Abouelaoualim, A., Das, K.C., Faria, L., Manoussakis, Y., Martinhon, C., Saad, R.: Paths and trails in edge-colored graphs. Theoret. Comput. Sci. **409**(3), 497–510 (2008)
3. Andrews, E., Lumduanhom, C., Laforge, E., Zhang, P.: On proper-path colorings in graphs. J. Comb. Math. Comb. Comput. **97**, 189–207 (2016)
4. Bang-Jensen, J., Gutin, G.: Alternating cycles and paths in edge-coloured multigraphs: a survey. Discrete Math. **165**(166), 39–60 (1997)
5. Bang-Jensen, J., Gutin, G.: Alternating cycles and trails in 2-edge-coloured complete multigraphs. Discrete Math. **188**(1–3), 61–72 (1998)
6. Baran, P.: The beginnings of packet switching: some underlying concepts. IEEE Commun. Mag. **40**(7), 42–48 (2002)
7. Barish, R.D.: On the Number of k-Proper Connected Edge and Vertex Colorings of Graphs. Accepted for publication in Thai J, Math (2023)
8. Bondy, J.A., Murty, U.S.R.: Graph Theory with Applications. Macmillan Press: New York, NY, 1st edn. (1976)
9. Borozan, V., et al.: Proper connection of graphs. Discrete Math. **312**(17), 2550–2560 (2012)
10. Chartrand, G., Johns, G.L., McKeon, K.A., Zhang, P.: Rainbow connection in graphs. Math. Bohem. **133**(1), 85–98 (2008)
11. Chartrand, G., Johns, G.L., McKeon, K.A., Zhang, P.: The rainbow connectivity of a graph. Networks **54**(2), 75–81 (2009)
12. Chou, W.S., Manoussakis, Y., Megalakaki, O., Spyratos, M., Tuza, Z.: Paths through fixed vertices in edge-colored graphs. Mathématiques et Sciences Humaines **127**, 49–58 (1994)
13. Dell, H., Husfeldt, T., Marx, D., Taslaman, N., Wahlén, M.: Exponential time complexity of the permanent and the Tutte polynomial. ACM Trans. Algorithms **10**(4), 21:1–21:32 (2014)
14. Dell, H., Lapinskas, J.: Fine-grained reductions from approximate counting to decision. ACM Trans. Comput. Theory **13**(2), 8:1–8:24 (2021)
15. Diestel, R.: Graph Theory. GTM, vol. 173. Springer, Heidelberg (2017). https://doi.org/10.1007/978-3-662-53622-3
16. Ducoffe, G., Marinescu-Ghemeci, R., Popa, A.: On the (di)graphs with (directed) proper connection number two. Discrete Appl. Math. **281**, 203–215 (2020)
17. Edmonds, J.: Paths, trees, and flowers. Can. J. Math. **17**, 449–467 (1965)
18. Gerek, A., Fujita, S., Magnant, C.: Proper connection with many colors. J. Comb. **3**(4), 683–693 (2012)
19. Goldberg, L.A., Jerrum, M.: Inapproximability of the Tutte polynomial. Inform. Comput. **206**(7), 908–929 (2008)
20. Gu, R., Li, X., Qin, Z.: Proper connection number of random graphs. Theoret. Comput. Sci. **609**, 336–343 (2016)
21. Hale, W.K.: Frequency assignment: theory and applications. Proc. IEEE **68**(12), 1497–1514 (1980)
22. Huang, F., Li, X., Qin, Z., Magnant, C.: Minimum degree condition for proper connection number 2. Theoret. Comput. Sci. **774**, 44–50 (2019)

23. Huang, Z., Li, X.: Hardness results for three kinds of colored connections of graphs, pp. 1–23 (2020). arxiv.org/abs/2001.01948
24. Huang, Z., Li, X.: Hardness results for three kinds of colored connections of graphs. Theoret. Comput. Sci. **841**, 27–38 (2020)
25. Impagliazzo, R., Paturi, R.: On the complexity of k-SAT. J. Comput. Syst. Sci. **62**(2), 367–375 (2001)
26. Impagliazzo, R., Paturi, R., Zane, F.: Which problems have strongly exponential complexity? J. Comput. Syst. Sci. **63**(4), 512–530 (2001)
27. Jerrum, M.: Counting, sampling and integrating: algorithms and complexity. Lectures in Mathematics, ETH Zuerich, Birkhauser Verlag, Basel, Switzerland (2013)
28. Jerrum, M.R., Valiant, L.G., Vazirani, V.V.: Random generation of combinatorial structures from a uniform distribution. Theoret. Comput. Sci. **43**(2–3), 169–188 (1986)
29. Li, X., Magnant, C.: Properly colored notions of connectivity - a dynamic survey. Theory Appl. Graphs **0**(1), 1–16 (2015)
30. Liśkiewicz, M., Ogihara, M., Toda, S.: The complexity of counting self-avoiding walks in subgraphs of two-dimensional grids and hypercubes. Theoret. Comput. Sci. **304**(1–3), 129–156 (2003)
31. Lumduanhom, C., Laforge, E., Zhang, P.: Characterizations of graphs having large proper connection numbers. Discuss. Math. Graph Theory **36**(2), 439–453 (2016)
32. Manoussakis, Y.: Alternating paths in edge-colored complete graphs. Discret. Appl. Math. **56**(2–3), 297–309 (1995)
33. Menger, K.: Zur allgemeinen kurventheorie. Fundam. Math. **10**(1), 96–115 (1927)
34. Micali, S., Vazirani, V.V.: An $\mathcal{O}\left(\sqrt{|V|} \cdot |E|\right)$ algorithm for finding maximum matching in general graphs. In: Proceedings of the 21st Annual Symposium on Foundations of Computer Science (FOCS), pp. 17–27 (1980)
35. Mulder, H.M.: Julius Petersen's theory of regular graphs. Discrete Math. **100**(1–3), 157–175 (1992)
36. Petersen, J.: Die theorie der regulären graphs. Acta Math. **15**, 193–220 (1891)
37. Roberts, L.G.: Multiple computer networks and intercomputer communication. In: Proceedings of the 1st ACM Symposium on Operating System Principles (SOSP), pp. 3.1-3.6 (1967)
38. Szeider, S.: Finding paths in graphs avoiding forbidden transitions. Discret. Appl. Math. **126**(2–3), 261–273 (2003)
39. Tutte, W.T.: The factors of graphs. Can. J. Math. **4**, 314–328 (1952)
40. Tutte, W.T.: The method of alternating paths. Combinatorica **2**(3), 325–332 (1982)
41. de Werra, D., Gay, Y.: Chromatic scheduling and frequency assignment. Discrete Appl. Math. **49**(1–3), 165–174 (1994)
42. Yeo, A.: A note on alternating cycles in edge-coloured graphs. J. Comb. Theory, Ser. B **69**(2), 222–225 (1997)

Combinatorics and Computing

Strong Edge Coloring of Subquartic Graphs

Junlei Zhu[1](✉)📵 and Hongguo Zhu[2]📵

[1] College of Data Science, Jiaxing University, Jiaxing 314001, China
zhujl-001@163.com
[2] Department of Mathematics, Zhejiang Normal University, Jinhua 321004, China
zhuhongguo@zjnu.edu.cn

Abstract. A strong k-edge coloring of a graph G is a mapping $c : E(G) \to \{1, 2, 3, ..., k\}$ such that for any two edges e and e' with distance at most two, $c(e) \neq c(e')$. The strong chromatic index of G, written $\chi'_s(G)$, is the smallest integer k such that G has a strong k-edge coloring. In this paper, using color exchange method and discharging method, we prove that for a subquartic graph G, $\chi'_s(G) \leq 11$ if $mad(G) < \frac{8}{3}$, where $mad(G) = \max\{\frac{2|E(G)|}{|V(G)|}, H \subseteq G\}$.

Keywords: subquartic graph · strong edge coloring · maximum average degree

1 Introduction

To solve the Channel Assignment Problem in wireless communication networks, Fouquet and Jolivet [8] first introduced the notion of strong edge coloring in 1983. A strong k-edge coloring of a graph G is a mapping $c : E(G) \to \{1, 2, 3, \cdots, k\}$ such that $c(e) \neq c(e')$ for any two edges e and e' with distance at most two. The smallest integer k such that G has a strong k-edge coloring of G is called the strong chromatic index of G, written $\chi'_s(G)$. By greedy algorithm, it is easy to see that $2\Delta^2 - 2\Delta + 1$ is a trivial upper bound on $\chi'_s(G)$, where Δ is the maximum degree of G. However, it is NP-complete to decide wether $\chi'_s(G) = k$ holds for a general graph G [14]. In 1989, Erdős and Nešetřil [7] proposed the following important conjecture while studying the strong edge coloring of graphs.

Conjecture 1. [7] For any graph G with maximum degree Δ, $\chi'_s(G) \leq \frac{5}{4}\Delta^2$ if Δ is even, $\chi'_s(G) \leq \frac{5}{4}\Delta^2 - \frac{1}{2}\Delta + \frac{1}{4}$ if Δ is odd.

In [7], Erdős and Nešetřil constructed two classes of graphs satisfying $\chi'_s(G) = \chi'(G) = |E(G)|$ while $|E(G)|$ attains the upper bound in Conjecture 1. This illustrate that the upper bound is sharp if Conjecture 1 is true. Also, they asked a question: For a general graph G, is there any positive number ε such that $\chi'_s(G) \leq (2-\epsilon)\Delta^2$, where Δ is the maximum degree of G. As yet, there are many research results on strong edge coloring. For a graph G with sufficient large Δ, Molloy

W. Wu and J. Guo (Eds.): COCOA 2023, LNCS 14462, pp. 139–146, 2024.
https://doi.org/10.1007/978-3-031-49614-1_9

and Reed [15] proved that $\chi'_s(G) \leq 1.998\Delta^2$ using probabilistic methods. In the next decides, this result was improved to $1.93\Delta^2$ by Bruhn and Joos [4], $1.835\Delta^2$ by Bonamy, Perrett and Postle [3]. For graphs with small Δ, scholars also made a lot of research works. It is an obvious result that $\chi'_s(G) \leq 5 = \frac{5}{4}\Delta^2$ while $\Delta = 2$. For subcubic graphs, the above conjecture was verified by Andersen [1], and independently by Horák, Qing, Trotter [10]. For subquartic graphs, $\chi'_s(G) \leq 22$ was proven by Cranston [6] using algorithms. Huang, Santana and Yu [11] reduced 22 to 21. For graphs with $\Delta = 5$, Zang [18] confirmed that $\chi'_s(G) \leq 37$.

For graphs with maximum average degree restriction, there are also a mount of results. The maximum average degree of a graph G, written $mad(G)$, is the largest average degree of its subgraph. In other words, $mad(G) = \max\{\frac{2|E(H)|}{|V(H)|} : H \subseteq G\}$. In 2013, Hocquard [9] studied the strong chromatic index of subcubic graphs with maximum average degree and obtained the following theorem.

Theorem 1. *[9] Let G be a graph with $\Delta(G) = 3$.*

(1) If $mad(G) < \frac{7}{3}$, then $\chi'_s(G) \leq 6$;
(2) If $mad(G) < \frac{5}{2}$, then $\chi'_s(G) \leq 7$;
(3) If $mad(G) < \frac{8}{3}$, then $\chi'_s(G) \leq 8$;
(4) If $mad(G) < \frac{20}{7}$, then $\chi'_s(G) \leq 9$.

The given upper bound on $mad(G)$ in Theorem 1(1)(2)(4) is optimal since there exist subcubic graphs with $mad(G) = \frac{7}{3}$ (or $mad(G) = \frac{5}{2}, \frac{20}{7}$) and $\chi'_s(G) > 6$ (or $\chi'_s(G) > 7, 9$), see Fig. 1.

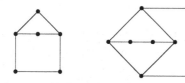

Fig. 1. $mad(G) = \frac{7}{3}$ (or $\frac{5}{2}, \frac{20}{7}$) and $\chi'_s(G) = 7$ (or $\chi'_s(G) = 8, 10$)

For subquartic graphs with bounded maximum average degree, Lv et al. [13] gave out the following theorem, which improved the corresponding upper bound on $mad(G)$ due to Bensmail et al. [2].

Theorem 2. *[13] Let G be a graph with $\Delta(G) = 4$.*

(1) If $mad(G) < \frac{61}{18}$, then $\chi'_s(G) \leq 16$;
(2) If $mad(G) < \frac{7}{2}$, then $\chi'_s(G) \leq 17$;
(3) If $mad(G) < \frac{18}{5}$, then $\chi'_s(G) \leq 18$;
(4) If $mad(G) < \frac{26}{7}$, then $\chi'_s(G) \leq 19$;
(5) If $mad(G) < \frac{51}{13}$, then $\chi'_s(G) \leq 20$.

Ruksasakchai and Wang [17] studied the strong edge coloring of graphs with $\Delta(G) \leq 4$ and $mad(G) < 3$ and obtained the following theorem.

Theorem 3. *[17] If G is a graphs G with maximum degree $\Delta \leq 4$ and $mad(G) < 3$, then $\chi'_s(G) \leq 3\Delta + 1$.*

For graphs with maximum degree 5 and bounded maximum average degree, Qin et al. [16] obtained the following theorem.

Theorem 4. *[16] Let G be a graph with $\Delta(G) = 5$.*

(1) If $mad(G) < \frac{8}{3}$, then $\chi'_s(G) \leq 13$;
(2) If $mad(G) < \frac{14}{5}$, then $\chi'_s(G) \leq 14$.

Additionally, Choi et al. [5] studied the strong edge coloring of graphs with maximum degree $\Delta \geq 7$ and bounded maximum average degree. They obtained a theorem as follows.

Theorem 5. *[5] Let G be a graph with maximum degree Δ.*

(1) If $\Delta \geq 9$ and $mad(G) < \frac{8}{3}$, then $\chi'_s(G) \leq 3\Delta - 3$;
(2) If $\Delta \geq 7$ and $mad(G) < 3$, then $\chi'_s(G) \leq 3\Delta$.

Recently, Li et al. [12] studied the strong edge coloring of graphs with maximum degree $\Delta \geq 6$ and bounded maximum average degree. The following theorem is given in [12].

Theorem 6. *[12] Let G be a graph with maximum degree Δ.*

(1) If $\Delta \geq 6$ and $mad(G) < \frac{23}{8}$, then $\chi'_s(G) \leq 3\Delta - 1$;
(2) If $\Delta \geq 7$ and $mad(G) < \frac{26}{9}$, then $\chi'_s(G) \leq 3\Delta - 1$.

In this paper, we further consider the strong edge coloring of subquartic graphs by using color exchange method and discharging method. We obtained the following theorem.

Theorem 7. *If G is a graph with $\Delta(G) = 4$ and $mad(G) < \frac{8}{3}$, then $\chi'_s(G) \leq 11$.*

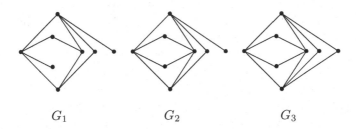

G_1 G_2 G_3

Fig. 2. Subquartic graphs.

G_1, G_2, G_3 in Fig. 2 are subquartic graphs, where $mad(G_1) = \frac{8}{3}$, $\chi'_s(G_1) = 10$; $mad(G_2) = \frac{20}{7}$, $\chi'_s(G_2) = 11$ and $mad(G_3) = 3$, $\chi'_s(G_3) = 12$ (we can take the graph obtained from G_1 by deleting two 1-vertices, G_2 by deleting the 1-vertex and G_3 as subgraphs, respectively). We do not know whether the upper bound $mad(G) < \frac{8}{3}$ in Theorem 7 is optimal. However, due to the graph G_3 in Fig. 2, we know that there exists a graph G with $\Delta(G) = 4$, $mad(G) = 3$ and $\chi'_s(G) = 12$.

For the strong edge coloring of subquartic graphs, Theorem 2 gives out some sufficient conditions for $\chi'_s(G) \leq 16$ (respectively 17,18,19,20). Theorem 3 indicates that any graph G with $\Delta(G) = 4$ and $mad(G) < 3$ satisfies $\chi'_s(G) \leq 13$. Therefore, Theorem 7 enriches the results of strong edge coloring for subquartic graphs.

2   Notations

All graphs considered here are finite undirected simple graphs. For a graph G, $V(G)$, $E(G)$, $\Delta(G)$ and $\delta(G)$ denote its vertex set, edge set, maximum degree and minimum degree respectively. For $v \in V(G)$, $d_G(v)$ (abbreviated by $d(v)$) denotes the degree of v in G. v is a i (or i^+, i^-)-vertex if $d(v) = i$ (or $d(v) \geq i$, $d(v) \leq i$). For a vertex v, a i-neighbor of v is a i-vertex in $N(v)$. A i_j-vertex is a i-vertex adjacent to exactly j 2-vertices. A 2-vertex is bad if it is adjacent to a 2-vertex, semi-bad if it is adjacent to a 3_2-vertex. A 2-vertex is good if it is neither bad nor semi-bad. For an edge e, $F(e)$ denotes the set of forbidden colors for it.

3   Proof of Theorem 7

Suppose G is a counterexample with minimum 2^+-vertices and then with minimum edges. Let H be the graph obtained from G by deleting all 1-vertices. Obviously, $H \subseteq G$ and then $mad(H) \leq mad(G) < \frac{8}{3}$. In the following, we first illustrate some properties of H.

Lemma 1. *H does not have vertices of degree 1.*

Proof. Suppose v is a 1-vertex in H and $uv \in E(H)$. Since H is the graph obtained from obtained from G by deleting all 1-vertices, $d_G(v) > 1$ and v has at least one 1-neighbor v_1 in G. Compared with G, $G - v_1$ has the same 2^+-vertices but fewer edges. By the minimality of G, $\chi'_s(G - v_1) \leq 11$. Note that in G, $|F(vv_1)| \leq 6$. Thus, vv_1 can be colored, which leads to a contradiction.

Lemma 2. *If $d_H(v) = 2$, then $d_G(v) = 2$.*

Proof. Suppose $d_G(v) > 2$. Then, v has at least one 1-neighbor v_1 in G. Compared with G, $G - v_1$ has fewer edges while the same 2^+-vertices. By the minimality of G, $\chi'_s(G - v_1) \leq 11$. Note that in G, $|F(vv_1)| \leq 9$. Thus, vv_1 can be colored, which leads to a contradiction.

Lemma 3. *If v is a 3_i-vertex in H, where $i \geq 1$, then $d_G(v) = 3$.*

Proof. Suppose $d_G(v) > 3$. Then, v has at least one 1-neighbor v' in G. Let v_1 be a 2-neighbor of v in H, By Lemma 2, $d_G(v_1) = 2$. Let $G' = G - v'$. Compared with G, G' has the same 2^+-vertices but fewer edges. By the minimality of G, $\chi'_s(G - v_1) \leq 11$. Note that in G, $|F(vv')| \leq 10$. Thus, vv' can be colored, which leads to a contradiction.

Lemma 4. *Every bad vertex in H is adjacent to a 4-vertex.*

Proof. Suppose v is a bad vertex in H and it is adjacent to a 2-vertex u and a 3^--vertex w. By Lemma 2, $d_G(u) = d_G(v) = 2$. Denote $N_G(u) = \{u_1, v\}$. Note that $2 \leq d_H(w) \leq 3$. If $d_H(w) = 2$, then $d_G(w) = 2$ by Lemma 2. If $d_H(w) = 3$, then by Lemma 3, $d_G(w) = 3$ since $d_H(v) = 2$. Let $G' = G - uv + ww_1$, where ww_1 is a pendent edge incident with w. Note that $3 \leq d_{G'}(w) \leq 4$ and G' has fewer 2^+-vertices than G. By the definition of maximum average degree, $mad(G') < 2$ if $mad(G) < 2$ and $mad(G') \leq mad(G) < \frac{8}{3}$ if $2 \leq mad(G) < \frac{8}{3}$. By the minimality of G, $\chi'_s(G') \leq 11$. Let c be a strong 11-edge coloring of G'. Note that in G, $|F(uv)| \leq 8$. If $c(uu_1) \neq c(vw)$, then uv can be colored, which is a contradiction. If $c(uu_1) = c(vw)$, then we first exchange the colors on pendent edges wv and ww_1 in G'. After that, uv can be colored, which leads to a contradiction.

Lemma 5. *H does not have 3_3-vertices.*

Proof. Suppose v is a 3_3-vertex in H and $N_H(v) = \{v_1, v_2, v_3\}$. By Lemma 2, $d_G(v_i) = 2$, $i = 1, 2, 3$. By Lemma 3, $d_G(v) = 3$. Let $G' = G - v$. Note that G' has fewer 2^+-vertices than G. By the minimality of G, $\chi'_s(G') \leq 11$. Note that in G, $|F(vv_i)| \leq 6$, $i = 1, 2, 3$, vv_i can be colored, which is a contradiction.

Lemma 6. *Every semi-bad vertex in H is adjacent to a 4-vertex.*

Proof. Suppose v is a semi-bad vertex in H and it is adjacent to a 3_2-vertex u and a 3^--vertex w. Let $N_H(u) = \{u_1, u_2, v\}$, where $d_H(u_1) = 2$ (see Fig. 3). By Lemma 2, $d_G(u_1) = d_G(v) = 2$. By Lemma 3, $d_G(u) = 3$. Note that $2 \leq d_H(w) \leq 3$ and $d_H(v) = 2$, we have $d_G(w) = d_H(w)$ by Lemma 2 and Lemma 3. Let $G' = G - uv + ww_1$, where ww_1 is a pendent edge incident with w. Note that G' has fewer 2^+-vertices than G, by the definition of maximum average degree, $mad(G') < 2$ if $mad(G) < 2$ and $mad(G') \leq mad(G) < \frac{8}{3}$ if $2 \leq mad(G) < \frac{8}{3}$. By

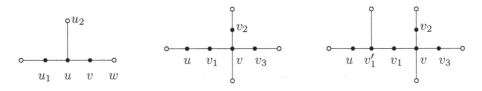

Fig. 3. .   Fig. 4. .   Fig. 5. .

the minimality of G, $\chi'_s(G') \leq 11$. Let c be a strong 11-edge coloring of G'. Erase on color on uu_1. Note that in G, $|F(uu_1)| \leq 9$, $|F(uv)| \leq 9$. If $c(uu_2) \neq c(vw)$, then uu_1, uv can be colored, which is a contradiction. If $c(uu_2) = c(vw)$, then we first exchange the colors on pendant edges wv and ww_1 in G'. After that, uu_1 and uv can be colored, which leads to a contradiction.

Lemma 7. *Let v be a 4_i-vertex in H, where $i \geq 3$. Then its 2-neighbors are all good vertices.*

Proof. Suppose that v_1, v_2, v_3 are 2-neighbors of v and at least one of them is not good. Without loss of generality, we assume that v_1 is not a good vertex. This implies that v_1 is adjacent to a 2-vertex or a 3_2-vertex.

If v_1 is adjacent to a 2-vertex u (see Fig. 4), then by Lemma 2, $d_G(v_i) = d_G(u) = 2$, $i = 1, 2, 3$. Let $G' = G - v_1$. Note that G' has fewer 2^+-vertices than G. By the minimality of G, $\chi'_s(G') \leq 11$. Note that in G, $|F(uv_1)| \leq 7$, $|F(vv_1)| \leq 9$, uv_1, vv_1 can be colored, which is a contradiction.

If v_1 is adjacent to a 3_2-vertex v'_1 and $u \neq v_1$ is the other 2-neighbor of v'_1 (see Fig. 5). By Lemma 2, $d_G(v_i) = d_G(u) = 2$, $i = 1, 2, 3$. By Lemma 3, $d_G(v'_1) = 3$. Let $G' = G - v_1$. Note that $G - v_1$ has fewer 2^+-vertices than G. By the minimality of G, $\chi'_s(G') \leq 11$. Note that in G, $|F(v_1v'_1)| \leq 9$, $|F(vv_1)| \leq 10$. Thus, $vv_1, v_1v'_1$ can be colored in order, which is a contradiction.

Proof of Theorem 7: We define weight function $w(v) = d(v)$ for each $v \in V(H)$ and we define five discharging rules R1-R5 as follows. Let $w'(v)$ be the final weight function while discharging finished. As we know, the sum weigh is fixed. However, we shall prove that $w'(v) \geq \frac{8}{3}$ for each $v \in V(H)$. This will lead to a contradiction as follow.

$$\frac{8}{3}|V(H)| \leq \sum_{v \in V(H)} w'(v) = \sum_{v \in V(H)} w(v) \leq mad(H)|V(H)| < \frac{8}{3}|V(H)|.$$

Discharging Rules:

R1 Each 4-vertex gives $\frac{2}{3}$ to each adjacent bad vertex.
R2 Each 4-vertex gives $\frac{1}{2}$ to each adjacent semi-bad vertex.
R3 Each 4-vertex gives $\frac{1}{3}$ to each adjacent good vertex.
R4 Each 3_2-vertex gives $\frac{1}{6}$ to each adjacent semi-bad vertex.
R5 Each 3_1-vertex gives $\frac{1}{3}$ to each adjacent good vertex.

In the following, we shall verify that $w'(v) \geq \frac{8}{3}$ for each $v \in V(H)$.
By Lemma 1, $\delta(H) \geq 2$.

- $d(v) = 2$

If v is bad, then by Lemma 4, v is adjacent to a 4-vertex. By R1, $w'(v) = 2 + \frac{2}{3} = \frac{8}{3}$.
If v is semi-bad, then by Lemma 6, v is adjacent to a 4-vertex. By R2 and R4, $w'(v) = 2 + \frac{1}{2} + \frac{1}{6} = \frac{8}{3}$.
If v is good, then by the definition of good vertex and Lemma 5, each neighbor of v is either 3_1-vertex or 4-vertex. By R3 and R5, $w'(v) = 2 + \frac{1}{3} \times 2 = \frac{8}{3}$.

- $d(v) = 3$

  By Lemma 5, v is 3_i-vertex, where $0 \leq i \leq 2$.
  If v is a 3_2-vertex, then by R4, $w'(v) \geq 3 - \frac{1}{6} \times 2 = \frac{8}{3}$.
  If v is a 3_1-vertex, then by R5, $w'(v) \geq 3 - \frac{1}{3} = \frac{8}{3}$.
  If v is a 3_0-vertex, then $w'(v) = w(v) = 3$.

- $d(v) = 4$

  If v is a 4_i-vertex, where $i \geq 3$, then by Lemma 7, the 2-neighbors of v are good. Thus, $w'(v) \geq 4 - \frac{1}{3} \times 4 = \frac{8}{3}$ by R3.
  If v is a 4_i-vertex, where $0 \leq i \leq 2$, then by R1-R3, $w'(v) \geq 4 - \frac{2}{3} \times 2 = \frac{8}{3}$.
  Therefore, for each $v \in V(H)$, $w'(v) \geq \frac{8}{3}$ and the proof of Theorem 7 is finished.   □

4 Further Considered Problems

Theorem 7 illustrates that $\chi'_s(G) \leq 11$ holds for any graph G with $\Delta(G) = 4$ and $mad(G) < \frac{8}{3}$. For the graph G_3 in Fig. 2, it satisfies that $\Delta(G_3) = 4$, $mad(G_3) = 3$ and $\chi'_s(G_3) = 12$. Then a question follows out naturally. What is the supremum M such that any graph G with $\Delta(G) = 4$ and $mad(G) < M$ satisfying $\chi'_s(G) \leq 11$?

Acknowledgement. This research was supported by National Natural Science Foundation of China under Grant Nos. 11901243, 12201569 and Qin Shen Program of Jiaxing University.

Declaration of Competing Interest. We declare that we have no conflicts of interest to this work. We also declare that we do not have any commercial or associative interest that represents a conflict of interest in connection with the work submitted.

References

1. Andersen, L.D.: The strong chromatic index of a cubic graph is at most 10. Discrete Math. **108**(1–3), 231–252 (1992)
2. Bensmail, J., Bonamy, M., Hocquard, H.: Strong edge coloring sparse graphs. Electron. Note Discrete Math. **49**, 773–778 (2015)
3. Bonamy, M., Perrett, T., Postle, L.: Colouring graphs with sparse neighbourhoods: bounds and applications. J. Combin. Theory Ser. B **155**, 278–317 (2022)
4. Bruhn, H., Joos, F.: A strong bound for the strong chromatic index. Combin. Probab. Comput. **27**(1), 21–43 (2018)
5. Choi, I., Kim, J., Kostochka, A.V., Raspaud, A.: Strong edge-colorings of sparse graphs with large maximum degree. European J. Combin. **67**, 21–39 (2018)
6. Cranston, D.W.: A strong bound edge-colouring of graphs with maximum degree 4 using 22 colours. Discrete Math. **306**, 2772–2778 (2006)
7. Erdős, P., Nešetřil, J., Halász, G.: Irregularities of Partitions, pp. 161–349. Springer, Berlin (1989). https://doi.org/10.1007/978-3-642-61324-1

8. Fouquet, J.L., Jolivet, J.L.: Strong edge-colorings of graphs and applications to multi-k-gons. Ars Combin. **16**, 141–150 (1983)

9. Hocquard, H., Montassier, M., Raspaud, A., Valicov, P.: On strong edge-colouring of subcubic graphs. Discrete Appl. Math. **161**(16–17), 2467–2479 (2013)

10. Horák, P., Qing, H., Trotter, W.T.: Induced matching in cubic graphs. J. Graph Theory **17**(2), 151–160 (1993)

11. Huang, M.F., Santana, M., Yu, G.X.: Strong chromatic index of graphs with maximum degree four. Electron. J. Combin. **25**(3), 3–31 (2018)

12. Li, X.W., Li, Y.F., Lv, J.B., Wang, T.: Strong edge-colorings of sparse graphs with $3\Delta - 1$ colors. Inform. Process. Lett. **179**, 106313 (2023)

13. Lv, J.B., Li, X.W., Yu, G.X.: On strong edge-coloring of graphs with maximum degree 4. Discrete Appl. Math. **235**, 142–153 (2018)

14. Mahdian, M.: On the computational complexity of strong edge-coloring. Discrete Appl. Math. **118**(3), 239–248 (2002)

15. Molloy, M., Reed, B.: A bound on the strong chromatic index of a graph. J. Combin. Theory Ser. B **69**(2), 103–109 (1997)

16. Qin, L.Z., Lv, J.B., Li, J.X.: Strong edge-coloring of some sparse graphs. Adv. Math. (China) **51**(1), 41–52 (2022)

17. Ruksasakchai, W., Wang, T.: List strong edge coloring of some classes of graphs. Australas. J. Combin. **68**, 106–117 (2017)

18. Zang C.Y.: The strong chromatic index of graphs with maximum degree Δ. arXiV1510.00785vl (2015)

Two Multicolor Ramsey Numbers
Involving Bipartite Graphs

Yan Li[1] and Ye Wang[2]

[1] University of Shanghai for Science and Technology, Shanghai 200093, China
[2] Harbin Engineering University, Harbin 150001, China
ywang@hrbeu.edu.cn

Abstract. For graphs G and H, the multicolor Ramsey number $r_{k,1}(G, H)$ is the minimum N such that any edge-coloring of K_N by $k+1$ colors contains a monochromatic G in the first k colors or a monochromatic H in the last color. In this note, we show the asymptotic upper bounds for $r_{k,1}(K_{m,n}; G)$ and $r_{k,1}(B_n; K_{t,s})$ if n is large, where G is any graph with at least one edge and B_n is a book.

Keywords: Multicolor Ramsey number · Bipartite graph · book

1 Introduction

For positive functions $f(n)$ and $g(n)$, we write $f(n) = O(g(n))$ if there is a constant $C > 0$ such that $f(n) \leq Cg(n)$, and $f(n) = \Theta(g(n))$ if $f(n) = O(g(n))$ and $g(n) = O(f(n))$. Denote by $f(n) = o(1)$ if $f(n) \to 0$ as $n \to \infty$.

For graphs G_i with $1 \leq i \leq k+1$, the multicolor Ramsey number $r(G_1, G_2, \ldots, G_{k+1})$ is defined as the minimum N such that any edge-coloring of K_N by $k+1$ colors contains a monochromatic G_i in color i, where $1 \leq i \leq k+1$. In particular, if $G_1 = G_2 = \ldots = G_k$, we write $r(G_1, G_2, \ldots, G_{k+1})$ as $r_{k,1}(G_1; G_{k+1})$.

For graphs G and H and integer $k \geq 2$, the Ramsey number $r_{k,1}(G; H)$ has been studied with special focus on the case of cycles and complete graphs first. Alon and Rödl [1] showed that $r_{k,1}(C_3; K_n) = \Theta(n^{k+1}\text{poly}\log n)$ and $r_{k,1}(C_{2m+1}; K_n) = \Omega(n^{1+\frac{k}{2m-1}}/(\log n)^{k+\frac{2k}{2m-1}})$. Xu and Ge [15] improved the cases of $r_{k,1}(C_5; K_n)$ and $r_{k,1}(C_7; K_n)$ by showing that $r_{k,1}(C_5; K_n) = \Omega((\frac{n}{\log n})^{\frac{3k}{8}+1})$ and $r_{k,1}(C_7; K_n) = \Omega((\frac{n}{\log n})^{\frac{2k}{9}+1})$. Zhang, Chen and Cheng [16] showed that $r_{k,1}(C_4; K_{1,n}) \leq n + \left\lceil k\sqrt{n + (k^2 + 2k - 3)/4} \right\rceil + \frac{k(k+1)}{2}$ for $n \geq 2$. Wang, Li and Li [13] showed that $r_{k,1}(K_{t,s}; K_{m,n}) \leq n + (1 + o(1))km(s - t + 1)^{1/t}n^{1-1/t}$ as $n \to \infty$. For more results on the multicolor Ramsey numbers, see, e.g., [3,6–12,14,17].

In this note, we show the asymptotic upper bounds for $r_{k,1}(K_{m,n}; G)$ and $r_{k,1}(B_n; K_{t,s})$ for large n as follows, where G is a graph with at least one edge and B_n is a book consisting of n triangles sharing a common edge.

Supported in part by NSFC (No. 12301451, 12101156).

Theorem 1. *Let G be a graph with $\chi = \chi(G) \geq 2$. For fixed positive integers k and m, it holds*

$$r_{k,1}(K_{m,n}; G) \leq (1 + o(1))k^m(\chi - 1)^m n$$

as $n \to \infty$.

Theorem 2. *For $k \geq 1$ and $s \geq t \geq 1$, it holds*

$$r_{k,1}(B_n; K_{t,s}) \leq (k + 2)!n,$$

for large n.

2   Proofs of Main Results

For graph G, the Turán number, denoted by $ex(N, G)$, is defined as the maximum number of edges of a graph on N vertices containing no G as a subgraph. A well-known argument called Erdős-Stone Theorem showed that for any fixed graph G with $\chi = \chi(G) \geq 2$,

$$ex(N, G) \leq \left(\frac{\chi - 2}{\chi - 1} + \epsilon \right) \binom{N}{2} \tag{1}$$

for large N, where the speed of $\epsilon \to 0$ can be at most $O(\frac{1}{\log N})$ as $N \to \infty$, see [2]. In particular, for $t \geq 3$,

$$ex(N, K_t) \leq \frac{(t - 2)}{2(t - 1)} N^2. \tag{2}$$

For complete bipartite graphs, Kövari, Sós and Turán [5] showed that

$$ex(N, K_{t,s}) \leq \frac{1}{2} \left[(s - 1)^{1/t} N^{2-1/t} + (t - 1)N \right]$$

for $s \geq t \geq 1$, and Füredi [4] proved

$$ex(N, K_{t,s}) \leq \frac{1}{2} \left[(s - t + 1)^{1/t} N^{2-1/t} + tN + tN^{2-2/t} \right]. \tag{3}$$

To simplify proofs, we shall change the bound slightly as

$$ex(N, K_{t,s}) \leq \frac{1}{2} \left[s^{1/t} N^{2-1/t} + (t - 1)N \right] \tag{4}$$

for $s \geq t \geq 1$.

Proof of Theorem 1. It is shown in (1) that $ex(N, G) \leq (\frac{\chi-2}{\chi-1} + \frac{c}{\log N})\binom{N}{2}$, where $c > 0$ is a constant. Let $N = k^m(\chi-1)^m n + \ell$ with $\ell = (1+o(1))cmk^m(\chi-1)^{m+1}\frac{n}{\log n}$. Consider an edge coloring of K_N by colors $1, 2, \cdots, k+1$. If

$$\binom{N}{2} > k \cdot ex(N, K_{m,n}) + ex(N, G),$$

then $r_{k,1}(K_{m,n}; G) \leq N$. Since the upper bound (4), it suffices to show that

$$\binom{N}{2} > \frac{k}{2}\left[n^{1/m}N^{2-1/m} + (m-1)N\right] + \left(\frac{\chi - 2}{\chi - 1} + \frac{c}{\log N}\right)\binom{N}{2},$$

equivalently

$$1 - \frac{k(\chi - 1)(m-1) + 1}{N} > \left(\frac{k^m(\chi - 1)^m n}{N}\right)^{1/m} + \frac{c}{\log N}(\chi - 1)(1 - \frac{1}{N}). \quad (5)$$

Note that

$$\left(\frac{k^m(\chi - 1)^m n}{N}\right)^{1/m} = \left(1 - \frac{\ell}{N}\right)^{1/m} = 1 - \frac{\ell}{mN} + \Theta\left(\frac{\ell^2}{N^2}\right),$$

and thus (5) becomes

$$\frac{\ell}{mN} \geq \frac{k(\chi - 1)(m-1) + 1}{N} + \frac{c(\chi - 1)}{\log N} + \Theta\left(\frac{\ell^2}{N^2}\right). \quad (6)$$

Since $\Theta\left(\frac{\ell^2}{N^2}\right) = o\left(\frac{1}{\log N}\right)$, if $\ell = (1 + o(1))cm(\chi - 1)\frac{N}{\log N} = (1 + o(1))cmk^m(\chi - 1)^{m+1}\frac{n}{\log n}$, then (6) holds for all large n and the claimed statement follows. $\square$

By the upper bounds (2) and (3), we obtain the following corollaries using the similar argument as in the proof of Theorem 1.

Corollary 1. *For $k, m \geq 1$ and $t \geq 3$, it holds*

$$r_{k,1}(K_{m,n}; K_t) \leq k^m n + c,$$

where $c = c(k, m, t) > 0$ is a constant.

Corollary 2. *For $k, m \geq 1$ and $s \geq t \geq 1$, it holds*

$$r_{k,1}(K_{m,n}; K_{t,s}) \leq k^m n + (1 + o(1))m(s - t + 1)^{1/t}(k^m n)^{1-1/t}$$

as $n \to \infty$.

In the following proof, when we color edges of K_N by $k + 1$ colors, we shall write the monochromatic graph induced by edges in color i as H_i for $1 \leq i \leq k + 1$. For a vertex v, we denote by $d_i(v)$ the degree of v in the graph H_i. We extend the notation of $r_{k,1}(G, H)$ slightly as follows. For graphs G, H and F, and integers $k, p, s \geq 1$, we write $r_{k,p,s}(G; H; F)$ for the Ramsey number $r(G, \ldots, G, H, \ldots, H, F \ldots, F)$, where the numbers of G, H and F are k, p and s, respectively. Denote by $e(G)$ the number of edges of graph G.

Proof of Corollary 2. Let $N_i = (1 + o(1))\left[i!(1 + \sum_{j=1}^{i} \frac{1}{j!})(k - 1) + 1\right]n$ as $n \to \infty$. We shall show that $r_{i,k-i,1}(B_n; K_{1,n}; K_{t,s}) \leq N_i$ for $1 \leq i \leq k$ by induction.

For $i = 1$, we shall show that $r_{1,k-1,1}(B_n; K_{1,n}; K_{t,s}) \leq N_1 = (1+o(1))(2k - 1)n$ as $n \to \infty$. Note that by Corollary 2, we have $r_{k,1}(K_{1,n}; K_{t,s}) \leq N_0 =$

$(1 + o(1))kn$. Consider an edge coloring of K_{N_1} by $k + 1$ colors. It is easy to see that $d_1(v) \leq r_{k,1}(K_{1,n}; K_{t,s}) - 1 \leq N_0 - 1$ for any vertex v, otherwise there exists a B_n in H_1, or a graph $K_{t,s}$ in H_{k+1}, or a star $K_{1,n}$ in some H_i with $2 \leq i \leq k$. Then

$$e(H_1) = \frac{1}{2} \sum_v d_1(v) \leq \frac{1}{2} N_1(N_0 - 1).$$

Hence

$$\sum_{j=2}^{k+1} e(H_i) = \binom{N_1}{2} - e(H_1) \geq \frac{N_1(N_1 - N_0)}{2}.$$

Therefore, there exists a B_n in H_1, or a graph $K_{t,s}$ in H_{k+1}, or a star $K_{1,n}$ in some H_i with $2 \leq i \leq k$ if

$$\frac{N_1(N_1 - N_0)}{2} > (k - 1)\frac{N_1(n - 1)}{2} + ex(N_1, K_{t,s}).$$

Equivalently, by (4), we only need to show

$$\frac{N_1(N_1 - N_0)}{2} > (k - 1)\frac{N_1(n - 1)}{2} + \frac{1}{2}\left[s^{1/t}N_1^{2-1/t} + (t - 1)N_1\right],$$

which is

$$N_1 - N_0 > (k - 1)(n - 1) + s^{1/t}N_1^{1-1/t} + t - 1.$$

So we can take $N_1 = (1 + o(1))(2k - 1)n$ as claimed.

Now we assume that $r_{i,k-i,1}(B_n; K_{1,n}; K_{t,s}) \leq N_i = (1 + o(1))\left(i!(1 + \sum_{j=1}^{i} \frac{1}{j!})(k - 1) + 1\right)n$ as $n \to \infty$, and we shall prove it holds for $i + 1$. Consider an edge coloring of $K_{N_{i+1}}$ by $k + 1$ colors. It is easy to see that $d_j(v) \leq r_{i,k-i,1}(B_n; K_{1,n}; K_{t,s}) - 1$ for any vertex v and $1 \leq j \leq i + 1$, otherwise there exists a B_n in some H_j with $1 \leq j \leq i + 1$, or a star $K_{1,n}$ in some H_j with $i + 2 \leq j \leq k$, or a graph $K_{t,s}$ in H_{k+1}. Then

$$e(H_j) = \frac{1}{2} \sum_v d_j(v) \leq \frac{1}{2} N_{i+1}(N_i - 1),$$

for $1 \leq j \leq i + 1$. Hence

$$\sum_{j=i+2}^{k+1} e(H_i) = \binom{N_{i+1}}{2} - \sum_{j=1}^{i+1} e(H_i) \geq \binom{N_{i+1}}{2} - \frac{i+1}{2} N_{i+1}(N_i - 1).$$

Therefore, there exists a B_n in some H_j with $1 \leq j \leq i + 1$, or a star $K_{1,n}$ in some H_j with $i + 2 \leq j \leq k$, or a graph $K_{t,s}$ in H_{k+1} if

$$\frac{N_{i+1}(N_{i+1} - (i + 1)N_i + i)}{2} > (k - i - 1)\frac{N_{i+1}(n - 1)}{2} + ex(N_{i+1}, K_{t,s}).$$

Equivalently, we only need to show

$$\frac{N_{i+1}(N_{i+1} - (i + 1)N_i + i)}{2} > \frac{(k - i - 1)N_{i+1}(n - 1)}{2} + \frac{1}{2}\left[s^{1/t}N_{i+1}^{2-1/t} + (t - 1)N_{i+1}\right],$$

which is

$$N_{i+1} \geq (i+1)N_i + (k-i-1)n + s^{1/t}N_{i+1}^{1-1/t} + t - k.$$

As $N_{i+1}^{1-1/t} = o(n)$, we can take

$$N_{i+1} = (1+o(1))(i+1)N_i + (k-i-1)n$$

$$= (1+o(1))\left[(i+1)!(1+\sum_{j=1}^{i+1}\frac{1}{j!})(k-1)+1\right]n$$

as claimed. Thus we have $N_k = \left[k!(1+\sum_{i=1}^{k}\frac{1}{i!})(k-1)+1+o(1)\right]n \leq (k+2)!n$, completing the proof. $\square$

References

1. Alon, N., Rödl, V.: Sharp bounds for some multicolor Ramsey numbers. Combinatorica **25**, 125–141 (2005)
2. Bollobás, B., Erdős, P.: On the structure of edge graphs. Bull. London. Math. Soc. **5**, 317–321 (1973)
3. Conlon, D., Fox, J., Sudakov, B.: Recent developments in graph Ramsey theory. Surv. Combinatorics **424**, 49–118 (2015)
4. Füredi, Z.: New asymptotics for bipartite Turán numbers. J. Comb. Theory Ser. A **75**, 141–144 (1996)
5. Kővári, T., Sós, V.T., Turán, P.: On a problem of K. Zarankiewicz. Colloq. Math. **3**, 50–57 (1954)
6. Lenz, J., Mubayi, D.: Multicolor Ramsey numbers for complete bipartite versus complete graphs. J. Graph Theory **77**, 19–38 (2014)
7. Li Y., Lin Q.: Elementary methods of graph Ramsey theory. Springer (2022). https://doi.org/10.1007/978-3-031-12762-5
8. Li, Y., Li, Y., Wang, Y.: Multicolor Ramsey numbers of bipartite graphs and large books. Graphs Combin. **39**, 21 (2023)
9. Liu, M., Li, Y.: Ramsey numbers involving an odd cycle and large complete graphs in three colors. Graphs Combin. **38**, 182 (2022)
10. Omidi, G., Raeisi, G.: A note on Ramsey number of stars-complete graphs. Europ. J. Combin. **32**, 598–599 (2011)
11. Omidi, G., Raeisi, G., Rahimi, Z.: Stars versus stripes Ramsey numbers. Europ. J. Combin. **67**, 268–274 (2018)
12. Wang, L.: Some multi-color Ramsey numbers on stars versus path, cycle or wheel. Graphs Combin. **36**, 515–524 (2020)
13. Wang, Y., Li, Y., Li, Y.: Ramsey numbers of several $K_{t,s}$ and a large $K_{m,n}$. Discrete. Math. **345**, 112987 (2022)
14. Wang Y., Song Y., Li Y., Liu M.: Multi-color Ramsey numbers of two bipartite graphs. submitted
15. Xu, Z., Ge, G.: A note on multicolor Ramsey number of small odd cycles versus a large clique. Discrete Math. **345**, 112823 (2022)
16. Zhang, X., Chen, Y., Cheng, T.: On three color Ramsey numbers $R(C_4, C_4, K_{1,n})$. Discrete Math. **342**, 285–291 (2019)
17. Zhang, X., Chen, Y., Cheng, T.: Bounds for two multicolor Ramsey numbers concerning quadrilaterals. Finite Fields Appl. **79**, 101999 (2022)

Mechanism Design for Time-Varying Value Tasks in High-Load Edge Computing Markets

Qie Li, Zichen Wang, and Hongwei Du[✉]

School of Computer Science and Technology, Harbin Institute of Technology (Shenzhen), Shenzhen 518055, China
hongwei.du@ieee.org

Abstract. A large number of computing task requests are generated by user terminals during peak hours in high-demand areas, but the resource capacity of edge servers is limited. It is necessary to design appropriate resource allocation and pricing mechanisms to address this resource competition dilemma. This paper proposes an auction-based mechanism called GMPO from an economic perspective. A market where multiple buyers and sellers compete with each other is considered, and the auction mechanisms is used to prevent these entities from falsely reporting information. As an extension of the concept of the age of information, the value of delay-sensitive computing tasks will decrease over time. This paper allocates resources greedily according to defined priorities and charge based on critical prices. The experiment results demonstrate that the proposed mechanism can effectively improve social welfare and guarantee the economic properties of auctions.

Keywords: Auction theory · Edge computing · Resource allocation

1 Introduction

Compared with traditional remote centralized cloud computing, edge computing (EC) has lower transmission delay because edge servers are closer to user equipments (UEs). With the development of the Internet of Things (IoT) and other technologies, there will be a massive amount of time-sensitive computationally intensive tasks that need to be offloaded to edge resource providers (ERPs) for processing. For example, these ERPs can be Google, AT&T, etc. The UEs can be smartphones, UAVs or smart vehicles etc. However, EC servers may not be able to fulfill a large number of user requests simultaneously during peak periods of resource shortage. Therefore, ERPs tend to prioritize UEs with higher cost-effectiveness, while UEs may be willing to pay more to enhance their competitive advantage and ensure a better service experience.

Some researches focus on the resource allocation problems between ERPs and UEs. Zhang et al. [1] proposed a decentralized multi-ERP resource allocation to maximize the total profit of all ERPs. Zeng et al. [2] used an improved matching algorithm to associate UEs and ERPs to maximize the final profit of ERPs. Chen et al. [3] proposed a stable matching algorithm for the cooperation and competition between UAVs and unmanned ground vehicles. But these matching algorithms cannot prevent buyers and sellers from falsely reporting information. As a reliable mechanism, auction theory [4] has been applied to many edge scenarios. Huang et al. [5] proposes a multi-dimensional dynamic programming and equivalent package algorithm based on auction theory, which allocates backhaul capacity and cache space for streams on ERPs. Yang et al. [6] used a multi-round sequential combinatorial auction mechanism to transform the matching problem into a multi-dimensional grouping knapsack problem, which can be efficiently solved using dynamic programming.

Age of information (AoI) [7] was originally defined as the time elapsed from the generation of the current information at the source until the reception of decision feedback. As a more reasonable indicator, AoI related models are widely used in real-time computing [8]. Lv et al. [9] proposed an auction framework with dynamic programming and preemption factors to minimize weighted AoI. Chen et al. [10] considered a multi-buyer and multi-seller market with stackelberg method to solve resource allocation problems, and developed an iterative algorithm to find equilibrium prices. However, these researches lack consideration for market with multidimensional heterogeneous resources and multiple-entities participation.

This paper proposes a mechanism called Greedy Method with Priority Order (GMPO) based on auction theory to address the resource allocation and pricing problem between UEs and ERPs. The contributions of this paper are as follows:

- This paper considers a competitive EC market with multiple buyers and sellers and uses auction mechanisms to prevent entities from lying. As an extension of AoI, the value of delay-sensitive tasks will decrease over time.
- This paper proposes the algorithm called Greedy Method with Priority Order (GMPO). The defined priority ensures that the winner decision process satisfies monotonicity, and the final transaction price is determined by the defined critical competitor. Experiments have shown that this algorithm can achieve higher social welfare in high-load environments. We have proven that the algorithm meets the economic properties of auctions.
- This paper compares the proposed GMPO mechanism with the GMMU, FCFS, and Random methods. The experimental results show that GMPO effectively improves the social welfare of the entire system.

The paper is organized as follows: Sect. 2 introduces the system model. Section 3 focuses on the detailed design of GMPO, including resource allocation strategies, winner determination algorithms, and pricing mechanisms. In Sect. 4, we give the simulation results on social welfare and the income of edge service providers. Finally, we conclude this paper in Sect. 5.

2   System Model and Problem Statement

As Fig. 1 shows, UEs use sealed bidding to upload their time-varying value and the number of requested resources to the auctioneer. ERPs provide their available bandwidth, CPU computing resources, and cost functions to the auctioneer. The auctioneer centrally combines this information to determine how ERPs and UEs are matched and how to allocate resources and determine the payment. Consider an area where M sellers ERP=$\{1,2,...,M\}$ are evenly distributed to provide bandwidth and computing resources to users, and N buyers U=$\{1,2,...,N\}$ are UEs who sealed bidding for performing computing tasks. Assuming that the tasks that each user i $(i \in U)$ needs to perform do not overlap, the user i mentioned below is equivalent to task i. If a user has multiple tasks to execute, these tasks can be abstractly represented as multiple virtual users. Divide a period of time into consecutive L time slots T=$\{1,2,...,L\}$. User i requests services from ERPs within the communication range at a certain time t, and the bidding information is represented as:

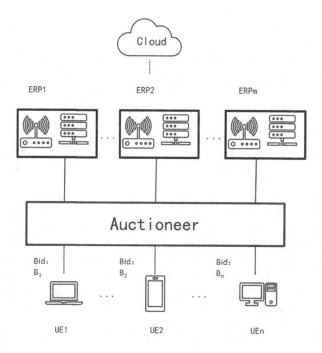

Fig. 1. System Architecture.

$$B_i(t) = \{t_{a,i}, s_i, g_i, v_i(t), DDL_i\}, \forall i \in U, \forall t \in T \tag{1}$$

where $t_{a,i}$ is the arrival time of task i, s_i is the data size of task i (bit), g_i is the number of CPU cycles required for task i, $v_i(t)$ is the time-varying value of task i. DDL_i is the time elapsed from task i's generation to the time when $v_i(t) = 0$.

The time-varying value [9] of the task is simply represented by a monotone and nonincreasing linear function:

$$v(t) = \begin{cases} v_0 - kt, \, t_a \leq t \leq DDL \\ 0, \qquad\quad \text{otherwise} \end{cases} \tag{2}$$

where v_0 is the initial value of the task, and k is derived from the d of $B(t)$.

2.1   Resource Model of the System

Radio Resources: This paper uses the Rayleigh fading model to describe the wireless channel. If ERP j allocated bandwidth ϕ_{ij}(HZ) to user i during communication, then according to Shannon's theorem, the data transmission rate (bits/s) can be expressed as:

$$r_{ij} = x_{ij}\phi_{ij}\Lambda \log_2 \left(1 + \frac{P_i h_{ij}^u}{\sigma_j^2} \right) \tag{3}$$

where $x_{ij} \in \{0,1\}$ indicates that ERP j communicates with user i, Λ is the duration of one time slot, P_i is wireless transmission power of user i, h_{ij}^u is the uplink channel gain between user i and ERP j, σ_j^2 is the noise power of ERP j .
The corresponding transmission delay from user i to ERP j is:

$$T_{ij}^{\text{trans}} = \left\lceil \frac{s_i}{r_{ij}} \right\rceil \tag{4}$$

Computing Resources: The computing resources requested by user i are the number of CPU cycles required to complete the task. Assuming that the CPU frequency allocated by ERP j to user i is f_{ij}, the computing latency is:

$$T_{ij}^{exe} = \left\lceil \frac{g_i}{f_{ij}\Lambda} \right\rceil \tag{5}$$

The Cost Function of Resources: The cost of wireless resources held by the seller ERP j is fixed and recorded as c_j^b. The cost of computing resources varies with changes in CPU power consumption, denoted as $c_j^f(z_j^f(t))$ [11], $z_j^f(t)$ representing the CPU frequency usage of ERP j at time t, $p_0^f(t)$ representing the lowest threshold of frequency cost, $\gamma_j^f, \varepsilon_j^f$ are power consumption parameter and related to Dynamic Voltage and Frequency Scaling (DVFS) technology.

$$c_j^f(z_j^f(t)) = \begin{cases} p_0^f + \gamma_j^f z_j^f(t)^{1+\varepsilon_j^f}, & \text{if } z_j^f \in \left[0, C_j^f\right] \\ +\infty, \quad \text{otherwise} \end{cases} \tag{6}$$

Because the data s_i is very small compared to ERP j's total storage resources, the cost of storage is ignored here.

2.2   Utility Model of the System

The Utility of Buyer: If W_u is the set of users who won the auction at time t and the completion time of the task $i \in W_u$ is t', then the utility of the winning buyer i in the auction is the difference between the task value and the payment p_i. The utility of the unselected buyer in the auction is 0.

$$util_i = \begin{cases} v_i(\Delta t) - p_i, & \forall t \in T, i \in W_u \\ 0, & \text{otherwise} \end{cases} \tag{7}$$

The Utility of Seller: The utility of seller ERP j in auctions is its income minus cost.

$$util_j = \sum_{t \in T} \sum_{i \in W_u} x_{ij} \left(p_i - c_j^f \left(z_j^f(t) \right) f_{ij} - c_j^b \phi_{ij} \right), \quad \forall j \in ERP, x_{ij} \in \{0,1\} \tag{8}$$

System Utility: The utility of the entire system is the sum of the utility of all buyers and sellers participating in the auction:

$$util_{sys} = \sum_{t \in T} \left(\sum_{i \in W_u} util_i + \sum_{j \in ERP} util_j \right) \tag{9}$$

2.3   Optimization Goals for Maximizing Social Welfare

We define social welfare as the total utility of the system, and design the optimization objective of the problem to maximize social welfare:

$$\text{maximize} \sum_{t \in T} \sum_{i \in W_u} \left(v_i(\Delta t) - \sum_{j \in ERP} x_{ij} \left(c_j^f \left(z_j^f(t) \right) f_{ij} + c_j^b \phi_{ij} \right) \right) \tag{10}$$

$$\text{s.t. } z_j^f(t) \in [0, F_j], \forall t \in T, j \in ERP \tag{11}$$

$$\sum_{i \in W_u} x_{ij} \phi_{ij} \le W_j, \forall t \in T, j \in M \tag{12}$$

$$\sum_{i \in W_u} x_{ij} f_{ij} \le z_j^f(t), \forall t \in T, j \in M \tag{13}$$

$$x_{ij} \in \{0, 1\} \tag{14}$$

The constraint (11) indicates that the current CPU frequency usage of ERP j does not exceed the total amount of its own computing frequency. (12) indicates that bandwidth allocated by ERP j to users does not exceed the total bandwidth it has. (13) indicates that the calculation frequency allocated by ERP j to users does not exceed the current CPU frequency usage. The x_{ij} in (14) represents the decision variable, $x_{ij} = 1$ indicates that ERP j will allocate its resources to user i. This problem is an NP-hard mixed integer optimization problem.

3   The Greedy Auction Mechanism

In this section, the Greedy Method with Priority Order (GMPO) is presented. The appendix demonstrated that GMPO satisfies the economic nature of auctions.

3.1   Resource Allocation Strategy

Using $\Gamma_{ij} = \langle f_{ij}, \phi_{ij} \rangle$ represent that ERP j allocate CPU frequency resources f_{ij} and bandwidth ϕ_{ij} to the winning user i. The different resource allocation schemes affect the completion time of tasks, and the completion time of tasks is closely related to their time-varying value. The system utility corresponding to allocation plan Γ_{ij} is:

$$\text{utils } _{\Gamma_{ij}} = v_i(\Delta t) - c_j^f \left(z_j^f(t) \right) f_{ij} - c_j^b \phi_{ij} \tag{15}$$

$$\Delta t = T^{\text{trans}} + T^{\text{exe}} \tag{16}$$

Then we have:

$$utils_{\Gamma_{ij}} = v_{0,i} - k \frac{S_i}{\phi_{ij} \Lambda \log_2 \left(1 + \frac{P_i h_{ij}^u}{\sigma_j^2} \right)} - k \frac{w_i}{\Lambda f_{ij}} - c_j^{\text{f}} \left(z_j^{\text{f}}(t) \right) f_{ij} - c_j^b \phi_{ij} \tag{17}$$

The resource allocation plan is based on the following idea: If the given task i can be transmitted or calculated within a one-time slot, and the remaining resources of ERP j is sufficient to complete the allocation, then resource allocation will be carried out; otherwise, no allocation will be carried out. Because the allocation schemes of bandwidth and frequency do not affect each other, they can be decoupled into the following two issues:

Determine the allocated frequency resources:

$$* f_{ij} = \begin{cases} \frac{g_i}{\Lambda}, & \frac{g_i}{\Lambda} <= F_{j\_\text{remain}} \\ 0, & \text{otherwise} \end{cases} \tag{18}$$

Determine the allocated bandwidth resources:

$$* \phi_{ij} = \begin{cases} \frac{s_i}{\Lambda \log_2 \left(1 + \frac{P_i h_{ij}^u}{\sigma_j^2} \right)}, & \frac{s_i}{\Lambda \log_2 \left(1 + \frac{P_i h_{ij}^u}{\sigma_j^2} \right)} <= BW_{j\_\text{remain}} \\ 0, & \text{otherwise} \end{cases} \tag{19}$$

If the utility after allocation is greater than 0, resources will be allocated according to the plan, otherwise, no allocation will be made.

3.2   GMPO-W Winner Decision

In order to maximize social welfare, the priority function D_{ij} is introduced to represent the cost-effectiveness of user i using ERP j to perform tasks, which is the ratio of task value to resource cost consumed by user i:

$$D_{ij} : \frac{v_i}{\sqrt{c_j^b * \varphi_{ij} + c_j^f * f_{ij}}} \tag{20}$$

Algorithm 1: GMPO-W winner decision and resource allocation algorithm

---

Input: U, ERP, F_j, BW_j
$B_i(t) = \{t_{a,i}, s_i, g_i, v_i(t), DDL_i\}, \forall i \in U, \forall t \in T$
$c_j^b, c_j^f \left(z_j^f(t)\right) \quad \forall j \in ERP, t \in T$
Output: $W_u, util_{sys}$

1 initialization $N_a \leftarrow \varnothing, W_u \leftarrow \varnothing, \forall t \in T, util_{sys} = 0$;
2 **for** $t \in T$ **do**
3   **for** $j \in ERP$ **do**
4 | Update the resource cost function at the current time c_j^f;
5   **end**
6   **for** $i \in U$ **do**
7   **if** $t_{a,i} \leq t \leq t_{a,i} + DDL_i$ *and* $i \notin W_u$ **then**
8              $N_a \leftarrow N_a \cup \{i\}$;
9   Calculate the ERP set of user i within the communication range M_i;
10   **end**
11   **end**
12   **for** $i \in N_a, j \in M_i$ **do**
13 | $D_{ij} \leftarrow\; < B_i(t), \Gamma_{ij}, c_j^b, c_j^f \left(z_j^f(t)\right) >$;
14   **end**
15   **while** *Priority list D is not empty* **do**
16         $\langle i, j \rangle \leftarrow \operatorname{argmax} D$;
17   **if** $i \notin W_u$ *and* $\Gamma_{ij} \neq \varnothing$ **then**
18             $W_u \leftarrow W_u \cup \{i\}$, utilsys$+ = utils_{\Gamma_{ij}}, N_a \leftarrow N_a/\{i\}$
19   ERP j allocate resources as Γ_{ij} and update remaining resource capacity $F_{j\_remain}, BW_{j\_remain}$;
20   **end**
21   **end**
22 **end**
23 **return** $W_u, util_{sys}$;

The greater the D_{ij} is, the greater the contribution of the $\langle i, j \rangle$ pair to social welfare. Sort D_{ij} by nonincreasing order and allocate resources accordingly and define the partial order relationship $\succ$ as follows:

$$B_{i_1} \succ B_{i_2} \Leftrightarrow \left(v^{i_1} > v^{i_2}\right) \cap \left(\phi_{i_1} < \phi_{i_2}\right) \cap \left(f_{i_1} < f_{i_2}\right) \tag{21}$$

If $B_{i_1} \succ B_{i_2}$, The time-varying task value of i_1 is higher than that of i_2, and the cost of resource consumption is lower. The partially ordered set satisfies monotonicity.

Based on the above definition, this article designs a winner determination and resource allocation algorithm as Agorithm 1. At the beginning of each moment, update the frequency cost function of ERP j based on the existing CPU resource usage (Line 4). Next, add the user requests that have not timed out to the active user set N_a(Line 8). Calculate the value of priority D_{ij} between users and their reachable servers in the active user set N_a (Line 13), traverse D_{ij} in nonincreasing order, select the UE-ERP pair corresponding to the highest priority, i.e. $\langle i, j \rangle$ pairs, and determine whether single slot resource allocation Γ_{ij} can be used for resource allocation (Line 17). If the allocation is possible, add users to the winner set W_u and update the system utility and the remaining resource capacity of ERP. The result returns the winner set W_u and system utility $util_{sys}$.

3.3   GMPO-P Price Determination Mechanism

Determine the payment price p_i of the winning user $i \in Wu$ by searching its critical competitors.

Algorithm 2: GMPO-P Price determination mechanism for winning users

---

Input: $U, ERP, D, \Gamma_{ij}, c_j^b, c_j^f \left(z_j^f(t) \right)$ $\forall i \in U, \quad \forall j \in ERP$

Output: p_i

1 $p_0 \leftarrow c_i^b * b_i + c_j^f \left(z_j^f(t) \right) * f_i$;

2 Traverse the global priority list D in non increasing order, get buyer q and seller s;

3 **for** $q \in U\,and\,q \neq i$ **do**

4 $\quad$ Allocates bandwidth and frequency resources for q based on Γ_{ij}

5 $\quad$ **if** $s = j$ *and there is not enough resources on ERP j for buyer i after allocation* **then**

6 $\quad\quad$ $cc(i) \leftarrow q$

7 $\quad\quad$ $p_i \leftarrow \max \left(D_{cc(i)j} * \sqrt{c_j^b * \varphi_{ij} + c_j^f * f_{ij}}, p_0 \right)$

8 $\quad$ **end**

9 $\quad$ **if** $p_i < p_0$ **then**

10 $\quad\quad$ $p_i \leftarrow p_0$

11 $\quad$ **end**

12 **end**

13 **return** p_i

Definition 1. *Critical Competitors*
Sort priority D by the nonincreasing order and allocate resources corresponding to the $\langle i, j \rangle$ pairs. If there is a user $q(q \neq i)$ for the remaining user $U/\{i\}$ excluding the winning user $i \in W_u$. If after allocating corresponding resources to user q on ERP j, user i will no longer be able to be allocated to resources on ERP j, then user q is a critical competitor of user i, denoted as cc(i).

As shown in Algorithm 2. The critical price p_i that user i should pay on ERP j is derived from the priority $D_{cc(i)j}$:

$$\frac{p_i}{\sqrt{c_j^b * \varphi_{ij} + c_j^f * f_{ij}}} >= D_{cc(i)j}, \ (p_i >= p_0) \tag{22}$$

p_0 is the cost of resources, and the above formula can be written as:

$$p_i = \max \left(D_{cc(i)j} * \sqrt{c_j^b * \varphi_{ij} + c_j^f * f_{ij}}, p_0 \right) \tag{23}$$

Theorem 1. *The mechanism proposed in this thesis is truthful, individual ratio national, budget balance, and can be completed within polynomial time.*

Proof. (Individual Rationality) Users only agree to execute tasks when the value of the task exceeds the cost of payment, while ERP only provides resources when the revenue exceeds the cost. Both will make rational decisions to ensure their utility is not negative.

(Budget Balance) The total payment amount of all users is greater than or equal to the total fee amount of all ERP, satisfying the weak budget balance.

(Polynomial Time Complexity) The auction mechanism consists of the winner decision algorithm GMPO-W and the payment mechanism GMPO-P. The time complexity of the winner determining algorithm GMPO-W is $O(nm + nm \log(nm))$, the running time of the payment mechanism GMPO-P is $O(nm)$, but it needs to be called $O(n)$ times, so the total running time of the GMPO mechanism is $O\left(n^2m + nm \log(nm)\right) \leq O\left(n^2m^2\right)$. Satisfy polynomial time complexity.

(Truthfulness) If user i cannot obtain higher profits by lying with B_i, then the truthfulness is satisfied. According to Myerson theorem [12], auction mechanisms that satisfy both monotonicity and critical price are strategy-proof.

4   Evaluation Results

4.1   Experimental Settings

Table 1. Experimental simulation parameters

Parameter	Value
Inter-base station distance	300 m
Base station coverage radius	450 m
Total CPU frequency of each base station	20 GHz
Total radio frequency of each base station	10 MHz
Data size of each task	0.5–1.5 MB
Computational workload of each task	1–4G[cycles]
Deadline of each task	3–9 s
Initial value of each task	50–150
Wireless transmission power of each task	1.5 W
Noise power	10^{-9}W

Consider a dense deployment of the cellular network in an area of 600 m $\times$ 600 m, consisting of $M = 9$ uniform-distributed ERP and $U = 2000$ randomly-distributed task requests. The bandwidth cost function of ERP j is $c_j^b = 2$, and the frequency cost function is $p_0^f \sim u[5, 10]$, $\gamma_j^f \sim u[0.4, 0.6]$, $\varepsilon_j^f \sim u[1.8, 2.2]$. Adopting Rayleigh fading channel model, where the channel gain from user i to ERP j is $h_{ij}^u = A_u d_{ij}^{-3} r_{ij}$, the antenna gain is $A_u = 6.25 * 10^{-4} (-32 \text{ dB})$, d_{ij} is the distance from user i to ERP j, -3 is the path loss exponent, and r_{ij} is a random variable following a Rayleigh distribution. Other parameters are shown in Table 1.

Each experiment lasts for $L = 20$ time slots, with each time slot $\Lambda = 1$ second. We adopt the average value of 100 experiments as the final result. The proposed GMPO algorithm is compared with the following mechanisms:

- Greedy Method with Maximum Utility (GMMU): Similar to GMPO, GMMU adopts a priority strategy with social welfare in a non-increasing order and determines the payment based on the critical price.
- Random Allocation: Resources are randomly allocated to arriving users using a single time-slot resource allocation strategy. Instead of the ctitical price, the charging price is half of the sum of the task completion value and its resource cost.
- First Come First Served (FCFS): Resources are allocated to arriving users in order using a single time-slot resource allocation strategy. The charging price is also half of the sum of the task completion value and its resource cost.

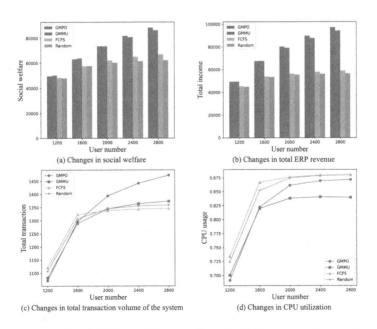

(a) Changes in social welfare

(b) Changes in total ERP revenue

(c) Changes in total transaction volume of the system

(d) Changes in CPU utilization

Fig. 2. Fix the scale of ERP to M=9, and gradually increase the number of task requests from 1200 to 2800

4.2   Numerical Results

The scale of ERP is initially set to $M = 9$, and the number of task requests is gradually increasing from 1200 to 2800. It can be seen in Fig. 2(a) and 2(b) that as the number of users increases, the total social welfare and total ERPs' revenue obtained by the four methods all increase. The results of GMPO and GMMU are close, with GMPO slightly better than GMMU. And their social welfare and total system income are much higher than those of FCFS and Random algorithms that do not adopt priority and critical pricing mechanisms.

It can be seen in Fig. 2(c) and 2(d) that when ERPs' resources are sufficient, as the number of users increases, the total transaction volume of the system increases and CPU utilization increases. When ERPs' resources enter a state of oversupply, the number of users further increases, and the total transaction volume of the system stabilizes after CPU utilization reaches saturation. At this point, the system transaction volume achieved by GMPO is higher than that of other algorithms, which means that when resources are scarce, the GMPO algorithm can achieve higher user satisfaction.

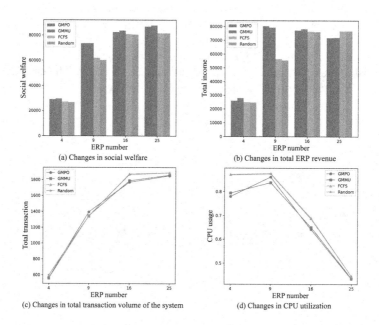

Fig. 3. Fix the number of task requests as U=2000, and the scale of ERP gradually increases from 4 to 25

Further, the number of task requests is fixed as $U = 2000$, and the scale of ERP gradually increases from 4 to 25. It can be seen in Fig. 3(a) that as the scale of ERP increases, more and more users are able to receive services. The social welfare of GMPO and GMMU are close, and still higher than that of FCFS and Random algorithms. Taking all three factors into consideration, it can be seen from Fig. 3(b), 3(c) and 3(d) that after the ERPs' scale changed from 9 to 25, the resource utilization situation changed from demand exceeding supply to supply exceeding demand. The transaction volume of the four algorithms is similar, but GMMU and GMPO will choose users who are more conducive to increasing social welfare for services, so the transaction volume is slightly lower than FCFS and Random algorithms.

With the growth of scale and transaction volume, the total ERPs' revenue of FCFS and Random gradually increases. But the ERPs' revenue of GMMU and GMPO first increases and then decreases. This suggests that when ERPs' resources are in short supply, UEs' prices are determined by their critical competitors. When ERPs' resources are in oversupply, the price charged is the cost price. However, the charging price of FCFS and Random is always half of the sum of the value of the task and its resource cost, which is higher than the cost price. This result reflects that the proposed auction mechanism is more suitable for scenarios where resource supply exceeds demand, but can still achieve higher social welfare in situations where resource supply exceeds demand.

5   Conclusion

This paper has considered an edge computing market where multiple buyers and sellers compete with each other in a high-load environment. GMPO was proposed based on auction theory. The winner selection algorithm GMPO-W is used to select the winning UE-ERP pairs. The defined priority ensures the monotonicity of the selection process; The GMPO-P mechanism is designed to determine the payment price for the winning user based on critical prices. The experimental results show that GMPO can effectively improve the social welfare of the entire system and maintain some economic attributes of auction.

Acknowledgment. This work is supported by National Natural Science Foundation of China (No. 62172124). It was also supported by the Shenzhen Basic Research Program (Project No. JCYJ20190806143011274).

References

1. Zhang, C., Du, H., Ye, Q., Liu, C., Yuan, H.: DMRA: a decentralized resource allocation scheme for multi-SP mobile edge computing. In: 2019 IEEE 39th international conference on distributed computing systems (ICDCS), pp. 390–398. IEEE(2019)
2. Zeng, G., Zhang, C., Du, H.: An efficient mechanism for resource allocation in mobile edge computing. In: Wu, W., Zhang, Z. (eds.) COCOA 2020. LNCS, vol. 12577, pp. 657–668. Springer, Cham (2020). https://doi.org/10.1007/978-3-030-64843-5_44
3. Chen, W., Su, Z., Xu, Q., Luan, T.H., Li, R.: VFC-based cooperative UAV computation task offloading for post-disaster rescue. In: IEEE INFOCOM 2020-IEEE Conference on Computer Communications, pp. 228–236. IEEE. (2020)

4. Qiu, H., et al.: Applications of auction and mechanism design in edge computing: a survey. IEEE Trans. Cogn. Commun. Netw. **8**(2), 1034–58 (2022)
5. Hung, Y.H., Wang, C.Y., Hwang, R.H.: Optimizing social welfare of live video streaming services in mobile edge computing. IEEE Trans. Mob. Comput. **19**(4), 922–34 (2019)
6. Yang, S.: A task offloading solution for internet of vehicles using combination auction matching model based on mobile edge computing. IEEE Access. **8**, 53261–73 (2020)
7. Kaul, S., Yates, R., Gruteser, M.: Real-time status: how often should one update?. In: 2012 Proceedings IEEE INFOCOM, pp. 2731–2735. IEEE. (2012)
8. Yates, R.D., Sun, Y., Brown, D.R., Kaul, S.K., Modiano, E., Ulukus, S.: Age of information: an introduction and survey. IEEE J. Sel. Areas Commun. **39**(5), 1183–210 (2021)
9. Lv, H., Zheng, Z., Wu, F., Chen, G.: Strategy-proof online mechanisms for weighted AoI minimization in edge computing. IEEE J. Sel. Areas Commun. **39**(5), 1277–92 (2021)
10. Chen, Y., Li, Z., Yang, B., Nai, K., Li, K.: A Stackelberg game approach to multiple resources allocation and pricing in mobile edge computing. Futur. Gener. Comput. Syst. **108**, 273–87 (2020)
11. He, X., Shen, Y., Ren, J., Wang, S., Wang, X., Xu, S.: An online auction-based incentive mechanism for soft-deadline tasks in collaborative edge computing. Futur. Gener. Comput. Syst. **137**, 1–3 (2022)
12. Myerson, R.B.: Optimal auction design. Math. Oper. Res. **6**(1), 58–73 (1981)

Computing Random r-Orthogonal Latin Squares

Sergey Bereg$^{(\boxtimes)}$

Department of Computer Science, Erik Jonsson School of Engineering and Computer
Science, University of Texas at Dallas, Richardson, USA
besp@utdallas.edu

Abstract. Two Latin squares of order n are r-orthogonal if, when superimposed, there are exactly r distinct ordered pairs. The spectrum of all values of r for Latin squares of order n is known. A Latin square A of order n is r-self-orthogonal if A and its transpose are r-orthogonal. The spectrum of all values of r is known for all orders $n \neq 14$. We develop randomized algorithms for computing pairs of r-orthogonal Latin squares of order n and algorithms for computing r-self-orthogonal Latin squares of order n.

Keywords: Latin squares · Latin rectangles · r-orthogonal Latin squares

1 Introduction

For a positive integer n, a Latin square of order n is an $n \times n$ array filled with n different symbols, each occurring exactly once in each row and exactly once in each column. For a positive integer n, let $[n]$ denote the set $\{1, 2, \ldots, n\}$. We denote the (i, j)-th element of a Latin square A by $A_{i,j}$. In this paper, we assume that the symbols in a Latin square of order n are in $[n]$, i.e. $A_{i,j} \in [n]$ for all $i, j \in [n]$.

Two Latin squares A and B of order n are said to be *orthogonal* if every ordered pair of symbols occurs exactly once among the n^2 pairs $(A_{i,j}, B_{i,j}), i \in [n], j \in [n]$. Orthogonal Latin squares were studied in detail by Leonhard Euler see Fig. 1. Orthogonal Latin squares exist for all orders $n > 1$ except $n = 2, 6$.

$A\alpha$	$B\delta$	$C\beta$	$D\gamma$
$B\gamma$	$A\beta$	$D\delta$	$C\alpha$
$C\delta$	$D\alpha$	$A\gamma$	$B\beta$
$D\beta$	$C\gamma$	$B\alpha$	$A\delta$

Fig. 1. Orthogonal Latin squares using Euler's notation (one Latin square uses the first n upper-case letters from the Latin alphabet and the other uses the first n lower-case letters from the Greek alphabet). Orthogonal Latin squares also known as a Graeco-Latin square or Euler square.

© The Author(s), under exclusive license to Springer Nature Switzerland AG 2024
W. Wu and J. Guo (Eds.): COCOA 2023, LNCS 14462, pp. 165–174, 2024.
https://doi.org/10.1007/978-3-031-49614-1_12

If A and B are orthogonal Latin squares, then we say that A is an *orthogonal mate* of B and vice versa. In 1779 Euler [11] proved that for every even n there exists a Latin square of order n that has no orthogonal mate. van Rees [25] called such squares *bachelor* Latin squares. The existence of bachelor Latin squares of orders $n \equiv 1 \pmod 4$ was shown by Mann [21]. In 2006 Evans [12] and Wanless and Webb [27] completed the remaining case of Latin squares of orders $n \equiv 3 \pmod 4$.

Theorem 1 ([12,27]). *For any positive integer $n \notin \{1,3\}$ there exists a Latin square of order n that has no orthogonal mate.*

van Rees [25] has conjectured that the portion of bachelor Latin squares of order n tends to one as $n \to \infty$. However, this conjecture was based on a study of Latin squares of small orders. Computational results by Egan and Wanless [10] show that only a small proportion of Latin squares of orders 7, 8 and 9 possess orthogonal mates. McKay *et al.* [24] tested 10 million random Latin squares of order 10 and estimated that around 60% of Latin squares of order 10 have mates.

A set of Latin squares of the same order, all pairs of which are orthogonal is called a set of *mutually orthogonal Latin squares* (MOLS). The study of mutually orthogonal Latin squares is a subject that has attracted much attention due to applications in error correcting codes, cryptographic systems, compiler testing, and statistics (the design of experiments) [18].

The construction of MOLS is a notoriously difficult combinatorial problem, see for example [22,23]. In this paper we study the problem of computing r-orthogonal Latin squares. Two Latin squares of order n are r-*orthogonal* (r-OLS) if their superposition produces exactly r distinct ordered pairs. Belyavskaya [3–5] systematically treated the following question: For which integers n and r does a pair of r-orthogonal Latin squares of order n exist? The spectrum of r-orthogonal Latin squares of order n was determined by Colbourn and Zhu [8] (with few exceptions) and completed later by Zhu and Zhang [29].

Table 1. (a) Genuine exceptions of pairs of r-orthogonal Latin squares of order n. (b) Genuine exceptions of r-self-orthogonal Latin squares of order n.

n	Genuine exceptions of r
2	4
3	5, 6, 7
4	7, 10, 11, 13, 14
5	8, 9, 20, 22, 23
6	33, 36

(a)

n	Genuine exceptions of r
2	4
3	5, 6, 7, 9
4	6, 7, 8, 10, 11, 12, 13, 14
5	8, 9, 12, 16, 18, 20, 22, 23
6	32, 33, 34, 36
7	46

(b)

Theorem 2 (Zhu and Zhang [29]**).** *For any integer $n \geq 2$, there exists a pair of r-orthogonal Latin squares of order n if and only if $n \leq r \leq n^2$ and $r \notin \{n+1, n^2-1\}$ with the exceptions of n and r shown in Table 1(a).*

A Latin square which is orthogonal to its transpose is called a *self-orthogonal Latin square* (SOLS). It is known that self-orthogonal Latin squares exist for all orders $n \notin \{2,3,6\}$ [13]. We say that a Latin square A is *r-self-orthogonal* (r-SOLS) if A and its transpose A^T are r-orthogonal. The spectrum of r-SOLS has been almost completely decided by Xu and Chang [29] and Zhang [28], as shown in the following theorem.

Theorem 3 (Zhang [28]**).** *For any integer $n \geq 1$, there exists an r-self-orthogonal Latin square of order n if and only if $n \leq r \leq n^2$ and $r \notin \{n+1, n^2-1\}$ with 26 genuine exceptions of n and r shown in Table 1(b) and one possible exception of $(n,r) = (14, 14^2 - 3)$.*

In this paper we develop randomized algorithms for computing r-orthogonal Latin squares and r-self-orthogonal Latin squares. The benefit of this approach is that r-orthogonal Latin squares of large orders can be computed. This is due to the polynomial running time. Our experiments show that the difficult case is when r is close to n^2. An evidence of this is the open problem of finding a 193-self-orthogonal Latin square of order 14 [28] which is the only case left to complete the spectrum of r-self-orthogonal Latin squares.

Related work. Keedwell and Mullen [19] investigated the construction of sets of t Latin squares of a given non-prime-power order q which are as close as possible to being a mutually orthogonal set. Dinitz and Stinson [9] studied the problem of constructing sets of s Latin squares of order m such that the average number of different ordered pairs obtained by superimposing two of the s squares in the set is as large as possible. Arce-Nazario *et al.* [1] discussed some computational problems concerning the distribution of orthogonal pairs in sets of Latin squares of small orders.

2   Preliminaries

A *partial Latin square* of order n is an $n \times n$ array in which

(i) each entry is either empty or it contains an element from $[n]$, and
(ii) each of the symbols $1, 2, \ldots, n$ occurs at most once in each row and at most once in each column of the array. Completing partial Latin squares is NP-complete [7]. However some partial Latin squares can always be completed, for example, Latin rectangles.

Theorem 4 (M. Hall [14]**).** *Every $r \times n$ Latin rectangle, $0 \leq r \leq n$, can be completed to a Latin square of order n.*

The proof is based on the following theorem for SDR. A *system of distinct representatives*, or *SDR*, for a collection of finite sets $S_1, S_2, \ldots, S_m$, is a sequence $\langle s_1, s_2, \ldots, s_m \rangle$ of m distinct elements $s_i \in S_i$. Each s_i is called a *representative* of set S_i. For example, the set $\langle 2, 3, 1 \rangle$ is an SDR for the sets $S_1 = \{2\}$, $S_2 = \{1,3\}$ and $S_3 = \{1,2\}$.

Theorem 5 (P. Hall [15]). *Let $S_1, S_2, \ldots, S_m$ be a collection of m finite sets. Then an SDR for these sets exists if and only if, for all $k \in \{0, 1, \ldots, m\}$, $|S_{i_1} \cup S_{i_2} \cup \cdots \cup S_{i_k}| \geq k$, where the k sets $S_{i_1}, \ldots, S_{i_k}$ represent any collection of k sets chosen from the m sets $S_1, S_2, \ldots, S_m$.*

The proof of Theorem 4 uses a collection of sets $S_i, i \in [n]$, where S_i is the set of all $x \in [n]$ such that x does not occur in the column j of a $r \times n$ Latin rectangle A. It can be shown that these sets satisfy Hall's condition (Theorem 5): for all $k \in \{0, 1, \ldots, n\}$, the union of any k sets in the collection contains at least k elements. By Theorem 5, an SDR for these sets exists. Therefore Latin rectangle A can be extended to a $(r+1) \times n$ Latin rectangle by adding a row of the representatives. An SDR can be computed as a matching in a bipartite graph $G_A = (V_1, V_2, E)$ where $V_1 = \{v_1, \ldots, v_n\}, V_2 = \{u_1, \ldots, u_n\}, E = \{(v_i, u_j) \mid j \in S_i\}$. A Latin square can be computed using this step $n - r$ times.

3   Completing Orthogonal Latin Rectangles

In order to construct r-orthogonal Latin squares of order n, we apply a randomized algorithm called Algorithm A1 where

1. the first rows of two Latin squares A and B are computed as random permutations of $[n]$ and
2. the remaining rows of A and B are computed using a matching algorithm applied to bipartite graphs $G_{A'}$ and $G_{B'}$ and random order of sets S_i where A' and B' are the current Latin rectangles.

$$A = \begin{matrix} 1\,2\,7\,5\,6\,4\,3 \\ 6\,7\,4\,2\,3\,1\,5 \\ 3\,1\,6\,7\,2\,5\,4 \\ 5\,4\,3\,1\,7\,6\,2 \\ 4\,5\,2\,3\,1\,7\,6 \\ 2\,6\,1\,4\,5\,3\,7 \\ 7\,3\,5\,6\,4\,2\,1 \end{matrix} \qquad B = \begin{matrix} 4\,2\,5\,3\,7\,1\,6 \\ 3\,6\,2\,4\,5\,7\,1 \\ 7\,3\,6\,2\,1\,5\,4 \\ 6\,1\,7\,5\,4\,2\,3 \\ 2\,7\,3\,1\,6\,4\,5 \\ 5\,4\,1\,6\,2\,3\,7 \\ 1\,5\,4\,7\,3\,6\,2 \end{matrix}$$

Fig. 2. A pair of 53-orthogonal Latin squares of order 7 where all pairs (i, j) except $(2,7),(3,2),(3,4),(4,5),(4,7),(6,1),(7,3)$ appear if A and B are superimposed.

We implemented Algorithm A1 and run it for $n \in [5, 20]$, see for example 42-orthogonal Latin squares of order 7 in Fig. 2. The range of values of r for $n \in [5, 20]$ computed by Algorithm A1 is shown in Table 2. One can observe that the range is complete for $n = 5$ and $n = 6$ (by Theorem 2). For $n \geq 7$, the algorithm only found a subset of possible values of r for two r-orthogonal Latin squares. For example, it computed 36 values of r for $n = 10$ which is $\frac{33}{89} \cdot 100 \approx 37.07\%$ of all values. For $n = 20$, it computed 70 values of r which is $\frac{70}{379} \cdot 100 \approx 18.46\%$ of all values. So, one needs to develop an algorithm for computing r-orthogonal Latin squares of order n for large and small values of r. It also gives rise to an interesting problem.

Open problem. Let $r(A, B)$ be the *orthogonality* of two Latin squares A and B of order n, i.e. the number of distinct ordered pairs when A and B are superimposed. What is the expected value of $r(A, B)$ of two random Latin squares A and B of order n?

We computed approximately the expected value of $r(A, B)$ of two random Latin squares A and B using the program[1] developed by Paul Hankin [16] implementing an algorithm for generating random Latin squares by Jacobson and Matthews [17]. 10000 runs were used for $n = 5, \ldots, 12$ and 1000 runs were used for $n = 13, \ldots, 20$. The result is shown in Table 2 (column $r(A, B)$). For all $n \in \{5, 6, \ldots, 20\}$, the expected value of $r(A, B)/n^2$ is close to 0.63.

Table 2. Pairs of r-orthogonal Latin squares of order n computed by Algorithm A1 (column A1) and Algorithm A2 (column A2). No entry in column A2 means the same result as in column A1. Column $r(A, B)$ shows the average orthogonality of two random Latin squares using the program developed by Paul Hankin [16].

n	A1	A2	$r(A, B)$
5	5, 7, 10–19,21, 25	-	15.8478
6	6, 8–32, 34	-	22.7416
7	13–44	7, 9, 11–47,49	30.9792
8	25–53	8, 10–60	40.4537
9	35–65	9,11–77	51.2081
10	47–79	10, 12–93	63.2259
11	60–93	11,13–114	76.55
12	71–109	12, 14–135	91.0366
13	85–128	13,15–157	106.847
14	102–145	14, 16–183	124.075
15	117–166	15, 17–210	142.279
16	137–189	16,18–236	161.437
17	155–207, 209	17, 19–268	182.85
18	173, 176–233	18, 20–300	204.693
19	196–257, 259, 261	19, 21–334	228.004
20	218–286, 290	20,22–370	252.536

---

[1] It is based on the Java implementation described by Ignacio Gallego Sagastume https://github.com/bluemontag/igs-lsgp.

In order to extend the range of $r(A, B)$ computed by Algorithm A1, a plausible approach would be to solve the problem of extending two $k \times n$ Latin rectangles to two $(k+1) \times n$ Latin rectangles maximizing (minimizing) $r(A, B)$. This problem seems difficult and we consider an approach where one Latin rectangle is extended first and then the other Latin rectangle is extended by solving the following problem.

Problem MaxRAB (MinRAB). Let A be a $k \times n$ Latin rectangle, $k < n$, and let B be a $(k + 1) \times n$ Latin rectangle. Extend A to a $(k + 1) \times n$ Latin rectangle A' maximizing (resp. minimizing) $r(A', B)$.

Theorem 6. *Problems MaxRAB and MinRAB can be solved by $O(n^3)$ time algorithm.*

Proof. Let A be a $k \times n$ Latin rectangle and B be a $(k+1) \times n$ Latin rectangle. Let A_j denote the set of positive integers that do not occur in column j of A. Let $S = \{(A_{i,j}, B_{i,j}) \mid i \in [k], j \in [n]\}$. Construct a weighted bipartite graph $G = (V_1, V_2, E)$ where $V_1 = \{v_1, \ldots, v_n\}$, $V_2 = \{u_1, \ldots, u_n\}$, and $E = \{(v_i, u_j) \mid j \in A_i\}$. To complete the construction, we assign the weight to each edge $(v_i, u_j) \in E$ as

$$w(v_i, u_j) = \begin{cases} 1 & \text{if } (j, B_{k+1,i}) \notin S, \\ 0 & \text{if } (j, B_{k+1,i}) \in S. \end{cases}$$

Graph G has a perfect matching since Latin rectangle A can be extended to a $(k+1) \times n$ Latin rectangle by Theorem 4. In order to solve problem MaxRAB, we compute the maximum weight matching M which is the solution of the assignment problem. It can be found using the Hungarian method [6,20] in $O(n^3)$ time. We extend Latin rectangle A to a $(k+1) \times n$ Latin rectangle A' by setting $A'_{k+1,i} = j$ if (v_i, u_j) is an edge of the matching. It remains to prove that $r(A', B)$ is maximized.

Let A^* be an extension of A (i.e. A^* is a $(k + 1) \times n$ Latin rectangle such that the first k rows of A^* and A are the same) maximizing $r(A^*, B)$. Consider the permutation $A^*_{k+1,1}, A^*_{k+1,2}, \ldots, A^*_{k+1,n}$. Clearly, $A^*_{k+1,i} \in A_i$ for all $i \in [n]$. Therefore $M^* = \{(v_i, u_j) \mid i \in [n], j = A^*_{k+1,i}\}$ is a matching in graph G. Let B' be a $k \times n$ Latin rectangle obtained using first k rows of Latin rectangle B. Then $r(A^*, B) - r(A, B')$ is the number of pairs $(A^*_{k+1,i}, B_{k+1,i})$ in $[n]^2 \setminus S$. Therefore $r(A^*, B) - r(A, B') = w(M^*)$. Similarly $r(A', B) - r(A, B') = w(M)$. Since M is the solution of the assignment problem, $w(M) \geq w(M^*)$. Then

$$\begin{aligned} r(A', B) &= r(A, B') + w(M) \\ &\geq r(A, B') + w(M^*) \\ &= r(A^*, B). \end{aligned}$$

```
4 3 2 5 1        4 3 2 5 1        4 2 3 5 1
5 1 3 2 4        5 1 3 2 4        5 1 2 3 4
3 4 5 1 2        3 5 1 4 2        3 4 5 1 2
1 2 4 3 5        1 2 4 3 5        1 3 4 2 5
2 5 1 4 3        2 4 5 1 3        2 5 1 4 3
   (a)              (b)              (c)
```

Fig. 3. (a) A Latin square A. (b) Switching a row cycle in A between the third row and the fifth row. (c) Switching a symbol cycle in A on symbols 2 and 3.

Since $r(A^*, B) \geq r(A', B)$, we have $r(A^*, B) = r(A', B)$. So, Latin rectangle A' is optimal.

A similar argument can be used to solve problems MinRAB. The minimum weight matching can be applied on the same graph G. $\qquad\square$

By Theorem 6, the extension of Latin rectangle A is optimal. However, the extension of Latin rectangle B in our approach might be not the best. We attempt to optimize (maximize or minimize) $r(A, B)$ using cycle switches [2, 26].

Switching cycles. Consider distinct rows r and s in a Latin square A. Let $\pi_{r,s}$ be a permutation which maps $A_{r,i}$ to $A_{s,i}$. Clearly, $\pi_{r,s}$ is a derangement, i.e. it has no fixed points. Take any cycle of $\pi_{r,s}$ and let C be the set of columns involved in the cycle. Switching *row cycle C* in A is defined by

$$
A'_{i,j} = \begin{cases} A_{s,j} & \text{if } i = r \text{ and } j \in C, \\ A_{r,j} & \text{if } i = s \text{ and } j \in C, \\ A_{i,j} & \text{otherwise,} \end{cases}
$$

see an example in Fig. 3(b). A *column cycle* is a set of elements which forms a row cycle when the square is transposed. Switching a *symbol cycle* on two symbols a and b is achieved by replacing every occurrence of a in the cycle by b and vice versa, see an example in Fig. 3(c).

In our approach, we have Latin rectangles instead of Latin squares. Switching row cycles can be applied to Latin rectangles since the rows are full. We adapt column cycles to Latin rectangles as *column paths*. Consider distinct columns c and d in a Latin rectangle A. Let $f_{c,d}$ be a function that maps $A_{i,c}$ to $A_{i,d}$.

```
4 3 2 5 1        2 3 4 5 1        4 3 2 5 1
5 1 3 2 4        5 1 3 2 4        5 1 4 2 3
3 4 5 1 2        3 4 5 1 2        3 4 5 1 2
1 2 4 3 5        4 2 1 3 5        1 2 3 4 5
   (a)              (b)              (c)
```

Fig. 4. (a) A Latin rectangle A. (b) Function $f_{1,3}$ for A induces a cycle (3,5) and a path (1,4,2). A Latin rectangle after switching column path (1,4,2) in A is shown. (c) Switching a symbol path in A on symbols 3 and 4.

Start with a symbol $A_{i,c}$ in column c which is not present in column d. Do the following step for row i. Find $A_{i,d}$ in column c, say $A_{j,c} = A_{i,d}$ if it exists. Swap $A_{i,c}$ and $A_{i,d}$. If j is not found then stop; otherwise set $i = j$ and repeat. See an example in Fig. 4(b). We have $c = 1$ and $d = 3$. Start with symbol $A_{4,1} = 1$ which is not present in column 3 in Fig. 4(a). Function $f_{1,3}$ maps 1 to 4, 4 to 2. Swap $A_{4,1}$ and $A_{4,3}$, then swap $A_{1,1}$ and $A_{1,3}$. The result is shown in Fig. 4(b).

Symbol cycles may occur in Latin rectangles. We also apply *symbol paths* which are defined as follows. Consider distinct symbols a and b in a Latin rectangle A. Start with a symbol $A_{i,j} = a$ in column j such that symbol b is not present in column j, so (i,j) is the first cell in the symbol path. Repeat the following step. Find symbol $\{a, b\} \setminus A_{i,j}$ in row i, say $A_{i,j'}$. Append (i, j') to the path. Find symbol $\{a, b\} \setminus A_{i,j'}$ in column j'. If it does not exist, then stop. Suppose it exists, say $A_{i',j'}$. Then append (i', j') to the path, set $i = i', j = j'$ and repeat the step. When the path is computed, swap symbols a and b in the path, see an example in Fig. 4(c).

We implemented algorithm A2 using method from Theorem 6 combined with switching row/column/symbol cycles and column/symbol path. The results for $n = 7, \ldots, 20$ are shown in Table 2. Note that the low values of $r(A, B)$ are covered completely (by Theorem 2) and new values of $r(A, B)$ larger than the ones

Table 3. r-self-orthogonal Latin squares of order n computed by Algorithm A3 (column A3) and Algorithm A4 (column A4). No entry in column A4 means the same result as in column A3. Column $r(A, A^T)$ shows the average value of $r(A, A^T)$ of a random Latin square A using the program developed by Paul Hankin [16].

n	A3	A4	$r(A, A^T)$
5	5,7,10–11,13–15,17,19,21,25	-	11.2769
6	6,8–31	-	15.4191
7	7, 9–45, 47, 49	-	20.2193
8	8,10–58	8,10–62,64	25.9692
9	20–69	9,11–79,81	32.368
10	29, 33–84	10, 12–98	39.5512
11	42, 44–99	11, 13–119	47.4226
12	56, 59–117	12, 14–140	56.1379
13	72–133	13, 15–163	65.5507
14	83–84,86,88–152	14, 16–214	75.7708
15	101, 103–170, 172–173	15, 17–211	86.5367
16	118, 120–196	16, 18–242	98.443
17	141–216	17, 19–271	110.387
18	161–239	18, 20–307	123.722
19	178–266, 268, 272	19, 21–340	137.122
20	202–292	20, 22–375	152.056

computed by algorithm A1. The problem of computing all large values of $r(A, B)$ is quite difficult. Specifically, the problem of computing random orthogonal Latin squares (i.e. for $r(A, B) = n^2$) is very difficult.

4 Computing Self-orthogonal Latin Rectangles

We experiment with two algorithms using the algorithms developed in the previous section. If A is a $k \times n$ Latin rectangle then we assume that B is an $n \times k$ Latin rectangle $B = A^T$. For example, if a column cycle is applied to Latin rectangle A, it also affect B. We call the algorithms corresponding to algorithms A1 and A2, Algorithm A3 and A4, respectively. The range of values of r for $n \in [5, 20]$ computed by Algorithm A3 is shown in Table 3. One can observe that the range is complete for $5 \leq n \leq 9$ (by Theorem 3). In general, algorithms A3 and A4 cover more values of r comparing to algorithms A1 and A2.

It is also related to an interesting problem: What is the expected value of $r(A, A^T)$ where A is a random Latin square of order n? We computed approximately the expected value of $r(A, A^T)$ of a random Latin square A using the program developed by Paul Hankin [16] implementing an algorithm for generating random Latin squares by Jacobson and Matthews [17]. 10000 runs were used for $n = 5, \dots, 15$ and 1000 runs were used for $n = 16, \dots, 20$. The result is shown in Table 3 (see column $r(A, A^T)$). For example, the expected value of $r(A, A^T)/n^2$ is close to 0.395 for $n = 10$ and is close to 0.38 for $n = 20$.

References

1. Arce-Nazario, R.A., Castro, F.N., Córdova, J., Hicks, K., Mullen, G.L., Rubio, I.M.: Some computational results concerning the spectrum of sets of Latin squares. Quasigroups Related Syst. **22**, 159–164 (2014)
2. Asratyan, A., Mirumyan, A.: Transformations of Latin squares (Russian). Diskret. Mat. **2**, 21–28 (1990)
3. Belyavskaya, G.B.: r-orthogonal quasigroups I. Math. Issled. **39**, 32–39 (1976)
4. Belyavskaya, G.B.: r-orthogonal quasigroups II. Math. Issled. **43**, 39–49 (1977)
5. Belyavskaya, G.B.: r-orthogonal Latin squares. In: DL enes, A.K.J., editor, Latin Squares: New Developments, pp. 169–202 (Chapter 6). Elsevier (1992)
6. Burkard, R.E., Çela, E.: Linear assignment problems and extensions. In: Du, D., Pardalos, P.M., editors, Handbook of Combinatorial Optimization, pp. 75–149. Springer (1999). https://doi.org/10.1007/978-1-4757-3023-4_2
7. Colbourn, C.J.: The complexity of completing partial Latin squares. Discret. Appl. Math. **8**(1), 25–30 (1984)
8. Colbourn, C.J., Zhu, L.: The spectrum of r-orthogonal Latin squares. In: Colbourn, C.J., Mahmoodian, E.S. (eds.) Combinatorics Adv., pp. 49–75. Springer, US, Boston, MA (1995)
9. Dinitz, J.H., Stinson, D.R.: On the maximum number of different ordered pairs of symbols in sets of Latin squares. J. Comb. Des. **13**(1), 1–15 (2005)
10. Egan, J., Wanless, I.M.: Latin squares with restricted transversals. J. Comb. Des. **20**(7), 344–361 (2012)

11. Euler, L.: Recherche sur une nouvelle espéce de quarrés magiques. Leonardi Euleri Opera Omnia **7**, 291–392 (1923)
12. Evans, A.B.: Latin squares without orthogonal mates. Des. Codes Cryptogr. **40**(1), 121–130 (2006)
13. Finizio, N.J., Zhu, L.: Self-orthogonal Latin squares (SOLS). In: Colbourn, C.J., Dinitz, J.H., editors, The Handbook of Combinatorial Designs, pp. 211–219. Chapman/CRC Press (2007)
14. Hall, M.: An existence theorem for Latin squares. Bull. Amer. Math. Soc. **51**(6), 387–388 (1945)
15. Hall, P.: On representative of subsets. J. London Math. Soc. **10**, 26–30 (1935)
16. Hankin, P.: Generating random Latin squares (blog). https://blog.paulhankin.net/latinsquares/ (2019)
17. Jacobson, M.T., Matthews, P.: Generating uniformly distributed random Latin squares. J. Comb. Des. **4**(6), 405–437 (1996)
18. Keedwell, A.D., Dénes, J.: Latin squares and their applications. Elsevier (2015)
19. Keedwell, A.D., Mullen, G.L.: Sets of partially orthogonal Latin squares and projective planes. Discret. Math. **288**(1–3), 49–60 (2004)
20. Kuhn, H.W.: The Hungarian method for the assignment problem. Naval Res. Logist. Quart. **2**(1–2), 83–97 (1955)
21. Mann, H.B.: On orthogonal Latin squares. Bull. Amer. Math. Soc. **50**, 249–257 (1944)
22. Mariot, L., Formenti, E., Leporati, A.: Constructing orthogonal Latin squares from linear cellular automata. CoRR, abs/1610.00139 (2016)
23. Mariot, L., Gadouleau, M., Formenti, E., Leporati, A.: Mutually orthogonal Latin squares based on cellular automata. Des. Codes Cryptogr. **88**(2), 391–411 (2020)
24. McKay, B.D., Meynert, A., Myrvold, W.: Small Latin squares, quasigroups, and loops. J. Comb. Des. **15**(2), 98–119 (2007)
25. van Rees, G.H.J.: Subsquares and transversals in Latin squares. Ars Combin. **29B**, 193–204 (1990)
26. Wanless, I.M.: Cycle switches in Latin squares. Graphs Comb. **20**(4), 545–570 (2004)
27. Wanless, I.M., Webb, B.S.: The existence of Latin squares without orthogonal mates. Des. Codes Cryptogr. **40**(1), 131–135 (2006)
28. Zhang, H.: 25 new r-self-orthogonal Latin squares. Discret. Math. **313**(17), 1746–1753 (2013)
29. Zhu, L., Zhang, H.: Completing the spectrum of r-orthogonal Latin squares. Discret. Math. **268**(1–3), 343–349 (2003)

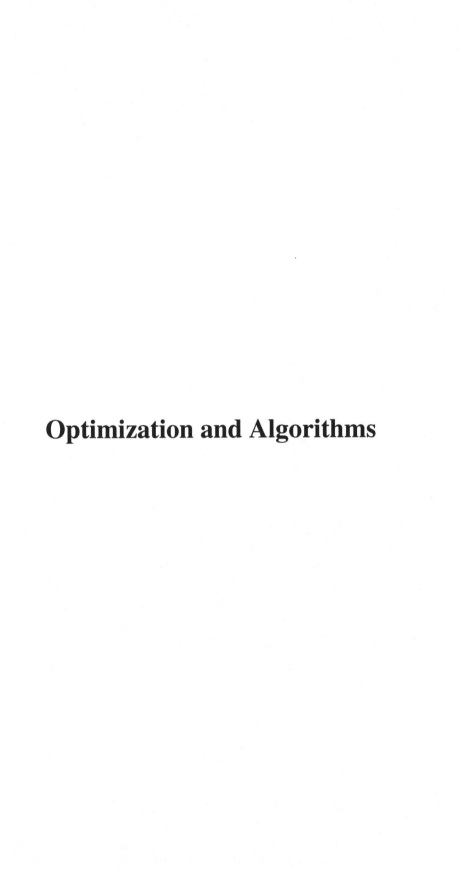

Optimization and Algorithms

A Two-Stage Seeds Algorithm
for Competitive Influence Maximization
Considering User Demand

Zhiheng You, Hongwei Du[(⊠)], and Ziwei Liang

School of Computer Science and Technology, Harbin Institute of Technology
(Shenzhen), Shenzhen 518055, China
`hongwei.du@ieee.org, ziweiliang@stu.hit.edu.cn`

Abstract. Competitive influence maximization (CIM) in online social
network has received widespread attention and research in recent years.
The traditional competitive influence maximization problem explores the
competition between multiple entities on the same network, aiming to
select a certain number of seeds for one of the entities to maximize the
spread of its influence. However, the latest competitive influence maximization researches ignore the impact of differences in user demand
on the spread process of competitive influence in real-world competitive relationships. Therefore, a novel propagation model called Competitive Linear Threshold Model considering User Demand (CLTMcUD) is
presented, which takes into account the difference in user demand for
two different brands of the same product category. For this propagation
model, a two-stage algorithm named Dual Influence Assessment based
on Community Structure (DIACS) algorithm is proposed, which utilizes
the characteristics of community structure and dual-influence of nodes
to select candidate seeds and maximize the influence of a competitor. We
test our algorithm on four real-world datasets and show that it outperforms state-of-the-art algorithms.

Keywords: Competitive influence maximization · User demand · Dual
influence · Community structure

1 Introduction

With the flourishing growth of online social networks, millions of people may now
engage with one another and produce record-breaking volumes of data, providing social networks a tremendous amount of potential. This potential has led
to an increasing interest in studying social networks, including analyzing their
structures and exploring the distribution of communities [14]. Social networks
can be regarded as complex networks composed of individuals and their relationships in society. They are important structures for information transmission and
dissemination between individuals. The process of information propagation in
social networks has great practical significance in viral marketing, personalized

recommendations, target advertising, and so on, all of which relate to the problem of maximizing the influence. The classic influence maximization problem's objective is to choose a group of k online social network members, or the "seed set", that maximizes the anticipated number of users who would be influenced by the information spread through the seed set [10,12].

In recent years, many scholars have studied the competitive influence maximization and propose different kinds of propagation models, including Competitive Linear Threshold model [7], Competitive Independent Cascade model [2], and so on [1,8,13]. In the real world, competition among different brands of the same product category is very common. Taking smartphones as an example, almost no one can live without them nowadays. Some people choose well-known brands such as iPhone for their high-quality experience while some choose more affordable brands that provide high cost-performance ratio. However, the exiting works have not taken this circumstance into consideration.

In this paper, we introduce the difference in user demand for two different brands of the same product category based on the CIM problem proposed by Bozorgi et al. [4], and explored the solution for the competition influence maximization problem under this scenario. The community structure in social networks has a significant impact on information dissemination and many scholars have conducted in-depth research on community-based impact maximization algorithms [6,15]. We also utilize the community structure and present a novel algorithm. We summarise the major contributions in this paper as follows:

1. We propose a new propagation model called Competitive Linear Threshold Model considering User Demand (CLTMcUD). In CLTMcUD, each node has the ability to affect its neighbors in a similar way to the conventional linear threshold model.
2. We present a competitive influence maximization method considering user demand, namely dual influence assessment based on community structure (DIACS) algorithm. In the DIACS algorithm, there are two steps for finding the seeds S_B that can ultimately maximize the influence of the competitor B against the already given seeds S_A of the competitor A.
3. we rely on four intricate real-world networks of varying sizes and complexity to verify the effectiveness and feasibility of the DIACS algorithm.

The rest of this paper is structured as follows. In Sect. 2, the competitive propagation model is introduced in detail, and the competitive influence maximization problem is defined and the corresponding algorithm is proposed. Section 3 shows and analyzes the experimental results of the DIACS algorithm. Section 4 presents the conclusion of this paper.

2   Propagation Model and Algorithm

2.1   Propagation Model

Decidable Competitive Model (DCM) propagation model was proposed by Bozorgi et al. [4], which can reflect some competitive phenomena in reality but ignores the competition relationship between entities of similar types. On this basis, we propose a model, CLTMcUD, which is suitable for the competition relationship between entities of similar types proposed in this paper.

Given a directed graph $G = (V, E, W, T)$, where V represents the set of all nodes, E represents the set of all directed edges, W represents the set of edge weights corresponding to the elements in the directed edge set E, and T represents the set of node tags corresponding to the elements in the node set V. We assume that there are two competing entities A and B in the network. The node in CLTMcUD spreads with discrete time steps, and its propagation rule is as follows:

(1) Each node v may be in one of four states, which are *inactive, thinking, active⁺*, and *active⁻*. Here, the *inactive* state represents that the node v is still in an inactive state, the *active⁺* state represents that the node v has chosen to accept entity A, while the *active⁻* state represents that the node v has chosen to accept entity B. While the *thinking* state represents that the node v has been activated for the first time but has not yet selected to accept any entity. And node v will choose to accept an entity and become *active⁺* or *active⁻* state after t timestamps after entering the *thinking* state people who want to buy a certain product often consider a certain amount of time before buying a product in real life.

(2) Each node v has a tag t_v which can take the values 1, 2, or 3. Here, we define that nodes with tag 1 have a higher demand for entity A, nodes with tag 2 have a higher demand for entity B, and nodes with tag 3 have equal demand for both entities. Nodes with less demand for an entity require an additional activation threshold $\theta_t (0 \leq \theta_t \leq 1)$ to be activated by that entity.

(3) Taking the node v with tag 1 as an example, when the total inbound propagation probability of node v satisfies:

$$\sum_{u \in N_{active+}^{in}} p_{u,v} \geq \theta_v \tag{1}$$

or

$$\sum_{u \in N_{active-}^{in}} p_{u,v} \geq \theta_v + \theta_t \tag{2}$$

The node v will switch to the *thinking* state from the *inactive* state, or it will stay in the *inactive* state otherwise. Nodes that enter the *thinking* state will choose the entity to accept after t timestamps, and the acceptance rules are as follows:

$$state_v = \begin{cases} active^+, & \text{if } \sum_{u \in A^+_{T+t}} p_{u,v} \geq \sum_{u \in A^-_{T+t}} p_{u,v} \\ active^-, & \text{if } \sum_{u \in A^+_{T+t}} p_{u,v} < \sum_{u \in A^-_{T+t}} p_{u,v} \text{ and } \sum_{u \in A^-_{T+t}} p_{u,v} \geq \theta_v + \theta_t \end{cases} \quad (3)$$

Here, A^+_{T+t} represents the set of nodes in the predecessor nodes of node v with status $active^+$ and A^-_{T+t} represents the set of nodes in the predecessor nodes of node v with status $active^-$.

(4) Any node can only be activated by one competing entity at the same time, and once it is activated by any entity, it cannot be activated by the other entity.

An example model is shown as Fig. 1:

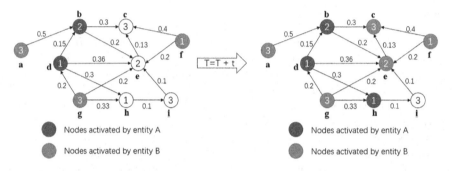

Fig. 1. A Example of Competitive Linear Threshold Model with User Demand

In this paper, we assume that there are two competing entities A and B in the network and make nodes marked as 1 have a higher demand for entity A, nodes marked as 2 have a higher demand for entity B, and nodes marked as 3 have an equal demand for both entities. For each node, we set the activation threshold of each node to $\theta_v = 0.3$. For example, as shown in Fig. 1, any entity that activates a node with low demand for itself requires an additional activation threshold $\theta_t = 0.3$. At time T, node c is marked as 3 and receives a propagation probability of 0.56 from entity A and 0.4 from entity B. Node e will be activated by entity B and enter the *thinking* state at $T + 1$ because $0.4 > 0.3$. Similarly, node c will be activated by entity B and enter the *thinking* state, and node h will be activated by entity A and enter the *thinking* state. After t timestamps, nodes c, e, and h will simultaneously choose which entity to accept. Although node e receive a propagation probability of 0.56 from entity A, which is greater than 0.4 from entity B, node e will accept entity B. The reason is node e is marked as 2, which means node e have a higher demand for entity B and will accept entity B with a less activation threshold θ_v. And the propagation from entity A is less than $\theta_v + \theta_v = 0.6$ so that node e cannot be activated by entity

A. Therefore, node e is activated by entity B. Similarly, node c will be activated by entity B, and node h will be activated by entity A. While node b cannot be activated by entity B because node b has already chosen to accept entity A.

2.2   Problem Formulation

We first introduce the definition of influence propagation score and some graph theory concepts. we present the definition and objective of the competitive influence maximization problem studied in this paper.

Definition 1. (Spread Influence Function f): We define the propagation influence of node v in CLTMcUD as $f(v)$, which represents the number of nodes that can be activated by node v according to the propagation rule we establish in the social network. The diffusion influence of a node set S is defined as the number of nodes that can be activated by simultaneously diffusing all nodes in the set until no additional nodes can be activated. The formula is expressed as follows:

$$f(S) = \sum_{v \in S} f(v) \tag{4}$$

Theorem 1. *Given a directed graph $G = (V, E, W, T)$, two competitors A, B and a seed set $S_A \subset V$ of A, the spread influence function f satisfies monotonicity and submodularity on the seed set $S_B \subset V$ in the CLTMcUD model.*

Proof. **Monotonicity:** According to the propagation rule of the CLTMcUD model mentioned in Sect. 2.1, each node that has reached the final state, i.e., either $active^+$ or $active^-$, will only propagate influence to its neighbors and will not change its own state again. Therefore, for any seed sets S_B, adding any node $v \in V \backslash (S_A \cup S_B)$ to S_B will always satisfy $f(S_B \cup \{v\}) \geq f(S_B)$. It means the function f satisfies monotonicity.

Submodularity: When function f satisfies the property of "diminishing marginal returns," we refer to it as submodularity. We let $F(S)$ represent the set of nodes that the set S can activate and suppose S can activate node v if and only if S can reach v within time t. For any set $S_B \subseteq T \subset V$ and any node $u \in V \backslash T$, for any node $v \in F(T \cup \{u\}) \backslash F(T)$, we set that $T \cup \{u\}$ can reach v within time t, but T cannot. Therefore, starting from node u must be able to reach v within time t. Because $S_B \subseteq T$, we can easily know S_B cannot reach v within time t. It means $v \in F(S_B \cup \{u\}) \backslash F(S_B)$, i.e. $F(T \cup \{u\}) \backslash f(T) \subseteq f(S_B \cup \{u\}) \backslash f(S_B)$. Then we can infer that $f(T \cup \{u\}) - f(T) \leq f(S_B \cup \{u\}) - f(S_B)$, which illustrates that f satisfies the submodularity.

We have proven that the influence spread function f satisfies monotonicity and submodularity, which means that greedy algorithms can be used on the CLTMcUD model and obtain an approximation ratio of $1 - \frac{1}{e}$.

In order to quickly assess the impact of nodes in the network, we use the Topology Importance (TI) value to rank the nodes. Here, we use $N^{in}(v)$ [16] to denote the set of internal neighbors of node v which are within the same community, and $N^{ex}(v)$ to represent the set of external neighbors of node v which are in other communities. In addition, since our CLTMcUD model is an improvement over the traditional linear threshold (LT) model, each activated node will continue to have an influence on its neighbors. Therefore, we use Eigenvector Centrality(EC) to evaluate the importance of a node in the network. This is important because EC can reflect the importance of a node and its neighbors, which is why we include it in the TI value. Next, we provide the definition of TI:

Definition 2. (Topology Importance, TI): TI consists of three components, that is used to evaluate the impact of a node within a community and across different communities. The calculation formula of TI is expressed as follows:

$$TI_v = |N^{in}(v)| + \frac{|N^{ex}(v)| - Min_{u_i \in V}(|N^{ex}(u_i)|)}{Max_{u_i \in V}(|N^{ex}(u_i)| - Min_{u_i \in V}(|N^{ex}(u_i)|)} + EC_v \quad (5)$$

The first parameter $|N^{in}(v)|$ represents the number of nodes in the set of internal neighbors of node v. And the second parameter is the normalized value of the set of external neighbors of node v. The third is the eigenvector centrality of node v.

Definition 3. (Competitive Influence Maximization considering User Demand, CIMcUD): Given a directed social network $G = (V, E, W, T)$ and two competitively similar entities A and B, where V represents the set of all nodes in the network, E represents the set of all edges in the network, W represents the corresponding weight set on edge set E, and the weight w_i on node v_i is used to represent the probability of propagation influence between two nodes, with w_i being a real number between 0 and 1. T represents the set of tags corresponding to each point in the network, with possible values of label t_i for node vi being 1, 2, or 3. In this paper, the seeds S_A of entity A has been given and S_B of entity B are selected from the remaining nodes after entity A has chosen its own seed nodes. Specifically, entity B selects k different nodes as its seeds, which is the same as the number of seeds selected by A. The objective of CIMcUD is to maximize the spread influence function $f(S_B)$ for entity B.

2.3 Algorithm Consider Dual Influence Assessment Based on Community Structure

Based on the CIMcUD problem, the seed selection algorithm of this paper will be influenced to some extent by the nodes already selected by competitors. Traditional CIM approaches may use community-based greedy algorithms, but they ignore the lasting influence of nodes on their neighboring nodes in the LT model [4]. Therefore, we design the DIACS algorithm, a two-stage algorithm. The first stage is a filter stage, which aims to quickly select a subset of nodes from all nodes as candidate seeds. The second stage is a greedy stage, which selects a group of seed nodes with the highest benefits through the marginal benefits coefficient. We now introduce the detailed algorithms for each step shown in Algorithm 1.

Algorithm 1. Dual Influence Assessment based on Community Structure Algorithm

---

Require: Graph G, Competitive Seeds S_A, Number of Seeds k
Ensure: Seeds $\{S_B\}$
1: $C \Leftarrow Louvain(G)$
2: Calculate the TI of all nodes
3: Select top-30% TI node as the $S_{candidate}$
4: **for** each $v \in C \cap v \in S_{candidate}$ **do**
5:     Calculate a CELF_list of Inf_{p1}, Inf_{p2}
6:     set $iter_v = 0$
7: **end for**
8: Select the node n with max Inf_{p2} with max Inf_{p1}
9: remove n from CELF_list
10: **while** $len(S_B) < k$ **do**
11:     Select the node n with max Inf_{p2} with max Inf_{p1}
12:     **for** each seed $s \in S_B$ **do**
13:         **if** s, n belong to the same community **then**
14:             **if** $iter_s == iter_n$ **then**
15:                 Add n into S_B and delete n from CELF_list
16:                 **break**
17:             **end if**
18:             Update $Inf_{p1}, Inf_{p2}, iter_n$ of n in CELF_list
19:         **end if**
20:     **end for**
21:     **if** n not belong to any community of the node in S_B **then**
22:         Add n into S_B and delete n from CELF_list
23:     **end if**
24: **end while**
25: **return** S_B

The two-stage algorithm is based on community structure and we use Louvain algorithm [3] to partition the social network into communities, which is based on modularity optimization and can be applied to networks of different types and sizes. The formula of the calculation of modularity is as follow [3]:

$$Q = \frac{1}{2m} \sum_{i,j} \left(A_{ij} - \frac{k_i k_j}{2m} \right) \delta(c_i, c_j) \tag{6}$$

By using the Louvain algorithm, we can obtain non-overlapping communities, which ensures that each node has a unique TI value during the filter stage.

The filter stage is shown in Line 2 to Line 3, where we calculate TI values for all nodes on the community-by-community basis mentioned before. By calculating the TI value of each node, we can quickly sort the nodes and select the top 30% nodes with the highest TI values as candidate seeds. In the next greedy phase, we will only assess the dual-influence value on the candidate seeds instead of all nodes.

Line 4 to Line 24 show the greedy stage, which first establish a CELF list [11] according to the dual-influence values of candidate seeds. The dual-influence consists of two parts, Inf_{p1} and Inf_{p2}. The value of $Inf_{p1}(v)$ represents the number of nodes that the node v can activate directly. The calculation method of Inf_{p1} is the same as the marginal benefit function considered by traditional greedy algorithms. It can help us understand the direct influence that can be generated when adding node v to the seed set. Based on this, we propose the Inf_{p2} parameter to evaluate the second-level influence of nodes. In the LT propagation model, each activated node will continue to influence its neighbors until the end of the propagation. Therefore, in addition to the neighbors that can be directly activated by the node, we also consider its influence on neighbors that have not yet been activated, which we refer to as the second-level influence inf_{p2}. In other words, the second-level influence is the influence of all nodes eventually activated by the set S on their neighbors that are in the *inactive* or *thinking* state. The formulas for calculating Inf_{p1} and Inf_{p2} are shown below:

$$inf_{p_1}(S) = \sum_{v \in S} inf_{p_1}(v) \tag{7}$$

$$inf_{p_2}(S) = \sum_{v \in inf_{p_1}(S)} \frac{inf_{v,a}}{\theta_a} \tag{8}$$

Where node a is a node in the *inactive* or *thinking* state, $inf_{v,a}$ represents the probability of node v propagating to node a, and θ_a represents the activation threshold for node a to be activated by the entity represented by node v.

After that, we built a CELF table based on the dual-influence of candidate seeds, and each time choose to add to the seed set S_2 the node with the highest dual-influence. It should be emphasized that we initially check the node with the current maximum dual-influence while choosing seeds, and if it belongs to the same community as the nodes in the current seed set, we update its dual-influence value and iteration flag $iter_n$ in the CELF table. The selected node is added to the seeds only when its iteration flag is equal to the current iteration number, which is showed from Line 12 to Line 19. When the node with the current maximum dual-influence does not belong to the same community as the nodes in the current seed set, it is directly added to the seeds. Repeat the above procedure until the same number of seeds as S1 has been selected.

3   Experiment

3.1   Dataset

We validate the algorithm on four real social networks, namely Wiki-Vote, CA-HepPh, Slashdot, and Amazon. The relevant information of the four datasets is shown in Table 1, where #nodes and #edges respectively represent the number of nodes and edges in the dataset and ave-degree represents the average degree of nodes in the dataset. All four datasets are obtained from the website[1]

Table 1. Real-world datasets

	Wiki-Vote	CA-HepPh	Slashdot	Amazon
#nodes	7115	12008	82168	262111
#edges	103689	118521	948464	1234877
ave-degree	29.11	19.71	23.06	2.77

3.2   Experiment Design

In this paper, we set the additional propagation influence θ_t to 0.5 for an entity to activate a node that has a lower demand for it. The in-degree of each node is defined as $in-degree$, and the propagation probability of each edge is $\frac{1}{in-degree}$. In addition, we stipulate that each node which enters the *thinking* state will select the entity it receives after 7 timestamps and propagate its influence [4]. We will confirm the viability and effectiveness of the algorithm through the following experiments:

Experiment 1 (Spread Influence): We stipulate the method of selecting seeds and the number of seeds for u_1. Here, we use the high-degree algorithm [9] to select the seeds for u_1. Then we compare the spread influence generated by the same number of seeds selected by u_2, which we present, using DIACS algorithm and other different seeds selection algorithms.

Experiment 2 (Running Time): Similarly, we use high-degree algorithm to select the seeds for u1 and set the number of seeds as u1. Then we compare the running time required for seeds selection of the DIACS algorithm with that of other algorithms.

We select the following algorithms for comparative experiments:

Random Algorithm, which randomly selects seeds.

High-Degree Algorithm, which selects the top-ranked nodes based on their degree as seeds.

PageRank Algorithm [9] , which selects the top-ranked nodes based on their PageRank values as seeds.

---

[1] http://snap.stanford.edu/.

DegreeDiscount Algorithm [5] , which measures the influence of nodes based on their degree, calculates the DegreeDiscount value for each node by averaging the degrees of its neighbors, and selects the top-ranked nodes based on their DegreeDiscount values as seeds.

CI2 Algorithm [4] , which first partitions the social network into communities and subsequently utilizes a local greedy algorithm to identify the node with the maximum marginal benefit in each community. It selects the node with the highest marginal benefit among all communities as the seed and updates the node with the maximum marginal benefit in the selected community.

3.3   Experiment Results

Spread Influence. As we can see in Fig. 2, overall, the DIACS algorithm consistently achieves propagation results comparable to the CI2 algorithm, and even outperforms the CI2 algorithm in most cases on the four datasets. The CI2 algorithm is a simple greedy algorithm based on community structure, and its theoretically achievable results are already locally optimal. Furthermore, the DIACS algorithm also performs better in terms of spread influence than other heuristic algorithms.

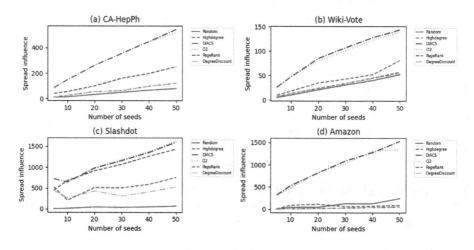

Fig. 2. Spread influence of different algorithms to select seeds on four datasets

As the Table 1 shows, the CA-HepPh and Wiki-Vote datasets have around ten thousand nodes, with an average degree of up to twenty or thirty. As shown in Fig. 2(a)(b), as the number of seeds increases, the propagation influence generated by the seeds selected by the DIACS algorithm is increasingly higher compared to other algorithms. This indicates that when the node average degree of the dataset is high, the dual-influence function we used can achieve better propagation influence when there are more seeds. However, Fig. 2(c) shows that in the Slashdot dataset, which also has a high average node degree, the DIACS algorithm can only achieve propagation influence results similar to or even lower than the CI2 algorithm. We believe that the reason is the number of selected seeds is not sufficient, resulting in the average performance of the dual-influence function. When the dataset, such as Amazon, has many nodes but a low average degree, the activation of a node is often affected by the difference in the probability of propagation on a single edge. This makes the second influence of the dual-influence function more important. Therefore, as shown in Fig. 2(d), when the number of seeds is 50, the DIACS algorithm's propagation influence is far superior to other algorithms. Therefore, the DIACS algorithm demonstrates certain advantages in terms of spread influence.

Running Time. As shown in the 3, although the greedy phase of the DIACS algorithm results in a similar growth rate of running time to the CI2 algorithm that uses a simple greedy algorithm, the DIACS algorithm is still always faster than the CI2 algorithm, which achieves similar propagation results in four datasets. Specifically, the DIACS algorithm is between $\frac{1}{2}$ and $\frac{1}{3}$ faster than the CI2 algorithm, indicating that the quality of candidate seeds selected during the filter stage of the DIACS algorithm is superior.

The DIACS algorithm demonstrates superiority as it achieves comparable or even better propagation results than the CI2 algorithm while being faster.

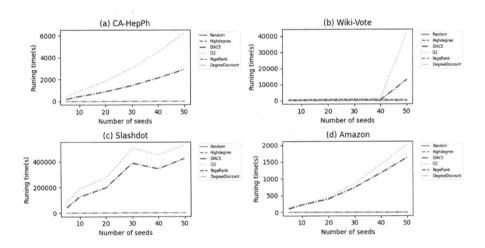

Fig. 3. Running time of different algorithms to select seeds on four datasets

4   Conclusion

In this paper, we study the competitive influence maximization problem considering user demand and dual influence, and propose a competitive influence diffusion model named CLTMcUD that takes user demand into account to simulate the competition of dual entities under different user preferences. Then, we provide a two-stage dual influence algorithm named DIACS based on community structure. This algorithm first selects candidate seeds in the filter stage after partitioning the social network into non-overlapping communities, and then chooses the final seeds in the greedy stage. Finally, we validate the effectiveness and feasibility of the algorithm on 4 real-world datasets, demonstrating that our approach surpasses currently available state-of-the-art approaches.

Acknowledgment. This work is supported by National Natural Science Foundation of China (No. 62172124). It was also supported by the Shenzhen Basic Research Program (Project No. JCYJ20190806143011274).

References

1. Ali, K., Wang, C.Y., Chen, Y.S.: Leveraging transfer learning in reinforcement learning to tackle competitive influence maximization. Knowl. Inf. Syst. **64**(8), 2059–2090 (2022)
2. Bharathi, S., Kempe, D., Salek, M.: Competitive influence maximization in social networks. In: Internet and Network Economics: Third International Workshop, WINE: San Diego, CA, USA, December 12–14, 2007. Proceedings 3. Springer, Heidelberg, pp. 306–311 (2007)
3. Blondel, V.D., Guillaume, J.L., Lambiotte, R., et al.: Fast unfolding of communities in large networks. J. Stat. Mech: Theory Exp. **2008**(10), P10008 (2008)
4. Bozorgi, A., Samet, S., Kwisthout, J., et al.: Community-based influence maximization in social networks under a competitive linear threshold model. Knowl.-Based Syst. **134**, 149–158 (2017)
5. Chen, W., Wang, Y., Yang, S.: Efficient influence maximization in social networks. In: Proceedings of the 15th ACM SIGKDD International Conference on Knowledge Discovery and Data Mining, pp. 199–208 (2009)
6. Guo, J., Wu, W.: Influence maximization: seeding based on community structure. ACM Trans. Knowl. Disc. Data (TKDD) **14**(6), 1–22 (2020)
7. He, X., Song, G., Chen, W., et al.: Influence blocking maximization in social networks under the competitive linear threshold model. In: Proceedings of the 2012 SIAM International Conference on Data Mining. Society for Industrial and Applied Mathematics, pp. 463–474 (2012)
8. Huang, H., Meng, Z., Shen, H.: Competitive and complementary influence maximization in social network: a follower's perspective. Knowl.-Based Syst. **213**, 106600 (2021)
9. Kempe, D., Kleinberg, J., Tardos, É.: Maximizing the spread of influence through a social network. In: Proceedings of the Ninth ACM SIGKDD International Conference on Knowledge Discovery and Data Mining, pp. 137–146 (2003)
10. Krömer, P., Nowaková, J.: Guided genetic algorithm for the influence maximization problem. In: Computing and Combinatorics: 23rd International Conference, COCOON: Hong Kong, China, August 3–5, 2017, Proceedings 23. Springer International Publishing, pp. 630–641 (2017)
11. Leskovec, J., Krause, A., Guestrin, C., et al.: Cost-effective outbreak detection in networks. In: Proceedings of the 13th ACM SIGKDD International Conference on Knowledge Discovery and Data Mining, pp. 420–429 (2007)

12. Li, Y., Fan, J., Wang, Y., et al.: Influence maximization on social graphs: a survey. IEEE Trans. Knowl. Data Eng. **30**(10), 1852–1872 (2018)
13. Liang, Z., He, Q., Du, H., et al.: Targeted influence maximization in competitive social networks. Inf. Sci. **619**, 390–405 (2023)
14. Naderipour, M., Fazel Zarandi, M.H., Bastani, S.: Fuzzy community detection on the basis of similarities in structural/attribute in large-scale social networks. Artif. Intell. Rev. **55**, 1373–1407 (2021). https://doi.org/10.1007/s10462-021-09987-x
15. Qiu, L., Jia, W., Yu, J., et al.: PHG: a three-phase algorithm for influence maximization based on community structure. IEEE Access **7**, 62511–62522 (2019)
16. Xie, X., Li, J., Sheng, Y., et al.: Competitive influence maximization considering inactive nodes and community homophily. Knowl.-Based Syst. **233**, 107497 (2021)

Practical Attribute-Based Multi-keyword Search Scheme with Sensitive Information Hiding for Cloud Storage Systems

Jie Zhao[1], Hejiao Huang[1]($\boxtimes$), Yongliang Xu[2], Xiaojun Zhang[3], and Hongwei Du[1]

[1] School of Computer Science and Technology, Harbin Institute of Technology (Shenzhen), Shenzhen 518055, China
zhaojswpu2017@163.com, {Huanghejiao,hwdu}@hit.edu.cn
[2] School of Mathematics and Statistics, Fuzhou University, Fuzhou 350108, China
xylwork@yeah.net
[3] School of Computer Science, Research Center for Cyber Security, Southwest Petroleum University, Chengdu 610500, China
zhangxjdzkd2012@163.com

Abstract. Attribute-based multi-keyword search (ABMKS) facilitates searching with fine-grained access control over outsourced ciphertexts. However, two critical issues impede wide application of ABMKS. Firstly, the majority of ABMKS schemes have suffered huge computation and communication costs in the process of ciphertexts matching and transmission. Secondly, the contents of data file containing sensitive information are encrypted as a whole, and data users with varying roles should have different access rights to the ciphertext returned by cloud, thereby preventing sensitive information in data files from being leaked to semi-trusted data users. In this paper, we tackle the issue of content access rights by introducing sensitive information hiding, a novel concept in the field of attribute-based keyword search. Specifically, we propose a practical multi-keyword search scheme with sensitive information hiding by integrating a modified blindness filtering technique into ciphertext policy attribute-based encryption under the multi-keyword search model. To minimize communication costs in the ciphertext transmission process, we utilize a super-increasing sequence to aggregate multiple blinding data blocks into a single ciphertext. The ciphertext can be recovered by using a recursive algorithm. Security analysis proves that our scheme is provably secure within the random oracle model, it guarantees keyword secrecy and selective security against chosen-keyword attacks. Performance evaluations demonstrate that our scheme surpasses state-of-the-art ABMKS schemes, making it highly suitable for cloud storage systems.

Keywords: Attribute-based multi-keyword search · Sensitive information hiding · Blindness filtering technique · Super-increasing sequence · Cloud storage systems

© The Author(s), under exclusive license to Springer Nature Switzerland AG 2024
W. Wu and J. Guo (Eds.): COCOA 2023, LNCS 14462, pp. 190–202, 2024.
https://doi.org/10.1007/978-3-031-49614-1_14

1   Introduction

Cloud computing has gained significant attention for its notable benefits like extensive storage capacity and flexible resource management [1]. However, the emergence of security and privacy issues has raised concerns among users, hindering their enjoyment of cloud storage and computing services [2]. To deter potential adversary attacks from posing a threat to data privacy, users often encrypt their data before uploading it to the cloud servers. While ensuring data confidentiality is crucial in cloud, encryption mechanisms inherently impose certain limitations on data availability [3,4].

Searchable encryption (SE) [5] enables users to search encrypted data files in the cloud based on user-specified keywords, just like searching on plaintext datasets. The original idea of SE was introduced by Song et al. [6], where symmetric keys were employed to construct keyword indexes and search trapdoors. Goh [7] formalized a security model for symmetric SE and proposed a secure searchable symmetric encryption scheme by leveraging a Bloom filter within this defined framework. Wang and Cao [8] proposed an effective order-preserving symmetric encryption mechanism specifically designed to facilitate ranked-keyword search. To address the secret key distribution problem inherent in symmetric SE, a new concept about public-key encryption based on keyword search (PEKS) was proposed by Boneh et al. [9]. Building upon the work of Boneh et al. [9], Cui et al. [10] proposed a novel approach to key-aggregate searchable encryption, which supports group data files sharing. Zhang and Huang [11] developed a biometric identity-based keyword search scheme using lattice-based technique over outsourced ciphertext, enhancing the security of ciphertext retrieval. Nevertheless, most of the previous SE schemes primarily operate in a "one-to-one" search mode and lack support for more expressive data sharing.

Fortunately, the attribute-based encryption (ABE) mechanism was developed by Sahai et al. [12], which provides a novel "one-to-many" ciphertext sharing paradigm for cloud users. To realize keyword search with fine-grained access control over ciphertext, Su et al. [13] put forward an attribute-based keyword search (ABKS) scheme in cloud storage. In ABKS, the access structure is determined by the data owner (DO) and embedded into the ciphertext. Only when the attribute set of data user (DU) satisfies the requirements of the access structure and the keyword within the search trapdoor matches the keyword in the index, the DU can get the relevant search results. Subsequently, Bao et al. [2], Zheng et al. [14], and Huang et al. [15] put forward an attribute-based keyword search scheme with distinct functional characteristics in the cloud-assisted healthcare systems. Most existing ABKS schemes lack the capability to facilitate multi-keyword search. This limitation arises from the insufficient power of a single keyword, resulting in an abundance of irrelevant search results and reducing search efficiency. To address this issue, Chen et al. [16] introduced a dual server model for attribute-based multi-keyword rank search. By combining a modified homomorphic MAC and a conjunctive keyword technique, Wen et al. [17] constructed a scheme for security searching multi-keyword search while maintaining privacy. Miao and Ma [18] presented a multi-keyword search scheme with attribute comparisons using 0-encoding and 1-encoding. However, these schemes incur huge communication and computation costs in the process of message transmission and ciphertext matching. Thus, how to design an effective multi-keyword searchable encryption mechanism in cloud storage systems is a crucial demand [19].

In addition, although existing ABKS schemes [3,14,20,21] have achieved "one-to-many" or "many-to-many" search mode with cloud access control, they mainly focus on matching user queries with ciphertexts encrypted by multiple distributed data users, overlooking the management of content rights over the returned ciphertext for data users with different roles. In some applications, data files owned by the DO often contain individual sensitive information that necessitates distinct access privileges for various data users [5]. For example, in a cloud-assisted healthcare sharing system, the patient acts as the data owner, doctors serve as the full-trusted (internal) searchers, and researchers as the semi-trusted (external) searchers. Nonetheless, considering the varying data user rights, the patient requires that doctors have access to the entire EMR, including sensitive information. On the contrary, researchers should only be granted the authority to access these shared contents within the EMR, while excluding sensitive information. Wu and Srivastava [22] put forward two structural models based on genetic algorithm, which filter user's sensitive information based on different thresholds of sensitive framework, but they do not realize fine-grained access control of outsourced ciphertext. Hence, it becomes urgent to develop effective methods for implementing a two-layer access control mechanism in ABKS, thereby ensuring comprehensive content rights management.

To address outsourced ciphertexts sharing issues and provide aforementioned functionalities, in this paper, we proposed ABMKS-SIH, a practical scheme for attribute-based multi-keyword search with sensitive information hiding in cloud storage systems. The main contributions of this work are elaborated below:

1. We combine a modified blindness filtering technique and block ciphers to realize the sensitive information hiding and other non-sensitive content sharing in data files. Any data user with a sufficient number of valid attributes can execute the keyword search operation of outsourced ciphertext by using the extracted keyword set, but only full-trusted DUs are granted access to the sensitive information in response ciphertext.
2. ABMKS-SIH leverages a super-increasing sequence to aggregate multiple blinding data blocks into a single ciphertext. This approach not only significantly reduces the communication overhead between two logical entities in the system but also facilitates the rapid recovery of the original data files by using a recursive algorithm. As such it enhances the overall search experience for users.
3. We formalize the security model of ABMKS-SIH and provide its corresponding security proofs based on the hardness assumption of decisional Diffie-Hellman (DDH) within the random oracle model. ABMKS-SIH guarantees keyword secrecy and selective security against chosen-keyword attack [14].

2   Problem Formulation and Preliminaries

2.1   System Model

The system model is depicted in Fig. 1 comprises the trusted authority (TA), data owner (DO), data users (DUs), and cloud server (CS). TA is in charge of determining public parameters $Para$ and master secret key Msk, and assigns the $Para$ to other entities in the system. According to the attribute set S of DU, it can generate an attribute key SK. DO is the original data owner who has numerous data files to outsource to CS. To enhance the data privacy protection and ensure efficient ciphertext search, DO first hides the sensitive information in data file, then constructs a ciphertext index based on the extracted keyword set and predefined access policies. Finally, DO uploads the storage data (i.e., ciphertext and secure index) to CS. DUs is the collective term for all data searchers in the system, including the internal searchers and external searchers. A new data user with attribute set S registers with TA to obtain the corresponding attribute key SK. When executing the ciphertext retrieval in the cloud, he/she can generate a secure search trapdoor based on his/her interests and send it to CS. CS, managed by the cloud server provider, provides cloud users with convenient storage services and rich computing resources. Upon receiving the search trapdoor from DUs, CS checks whether the DUs have permission to access the storage data of DO. If not, CS outputs ⊥; otherwise, it executes the ciphertext matching operation and responds the corresponding search results to DUs.

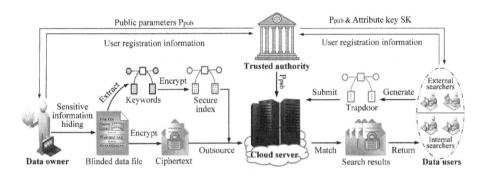

Fig. 1. System model of ABMKS-SIH.

2.2   Threats Model and Design Goals

According to the proposed ABMKS-SIH system model, we mainly consider threats from three different several angles: honest-but-curious CS and semi-trusted internal searchers. The CS follows the agreed-upon protocol to perform the ciphertext retrieval tasks, but it is awfully curious about the keywords over secure indexes or the keywords over search trapdoors. The external searchers may try to derive the sensitive information about the DO from the search results. Even worse, they impersonate other DUs to learn valuable information about the target DUs, seeking unauthorized data access.

The proposed ABMKS-SIH scheme seeks to reach the following goals: (1) *Data Confidentiality*. It should prevent the CS and other malicious DUs from accessing the content of outsourced storage data. The external searchers should not learn sensitive information about the DO from the response search results. (2) *Keyword Security*. The CS should not have the ability to identify keywords from the secure index or search trapdoors, nor link secure index or search trapdoors sent by the same DO/DUs, even if they have the same keywords. (3) *Dual Access Control*. The DO can define fine-grained access policies for the outsourced data file with sensitive information hiding. CS can execute the task of multi-keyword search matching if the attributes of DUs meet DO's access policies. External searchers only have permission to access the shared content on the data file, excluding the sensitive information. (4) *High Performance*. Ensuring high performance of ABMKS-SIH has always been a pressing demand for cloud storage systems. Particularly, maintaining an appropriate trapdoor size, regardless of the number of searching keywords, will minimize the network delay between CS and DUs. The Computation costs of ciphertext search matching and decryption should be kept as low as possible.

2.3   Preliminaries

Bilinear Pairing. Given two multiplicative cyclic groups $\mathbb{G}$ and $\mathbb{G}_T$, both having a prime order q. The generator of $\mathbb{G}$ is denoted as g. Let $\hat{e}$ be a bilinear pairing $\hat{e} : \mathbb{G} \times \mathbb{G} \rightarrow \mathbb{G}_T$, which should satisfies: *Bilinearity*: $\forall a, b \in \mathbb{Z}_q$, $\hat{e}(g^a, g^b) = \hat{e}(g, g)^{ab}$; *Non-degeneracy*: $\hat{e}(g, g) \neq 1$; *Computability*: $\hat{e}$ can be effectively calculated.

Decisional Diffie-Hellman (DDH) Assumption. Given a quintuple $(g, g^a, g^b, g^{ab}, g^\chi) \in \mathbb{G}^5$ for unknown random values $a, b, \chi \in \mathbb{Z}_q^*$, the goal of DDH problem in the group $\mathbb{G}$ is to distinguish whether g^{ab} or g^χ from the quintuple. For any probabilistic polynomial-time (PPT) adversary $\mathcal{A}$, the probability of figuring out the DDH problem is considered negligible, This can be expressed as $Adv_{\mathbb{G},\mathcal{A}}^{DDH}(\varsigma) = |Pr[\mathcal{A}(q, \mathbb{G}, g, g^a, g^b, g^{ab}) = 1] - Pr[\mathcal{A}(q, \mathbb{G}, g, g^a, g^b, g^\chi) = 1]| \leq \epsilon$, where ϵ is a negligible advantage.

Super-Increasing Sequence. Super-increasing sequence is a key technique for handling multi-dimensional data. It is a sequence of positive integers $\{\rho_1 = 1, \rho_2, \cdots, \rho_n\}$, $\forall i, j \in [1, n], i < j, \rho_j > \sum_{i=1}^{j-1} \rho_i$. Here, each element ρ_i $(1 \leq i \leq n)$ surpasses the cumulative total of all preceding elements within the sequence.

Access Policy Tree. An access tree can embody the access structure $\mathcal{T}$ [20]. This structure involves two types of nodes: leaf nodes *lns* and non-leaf nodes *nlns*, Each leaf node $x in lns$ within $\mathcal{T}$ is linked to an attribute $att(x)$, and the non-leaf node $\nu \in nlns$ denotes a threshold gate (k_ν, Num_ν). Here, $k_\nu \in [1, n]$ denotes the threshold value of ν and Num_ν indicates the number of children in ν. When $k_\nu = 1$, the threshold gate function as an *Or* gate, whereas it behaves as an *And* gate when $k_x = Num_x$. For each inner node ν in the $\mathcal{T}$, it generates a polynomial Γ_ν with a degree $k_\nu - 1$ from top to bottom. If the ν is the $\mathcal{T}$'s root node R, $\Gamma_{\nu=R}(0) = s$, where s is a secret value chosen for the $\mathcal{T}$. Otherwise, $\Gamma_\nu(0)$ is set to $\Gamma_{parent(\nu)}(index(\nu))$. Here, $parent(\nu)$ denotes the parent of ν, and $index(\nu)$ represents the sequence number associated with ν in the $\mathcal{T}$. Let $\Gamma(S, \mathcal{T}_\nu) = 1$ indicate that the attribute set S of $att(x)$ meets the access policies of subtree $\mathcal{T}_\nu$. For a non-leaf node ν, at least k_x children return 1 if $\Gamma(S, \mathcal{T}_\nu) = 1$ holds by recursive calculation [3]. For a leaf node x, if $\Gamma(S, \mathcal{T}_x) = 1$ holds, than $att(x) \in S$. As a result, a data user DU with the attribute set S should satisfy the minimum requirements of the $\mathcal{T}$ when he/she intends to access a ciphertext generated based on access policy.

3   The Proposed ABMKS-SIH Scheme

Now we describe the concrete construction of ABMKS-SIH, consisting of a combination of five algorithms described as follows.

Setup$(1^\varsigma) \rightarrow (P_{pub}, Msk)$: Input a security parameter ς, TA define a bilinear mapping $\widehat{e}: \mathbb{G} \times \mathbb{G} \longrightarrow \mathbb{G}_T$, where $\mathbb{G}$ and $\mathbb{G}_T$ are two multiplicative cyclic groups with the same prime order q. Here, g serves as a generator in $\mathbb{G}$. TA selects three distinct values $\alpha, \beta, \varpi \leftarrow \mathbb{Z}_q^*$ and proceeds to calculate $A = \widehat{e}(g, g)^\alpha$, $B = g^\beta$, and $W = g^{\alpha\varpi}$. Then, it determines three collision-resistant functions: $H_0 : \{0,1\}^* \longrightarrow \mathbb{G}$, $H_1 : \mathbb{G}_T \longrightarrow \mathbb{Z}_q^*$, and $H_2 : \{0,1\}^* \longrightarrow \mathbb{Z}_q^*$, respectively. TA defines a pseudorandom function $Prf : \{1, 2, \cdots, n\} \times \mathbb{Z}_q^* \rightarrow \mathbb{Z}_q^*$ and selects a secure symmetric encryption algorithm pair Enc/Dec. TA generates a super-increasing sequence with n positive integers $\{\nu_1 = 1, \nu_2, \cdots, \nu_n\}$. As for $\forall \imath, \jmath \in [1, n], \jmath > \imath$, the coefficients need to satisfy $q > \nu_\jmath > \sum_{\imath=1}^{\jmath-1} \nu_\imath \cdot \varphi_\imath \cdot n$, where $\varphi_\imath$ denotes the upper bound of the $\imath$-th data block. Finally, TA sets $\widetilde{\mu} = \{\nu_1, \nu_2, \cdots, \nu_n\}$, and issues all public parameters P_{pub} and saving the master secret key Msk as:

$$\left(\begin{array}{c} Para = \{(\widehat{e}, \mathbb{G}, \mathbb{G}_T, g, q), A, B, W, H_0 \sim H_2, Prf, Enc/Dec, \widetilde{\mu}\}, \\ Msk = (g^\alpha, \beta, \varpi) \end{array} \right). \quad (1)$$

KeyGen$(Para, Msk, S) \rightarrow SK$: TA takes the public parameters $Para$, master secret key Msk, and an attribute set S of DUs as inputs, it then generates the attribute key SK for the DUs using the Eq. (2). Particularly, the TA needs to randomly select a value $r \leftarrow \mathbb{Z}_q^*$ for DUs. As for each attribute value $j \in S$ of the DUs, it chooses a number $r_j \leftarrow \mathbb{Z}_q^*$ randomly. Finally, TA returns the attribute key $SK = (K_1, K_2, \forall j \in S : K_j, K_j')$ to the DUs by a secure channel.

$$SK = \left(\begin{array}{c} K_1 = g^{\alpha(1+r)}, \ K_2 = g^{\alpha(r-\beta\varpi)}, \\ \{\forall j \in S : K_j = g^{\alpha r} \times H_0(j)^{r_j}, \ K_j' = g^{r_j}\} \end{array} \right). \quad (2)$$

IndexGen$(Para, F, Prf, \kappa, \widetilde{\mu}, \mathcal{T}, Enc) \rightarrow \Omega$: The DO takes the public parameters $Para$, a data file F, the pseudorandom function Prf with a seed secret key κ, the super-increasing sequence $\widetilde{\mu}$, a certain access policy $\mathcal{T}$, and a symmetric encryption algorithm Enc as inputs. Then it generates the secure index of F by executing the following algorithm steps.

1. Given a data file $F = \{f_1, f_2, \cdots, f_n\}$ with the file identifier $Fname \in \mathbb{Z}_q^*$, the DO first blinds these data blocks corresponding to sensitive information in F by computing the following Eq. (3).

$$f_i^* = f_i \oplus \xi_i, \quad \xi_i \leftarrow Prf_\kappa(i, Fname), \tag{3}$$

where $f_i^* \neq f_i$ if and only if $i \in [1, n]$ and $i \in \overline{K}$. Here, $\overline{K}$ represents the index set corresponding to the sensitive information of F; otherwise, $f_i^* = f_i$. The κ is shared between DO and internal searchers. Finally, the data file $F = \{f_1, f_2, \cdots, f_n\}$ is blinded as $F^* = \{f_1^*, f_2^*, \cdots, f_n^*\}$.

2. To minimize the communication cost in the process of data file outsourcing, the DO exploits the super-increasing sequence $\tilde{\mu}$ to aggregate the n blinding data blocks into a single ciphertext value $\widehat{F}$ below:

$$\widehat{F} = \nu_1 f_1^* + \nu_2 f_2^* + \cdots + \nu_n f_n^* = \sum_{i=1}^{i=n} \nu_i f_i^* \in \mathbb{Z}_q^*. \tag{4}$$

3. The DO selects a random element $\partial \leftarrow \mathbb{G}_T$ and computes $\zeta = H_1(\partial)$. Then, it chooses a random value $s_0 \leftarrow \mathbb{Z}_q^*$ as the secret value of root node R in $\mathcal{T}$ and encrypts $\widehat{F}$ as:

$$C = Enc_\zeta(\widehat{F}), \quad C_1 = g^{s_0}, \quad C_2 = \partial \cdot A^{s_0} = \partial \cdot \widehat{e}(g, g)^{\alpha s_0}. \tag{5}$$

4. Let the leaf node set as Y in access policy $\mathcal{T}$, for each leaf node $y \in Y$, the DO calculates the ciphertext as:

$$\forall y \in Y : C_y = g^{\Gamma_y(0)}, \quad C_y' = H_0(att(y))^{\Gamma_y(0)}, \tag{6}$$

5. The DO extracts the keyword set $KW = \{kw_1, kw_2, \cdots, kw_m\}$ from the binding data file F^*. After that, it generates the secure index $I_{F^*} = \{I_1, I_2, \cdots, I_m\}$ by computing the following Eq. (7).

$$\forall kw_\ell \in KW : I_\ell = B^{s_0} \cdot g^{H_2(kw_\ell)} = g^{\beta s_0 + H_2(kw_\ell)}. \tag{7}$$

Finally, the DO sets $C_{F^*} = \{C, C_1, C_2, \forall y \in Y : C_y, C_y'\}_{Fname}$ and uploads the storage data $\Omega = (C_{F^*}, I_{F^*})$ to CS.

TrapGen$(Para, SK, KW') \rightarrow TK$: The DUs input the public parameters $Para$, the attribute key SK, and a queried keyword set $KW' = \{kw_1', kw_2', \cdots, kw_v'\}$, where $KW' \subseteq KW$. Then, it selects a random value $\eta \leftarrow \mathbb{Z}_q^*$ and outputs a search trapdoor TK by computing the following Eq. (8).

$$TK = \begin{pmatrix} \widehat{D} = \prod_{i=1}^{i=\ell} g^{\eta H_2(kw_i')}, D_1 = K_2^\tau = g^{\alpha(r - \beta \varpi)\eta}, D_2 = W^\eta = g^{\alpha \varpi \eta}, \\ \{\forall j \in S : D_j = K_j^\eta = g^{\alpha r \eta} \cdot H_0(j)^{\eta r_j}, D_j' = K_j'^\eta = g^{\eta r_j}\} \end{pmatrix}. \tag{8}$$

Finally, DUs submit the search trapdoor TK to the CS and keeps η in secret.

CiphertextSearch$((Para, \Omega, TK) \rightarrow \Lambda)$: The CS inputs the public parameters $Para$, DUs's search trapdoor TK, and the storage data Ω. Then, it checks whether the attribute set S of DUs satisfies the access tree $\mathcal{T}$. If not, the CS suspends the ciphertext search testing process and outputs $\bot$; otherwise, it outputs the search result Λ by executing the following algorithm steps.

1. Let y represent a leaf node in $\mathcal{T}$, with j denoting the attribute associated with y. The CS assigns j the attribute of y by setting $j = att(y)$. If $j \in S$, an $Error$ is emitted, resulting in $E_y = \bot$; otherwise, the CS calculates E_y as

$$E_y = \frac{\widehat{e}(D_j, C_y)}{\widehat{e}(D_j', C_y')} = \widehat{e}(g,g)^{\alpha r \eta \Gamma_y(0)}. \tag{9}$$

2. If y is a non-leaf node within $\mathcal{T}$ and x is one of its child nodes, the CS can compute $E_x = \widehat{e}(g,g)^{\alpha r \eta \Gamma_x(0)}$. Let S_y be a random set of child nodes x with a size of z, such that $E_x \neq \bot$. In case there is no such set, CS outputs $\bot$ by emitting an $Error$; otherwise, it computes E_y as

$$E_y = \prod_{x \in S_y} E_x^{\Delta_{z,S_y'}(0)} = \prod_{x \in S_y} \left(\widehat{e}(g,g)^{\alpha r \eta (q_{parent(y)}(index(y)))}\right)^{\Delta_{z,S_y'}(0)}$$
$$= \prod_{x \in S_y} \left(\widehat{e}(g,g)^{\alpha r \eta q_y(z)}\right)^{\Delta_{z,S_y'}(0)} = \widehat{e}(g,g)^{\alpha r \eta \Gamma_y(0)}, \tag{10}$$

where $\Delta_{j,S_y'}(0)$ denotes the Lagrange coefficient, $z = index(x)$ indicates that x is the z-th child node of y, and $S_y' = \{index(x)\}_{\forall x \in S_y}$.

3. Only when the attribute set S meets the requirement of $\mathcal{T}$, the root node R can be computed as $E_R = \widehat{e}(g,g)^{\alpha r \eta \Gamma_R(0)} = \widehat{e}(g,g)^{\alpha r \eta s_0}$. CS performs the ciphertext search testing by verifying

$$\widehat{e}(D_1, C_1) \cdot \widehat{e}(\prod_{i=1}^{i=\ell} I_i, D_2) \stackrel{?}{=} E_R \cdot \widehat{e}(\widehat{D}, W). \tag{11}$$

If the above ciphertext search testing Eq. (11) does not hold, it means that no valid ciphertext is found and the CS outputs $\bot$; otherwise, it returns the search result $\Lambda = \{C, C_1, C_2, E_R\}_{Fname}$ to the DUs.

CiphertextDecrypt$((Para, \psi, Dec, \eta, \widetilde{\mu}, Prf, \kappa) \to F^*/F)$: DUs utilize the following inputs: public parameters $Para$, search result ψ, symmetric decryption algorithm Dec, super-increasing sequence $\widetilde{\mu}$, random value η, and pseudorandom function Prf with a seed secret key κ. It outputs the blinding data file F^* or original data file F by performing the following algorithm steps.

1. The DUs recover the random element ∂ by computing

$$C_2 \cdot \frac{(E_R)^{1/\eta}}{\widehat{e}(K_1, C_1)} = \partial \cdot \widehat{e}(g,g)^{\alpha s} \cdot \frac{(\widehat{e}(g,g)^{\alpha r \eta s})^{1/\eta}}{\widehat{e}(g^{\alpha(1+r)}, g^s)} = \partial. \tag{12}$$

2. The DUs compute $\zeta = H_1(\partial)$ and make it as the secret key of symmetric decryption algorithm. After that, it recovers the $\widehat{F}$ by computing

$$\widehat{F} = Dec_\zeta(C) = Dec_\zeta(Enc_\zeta(\widehat{F})). \tag{13}$$

3. Based on the $\widehat{F}$ and the super-increasing sequence $\widetilde{\mu} = \{\nu_1, \nu_2, \cdots, \nu_n\}$, the DUs can exploit the recursive method $f_i^* = (\widehat{F} - (\widehat{F} \bmod \nu_i))\nu_i^{-1}$ to retrieve $F^* = \{f_1^*, f_2^*, \cdots, f_n^*\}$.

4. Furthermore, if the DUs are the internal searchers, they can recover all of the original contents F from F^* by utilizing the pseudorandom function Prf with the secret seed key κ. Specifically, the internal searchers compute $\xi_i \leftarrow Prf_\kappa(i, Fname)$ and $f_i = f_i^* \oplus \xi_i$, where $i \in [1, n], i \in \overline{K}$. Hence, the blinding data file $F^* = \{f_1^*, f_2^*, \cdots, f_n^*\}$ is decrypted as $F = \{f_1, f_2, \cdots, f_n\}$.

4   Security Analysis of ABMKS-SIH

4.1   Security Model

The proposed ABMKS-SIH scheme presents a formal security model of the selective CKA game and keyword secrecy game.

4.2   Security Proofs

The security of the ABMKS-SIH scheme can be rigorously established through the following two theorems.

Theorem 1. *ABMKS-SIH demonstrates selective secure against chosen-keyword attacks under the DDH problem, where $H_0(\cdot)$ and $H_2(\cdot)$ are treated as a random oracle machine and a collision-resistant hash function, respectively.*

Theorem 2. *ABMKS-SIH guarantees keyword secrecy within the framework of the random oracle model. Within this model, $H_0(\cdot)$ and $H_2(\cdot)$ are treated as random oracle machine and a collision-resistant hash function, respectively.*

5   Comparison and Evaluation

In this section, we evaluate a comprehensive performance of our scheme with state-of-the-art ABKRS-KGA scheme [16] and PAB-MKS scheme [18] from the perspective of computation communication costs. The experimental simulations are conducted using the C language and the pairing-based cryptography (PBC) library[1]. All implementations are executed on an Ubuntu 22.04 Linux system (x86-64) with an Intel(R) Core(TM) i9-12900K CPU operating at 3200 MHz and 31.00 GB of RAM. To ensure accuracy, we average the results from 30 trials.

We first specify the following notations of cryptographic operation: T_{bp}, T_{ex}, T_{en}, T_{mu}, T_H, and T_h represent the execution time of a bilinear pairing, a modular exponentiation operation, a symmetric encryption operation, a multiplication

---

[1] The PBC library defines the Type A curve as $E(F_p) :< y^2 = x^3 + x >$. It includes two multiplicative cyclic groups $\mathbb{G}$ and $\mathbb{G}_T$, both of order q, which are subgroups of $E(Fp)$. Two large primes p and q have sizes of 512 bits and 160 bits, respectively.

operation, a hash-to-point operation, and a general hash operation, respectively. The bit length of elements in the groups $\mathbb{G}$ and $\mathbb{G}_T$ is denoted by $|G|$ and $|G_T|$, respectively. $|E|$ stands for the symmetric encryption algorithm such as AES-256. For comparison convenience, let's use the variables n, m, ℓ, θ, and ϑ to represent different quantities: $n = 10000$ stands for the number of data blocks in the data file F, $m = 50$ stands for the number of keywords in the index, $\ell = 20$ refers to the number of keywords in the trapdoor, $\theta \in [1, 100]$ refers to the number of attributes in $\mathcal{T}$, and $\vartheta \in [1, 50]$ stands for the number of attributes in the DU. Table 1 presents the computational costs of various schemes, while Fig. 2 (a)-(d) illustrates the corresponding implementation result of the computation costs comparison. Specifically, according to Fig. 2 (a), it can be observed that the computational costs of all schemes exhibit a linear increase as the number of submitted attributes for the *KeyGen* algorithm grows. Our scheme and PAB-MKS [18] is more efficient than ABKRS-KGA [16]. In Fig. 2 (b), our scheme does incur lower computational costs compared to the ABKRS-KGA [16] scheme but higher than PAB-MKS [18] when performing the *IndexGen*. This is because in our ABMKS-SIE scheme, the data blocks corresponding to sensitive information require a certain amount of time to undergo blinding encryption during the process. The implementation results of trapdoor generation are depicted in Fig. 2 (c), demonstrating the superior efficiency of ABMKS-SIE compared to other related schemes. From the Fig. 2 (d), we can get that computation costs are sensitive to the size of the submitted attribute ϑ, and the primary reason is that the size of ϑ determines the iteration times of CS.

Optimizing end-to-end network latency is a crucial factor that impacts the deployment of ABMKS-SIE. In this paper, we compare ABMKS-SIE along with related schemes such as ABKRS-KGA [16] and PAB-MKS [18] in terms of communication cost. Table 2 presents the communication costs of various schemes. The corresponding implementation results of the communication costs comparison are exhibited in Fig. 2 (e)-(f). For the Fig. 2 (e), we can obtain that ABKRS-KGA [16] and PAB-MKS [18] require over 100 times more communication cost than our scheme, when the leaf node is $\theta = 100$. This demonstrates

Table 1. The Computation Costs in Related Schemes

Algorithms	ABKRS-KGA [16]	PAB-MKS [18]	ABMKS-SIH
KeyGen	$(4T_{ex} + T_H + T_{mu})\vartheta$ $+ T_{ex}$	$(3T_{ex} + T_H + T_{mu})\vartheta$ $+ 2T_{ex} + 3T_{mu}$	$(3T_{ex} + T_H + T_{mu})\vartheta$ $+ 2T_{ex} + 2T_{mu}$
IndexGen	$(2T_{ex} + T_H + T_{mu})\theta +$ $(3T_{ex} + T_H + T_{mu})m$ $+ 2T_{ex}$	$(2T_{ex} + T_H)\theta +$ $(T_{ex} + T_H + T_{mu})m$ $+ 3T_{ex} + nT_{en}$	$(2T_{ex} + T_H)\theta +$ $(T_{ex} + T_h + T_{mu})m$ $+ T_{ex} + T_{bp} + T_{en}$
TrapGen	$T_{ex}\vartheta + 2T_{ex} +$ $(2T_{ex} + T_H + T_{mu})\ell$	$2T_{ex}\vartheta + 2T_{ex} +$ $(2T_{ex} + T_H + T_{mu})\ell$	$2T_{ex}\vartheta + 3T_{ex} + T_h\ell$
CipertextSearch	$(T_{bp} + T_{ex} + T_{mu})\vartheta$ $+ (2T_{bp} + 5T_{ex})\ell$ $+ 2T_{bp} + 2T_{ex}$	$(T_{bp} + T_{ex} + T_{mu})\vartheta$ $+ (T_{ex} + 2T_{mu})\ell$ $+ 7T_{bp} + 2T_{mu}$	$(T_{bp} + T_{ex} + T_{mu})\vartheta$ $+ (T_{ex} + T_{mu})\ell$ $+ 6T_{bp} + 2T_{mu}$

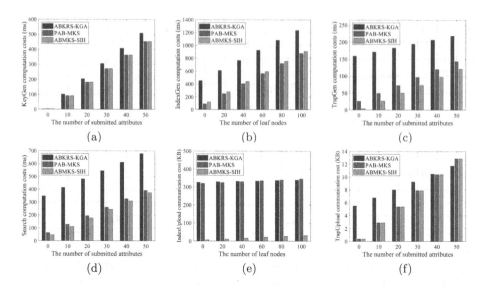

Fig. 2. The performance comparison between our scheme and related schemes: Computation costs of (a) *KeyGen*, (b) *IndexGen*, (c) *TrapGen*, (d) *Search*. Communication costs of (e) Secure Index Uploading, (f) Search Trapdoor Uploading.

that the super-increasing sequence indeed greatly reduces the communication cost of index outsourcing phase. From the Fig. 2(f), we can obtain that our scheme and PAB-MKS [18] are lower than ABKRS-KGA [16] when the number of submitted attributes is less than 40, however, as the number of attributes submitted by the DU continues to increase, the experimental results show the opposite trend. This is attributed to the fixed number of 20 keywords in the trapdoor that we maintained during the experimental simulations. Therefore, the proposed ABMKS-SIE scheme is still feasible in practical applications.

Table 2. The Communication Costs (Communi-Costs) in Related Schemes

Schemes	IndexUpload Communi-Costs	TrapUpload Communi-Costs
ABKRS-KGA [16]	$|G_1|\theta + |E|n + 2|G_2| + 2|G_1|m$	$|G_1|\vartheta + |G_2| + 2|G_1| + |G_2|\ell$
PAB-MKS [18]	$2|G|\theta + |E|n + 3|G| + |G|m$	$2|G|\vartheta + 3|G|$
ABMKS-SIH	$2|G|\theta + |E| + |G| + |G_T| + |G|m$	$2|G|\vartheta + 3|G|$

Notes. In the context of an asymmetric bilinear parting, the bit length of elements in groups $\mathbb{G}_1$ and $\mathbb{G}_2$ is denoted by $|G_1|$ and $|G_2|$, respectively.

6   Conclusion

In this paper, we proposed ABMKS-SIH, a practical scheme for attribute-based multi-keyword search with sensitive information hiding in cloud storage systems.

ABMKS-SIH enables data users multiple keywords to accurately retrieve the ciphertext with high efficiency in cloud. Meanwhile, it manages the content rights of the returned search ciphertext for data users with varying roles. Full-trusted searchers are granted the right to access the whole original data file including sensitive information. Conversely, the semi-trusted searchers are only allowed access to shared contents in data file excluding sensitive information. We conduct a formal security analysis and evaluate the performance of communication and computation costs with state-of-the-art schemes.

Acknowledgments. This work is supported by the Shenzhen Science and Technology Program under Grant No. GXWD20220817124827001, and No. JCYJ202-10324132406016.

References

1. Gan, Q., Wang, X., Huang, D., Li, J., Zhou, D., Wang, C.: Towards multi-client forward private searchable symmetric encryption in cloud computing. IEEE Trans. Serv. Comput. **15**(6), 3566–3576 (2022)
2. Bao, Y., Qiu, W., Tang, P.: Efficient, revocable, and privacy-preserving fine-grained data sharing with keyword search for the cloud-assisted medical IoT system. IEEE J. Biomed. Health Inform. **26**(5), 2041–2051 (2021)
3. Wang, M., Miao, Y., Guo, Y., Huang, H., Wang, C., Jia, X.: Aesm^2 attribute-based encrypted search for multi-owner and multi-user distributed systems. IEEE Trans. Parallel Distrib. Syst. **1**(34), 92–107 (2023)
4. Zhao, J., Zheng, Y., Huang, H., Wang, J., Zhang, X., He, D.: Lightweight certificateless privacy-preserving integrity verification with conditional anonymity for cloud-assisted medical cyberCphysical systems. J. Syst. Architect. **138**, 102860 (2023)
5. Xu, C., Wang, N., Zhu, L., Sharif, K., Zhang, C.: Achieving searchable and privacy-preserving data sharing for cloud-assisted E-healthcare system. IEEE Internet Things J. **5**(6), 8345–8356 (2019)
6. Song, D., Wagner, D., Perrig, A.: Practical techniques for searches on encrypted data. In: Proceedings of IEEE Symposium on Security and Privacy, pp. 44–55. IEEE, Berkeley (2000). https://doi.org/10.1109/SECPRI.2000.848445
7. Goh, E. J.: Secure indexes. Cryptology ePrint Archive, pp. 1–18 (2003)
8. Wang, C., Cao, N., Li, J., Ren, K., Lou, W.: Secure ranked keyword search over encrypted cloud data. In 2010 IEEE 30th International Conference on Distributed Computing Systems ICDCS, pp. 253–262. IEEE (2011). https://doi.org/10.1109/SECPRI.2000.848445
9. Boneh, D., Di Crescenzo, G., Ostrovsky, R., Persiano, G.: Public key encryption with keyword search. In: Cachin, C., Camenisch, J.L. (eds.) EUROCRYPT 2004. LNCS, vol. 3027, pp. 506–522. Springer, Heidelberg (2004). https://doi.org/10.1109/ICDCS.2010.34
10. Cui, B., Liu, Z., Wang, L.: Key-aggregate searchable encryption (KASE) for group data sharing via cloud storage. IEEE Trans. Comput. **65**(8), 2374–2385 (2015)
11. Zhang, X., Huang, C., Gu, D., Wang, H.: BIB-MKS: post-quantum secure biometric identity-based multi-keyword search over encrypted data in cloud storage systems. IEEE Trans. Serv. Comput. **16**(1), 122–133 (2023)

12. Sahai, A., Waters, B.: Fuzzy identity-based encryption. In: Cramer, R. (ed.) EURO-CRYPT 2005. LNCS, vol. 3494, pp. 457–473. Springer, Heidelberg (2005). https://doi.org/10.1007/11426639_27

13. Sun, W., Liu, X., Lou, W., Hou, Y. T., Li, H.: Catch you if you lie to me: efficient verifiable conjunctive keyword search over large dynamic encrypted cloud data. In 2015 IEEE Conference on Computer Communications (INFOCOM), pp. 2110–2118. IEEE (2015). https://doi.org/10.1109/INFOCOM.2015.7218596

14. Zheng, Q., Xu, S., Ateniese, G.: VABKS: verifiable attribute-based keyword search over outsourced encrypted data. In: IEEE INFOCOM 2014-IEEE Conference on Computer communications, pp. 522–530 (2014). https://doi.org/10.1109/INFOCOM.2014.6847976

15. Huang, Q., Yan, G., Yang, Y.: Privacy-preserving traceable attribute-based keyword search in multi-authority medical cloud. IEEE Trans. Cloud Comput. **11**(1), 678–691 (2023)

16. Chen, Y., Li, W., Gao, F., Wen, Q., Wang, H.: Practical attribute-based multi-keyword ranked search scheme in cloud computing. IEEE Trans. Serv. Comput. **2**(15), 724–735 (2022)

17. Wan, Z., Deng, R.H.: VPSearch: achieving verifiability for privacy-preserving multi-keyword search over encrypted cloud data. IEEE Trans. Depend. Secure Comput. **15**(6), 1083–1095 (2018)

18. Miao, Y., Ma, J., Liu, X., Li, X., Liu, Z., Li, H.: Practical attribute-based multi-keyword search scheme in mobile crowdsourcing. IEEE Internet Things J. **5**(4), 3008–3018 (2018)

19. Liu, J., Wu, M., Sun, R., Du, X., Guizani, M.: BMDS: a blockchain-based medical data sharing scheme with attribute-based searchable encryption. ICC 2021, IEEE International Conference on Communications, pp. 14–23. IEEE (2021). https://doi.org/10.1109/ICC42927.2021.9500966

20. Liu, Z., Liu, Y., Xu, J., Wang, B.: Privacy-preserving attribute-based multi-keyword search encryption scheme with user tracing. In: Vaidya, J., Zhang, X., Li, J. (eds.) CSS 2019. LNCS, vol. 11983, pp. 382–397. Springer, Cham (2019). https://doi.org/10.1007/978-3-030-37352-8_34

21. Yin, H., Zhang, W., Deng, H., Qin, Z., Li, K.: An attribute-based searchable encryption scheme for cloud-assisted IIoT. IEEE Internet Things J. **10**(12), 11014–11023 (2023)

22. Wu, J.M.T., Srivastava, G., Jolfaei, A., Fournier-Viger, P., Lin, J.C.W.: Hiding sensitive information in eHealth datasets. Futur. Gener. Comput. Syst. **1**(117), 169–180 (2021)

Testing Higher-Order Clusterability on Graphs

Yifei Li[1,2]([✉]) [ID], Donghua Yang[1] [ID], and Jianzhong Li[2] [ID]

[1] Harbin Institute of Technology, Harbin, Heilongjiang, China
yf.li@stu.hit.edu.cn, yang.dh@hit.edu.cn
[2] Shenzhen Institute of Advanced Technology, Chinese Academy of Sciences,
Shenzhen, China
lijzh@hit.edu.cn

Abstract. Analysis of higher-order organizations, usually small connected subgraphs called motifs, is a fundamental task on complex networks. This paper studies a new problem of testing higher-order clusterability: given query access to an undirected graph, can we judge whether this graph can be partitioned into a few clusters of highly-connected motifs? This problem is an extension of the former work proposed by Czumaj et al. (STOC' 15), who recognized cluster structure on graphs using the framework of property testing. In this paper, a good graph cluster on high dimensions is first defined for higher-order clustering. Then, query lower bound is given for testing whether this kind of good cluster exists. Finally, an optimal sublinear-time algorithm is developed for testing clusterability based on triangles.

Keywords: Higher-order Clustering · Property Testing · High Dimensional Expander · Spectral Graph Theory

1 Introduction

1.1 Motivation

In many real-world systems, interactions and relations between entities are not pairwise, but occur in higher-order organizations that are usually small connected patterns denoted as motifs, including triangles, wedges, cliques, etc. Some researches focus on higher-order clustering [2], which captures connected motifs into cohesive groups while motifs between different groups have few connections. Authors in [2] gave an example of clustering based on a particular triangle motif, which correctly represents three well-known aquatic layers in Florida Bay foodweb. Higher-order clustering has been widely applied in social network analysis [16], gene regulation [8] and neural networks [6]. However, graphs such as

---

This work was supported by the National Natural Science Foundation of China under grants 61832003, Shenzhen Science and Technology Program (JCYJ202208181002205012) and Shenzhen Key Laboratory of Intelligent Bioinformatics (ZDSYS20220422103800001).

W. Wu and J. Guo (Eds.): COCOA 2023, LNCS 14462, pp. 203–214, 2024.
https://doi.org/10.1007/978-3-031-49614-1_15

Actors [1] and Coauthoring [18], which are nearly bipartite, are not suitable for higher-order clustering based on triangles or cliques. Therefore, it is important to judge whether the given graph is suitable for clustering based on the specified motif. A helpful method is to use property testing [19], which is a framework that decides whether an object has a specific property or is "far" from objects having this property. However, none of the former property testing on graphs considered higher-order motifs.

In this paper, we develop a new framework of testing whether a given graph is higher-order clusterable, which is compatible with the low-order testing problem given by Czumaj et al. [4]. First, what is a good high-dimension cluster is defined on undirected graphs. Requirements of the high-dimension cluster wouldn't violate the topological structure in lower-dimension. The problem of testing higher-order clusterability is then proposed. It asks whether there exists a good high-dimension cluster or is far from having that kind of cluster. Finally, A sublinear-time algorithm for testing triangle-based clusterability is developed, which reaches the lower bound, $\Omega(\sqrt{n})$, and is nearly optimal.

1.2   Related Work

This section reviews some previous researches and analyze their shortages or differences compared to the work in this paper.

Previous Work on Higher-Order Graph Clustering. Earlier researches on higher-order clustering is related to hypergraph partitioning [14]. Benson et al. [2] proposed a generalized framework with motif conductance that could cluster higher-order connectivity patterns. They implemented an algorithm without suffering hypergraph fragmentation and its time complexity is bounded by the number of motifs. Tsourakakis et al. [21] shared the same contribution in parallel that a weighted graph can be used to replace the hypergraph in motif-based clustering. Li et al. [17] proposed an edge enhancement approach to solve the issue in isolated nodes.

However, there solutions have some drawbacks. First, whether there exists triangle expanders that are not edge expanders is still unknown. It means that triangle-based clustering in [21] could still violate lower-order cluster structure. In addition, time complexity of higher-order clustering are bounded by the number of motifs, which is $\Omega(n^{3/2})$ if motif is triangle and is far from sublinear. Furthermore, they do not consider whether the given graph is suitable for higher-order clustering. As a result, an inappropriate clustering would suffer severe computation cost on huge graphs.

Previous Work on Graph Property Testing. Framework on testing graph properties was first proposed by Goldreich and Ron [11], who present an alternative model that each query on bounded-degree graph returns a vertex with one indexed neighbor. They showed that testing whether a graph is an expander requires $\Omega(\sqrt{n})$ queries under this model. Latter work [5,12] provided optimal algorithms that reach this lower bound. Czumaj et al. [4] defined (k,ϕ)-clusterable graphs that can be partitioned into k clusters with requirements on both internal and external conductance. They maintained a logarithmic gap between two conductance so that testing clusterability is equivalent to testing expansion when $k = 1$. Chiplunkar et al. [3] eliminated the logarithmic gap at the cost of rising the lower bound to $\Omega(n^{1/2+O(\epsilon)})$. Gluch et al. [10] designed a clustering oracle that allows fast query access and proposed an optimal algorithm.

All the above testers consider the property of low-order expansion and fail to unravel higher-order organizations such as dense cluster of triangles. Furthermore, these testers adopt simple or lazy random walk that starts from vertices, which cannot catch information of triangles or k-cliques. Therefore, they cannot be easily extended to learning clusterability of higher order motifs.

1.3   Contributions

Specifically, contributions are summarized as follows:

1. Problem of testing higher-order clusterability based on a new definition of high-dimension cluster.
2. Proof that the redefined problem is compatible with the original one defined by Czumaj et al. [4]
3. An $\Omega(\sqrt{n})$ query lower bound of testing higher-order clusterability.
4. A sublinear-time algorithm for testing triangle-based clusterability, which reaches the lower bound with neighbor query oracle.

1.4   Organization of the Paper

Section 2 provides preliminary and statement of testing higher-order clusterability on bounded degree graphs. Section 3 establishes relationship between higher-order clusterability and counterpart, and then gives a query lower bound. Section 4 proposes algorithms for testing triangle clusterability and analysis of correctness and running time. Section 5 gives a summary on the whole paper and presents the future work.

2   Preliminary and Problem Statement

2.1   Testing Clusterability on Bounded-Degree Graphs

Here is a brief review on the problem of testing graph cluster structure. Let $G = (V, E)$ be an undirected graph. $deg(v)$ denotes the degree of vertex v. For two non-empty vertex sets S and C, $S \subset C \subseteq V$, let $Vol(S) = \Sigma_{v \in S} deg(v)$ denote the volume of set S. The *external conductance* [13] of S on C is defined as

$$\Phi_C(S) = \frac{|E(S, C \backslash S)|}{\min\{Vol(S), Vol(C \backslash S)\}},$$

where $E(S, C\backslash S)$ is the set of edges with two endpoints contained in set S and $C\backslash S$ respectively. In addition, $G[C]$ denotes the *induced graph* whose vertex set is C and whose edge set consists of all edges with both endpoints included in C. Then the *internal conductance* is defined as

$$\Phi(G[C]) = \min_{\substack{\emptyset \neq S \subset C \\ |S| \leq |C|/2}} \frac{|E(S, C\backslash S)|}{Vol(S)}.$$

Since $\Phi_C(S) = \Phi_C(C\backslash S)$, usually only vertex sets S with $|S| \leq |V|/2$ are considered for convenience. The definition of characterizing the cluster structure of undirected graph is shown as follows,

Definition 1 *((k, ϕ_{in}, ϕ_{out})-cluster* [9]*). Given an undirected graph $G(V, E)$ with parameters k, ϕ_{in}, ϕ_{out}, find an h-partition $\mathbb{P}$ of V, $\mathbb{P} = (P_1, P_2, \ldots, P_h)$ with $1 \leq h \leq k$, and for each i, $1 \leq i \leq h$, $\Phi(G[P_i]) \geq \phi_{in}$ and $\Phi_G(P_i) \leq \phi_{out}$.*

In the property testing framework, G is given as a *neighbor query oracle*. When given an index pair (v, i), the oracle returns the predetermined ith neighbor of vertex v if i doesn't exceed the degree of v, otherwise it would return *NULL*.

Definition 2 *(Testing $(k, \phi_{in}, \phi_{out}, \epsilon)$-clusterability)* [4]*. Given a neighbor oracle access to graph $G(V, E)$ with maximum degree at most d_{max} and parameters $k, \phi_{in}, \phi_{out}, \epsilon$, in which ϕ_{in}, ϕ_{out} satisfy $\phi_{out} = O(\frac{\epsilon^4}{\log n}\phi_{in}^2)$, with probability at least $2/3$,*

- *accept if there exists a $(k, \phi_{in}, \phi_{out})$-cluster on G,*
- *reject if G is ϵ-far from having a $(k, \phi_{in}, \phi_{out})$-cluster,*

where ϵ-far means that G cannot be accepted by modifying (inserting or deleting) no more than $\epsilon d_{max} n$ edges.

Authors in [4] gave a detailed explanation on why they chose a logarithmic gap between ϕ_{in}^2 and ϕ_{out}. This paper maintains the gap of [4] in testing higher-order cluster structure and show that the 1-dimension cluster is compatible with $(k, \phi_{in}, \phi_{out})$-cluster in Theorem 1.

2.2   Testing Higher-Order Clusterability on Bounded-Degree Graphs

A few concepts on simplicial complex would be introduced before showing the Definition 6 of higher-order cluster and the Problem 7 of testing higher-order clusterability. These concepts could help us understand the graph in a high dimensional view. A *d-simplex* [20] is the simplest geometric figure in d dimension, e.g., point (0-simplex), line segment (1-simplex), triangle (2-simplex) and tetrahedron (3-simplex). A *d-simplicial complex* X is a collection of sets constructed by gluing together simplices with maximal dimension d. X should satisfy a closure property that for any simplex $\sigma \in X$, all of its subsets $\tau \subset \sigma$ are also in X. σ is denoted as a *face* of X. Dimension of a face $dim(\sigma)$ equals to the number of vertices in it minus 1, i.e., $dim(\sigma) = |\sigma| - 1$. Empty set satisfies $\emptyset \in X$ with dimension -1 to keep closure. Other definitions are shown as follows:

- *i-faces* X_i is a set of all faces with dimension i.
- *i-cochain* $C(i)$ is a subset of X_i. Space of i-cochain is $S^i(X)$.
- *Degree of face* $deg_d(\sigma)$ is the number of d-dimension faces that contain σ.
- *Volume of i-cochain* $Vol_d(C(i)) = \sum_{\tau \in C(i)} deg_d(\tau)$.
- *Norm of i-cochain* $\|C(i)\|_d = \frac{Vol_d(C(i))}{Vol_d(X_i)}$.
- *Adjacent i-dimension faces* $a \sim b$ means there exists a face $\tau \in X_{i+1}$ that $a, b \subset \tau$.
- *Induced $(i+1)$-subcomplex* $C(i)[X_{i+1}] = (C(i), \{\sigma \in X_{i+1} | \exists \tau \in \sigma : \tau \in C(i)\})$.

Kaufman and Mass [15] proposed a high dimensional expander as follows,

Definition 3 *(Colorful Expander* [15]*). Let X be a d-dimension simplicial complex. X is an ϵ-colorful expander, $\epsilon > 0$, if for any i-cochain $C(i) \in S^i(X), 0 \le i < d, 0 < \|C(i)\|_d \le 1/2$,*

$$\frac{\|\mathbb{F}(C(i), X_i \backslash C(i))\|_d}{\|C(i)\|_d} \ge \epsilon,$$

*where $\mathbb{F}(C(i), X_i \backslash C(i))$ is the **expander face** (similar to cut on graphs) that is defined as*

$$\mathbb{F}(C(i), X_i \backslash C(i)) = \{\sigma \in X_{i+1} | \exists \tau, \tau' \subset \sigma : \tau \in C(i), \tau' \in X_i \backslash C(i)\}.$$

Similar to the internal and external conductance on undirected graphs, a normalized version of conductance is extended to simplicial complex.

Definition 4 *(Normalized External Conductance). Let X be a d-dimension simplicial complex, $d \ge 1, 0 \le i < d$, $C(i)$ and $S(i)$ are both i-cochains, $\emptyset \ne S(i) \subset C(i) \subseteq X_i$, the normalized external conductance of $S(i)$ on $C(i)$ equals to*

$$\Psi_{d,C(i)}\{S(i)\} = \frac{\|\mathbb{F}(S(i), C(i) \backslash S(i))\|_d}{\min\{\|S(i)\|_d, \|C(i) \backslash S(i)\|_d\}}.$$

Definition 5 *(Normalized Internal Conductance). Let X be a d-dimension simplicial complex, $d \le 1, 0 \le i < d$, $C(i)$ is an i-cochain, $\emptyset \ne C(i) \subseteq X_i$. The normalized internal conductance of $C(i)$ is*

$$\Psi_d(C(i)[X_{i+1}]) = \min_{\substack{\emptyset \ne S(i) \subset C(i) \\ Vol_d(S(i)) \le Vol_d(C(i))/2}} \frac{\|\mathbb{F}(S(i), C(i) \backslash S(i))\|_d}{\|S(i)\|_d}.$$

The final step is to establish a unique mapping from simple undirected graph to d-dimension simplicial complex, which is easy to implement since the process can be seen as dimension raising.

Lemma 1. *Given an undirected graph $G(V,E)$ and integer $d > 1$, there exists a unique d-dimension simplicial complex $X^d(G) = \{X_0(G), X_1(G), X_2(G), \ldots, X_d(G)\}$ that satisfies $X_0(G) = V, X_1(G) = E$, for each i, $1 < i \leq d$,*

$$X_i(G) = \{\bigcup(s_1, s_2, \ldots, s_{i+1}) | s_j, s_k \in X_{i-1}(G) : s_j \sim s_k, \forall 1 \leq j < k \leq i+1\}.$$

More generally, $X^d(G)$ is constructed by gluing together all i-cliques (triangles when $i = 3$) to be its $(i-1)$-faces. The formal definition of high-dimension cluster that mentioned in the abstract is as follows,

Definition 6 (*d-dimension $(k, \psi_{in}, \psi_{out})$-cluster*). *Given an undirected graph $G(V,E)$ with parameters $d, k, \psi_{in}, \psi_{out}$, find an h-partition $\mathbb{P}$ of V, $\mathbb{P} = (P_1, P_2, \ldots, P_h)$ with $1 \leq h \leq k$, and for each i, r, $1 \leq i \leq h, 0 \leq r < d$, $\Psi_d(X_r(G[P_i])[X_{r+1}(G)]) \geq \psi_{in}$ and $\Psi_{d, X_r(G)}(X_r(G[P_i])) \leq \psi_{out}$.*

The problem of testing higher-order clusterability is defined as follows,

Definition 7 (*Testing d-dimension $(k, \psi_{in}, \psi_{out}, \epsilon)$-clusterability*). *Given a neighbor oracle access to graph $G(V,E)$ with maximum degree at most d_{max} and parameters d, k, $\psi_{in}, \psi_{out}, \epsilon$, in which ψ_{in}, ψ_{out} satisfies $\psi_{out} = O(\frac{\epsilon^4}{\log n}\psi_{in}^2)$, with probability at least 2/3,*

- ***accept*** *if there exists a d-dimension $(k, \psi_{in}, \psi_{out})$-cluster on G,*
- ***reject*** *if G is ϵ-far from having a d-dimension $(k, \psi_{in}, \psi_{out})$-cluster,*

where ϵ-far denotes G cannot be accepted by modifying (insertion or deletion) no more than $\epsilon d_{max} n$ edges.

3   Analysis of Compatibility and Lower Bound

3.1   Compatibility with Framework of Testing Clusterability

The relationship between 1-dimension $(k, \psi_{in}, \psi_{out})$-partiton and $(k, \phi_{in}, \phi_{out})$-partiton [9] is shown as follows,

Theorem 1. *1-dimension $(k, \psi_{in}, \psi_{out})$-cluster is equivalent to $(k, \frac{\psi_{in}}{2}, \frac{\psi_{out}}{2})$-cluster on undirected graph.*

Proof. For 1-dimension $(k, \psi_{in}, \psi_{out})$-partiton, $1 \leq i \leq h$, $X_0 = V$ and $X_1 = E$, so $X_0(P_i[G])[X_1(G)] = P_i[G]$, $X_0(P_i[G]) = P_i$ and $X_0(G) = V$. Therefore,

$$\psi_{out} \geq \Psi_{d, X_r(G)}(X_r(P_i[G])) = \frac{|E(P_i, V \backslash P_i)|/|E|}{Vol(P_i)/Vol(V)} = 2\Phi_V(P_i).$$

Similarly, $\psi_{in} \leq \Psi_d(X_0(P_i[G])[X_1(G)]) = 2\Phi(P_i[G])$. The proof is finished by combining these two inequalities.

According to Theorem 1, algorithms for testing 1-dimension $(k, \psi_{in}, \psi_{out}, \epsilon)$-clusterability can also test $(k, \frac{\psi_{in}}{2}, \frac{\psi_{out}}{2}, \epsilon)$-clusterability in [4].

3.2 Compatibility of High-Dimension $(k, \psi_{in}, \psi_{out})$-Cluster

This section mainly deals with undirected graphs without outliers, which means all vertices or edges are contained in at least one triangle or d-clique. It is natural since if the graph has outliers, they can be eliminated without affecting quality of higher-order clustering. Following definition is necessary to prove compatibility,

Definition 8 *(Induced i-graph [15])*. *Given a d-dimension simplicial complex X. For any i with $0 \leq i < d$, the i-graph $G_i(V_i, E_i)$ satisfies,*

1) Every i-dimension face τ in X_i is corresponding to a unique vertex $V(\tau)$.
2) There is an edge between the corresponding vertex for any two adjacent i-dimension faces τ, τ', i.e., $E_i = \{(V(\tau), V(\tau'))|\tau \sim \tau'\}$

Generally speaking, induced i-graph is a dimensional reduction that maps the complex constructed by two i-faces (X_i, X_{i+1}) to an undirected graph. Corresponding to the graph without outliers, *pure simplicial complex* X is adopted that for any face $\tau \in X$ with $dim(\tau) < dim(X)$, there exists a face $\sigma \in X$, $dim(\sigma) = dim(X)$, such that $\tau \subset \sigma$. Then the following lemma holds,

Lemma 2. *Let X be a pure d-dimension simplicial complex. Given t that satisfies $1 \leq t < d$, for any i-cochain $C(i)$ that satisfies $0 \leq i < t$, the external conductance is equal to $\frac{\|\mathbb{F}(C(i), X_i \backslash C(i))\|_d}{\|C(i)\|_d}$.*

Lemma 3. *Let X be a pure d-dimension ϵ-colorful expander, then for any t that $1 \leq t < d$, X must be a t-dimension ϵ-colorful expander.*

Through the above two lemmas, it can be proved that if there exists a good cluster in high dimension, it is exactly a good cluster in lower dimension.

Theorem 2. *Given an undirected graph $G(V, E)$ without outliers, if h-partition $\mathbb{P}$ is a d-dimension $(k, \psi_{in}, \psi_{out})$-cluster with $d \geq 2, 1 \leq h \leq k$, then $\mathbb{P}$ must be a t-dimension $(k, \psi_{in}, \psi_{out})$-cluster for all t that satisfies $1 \leq t \leq d - 1$.*

3.3 Lower Bound of Testing Higher-Order Clusterability

Theorem 3. *With neighbor query oracle access, testing d-dimension $(k, \psi_{in}, \psi_{out}, \epsilon)$-clusterability on bounded-degree graph with neighbor query oracle has a lower bound $\Omega(\sqrt{n})$.*

Proof. Consider the special case when $k = 1$. The origin testing problem would reduce to testing d-dimension ψ-colorful expansion, while any $\psi_{out} > 0$ could be satisfied. Consider an undirected graph G without outliers, which means a pure d-dimension simplicial complex X can be constructed on it. According to Lemma 3 and Theorem 1, if X is a pure d-dimension ψ_{in}-colorful expander, (X_0, X_1) should be a 1-dimension ψ_{in}-colorful expander, which means G is a normal $\frac{\psi_{in}}{2}$-expander. Goldreich and Ron [11] proved that testing expansion on bounded degree graphs with has an $\Omega(\sqrt{n})$ lower bound. Suppose that there exists an algorithm that can test d-dimension $(1, \psi_{in}, \psi_{out})$-clusterability in $o(\sqrt{n})$ queries, it can also answer the expansion test in $o(\sqrt{n})$ queries, which is a contradiction. To conclude, query lower bound of testing higher-order clusterability is $\Omega(\sqrt{n})$.

In the next section, we would give an approach on triangle-based clusterability that could reach this lower bound.

Algorithm 1: 2-dimension Random Walk (2DRW)

Input: Initial vertex v_0 or edge e_0, length l.
Output: $(v_0, v_1, \ldots, v_l)$ if input is vertex; $(e_0, e_1, \ldots, e_l)$ if input is edge.

1 **for** *Step* $t \in [0, l-1]$ **do**
2   **if** *Move from vertex v_t* **then**
3   **for** *Each neighbor u_t of v_t* **do**
4   Search all neighbors of u_t and v_t;
5   Set the number of common neighbors $c(u_t)$ to u_t;
6   Choose u_t' with probability $\frac{c(u_t')}{\sum_{u_t \sim v_t} c(u_t)}$ as v_{t+1} and move to it.;
7   **else if** *Move from edge $e_t = (x_t, y_t)$* **then**
8   Search all neighbors of x_t and y_t;
9   **for** *Each common neighbor z_t* **do**
10   Put edges (x_t, z_t) and (y_t, z_t) into candidate set $C(e_t)$;
11   Choose e_{t+1} from set $C(e_t)$ uniformly at random and move to it;

---

4 Algorithm of Testing Triangle-Based Clusterability

4.1 Design of Triangle-Based k-Cluster Tester

This section would give an example how to recognize triangle-based clusterability in sublinear-time with neighbor query oracle access. High-order random walk, which is used to catch information of network motifs, would be invoked in our algorithm. Related definition is shown as follows,

Definition 9 (High-order Random Walk [15]). *Given a simplicial complex X with d-dimension higher than i, the i-dimension high-order random walk W_i starts from an initial i-dimension face $\tau_0 \in X_i$. Then let τ_t be the position W_i stays after t steps. Choose τ_{t+1} as follows*

1) *Choose an $(i+1)$-dimension $\sigma_t \supset \tau_t$ with probability proportional to its degree $deg_d(\sigma_t)$.*
2) *Uniformly choose an i-dimension face $\tau_{t+1} \subset \sigma_t, \tau_{t+1} \neq \tau_t$ at random and move to it.*

W_i stops at τ_t if no σ_t or τ_{t+1} exists.

The exact probability for moving from τ to τ', where $\tau \sim \tau'$, is as follows:

$$Pr[\tau_{t+1} = \tau' | \tau_t = \tau] = \frac{deg_d(\tau \cup \tau')}{\sum_{\tau'' \sim \tau} deg_d(\tau \cup \tau'')}$$

Generally speaking, high-order random walk is an up-down Markov chain that moves on the induced i-subcomplex (X_i, X_{i+1}). Also, this random walk is equivalent to simple random walk on induced i-graph [15] with probability distribution $\pi_0, \pi_1, \cdots \in \mathbb{R}^{|X_i|}$ and $\pi_{t+1} = \pi_t \cdot \tilde{A}_i$, where $\tilde{A}_i$ is the normalized adjacency matrix of the i-graph. Thus, the distribution becomes stable when it equals to one of the eigenvectors of $\tilde{A}_i$. A complex with high expansion should satisfy that any high-order random walk converges rapidly to the uniform distribution.

However, neighbor query oracle cannot directly catch i-dimension face, so it is necessary to simulate this process by using more queries for each moving step. A 2-dimension random walk sampler in 1 is implemented for testing triangle-based clusterability. Given a vertex or edge as input, this sampler could perform the same up-down walk as that on the induced 0-graph and 1-graph. Transition probability is proportional to the degree of the pass edge or triangle. Note that if no common neighbor exists, which means it is an outlier, the sampler would stop here as an endpoint.

Algorithm 2: Triangle-based k-cluster tester

Input: Query oracle of undirected bounded-d_{max} graph $G(V, E)$, maximum cluster k, error ϵ
Output: Decision **Accept** or **Reject**

1 Sample a set S_0 of s vertices independently and uniformly at random with query oracle;
2 For each $v \in S$, perform m times of lazy $2DRW(v, l)$ and calculate the distribution π_u^l of the endpoints;
3 **if** k-cluster-test$(\pi^l, |V|, k, s, m, \theta, \delta, \epsilon)$ *rejects* **then**
4 | Abort and return *Reject*;
5 **else**
6 | Sample a set S_1 of s edges independently and uniformly at random with edge sampler $S(G, \eta)$;
7 | For each $e \in S$, perform m times of lazy $2DRW(e, l)$ and calculate the distribution π_e^l of the endpoints;
8 | return k-cluster-test$(\pi^l, |E|, k, 2s, m, \theta, \delta, \epsilon)$;

Here we briefly introduce our algorithm in 2. Similar to the approach in [4] and [3], the algorithm embeds samples of vertices or edges into points on Euclidean spaces and cluster them based on the estimates of Euclidean distances. There are two main differences between our method and former ones. First, it is a two-step approach with *k-cluster-tester* 3 that tests whether the distribution vectors can be embedded into no more than k clusters on Euclidean space. Second, simulated high-order random walks, which is promised to converge rapidly in high-dimension expander, is performed to estimate distribution of endpoints that reveals the similarities to each other. Note that *lazy random walk* means with the probability $1/2$ for each step, the walk stay at the current vertex or edge. The edge sampler [7] returns an edge that is n uniformly at random with bias η. Running time of the edge sampler is $O(\frac{n}{\sqrt{m}})$, which is $O(\sqrt{n})$ on bounded-degree graphs. Our algorithm would use the same configuration as that in [3] that $\eta = \frac{1}{2}$ and number of edge samples would be doubled.

Algorithm 3: k-cluster-test

   Input: Distribution of endpoints π^l, maximum set size n, maximum cluster k, sample size s, number of each distribution m, parameters θ, δ, ϵ

   Output: Decision **Accept** or **Reject**

1 Similarity Graph $H = (\emptyset, \emptyset)$;

2 For each $u \in S$, if $l_2^2\text{-}norm(\pi_v^l, \theta, m, s)$ rejects, abort and return **Reject**;

3 For each pair of $u, v \in S$, if $l_2\text{-}distribution(\pi_u^l, \pi_v^l, m, s, \delta, \epsilon)$ accepts, then add an edge (u, v) to H;

4 If H contains more than k connected components, return **Accept**; Else, return **Reject**;

4.2 Correctness and Running Time Analysis

Now we prove the correctness of our algorithm. Since high order random walks are different from simple random walk, it is essential to make sure that distributions of endpoints converge as the input of *k-cluster-test*.

Lemma 4 (Mixing Rate [15]). *Given an undirected graph $G(V, E)$, $\tilde{A}$ is its normalized adjacency matrix, $1 = \alpha_1 \geq \alpha_2 \geq \cdots \geq \alpha_n \geq -1$ the eigenvalues of $\tilde{A}$ and $\alpha = \max\{|\alpha_2|, |\alpha_{|V|}|\}$. Then for any initial probability distribution $\pi_0 \in R^{|V|}$ and any $t \in \mathbb{N}$,*

$$\|\pi^t - \pi\|_2 \leq \sqrt{\frac{d_{max}}{d_{min}}}\alpha^t,$$

where π^t is the probability distribution after t steps of the random walk, π is the stationary distribution, $d_{max} = \max\limits_{v \in V}\{deg(v)\}$ and $d_{min} = \min\limits_{v \in V}\{deg(v)\}$.

We prove that a lazy 2-dimension random walk with 11 times number of the original steps is enough for *k-cluster-test*.

Lemma 5. *Given an undirected graph $G(V, E)$, mixing rate of lazy 2-dimension random walk is μ', for any initial probability distribution $\pi_0 \in R^{|V|}$ and any $t \in \mathbb{N}$,*

$$\|\pi^t - \pi\|_2 \leq \sqrt{\frac{d_{max}}{d_{min}}}\mu'^{11t},$$

where π^t is the probability distribution after t steps of the random walk, π is the stationary distribution, $d_{max} = \max\limits_{v \in V}\{deg(v)\}$ and $d_{min} = \min\limits_{v \in V}\{deg(v)\}$.

Then Theorem 4 can be deduced by the next lemma. In convenience, we set $\psi_{in} = \psi$ and $\psi_{out} = O(\epsilon^4 \psi^2 / \log(n))$. We say a given graph is 2-*dimension* (k, ψ)-*clusterable* if there exists a 2-dimension $(k, \psi_{in}, \psi_{out})$-cluster on it.

Lemma 6.   *Given the same constants* $c_{3.1}, c_{4.2}, c_{4.3}$ *as those in [4], set* $s = \frac{1536k \ln(18(k+1))}{\epsilon^2}, l = \frac{11max\{c_{4.2}, c_{4.3}\}k^4 \log(n)}{\psi^2}, m = 384c_{3.1}s\sqrt{skn} \ln s, \theta = \frac{288sk}{n}, \delta = \frac{1}{24s^2}$. *k-cluster-test accepts 2-dimension* (k, ψ)-*clusterable graph and rejects every graph* ϵ-*far from being 2-dimension* (k, ψ)-*clusterable with probability at least* $\frac{5}{6}$.

Then Theorem 4 holds since our algorithm invoke *k-cluster-test* twice.

Theorem 4 (*Correctness*). *With proper setting of the parameters, algorithm 2 can accept every 2-dimension* $(k, \psi_{in}, \psi_{out})$-*clusterable graph with probability at least* $\frac{2}{3}$ *and reject every graph* ϵ-*far from being 2-dimension* $(k, \psi_{in}, \psi_{out})$-*clusterable with probability at least* $\frac{2}{3}$.

Theorem 5 (*Running Time*). *With proper setting of parameters, triangle-based k-cluster tester runs in time* $O(\frac{\sqrt{n}k^7 d_{max}^3 (\ln k)^{7/2} \ln 1/\epsilon \ln n}{\psi_{in}^2 \epsilon^5})$.

Proof First, the algorithm generates a sample set of s vertices with query oracle and s edges with edge sampler. Next, the algorithm performs m random walks with step l for all s samples, while time for each step is $O(d_{max}^3)$. Then, the algorithm invoke l_2-*norm* tester, which has running time $O(m)$, for each sample in S. Finally, the algorithm invoke l_2-*distribution* tester with running time $O(m)$ for each pair of samples in S. To conclude, the total running time of the algorithm is $O(s\sqrt{n} + d_{max}^3 sml + sm + s^2m) = O(\frac{\sqrt{n}k^7 d_{max}^3 (\ln k)^{7/2} \ln 1/\epsilon \ln n}{\psi_{in}^2 \epsilon^5})$.

5   Summary and Future Work

In this work, a problem of testing higher order clusterability is proposed based on the new definition of high-dimension cluster. Besides, an algorithm for testing triangle-based clusterability, which reaches the proved lower bound, is designed. In the future, we would seek the lower bound when the logarithmic gap between normalized internal and external conductance is eliminated. We also plan to develop new algorithms for testing clique-based clusterability with more powerful query and sample oracles.

References

1. The internet movie database. http://www.imdb.com/
2. Benson, A.R., Gleich, D.F., Leskovec, J.: Higher-order organization of complex networks. Science **353**(6295), 163–166 (2016)
3. Chiplunkar, A., Kapralov, M., Khanna, S., Mousavifar, A., Peres, Y.: Testing graph clusterability: algorithms and lower bounds. In: 2018 IEEE 59th Annual Symposium on Foundations of Computer Science (FOCS), pp. 497–508. IEEE (2018)

4. Czumaj, A., Peng, P., Sohler, C.: Testing cluster structure of graphs. In: Proceedings of the Forty-Seventh Annual ACM Symposium on Theory of Computing, pp. 723–732 (2015)
5. Czumaj, A., Sohler, C.: Testing expansion in bounded-degree graphs. Comb. Probab. Comput. **19**(5–6), 693–709 (2010)
6. Duval, A., Malliaros, F.: Higher-order clustering and pooling for graph neural networks. In: Proceedings of the 31st ACM International Conference on Information & Knowledge Management, pp. 426–435 (2022)
7. Eden, T., Rosenbaum, W.: On sampling edges almost uniformly. In: 1st Symposium on Simplicity in Algorithms (2018)
8. Gama-Castro, S., et al.: Regulondb version 9.0: high-level integration of gene regulation, coexpression, motif clustering and beyond. Nucleic Acids Res. **44**(D1), D133–D143 (2016)
9. Gharan, S.O., Trevisan, L.: Partitioning into expanders. In: Proceedings of the Twenty-Fifth Annual ACM-SIAM Symposium on Discrete Algorithms, pp. 1256–1266. SIAM (2014)
10. Gluch, G., Kapralov, M., Lattanzi, S., Mousavifar, A., Sohler, C.: Spectral clustering oracles in sublinear time. In: Proceedings of the 2021 ACM-SIAM Symposium on Discrete Algorithms (SODA), pp. 1598–1617. SIAM (2021)
11. Goldreich, O., Ron, D.: Property testing in bounded degree graphs. In: Proceedings of the Twenty-Ninth Annual ACM Symposium on Theory of Computing, pp. 406–415 (1997)
12. Kale, S., Seshadhri, C.: An expansion tester for bounded degree graphs. SIAM J. Comput. **40**(3), 709–720 (2011)
13. Kannan, R., Vempala, S., Vetta, A.: On clusterings: Good, bad and spectral. J. ACM (JACM) **51**(3), 497–515 (2004)
14. Karypis, G., Kumar, V.: Multilevel k-way hypergraph partitioning. In: Proceedings of the 36th Annual ACM/IEEE Design Automation Conference, pp. 343–348 (1999)
15. Kaufman, T., Mass, D.: High dimensional random walks and colorful expansion. In: 8th Innovations in Theoretical Computer Science Conference (ITCS 2017). Schloss Dagstuhl-Leibniz-Zentrum fuer Informatik (2017)
16. Li, P., Dau, H., Puleo, G., Milenkovic, O.: Motif clustering and overlapping clustering for social network analysis. In: IEEE INFOCOM 2017-IEEE Conference on Computer Communications, pp. 1–9. IEEE (2017)
17. Li, P.Z., Huang, L., Wang, C.D., Lai, J.H.: Edmot: an edge enhancement approach for motif-aware community detection. In: Proceedings of the 25th ACM SIGKDD International Conference on Knowledge Discovery & Data Mining, pp. 479–487 (2019)
18. Newman, M.E., Watts, D.J., Strogatz, S.H.: Random graph models of social networks. Proc. Natl. Acad. Sci. **99**(suppl-1), 2566–2572 (2002)
19. Rubinfeld, R., Sudan, M.: Robust characterizations of polynomials with applications to program testing. SIAM J. Comput. **25**(2), 252–271 (1996)
20. Spanier, E.H.: Algebraic Topology. Springer, New York (1981). https://doi.org/10.1007/978-1-4684-9322-1
21. Tsourakakis, C.E., Pachocki, J., Mitzenmacher, M.: Scalable motif-aware graph clustering. In: Proceedings of the 26th International Conference on World Wide Web, pp. 1451–1460 (2017)

The 2-Mixed-Center Color Spanning Problem

Yin Wang[1,2], Yi Xu[3], Yinfeng Xu[1,2], and Huili Zhang[1,2(✉)]

[1] School of Management, Xi'an Jiaotong University, Xi'an 710049, Shaanxi,
People's Republic of China
`zhang.huilims@xjtu.edu.cn`
[2] State Key Lab for Manufacturing Systems Engineering, Xi'an 710049, Shaanxi,
People's Republic of China
[3] School of Economics and Management, Xi'an University of Technology, Xi'an
710054, Shaanxi, People's Republic of China

Abstract. Inspired by the applications in cloud manufacturing, we introduce a new 2-mixed-center version of the minimum color spanning problem, the first mixed-center model for color spanning problems to the best of our knowledge. Given a set P of n colored points on a plane, with each color chosen from a set C of $m \leq n$ colors, a 2-mixed-center color spanning problem determines the locations and radii of two disks to make the union of two disks contains at least one point of each color. Here, one center is called a *discrete center*, which is selected from P, while the other center is called a *continuous center*, which is selected from a plane. The objective is to minimize the maximum of three terms, i.e. the radii of the two disks and the distance between the two centers. We develop an exact algorithm to find the optimal solution in time complexity of $O(n^7, n^5 m^3 \log n)$. Furthermore, we propose a 2-approximation algorithm that reduces the time complexity to $O(nm \log n)$.

Keywords: Color-spanning · Mixed center problem · Voronoi diagram · Approximation algorithm

1 Introduction

The minimum color-spanning problem (CSP) focuses on finding a location that minimizes the cost to cover all types of facilities, such as hospitals, schools, etc. [16]. In the realm of cloud manufacturing involving m types of n suppliers, each type of suppliers provides unique component for production of personalized products to fulfill online orders.

To meet a wide range of personalized demands, a common approach involves establishing multiple centers. However, this often results in hefty construction cost and transfer cost between centers. To reduce some construction cost, we can enlist the help of the suppliers to locate the center instead of setting up all the centers from scratch. This gives rise to what we term as the mixed center spanning problem, specially the 2-mixed center color-spanning problem (abbreviated as 2-MCCSP) in this study. This problem includes a *discrete center* selected from

© The Author(s), under exclusive license to Springer Nature Switzerland AG 2024
W. Wu and J. Guo (Eds.): COCOA 2023, LNCS 14462, pp. 215–226, 2024.
https://doi.org/10.1007/978-3-031-49614-1_16

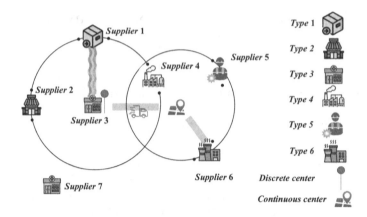

Fig. 1. An illustration of the mixed centers for cloud manufacturing. There are 7 suppliers of 6 different types. One disk centered at Supplier 3 covers suppliers of Type 1, Type 2, Type 3 and Type 4. The other disk with a continuous center covers the suppliers of Type 4, Type 5 and Type 6. The cost of 2-MCCSP here depends on the maximum of three terms, i.e. the radii of the two disks and the distance between centers. In this case, the maximum is the radius of discrete center on Supplier 3, which is indicated by the wave line here.

existing suppliers and a *continuous center* on a plane. These two centers collaborate to assemble the components from the nearest suppliers and produce the ordered products. To improve the manufacturing efficiency, it becomes essential to determine the optimal locations of the two centers to curtail transportation costs. This includes the cost of transporting components from the suppliers to the centers, as well as the inter-center transfer. Specifically, we aim to determine two disks spanning all types of suppliers, intending to minimize the maximum of the radii of the disks and the distance between two centers. See Fig. 1 as an example.

Our Contribution. To the best of our knowledge, we are the first to explore the mixed center model for color spanning problems. For n points and each of which has one of m colors, we propose an exact algorithm based on algorithms inspired by Abellanas et al. [2] and Banerjee et al. [4] which achieves the time complexity of $O(n^7, n^5 m^3 \log n)$. In addition, we develop an efficient $O(nm \log n)$-time algorithm based on Voronoi diagrams, and prove that its approximation ratio is 2.

Roadmap. The remaining parts of the paper are organized as follows. Related works are reviewed in Sect. 2. We introduce essential concepts, terminologies, and notations in Sect. 3. An exact algorithm for the 2-MCCSP is presented in Sect. 4, while an approximation algorithm is discussed in Sect. 5. We make a conclusion and discussion about future research in Sect. 6.

2   Literature Review

Our work explores the Minimum Color Spanning Problem (CSP) within the mixed center setting. CSP aims to identify smallest color spanning regions of various types. The concept was originally proposed by Abellanas et al. in 2001 [1]. It finds applications in areas such as location planning, statistical clustering, and pattern recognition [7]. This section will provide an overview of related works in this field.

Specific instances of CSPs, such as the minimum color spanning disk and minimum area axis-parallel square, can be solved in $O(mn \log n)$ time by computing the upper envelope of Voronoi surfaces, as indicated by Abellanas et al. in their 2001 work [1]. They also discussed the Farthest Color Voronoi Diagram (FCVD) in which the region of a c-colored site p includes all points in the plane for which c is the farthest color and the site p is the nearest c-colored site. The minimum color spanning disk can be determined by an $O(n^2 \alpha(m) \log m)$-time algorithm if the FCVD is given for its formulation and in time $O(m^3 n \log n)$ without FCVD [1], where $\alpha(\cdot)$ is the inverse of the Ackermann's function. For the minimum area axis-parallel square, Khanteimouri et al. [14] presented an efficient algorithm that runs in $O(n\log^2 n)$ time.

Abellanas et al. [2] further explored two additional CSPs, the narrowest color spanning corridor and the minimum area color spanning axis-parallel rectangle. Here a corridor is defined as an open region bounded by a pair of parallel lines that intersects the convex hull of the given n points. The former problem can be solved in $O(n^2 \alpha(m) \log m)$ time, while the latter one has a time complexity of $O(n(n-m) \log^2 m)$. Das et al. [7] revisited the narrowest color spanning corridor problem to $O(n^2 \log n)$ and also proposed an $O(n^3 \log m)$-time algorithm to solve the minimum area color spanning rectangle of arbitrary orientation.

The minimum diameter color spanning set (MDCS) problem was approached by Zhang et al. [23] using a brute-force $O(n^m)$-time algorithm. Fleischer and Xu continued to explore the MDCS problem in [8,9] and showed that it can be solved in polynomial time when the distance is measured under the L_1 and L_∞ metrics, while it is NP-hard for all other L_p metrics even in two dimensions. The approximability of the MDCS problem was further studied in a series of works [8,9,13,15], where inapproximability was established, and a constant ratio approximation algorithms were introduced.

Bereg et al. [5] investigated the FPT tractability of the related problems. Several researchers extended the color spanning model to include more geometric properties [12] and showed all these problems become NP-hard. In particular, an exact algorithm for 2 centers from given point set P with n points and k colors was offered by Banerjee et al. [4], running in $O(n^3, (n^2 k \log^2 n + \min\{mkn \log n, m^2 k^{\omega-2}\}) \log n)$ time, where $m = \min\{2^k, n^2\}$, ω denotes a word of length m over the alphabet $\{1, 2, \ldots, s^*\}$, and s^* denotes the maximal demand among all the colors.

The mixed center problem [21] involves finding m discrete centers from a given point set and $k - m$ centers on the plane to minimize the maximal radius, which was solved using Megiddo's algorithm ([17]) within complexity of

$O(n^2 \log n)$. Furthermore, following the work of Chen and Chen's work [6], Xu et al. [22] discussed a new relaxation algorithm suitable for both the uncapacitated continuous and discrete p-center problems, proving an important result about the optimal solution of the mixed minimax 2-center problem. The time complexity is $O(n^2 \log^2 n)$ time for covering all demands in their works.

More earlier, for the discrete 2-center problem, it was solved in $O(n^{\frac{4}{3}} \log^5 n)$ time [3,11] and then improved to near-linear time $O(n \log^3 n)$ by Sharir [18]. Instead of selecting from the given points for the discrete center, the continuous center is located at anywhere on given region. The continuous 2-center version was solved in $O(n^2)$ by Hershberger [11] and improved to $O(n\log^2 n)$ [19].

To date, to the best of our knowledge, no studies have adapted the mixed-center setting to the color spanning model. Our 2-approximation algorithm achieves the time complexity of $O(mn \log n)$, filling this gap in the literature of approximation algorithm with constant ratio.

3   Preliminary

Given a set of n points $P = \{p_1, p_2, \ldots, p_n\}$ in $\mathbb{R}^2$ with a coloring function $\phi(\cdot)$ assigning each point one of the m colors $C = \{c_1, c_2, \ldots, c_m\}$. To avoid naming confusions, we may refer to the points in P as *sites*. Let $P_j = \{p \in P \mid \phi(p) = c_j\}$, $j \in \{1, 2, \ldots, m\}$, signify all points in P_j of the same color c_j. We denote $I = (P, C, \bigcup P_j)$ as a specific instance of the problem.

For any two points $p, q \in \mathbb{R}^2$, let $d(p, q)$ measure the distance between them. We extend the notation as $d(p, c_j)$ or $d(p, P_j)$ to define the minimum distance between point p and sites of the color c_j, as per Eq. 1.

$$d(p, c_j) = d(p, P_j) = \min_{q \in P_j} d(p, q). \tag{1}$$

Let $q_{i,j}$ be the nearest site with color c_j to p_i, i.e. $d(p_i, c_j) = d(p_i, P_j) = d(p_i, q_{i,j})$.

The 2-MCCSP seeks two disks, where one disk is at a *discrete center* $p \in P$, and the other is at a *continuous center* o on a plane. Let $D(p, r(p))$ be a disk centering at a discrete center $p \in P$ with a radius $r(p)$. The colors covered by disk $D(p, r(p))$ are collected in a color subset C_p. Similarly, let $D(o, r(o))$ be a disk centering at a continuous center o with radius $r(o)$, let C_o be the set of colors covered by disk $D(o, r(o))$. We also use $r(p, C_p)$ and $r(o, C_o)$ as the radii of disks to cover C_p and C_o, respectively. It is required that $C_p \cup C_o = C$.

The operation cost includes the distribution cost between suppliers and center, as well as the transportation costs between the two centers. Therefore, This cost is formulated as the maximum of three terms: the radii of the two disks and the distance between the centers, as per Eq. 2. Summarizing, for an instance I, the 2-MCCSP problem aims at finding two disks $D(p, r(p))$ and $D(o, r(o))$ to minimize

$$f_I = f(p, o) = \max\{r(p), r(o), d(p, o)\}. \tag{2}$$

We denote (p^*, o^*) as the optimal locations for the mixed centers with $f_I^* = f(p^*, o^*) = \max\{r(p^*), r(o^*), d(p^*, o^*)\}$ representing the optimal cost. For any approximation algorithm $\mathcal{A}$, the approximation ratio $\alpha_{\mathcal{A}}$ is defined as per Eq. 3.

$$\alpha_{\mathcal{A}} = \sup_I \frac{\mathcal{A}(I)}{OPT(I)} = \sup_I \frac{f_I}{f_I^*} \tag{3}$$

4   The Exact Algorithm for 2-MCCSP

In this section, we propose an exact algorithm based on two key observations.

Observation 1. *A tight disk can be determined by two or three sites of different colors, where the sites with these colors only locate on the boundary of the disk.*

Observation 2. *For a given disk $D(p, r_p)$ (or $D(o, r_o)$) of 2-MCCSP, this disk is included in the optimal solution only when the centers are both within a single disk, or when o (or p) lies on the boundary of $D(p, r_p)$ (or $D(o, r_o)$).*

Note that this case, studied in [21, 22], where all sites of different colors (i.e. $|P_j| = 1, \forall j \in [n]$), is a special case of our research.

Employing these observations, we enumerate all potential optimal pairs of centers. As Observation 1 applied to both discrete and continuous disks, we first use the *discrete-continuous* process to obtain p and o, followed by the *continuous-discrete* process to retrieve o and p similarly. We utilize the algorithm of [4] to determine the p and algorithm of [1] to determine o, as detailed in Algorithm 1.

For simplicity, we refer to a pair or a triplet of points of different colors as a *candidate* in this section.

Algorithm 1: Exact Algorithm for 2-MCCSP

Data: a set C of m colors; a set P of n colored points that can be partitioned by colors into $P = P_1 \bigcup \ldots \bigcup P_m$;
a center $p \in P$ with radius $r(p, C_p)$, where C_p is the color
Result: $D(p^*, r(p^*))$, $D(o^*, r(o^*))$, $f(p^*, o^*) = \max\{r(p^*), r(o^*), d(p^*, o^*)\}$;

1 initialization;
2 For each $j \in \{1, 2, \ldots, m\}$, construct the Voronoi diagram VD_j for points in P_j;
3 **foreach** *candidate* **do**   /*continuous-discrete process /
4 | Compute continuous center o by algorithm of [1] with smallest radius enclosing the sites in a *candidate*;
5 | Save C_o as the colors covered by disk $D(o, r(o, C_o))$ and compute $C \setminus C_o$, $s(o) \leftarrow o$;
6 **end**
7 **foreach** $o \in s(o)$ **do**
8 | **foreach** $p \in D(o, r(o, C_o)) \cap P$ **do**   /*to determine p /
9 | | Compute $D(p, r(p, C \setminus C_o))$ based on VDs, and get $f(p, o)$;
10 | **end**
11 | $p_o \leftarrow \arg\min f(p, o)$;
12 **end**
13 $f^1(p, o) = \min_{o \in s(o)} f(p_o, o)$;
14 **foreach** $p \in P$ **do**   /*discrete-continuous process /
15 | Compute all *candidates* containing p by algorithm in [4] based on VDs;
16 **end**
17 **foreach** *candidate containing* p **do**
18 | Compute the minimal disk enclosing the points in the *candidate* centering on $p,$;
19 | Save C_p as the colors covered by disk $D(p, r(p, C_p))$;
20 | Compute all possible location of continuous center o by algorithm of [1] with minimum radius enclosing the *candidates* of colors in $C \setminus C_p$ /*to determine o /;
21 | **if** *one possible location of o is in* $D(p, r(p, C_p))$ **then**
22 | | $s(o) \leftarrow o$, and compute $f(p, o)$;
23 | **end**
24 | $o(p) \leftarrow \arg\min_{o \in s(o)} f(p, o)$ for each candidate containing p;
25 **end**
26 $f(p, o_p) = \min_{o_p} f(p, o(p))$;
27 $f^2(p, o) = \min_{p \in P} f(p, o_p)$;
28 $f(p^*, o^*) = \min\{f^1(p, o), f^2(p, o)\}$ **return** $D(p^*, r(p^*))$, $D(o^*, r(o^*))$, $f(p^*, o^*)$

Theorem 1. *Given a set of n points on the plane, each with one of m colors, the optimal solution of 2-MCCSP can be found with a time complexity of $O(n^7, n^5 m^3 \log n)$.*

Proof. Let n_j be the number of sites in P_j, $j \in \{1, 2, \ldots, m\}$. Since the time complexity of constructing VD_j is $O(n_j \log n_j)$ by Fortune's algorithm [10], Line 2 will take $O\left(\sum_{j=1}^{m} n_j \log n_j\right) = O(n \log n)$ time. The *continuous-discrete* process from Line 3 to Line 6, which involves all candidates of continuous centers, requires $O(m^3 n \log n)$ time [2]. The size of these candidates is $O(n^3)$(i.e. $|s(o)| = O(n^3)$) at most. Given that each discrete center in the third for-loop in Line 9 requires m times of point locations in VDs, it takes $O(mn^2 \log n)$ for at most n sites satisfied $D(o, r(o, C_o)) \cap P$. Hence, all candidates in $s(o)$ necessitate $O(n^5 m \log n)$ time to find paired optimal discrete centers, with the final selection of minimum $f(p, o)$ taking $O(n^3)$.

Similarly, Line 15 takes $O(n^3, n^2 m \log^2 n)$ time [4]. The time complexity of the last for-loop (from Line 17 to Line 25) mainly relies on the computing of Line 20 , which requires $O(m^3 n \log n)$ as per the Algorithm in [2]. The number of all candidates of continuous center paired with a given p could be at most $O(n^3)$. They all need to judge if $o \in D(p, r(p, C_p))$ and compute their distance to given p with a time complexity of $O(n^3)$ from Line 21 to Line 23. Thus, the total computation for n points requires $O(n^7, n^5 m^3 \log n)$ time by $O(n^3, n^2 m \log^2 n) + n \times n^3 \times (O(m^3 n \log n) + O(n^3) + O(n^3)) = O(n^7, n^5 m^3 \log n)$, considering all candidates of the discrete centers with a size of $O(n^3)$ from Line 14 to Line 27. The extremum selections require negligible time. In conclusion, the total computation and selection thus takes a total of $O(n^7, n^5 m^3 \log n)$ of time, i.e.

$$T(n, m) = O(nm \log n) + O(n^5 m \log n) + O(n^7, n^5 m^3 \log n) = O(n^7, n^5 m^3 \log n)$$

given that the number of colors m is no more than the number of points n. $\square$

5   A 2-Approximation Algorithm for 2-MCCSP

In this section, we introduce an 2-approximation algorithm, Algorithm 2, which reduces the time complexity to $O(mn \log n)$. The approximation ratio is validated in Theorem 3.

Similar to the exact algorithm, we also requires a Voronoi diagram of a set of sites. The central concept is as follows. For each site $p_i \in P$, we first determine the minimum radius $d_{i,m}$ that encompasses all colors based on Voronoi diagrams. We then locate the discrete center at the site $\bar{i}$ that has the smallsest radius among all sites, where $d_{\bar{i},m} = \min_{p_i \in P} d_{i,m}$. The continuous center is placed on the line between the discrete center and its farthest color or anywhere within the discrete disk, dependenting on the circumstance. The algorithm's specifics are outlined in Algorithm 2.

Algorithm 2: Approximation Algorithm for 2-MCCSP

Data: a set C of m colors; a set P of n colored points that can be partitioned by colors into $P = P_1 \dot\cup \ldots \dot\cup P_m$

Result: $D(p_{\bar{i}}, r(p_{\bar{i}}, C_{p_{\bar{i}}}))$, $D(o_{\bar{i}}, r(o_{\bar{i}}, C_{o_{\bar{i}}}))$ and $f(p_{\bar{i}}, o_{\bar{i}})$

1 Initialization;
2 For each $j \in \{1, 2, \ldots, m\}$, construct the Voronoi diagram VD_j for points in P_j;
3 **foreach** $p_i \in P$ **do**
4 $\quad$ Compute $d(p_i, c_j)$, $j = 1, \cdots, m$;
5 $\quad$ Obtain the farthest color distance $d_{i,m} = \max_{\{j=1,\cdots,m\}}\{d(p_i, c_j)\}$ and the farthest color $c_{i,m} = \arg\max_{\{j=1,\cdots,m\}} d(p_i, c_j)$, where $c_{i,m}$ is the color of $q_{i,m}$;
6 $\quad$ Obtain $d_{i,m-1} = \max_{c \in C \setminus c_{i,m}}\{d(p_i, c)\}$
7 **end**
8 $\bar{i} = \arg\min_{\{i=1,2,\cdots,n\}} d_{i,m}$;
9 **if** $d_{\bar{i},j} \neq d_{\bar{i},m}$ *for some* $j \in [m]$ **then**   /*Case 1 /
10 $\quad$ Put $o_{\bar{i}}$ on the line between $p_{\bar{i}}$ and $q_{\bar{i},m}$ satisfying that $d(o_{\bar{i}}, q_{\bar{i},m}) = \frac{d_{\bar{i},m} + d_{\bar{i},m-1}}{2}$;
11 $\quad$ **foreach** $c_j \in C$ **do**
12 $\quad\quad$ Compute $d(p_{\bar{i}}, c_j)$ and $d(o_{\bar{i}}, c_j)$;
13 $\quad\quad$ **if** $d(p_{\bar{i}}, c_j) \geq d(o_{\bar{i}}, c_j)$ **then**
14 $\quad\quad\quad$ $C_{o_{\bar{i}}} \leftarrow C_{o_{\bar{i}}} \cup c_j$;
15 $\quad\quad$ **else**
16 $\quad\quad\quad$ $C_{p_{\bar{i}}} \leftarrow C_{p_{\bar{i}}} \cup c_j$;
17 $\quad\quad$ **end**
18 $\quad$ **end**
19 $\quad$ $r(p_{\bar{i}}, C_{p_{\bar{i}}}) = \max_{c \in C_{p_{\bar{i}}}} r(p_{\bar{i}}, c)$;
20 $\quad$ $r(o_{\bar{i}}, C_{o_{\bar{i}}}) = \max_{c \in C_{o_{\bar{i}}}} r(o_{\bar{i}}, c)$;
21 $\quad$ $f(p_{\bar{i}}, o_{\bar{i}}) = \max\{r(p_{\bar{i}}, C_{p_{\bar{i}}}), r(o_{\bar{i}}, C_{o_{\bar{i}}}), d(o_{\bar{i}}, q_{\bar{i},m})\}$;
22 **else**   /*Case 2 /
23 $\quad$ Put $o_{\bar{i}}$ on anywhere within the disk $D(p_{\bar{i}}, d_{\bar{i},m})$;
24 $\quad$ $r(p_{\bar{i}}, C_{p_{\bar{i}}}) = d_{\bar{i},m}$ with $C_{p_{\bar{i}}} = C$;
25 $\quad$ $r(o_{\bar{i}}, C_{o_{\bar{i}}}) = \epsilon$, $C_{o_{\bar{i}}} = \phi$, where ϵ is a positive number that is small enough;
26 $\quad$ $f(p_{\bar{i}}, o_{\bar{i}}) = d_{\bar{i},m}$;
27 **end**
28 **return** $D(p_{\bar{i}}, r(p_{\bar{i}}, C_{p_{\bar{i}}}))$ and $D(o, r(o_{\bar{i}}, C_{o_{\bar{i}}}))$ the objective value $f(p_{\bar{i}}, o_{\bar{i}})$.

The location of o is determined from Line 9 to Line 27. It is divided into two cases as shown in Fig. 2, where Case 1 is in Fig. 2.a, and Case 2 is in Fig. 2.b.

Theorem 2. *Algorithm 2 runs in time* $O(nm \log n)$.

Proof. The time complexity of constructing a VD for n points is $O(n \log n)$ according to Fortune's algorithm [10]. Hence, Line 2 takes $O\left(\sum_{j=1}^m n_j \log n_j\right) = O(n \log n)$ time, where n_j represents the number of sites in P_j for $j \in \{1, 2, \ldots, m\}$. Each computation of $d(p_i, c_j)$ in Line 4 takes $O(\log n)$ time, and identifying the maximum $d_{i,m}$ and $d_{i,m-1}$ in Line 5 and 6 takes $O(m)$ time. Therefore, the total time from Line 3 to Line 7 is $O(nm \log n)$. The selection of discrete center $p_{\bar{i}}$ in Line 8 clearly requires linear time $O(n)$. Line 9

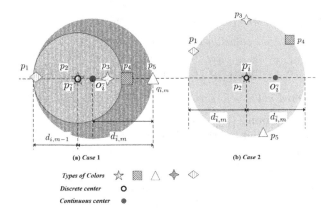

Fig. 2. Two instances of Case 1 and Case 2 respectively. $P = \{p_1, p_2, p_3, p_4, p_5\}$ and $C = \{c_1, c_2, c_3, c_4, c_5\}$, $n = 5$, $m = 5$, where $p_{\bar{i}}$ is on p_2, the colors are presented by different shapes. In particular, $d_{\bar{i},2} = d_{\bar{i},3} = d_{\bar{i},3} = d_{\bar{i},4} = 1$ in Case 2.

takes $O(m)$ time for check the condition, Case 1 from Line 10 to Line 22 takes $O(m \log n + m) = O(m \log n)$ time, Case 2 from Line 23 to Line 27 takes $O(1)$ time.

In conclusion, our algorithm has the time complexity of $O(nm \log n)$, i.e. $T(n, m) = O(n \log n) + O(nm \log n) + O(n) + O(m \log n) = O(nm \log n)$. □

Lemma 1. *The solution $(p_{\bar{i}}, o_{\bar{i}})$ by Algorithm 2 satisfies:*

$$f(p_{\bar{i}}, o_{\bar{i}}) = \max\{r(o_{\bar{i}}, C_o), r(p_{\bar{i}}, C_p), d(p_{\bar{i}}, o_{\bar{i}})\} \leq d_{\bar{i},m}$$

such that the covered color sets satisfied $C_p \cup C_o = C$.

Proof. There are two cases about the location rule to $o_{\bar{i}}$ in Algorithm 2.

- **Case 1:** $\forall j \in \{1, 2, \ldots, m\}, d_{\bar{i},j} \neq d_{\bar{i},m}$. The center $o_{\bar{i}}$ at the location between $p_{\bar{i}}$ and $q_{\bar{i},m}$ with $d(o_{\bar{i}}, q_{\bar{i},m}) = \frac{d_{\bar{i},m} + d_{\bar{i},m-1}}{2} \leq d_{\bar{i},m}$. Here, the radius of discrete disk has the inequality $r(p_{\bar{i}}, C_p) = \max_{j=\{1,2,\cdots,m\}} r(p_{\bar{i}}, c_j) \leq r(p_{\bar{i}}, C) = d_{\bar{i},m}$. At the same time, the radius of continuous disk $o_{\bar{i}}$ has

$$r(o_{\bar{i}}, C_o) = \max_{j=\{1,2,\cdots,m\}} r(o_{\bar{i}}, c_j) \leq r(o_{\bar{i}}, C) = \frac{d_{\bar{i},m} + d_{\bar{i},m-1}}{2} \leq d_{\bar{i},m},$$

where last equation is proved by Wang et al. [20]. Thus, we have $f(p_{\bar{i}}, o_{\bar{i}}) = \max\{r(p_{\bar{i}}, C_p), r(o_{\bar{i}}, C_o), d(p_{\bar{i}}, o_{\bar{i}})\} \leq d_{\bar{i},m}$.
- **Case 2:** $\forall j \in \{1, 2, \ldots, m\}, d_{\bar{i},j} = d_{\bar{i},m}$. We have the equation that $r(p_{\bar{i}}, C_p) = \max_{j=\{1,2,\cdots,m\}} r(p_{\bar{i}}, c_j) = r(p_{\bar{i}}, C) = d_{\bar{i},m}$. As $r(o_{\bar{i}}) = \epsilon$ and $d(p_{\bar{i}}, o_{\bar{i}}) \leq d_{\bar{i},m}$, we know that $f(p_{\bar{i}}, o_{\bar{i}}) = d_{\bar{i},m}$ in this case. We have

$$f(p_{\bar{i}}, o_{\bar{i}}) = \max\{r(p_{\bar{i}}, C_p), r(o_{\bar{i}}, C_o), d(p_{\bar{i}}, o_{\bar{i}})\} \leq d_{\bar{i},m}.$$

To sum up, $f(p_{\bar{i}}, o_{\bar{i}}) \leq d_{\bar{i},m}$. This proves the lemma. □

In the following, we will prove the lower bound of the optimal solution.

Lemma 2. *The optimal solution of 2-MCCSP satisfies:*

$$f(p^*, o^*) = \max\{r(o^*, C_o), r(p^*, C_p), d(o^*, p^*)\} \geq \frac{d_{\bar{i}, m}}{2}$$

such that the covered color sets satisfied $C_p \cup C_o = C$.

Proof. Let $i^* \leftarrow \arg\min_{i \in \{1,2,\dots,n\}} \max\{r(p_{i^*}), r(o_{i^*}, d(p_{i^*}), o_{i^*}))\}$. The main idea in this proof is based on the difference between definitions of $d_{i^*, m}$ and $d_{\bar{i}, m}$. It holds that $d_{\bar{i}, m} \leq d_{i^*, m}$ according to definition of $\bar{i}$.

We prove it by deriving a contradiction. Suppose that an optimal solution $f(p^*, o^*) = \max\{r(p^*, C_{p^*}), r(o^*, C_{o^*}), d(o^*, p^*)\} < \frac{d_{i^*, m}}{2}$ exists. It leads to that $r(o^*, C_{o^*}) < \frac{d_{i^*, m}}{2}$, $r(p^*, C_{p^*}) < \frac{d_{i^*, m}}{2}$ and $d(p^*, o^*) < \frac{d_{i^*, m}}{2}$. Let color of $q_{i^*, m}$ be color $c_{i^*, m}$, which is $d(p^*, q_{i^*, m}) = d(p^*, c_{i^*, m}) = d_{i^*, m}$. $c_{i^*, m}$ must be covered by at least one of the disk $D(p^*, r(p^*, C_{p^*}))$ and disk $D(o^*, r(o^*, C_{o^*}))$. Because of $r(p^*, C_{p^*}) < \frac{d_{i^*, m}}{2}$, $c_{i^*, m}$ must be covered $D(o^*, r(o^*, C_{o^*}))$. Let q_x be the nearest site of color $c_{i^*, m}$. we consider the following two cases.

- **Case 1:** $q_x \notin D(p^*, d_{i^*, m}) \cap P$. It means that q_x is out of $D(p^*, d_{i^*, m})$. See the instance of Fig. 3. (a) as an example. We can conduct that $d(p^*, q_x) > \frac{d_{i^*, m}}{2}$ because the $q_x \notin D(p^*, d_{i^*, m})$. Thus, we must have $d(o^*, q_x) \geq d(p^*, q_x) - d(p^*, o^*) > \frac{d_{i^*, m}}{2}$ because of the triangle inequality. It makes that $f(p^*, o^*) \geq r(o^*, C_{o^*}) \geq r(o^*, c_{i^*, m}) = d(o^*, q_x) > \frac{d_{i^*, m}}{2}$, which is in contradiction with the assumption $f(p^*, o^*) < \frac{d_{i^*, m}}{2}$.
- **Case 2:** $q_x \in D(p^*, d_{i^*, m}) \cap P$. It means that q_x is in $D(p^*, d_{i^*, m})$. See the instance of Fig. 3. (b) as an example. As the $c_{i^*, m}$ is covered by $D(o^*, r(o^*, C_{o^*}))$, $r(o^*, C_{o^*}) \geq d(o^*, c_{i^*, m}) = d(o^*, q_{i^*, m})$. Because of the triangle inequality $d(o^*, p^*) + d(o^*, q_{i^*, m}) > d_{i^*, m}$ and $d(o^*, p^*) < \frac{d_{i^*, m}}{2}$, $f(p^*, o^*) \geq r(o^*, C_{o^*}) \geq d(o^*, q_{i^*, m}) > \frac{d_{i^*, m}}{2}$, which is in contradiction with the assumption of $f(p^*, o^*) < \frac{d_{i^*, m}}{2}$.

Thus, the assumption of $f(p^*, o^*) < \frac{d_{i^*, m}}{2}$ does not hold, and we have $f(p^*, o^*) \geq \frac{d_{i^*, m}}{2} \geq \frac{d_{\bar{i}, m}}{2}$, in which the last inequality follows the definition of $\bar{i}$.

This proves the Lemma 2.   □

Theorem 3. *The approximation ratio of Algorithm2 for the 2-MCCSP is at most 2.*

Proof. The approximation ratio α can be conducted by

$$\alpha = \frac{f(p_{\bar{i}}, o_{\bar{i}})}{f(p^*, o^*)} = \frac{\max\{r(p_{\bar{i}}), r(o_{\bar{i}}), d(p_{\bar{i}}, o_{\bar{i}})\}}{\max\{r(p^*), r(o^*), d(p^*, o^*)\}}$$

$$\leq \frac{\max\{r(p_{\bar{i}}), r(o_{\bar{i}}), d(p_{\bar{i}}, o_{\bar{i}})\}}{\frac{d_{\bar{i}, m}}{2}} \leq \frac{d_{\bar{i}, m}}{\frac{d_{\bar{i}, m}}{2}} = 2,$$

where the first inequality comes from Lemma 2, and the last inequality follows from Lemma 1. This proves the theorem.   □

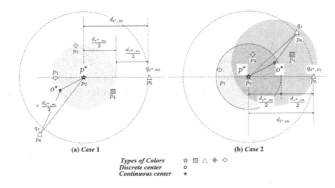

Fig. 3. Two cases in the proof of Lemma 2. Case 1 is about q_x out of $D(p^*, d_{i^*,m}) \cap P$, and Case 2 is that q_x in $D(p^*, d_{i^*,m}) \cap P$. In both instances, all sites are in set $P = \{p_1, p_2, p_3, p_4, p_5, p_6\}$, and all colors are in set $C = \{c_1, c_2, c_3, c_4, c_5\}$, where $n = 6$, $m = 5$, p^* is on p_2, $d_{i^*,m} = d(p_2, p_5)$.

6   Conclusions

Inspired by the concept of cloud manufacturing, we introduced a novel variant of Min-Cost Color-spanning Problems with Mixed Centers, termed as 2-MCCSP. This variant represents a mixed-center adaptation of the classic color-spanning disk problem. We presented an exact algorithm with a polynomial time complexity of $O(n^7, n^5 m^3 \log n)$, utilizing two existing algorithms for n points, where each point is assigned one of the given m colors ($m \leq n$). Moreover, we developed a 2-approximation algorithm, which significantly reduces the time complexity to $O(nm \log n)$.

Looking forward, it would be intriguing to extend our model to other cost formulations and color-spanning objects. This could pave the way for designing even more efficient and effective algorithms.

Acknowledgements. YW and YX are supported by the National Natural Science Foundation of China (NSFC) (Grant No. 71832001). YX is partially supported by the National Natural Science Foundation of China (NSFC) (No.72301209). HZ is partially supported by the National Natural Science Foundation of China (NSFC) (No. 72071157, No. 72192834).

References

1. Abellanas, M., et al.: Smallest color-spanning objects. In: auf der Heide, F.M. (ed.) ESA 2001. LNCS, vol. 2161, pp. 278–289. Springer, Heidelberg (2001). https://doi.org/10.1007/3-540-44676-1_23
2. Abellanas, M., et al.: The farthest color voronoi diagram and related problems. In: European Workshop on Computational Geometry, pp. 113–116 (2006)
3. Agarwal, P.K., Sharir, M., Welzl, E.: The discrete 2-center problem. In: Proceedings of the Thirteenth Annual Symposium on Computational Geometry, pp. 147–155 (1997)

4. Banerjee, S., Misra, N., Nandy, S.C.: Color spanning objects: algorithms and hardness results. Disc. Appl. Math. **280**, 14–22 (2020)
5. Bereg, S., Ma, F., Wang, W., Zhang, J., Zhu, B.: On some matching problems under the color-spanning model. Theor. Comput. Sci. **786**, 26–31 (2019)
6. Chen, D., Chen, R.: New relaxation-based algorithms for the optimal solution of the continuous and discrete p-center problems. Comput. Oper. Res. **36**(5), 1646–1655 (2009)
7. Das, S., Goswami, P.P., Nandy, S.C.: Smallest color-spanning object revisited. Int. J. Comput. Geom. Appl. **19**(05), 457–478 (2009)
8. Fleischer, R., Xu, X.: Computing minimum diameter color-spanning sets. In: International Workshop on Frontiers in Algorithmics, pp. 285–292 (2010)
9. Fleischer, R., Xu, X.: Computing minimum diameter color-spanning sets is hard. Inf. Process. Lett. **111**(21–22), 1054–1056 (2011)
10. Fortune, S.: A sweepline algorithm for voronoi diagrams. In: Proceedings of the Second Annual Symposium on Computational Geometry, pp. 313–322 (1986)
11. Hershberger, J.: A faster algorithm for the two-center decision problem. Inf. Process. Lett. **47**(1), 23–29 (1993)
12. Ju, W., Fan, C., Luo, J., Zhu, B., Daescu, O.: On some geometric problems of color-spanning sets. J. Comb. Optim. **26**(2), 266–283 (2013)
13. Kazemi, M.R., Mohades, A., Khanteimouri, P.: Approximation algorithms for color spanning diameter. Inf. Process. Lett. **135**, 53–56 (2018)
14. Khanteimouri, P., Mohades, A., Abam, M.A., Kazemi, M.R.: Computing the smallest color-spanning axis-parallel square. In: International Symposium on Algorithms and Computation, pp. 634–643 (2013)
15. Li, C., Fan, C., Luo, J., Zhong, F., Zhu, B.: Expected computations on color spanning sets. J. Comb. Optim. **29**(3), 589–604 (2015)
16. Mantas, I., Papadopoulou, E., Sacristán, V., Silveira, R.I.: Farthest color Voronoi diagrams: complexity and algorithms. In: Latin American Symposium on Theoretical Informatics, pp. 283–295 (2021)
17. Megiddo, N.: Linear-time algorithms for linear programming in r̂3 and related problems. SIAM J. Comput. **12**(4), 759–776 (1983)
18. Sharir, M.: A near-linear algorithm for the planar 2-center problem. In: Proceedings of the Twelfth Annual Symposium on Computational Geometry, pp. 106–112 (1996)
19. Wang, H.: On the planar two-center problem and circular hulls. Disc. Comput. Geom. **68**, 1–52 (2022)
20. Wang, Y., Xu, Y., Zhang, H., Tong, W.: Online k-color spanning disk problems (2023)
21. Xu, Y., Peng, J., Xu, Y.: The mixed center location problem. J. Comb. Optim. **36**, 1128–1144 (2018)
22. Xu, Y., Peng, J., Xu, Y., Zhu, B.: The discrete and mixed minimax 2-center problem. In: Lu, Z., Kim, D., Wu, W., Li, W., Du, D.-Z. (eds.) COCOA 2015. LNCS, vol. 9486, pp. 101–109. Springer, Cham (2015). https://doi.org/10.1007/978-3-319-26626-8_8
23. Zhang, D., Chee, Y.M., Mondal, A., Tung, A.K., Kitsuregawa, M.: Keyword search in spatial databases: towards searching by document. In: IEEE 25th International Conference on Data Engineering, pp. 688–699 (2009)

A Dynamic Parameter Adaptive Path Planning Algorithm

Guangyu Yao, Nan Zhang$^{(\boxtimes)}$, Zhenhua Duan, and Cong Tian

Institute of Computing Theory and Technology, and ISN Laboratory,
Xidian University, Xi'an 710071, China
`nanzhang@xidian.edu.cn`, {`zhhduan,ctian`}`@mail.xidian.edu.cn`

Abstract. Path planning in complex environments has always been a
focus of research for scholars both domestically and internationally. This
study addresses the challenge of path planning that combines obstacle
avoidance and optimal path searching in scenarios lacking prior knowledge. The proposed approach introduces a parameter dynamic adaptation strategy for path planning. Experimental investigations are conducted using grid-based maps, and the results demonstrate that the
method presented in this paper surpasses Q-learning and Sarsa algorithms in terms of comprehensive exploration, enhanced stability, and
quicker convergence speed.

Keywords: Path planning · Reinforcement learning · ε-greedy policy

1 Introduction

Path planning is essential in real-world production, optimizing processes, increasing efficiency, cutting costs, and enhancing safety and flexibility. Optimal path
planning and effective obstacle avoidance are critical challenges [7]. The objective is for agents to achieve the task of searching for a relatively optimal route
from the starting point to the destination, utilizing excellent performance metrics [6,13].

Path planning methods have a long history, and different algorithms yield
varying results under different constraints. The traditional Dijkstra's algorithm
[3]was introduced in 1956. This method was developed to address the single-
source shortest path problem. The A* algorithm [4] is a path planning algorithm
based on heuristic search. It combines cost and heuristic information, making it
quite efficient. Genetic algorithms [8] can also be employed to solve path planning
problems. Global search algorithms can find relatively optimal solutions in path
planning but may have slower convergence speeds.

This research is supported by National Natural Science Foundation of China under
Grant Nos. 62272359 and 62172322; Natural Science Basic Research Program of
Shaanxi Province under Grant Nos. 2023JC-XJ-13 and 2022JM-367.

W. Wu and J. Guo (Eds.): COCOA 2023, LNCS 14462, pp. 227–239, 2024.
https://doi.org/10.1007/978-3-031-49614-1_17

In many real-world applications, intelligent agents face increased uncertainty in their environments. This has led to the rise of reinforcement learning, gaining attention among researchers to address these challenges [12]. Reinforcement Learning [9,10] is fundamentally a machine learning approach that involves learning through a "trial-and-error" process. In [11], neural networks are combined with the Q-Learning algorithm from reinforcement learning to address path planning problems in diverse environments. [1] introduces a novel approach by combining greedy and Boltzmann probability selection strategies as a means to avoid getting trapped in local optima. In [5], heuristic knowledge is integrated into reinforcement learning, resulting in improved efficiency for path planning and obstacle avoidance within the Deep Q-learning Network (DQN) algorithm. In [2], by optimizing convolutional neural networks in a specific way, the DQN algorithm is successfully employed to navigate through 3D mazes using visual information.

The main focus of this study is the application of reinforcement learning methods in path planning issue. The primary objective of this research is to enable intelligent agents to navigate even in unknown and relatively intricate scenarios. We propose a path planning strategy that involves dynamic parameter adaptation. The proposed strategy demonstrates more comprehensive search, better convergence.

2 Two Reinforcement Learning Algorithms

Reinforcement Learning is a machine learning algorithm based on the interaction between an agent and its environment, aiming to maximize the cumulative long-term reward through a sequence of decisions. The fundamental principles are illustrated in Fig. 1.

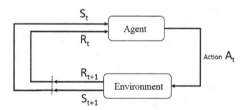

Fig. 1. Reinforcement Learning Diagram

Assuming the current step t, then S_t, S_{t+1}, A_t, R_t and R_{t+1} respectively represent the current state, the next state, the action, the current reward, and the next reward. The principle is to select the best action based on the current state and current reward, and generate the next reward and transition to the next state. Through learning an appropriate policy, the intelligent agent can make optimal action choices when facing different environmental states, with the aim of maximizing its expected cumulative reward.

2.1   Q-Learning Algorithm

The Q-learning algorithm is one of the most widely used techniques in reinforcement learning. Q represents the value associated with a state-action pair, representing the expected return when the agent takes action $a(a \in A)$ in state $s(s \in S)$. The environment provides an immediate reward r based on the agent's action a. Through continuous interaction between the agent and the environment, the algorithm updates the expected values based on the actual reward r obtained by taking action r in state s .

Q-learning employs a temporal difference (TD) method to update the value $Q_{(s_t, a_t)}$, and the update formula is as follows: $Q_{(s_t,a_t)} \leftarrow Q_{(s_t,a_t)} + \alpha(R + \gamma \max_{a'} Q_{(s_{t+1}, a')} - Q_{(s_t,a_t)})$. a' represents the action that yields the maximum value in the subsequent state. The pseudocode for the Q-learning algorithm is presented in Table 1.

Table 1. Q-learning Algorithm Pseudocode

Algorithm 2.1: Q-learning Algorithm
Initialization:
set the learning rate α; the discount factor γ; maximum number of iterations T; the value function $Q_{(s,a)}$
while Repeat for each episode **do**
Initialize the starting state s
while Repeat for each step of the episode **do**
Choose action a from state s
Take action a, observe reward R and new state s'
Update $Q_{(s,a)}$ using the Q-learning update rule:
$Q_{(s,a)} \leftarrow Q_{(s,a)} + \alpha(R + \gamma \max_{a} Q_{(s',a)} - Q_{(s,a)})$
Update state $s \leftarrow s'$
end while
end while

2.2   Sarsa Algorithm

The distinction between the Sarsa algorithm and the Q-learning algorithm lies in the different strategies for selecting the next action and the timing of updating the value function $Q_{(s,a)}$.

In Q-learning, the next action is determined using a greedy policy, selecting the action with the highest Q value. In contrast, Sarsa employs an ε-greedy policy. Additionally, the timing of $Q_{(s,a)}$ updates also varies between the two algorithms. In Q-learning, the update of $Q_{(s,a)}$ takes place immediately after executing the current action and selecting the next action. In Sarsa, the update of $Q_{(s,a)}$ occurs after executing both the current action and the selected next action.

The pseudocode for the Sarsa algorithm is presented in Table 2.

Table 2. Sarsa Algorithm Pseudocode

Algorithm 2.2: Sarsa Algorithm
Initialization:
set the learning rate α; the discount factor γ; maximum number of iterations T; the value function $Q_{(s,a)}$; The value of ε in the ε-greedy policy
while Repeat for each episode **do**
Initialize the starting state s
while Repeat for each step of the episode **do**
Choose action a from state s
Take action a, observe reward R and new state s'
Choose next action a' from new state s'
Update $Q_{(s,a)}$ using the Sarsa update rule:
$Q_{(s,a)} \leftarrow Q_{(s,a)} + \alpha(R + \gamma Q_{(s',a')} - Q_{(s,a)})$
Update state $s \leftarrow s'$ action $a \leftarrow a'$
end while
end while

3   DPARL Algorithm

The Dynamic Parameter Adaptive-Reinforcement Learning (DPARL) algorithm is a reinforcement learning approach with dynamically changing parameters. Specifically, it involves the dynamic adjustment of the ε-greedy policy in the Sarsa algorithm. Depending on the progression of the path search process, whether at different episodes or different time steps within the same episode, the probability of random exploration varies accordingly. This adaptive approach is aimed at enhancing the interaction efficiency between the agent and the environment, ultimately improving the performance of the original algorithm.

Consider the path planning problem on a grid map as an example. During the initial exploration phase, the agent lacks a comprehensive understanding of the environment, necessitating substantial random actions for exploration. Towards the end of exploration, the agent should become more targeted in its exploration, requiring adaptive adjustments to its perception range. In the original strategy, the value of ε remains fixed. Thus, altering the ε factor in stages to accommodate the needs of path planning is crucial. The aim is to have a higher-than-average likelihood of random exploration in the initial stage of the program and a lower-than-average likelihood towards the end.

The expression for the sigmoid function is as follows:

$$Sigmoid(x) = \frac{1}{1 + e^{-x}} \tag{1}$$

The graph of this function is depicted in Fig. 2. The graph illustrates that as x approaches negative infinity, the function value approaches 0 with a small slope. Conversely, as x tends towards positive infinity, the function value approaches 1 with a small slope. The function is bounded with values between 0 and 1. And the curve is smooth, continuously differentiable everywhere. Hence, the sigmoid function aligns with the desired probability reduction approach.

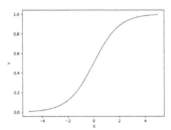

Fig. 2. Sigmoid Function

This paper proposes a path planning strategy called DPARL that improves Sarsa algorithm.

The ε-greedy policy is a commonly used trade-off strategy for exploration and exploitation in reinforcement learning. In the ε-greedy policy, ε is a positive number less than 1, typically between 0 and 1. The main idea of the strategy is that, during each action selection, the action with the highest estimated value is chosen with a probability of 1-ε, while random exploration is conducted with a probability of ε by selecting a random action. Therefore, the value of ε controls the degree of exploration. For example, if $\varepsilon = 0.1$, then in each decision-making step, there is a 90% probability of selecting the action with the highest estimated value, and a 10% probability of selecting a random action. The parameter ε can be adjusted using the following formula:

$$\varepsilon = \frac{\rho}{1 + e^{-(\frac{max\_iterations}{2} - n\_iter)}} + \phi \tag{2}$$

In the initial stage of the program, when $n\_iter - \frac{max\_iterations}{2}$ is at its minimum, $\frac{max\_iterations}{2} - n\_iter$ is at its maximum, $\varepsilon \approx \rho + \phi$ is at its maximum. As the program progresses, when $n\_iter - \frac{max\_iterations}{2} = 0$, $\frac{max\_iterations}{2} - n\_iter = 0$, $\varepsilon = \frac{\rho}{2} + \phi$. In the final stage of the program, when $n\_iter - \frac{max\_iterations}{2}$ is at its maximum, $\frac{max\_iterations}{2} - n\_iter$ is at its minimum, $\varepsilon \approx \phi$ is at its minimum.

Hence, by appropriately setting the parameters ρ and ϕ according to the specific problem, the desired probability reduction approach will be satisfied.

For a single episode's iteration cycle, ε can also be dynamically adjusted. When the agent is farther from the target location, set ε to a value higher than the average probability, promoting exploration. Conversely, when the agent is closer to the target location, set ε to a value lower than the average probability, focusing on exploitation. Using μ to represent the adaptive factor for the fluctuation of ε. As shown in the formula 3 (ε in formula 3 is the same as ε in formular 2):

$$\varepsilon_{episode} = \frac{\mu}{1 + e^{-[(row\_current + col\_current) - \frac{row+col}{2}]}} + \varepsilon - \frac{\mu}{2} \tag{3}$$

In a single episode iteration, when the agent is at the farthest distance from the target position, $(row\_current + col\_current) - \frac{row+col}{2}$ is at its maximum, $\varepsilon_{episode} = \varepsilon + \frac{\mu}{2}$ is at its maximum. This encourages exploration with a probability higher than the average ε. When the agent is closest to the target position, $(row\_current + col\_current) - \frac{row+col}{2}$ is at its minimum, $\varepsilon_{episode} = \varepsilon - \frac{\mu}{2}$ is at its minimum. This leads to exploration with a probability lower than the average ε.

The parameter dynamic adjustment path planning strategy adapts the random exploration probability, enhancing the efficiency of exploring the environment. This strategy better balances the exploration-exploitation trade-off, improves the interaction efficiency between the agent and the environment, and

Table 3. DPARL Algorithm Pseudocode

Algorithm 3.1: DPARL Algorithm
Initialization:
set the learning rate α; the discount factor γ; maximum number of iterations T; the adaptive factor ρ; the adaptive factor ϕ; the adaptive factor μ; the value function $Q_{(s,a)}$; the value of ε in the ε-greedy policy
while Repeat for each episode **do**
Initialize the starting state s
Update the value of ε
while Repeat for each step of the episode **do**
Update the value of $\varepsilon_{episode}$
Choose action a from state s through ε-greedy policy
Take action a, observe reward R and new state s'
Update $Q_{(s,a)}$ using the Sarsa update rule:
$Q_{(s,a)} \leftarrow Q_{(s,a)} + \alpha(R + \gamma Q_{(s',a')} - Q_{(s,a)})$
Update state $s \leftarrow s'$ action $a \leftarrow a'$
end while
end while

enhances the performance of the original algorithm. The pseudocode for the DPARL algorithm is presented in Table 3.

4   Experimental Setup and Experimental Results

4.1   Description of the Path Planning Problem

The experimental environment consists of maps constructed using grid cells, as depicted in Fig. 3. There are two maps: one composed of 5*5 grid cells and the other of 10*10 grid cells. Gray grid cells represent obstacles, white grid cells depict possible paths, red dots indicate the starting points, and blue dots mark the destination points. The agent begins from the starting coordinates and navigates to the destination point while passing only through white grid cells.

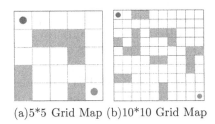

(a)5*5 Grid Map (b)10*10 Grid Map

Fig. 3. Grid Maps (Color figure online)

4.2   Modeling and Parameter Settings for the Path Planning Problem

4.2.1   State and Action Set
The starting grid cell serves as the initial state, the termination grid cell denotes the goal state, while the collection of all white grid cells constitutes the state set. The set of actions comprises valid actions after excluding illegal ones, defined as actions that would lead the agent to step outside the grid boundaries. The set of actions can be represented using coordinate transformations.

4.2.2   Q-Matrix
The matrix contains probabilities corresponding to state-action pairs.

4.2.3   Reward Function
When the agent moves out of the map boundaries or enters a gray grid cell, it receives a punitive reward of -10. When the agent reaches the target location, it receives a reward of 10. As the agent gets closer to the target location, it receives a reward of 0. If the agent moves away from the target location, it receives a punitive reward of –0.2.

4.2.4 Partial Parameter Settings

For a 5*5 map, the iteration number T is set to 100. For a 10*10 map, the iteration number T is set to 500. Learning rate α is 0.1. Discount factor γ is 0.9. Greedy policy parameter ε is 0.2. Dynamic adaptive factor ϕ is 0.15. ε adjustment factor μ is 0.05.

4.2.5 Experimental Environment

The experiments were conducted on a machine with 64 GB of RAM, using VMware Workstation 15 as the virtualization software. The operating system employed was Ubuntu 20.04, with 16 GB of virtual memory allocated. The programming language used for the experiments was Python 3.

4.2.6 Evaluation Metrics

(1) Runtime: The time taken for the algorithm to perform one path planning task. Shorter runtime indicates lower resource consumption of the algorithm.
(2) Average steps: The average length of paths generated by the algorithm in one path planning task.
(3) First detection of optimal path time: The time taken by the algorithm to first detect the expected optimal path in one path planning task.
(4) First detection of optimal path iteration number: The iteration number at which the algorithm first detects the expected optimal path in one path planning task.

4.3 Experimental Results and Analysis

The experiment designed two scenarios: a 5*5 grid map and a 10*10 grid map. Path planning was conducted using three different methods in each scenario. Optimal path diagrams were generated and the performance of the three methods was compared.

4.3.1 5*5 Grid Map Scenario

In Experiment Scenario 1, it's evident that there is more than one shortest path. Even among paths with the same shortest length, some algorithms might favor turns while others might prefer straight-line paths. In this experiment, each algorithm was trained 100 times, and the resulting path data was collected. The path that appeared most frequently was chosen as the optimal path for that algorithm. The optimal paths generated by these three algorithms in this experiment are displayed in Fig. 4, where the red dashed line represents the optimal path for each algorithm.

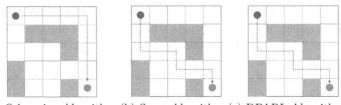

(a) Q-learning Algorithm (b) Sarsa Algorithm (c) DPARL Algorithm

Fig. 4. Three Different Algorithms' Optimal Paths (Color figure online)

From Fig. 4, it is apparent that all three algorithms can provide optimal paths. However, the Q-learning algorithm exhibits fewer turns, while the other two algorithms display more frequent turns, indicating that the Q-learning algorithm tends to favor straight-line exploration and is less likely to change its behavioral habits. Conversely, this suggests that the other two algorithms lean towards comprehensive exploration, resulting in a more evenly distributed behavior pattern. In addition to this, the paper also compares the performance of the three algorithms by calculating the average runtime, average step count, time taken for the first detection of the optimal path, and the number of iterations for the first detection of the optimal path over 100 runs. The results are summarized in Table 4.

Table 4. Performance Comparison of Three Algorithms (Averaged Over 100 Runs)

Algorithm	Q-learning	Sarsa	DPARL
Average Runtime (ms)	77.8	52.5	53.6
Average Step Count(steps)	17.6	14.6	13.5
Time for 1st Optimal Path(ms)	18.3	17.2	11.9
Iterations for 1st Path	12.5	6.2	4.6

From Table 4, it is evident that among the three algorithms, the Sarsa algorithm consumes the least amount of time and Q-learning requires the longest time. Looking at the average step count, both the DPARL and Sarsa algorithms exhibit similar step counts, both significantly lower than the Q-learning algorithm. Analyzing the time taken for the first detection of the optimal path, the DPARL algorithm demonstrates the shortest time and the Q-learning algorithm takes the longest time. In terms of the number of iterations required for the first optimal path, DPARL demands the fewest iterations, followed by Sarsa, and Q-learning requires the most iterations. This observation suggests that the DPARL algorithm is capable of finding the optimal path with fewer iterations and less time, showcasing excellent stability and a lower frequency of encountering invalid paths.

In Scenario 1, the path planning was performed using three different algorithms, and the reward curves for single training runs are depicted in Fig. 5. The horizontal axis represents the current iteration episodes, while the vertical axis represents the total reward obtained in each episode.

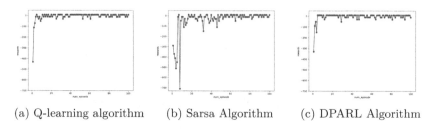

(a) Q-learning algorithm   (b) Sarsa Algorithm   (c) DPARL Algorithm

Fig. 5. The rewards curves for the three different algorithms

From Fig. 5, it can be observed that when using the DPARL algorithm, the agent is able to reach the optimal path earlier and more frequently. On the other hand, the Q-learning algorithm and Sarsa algorithm take longer to converge to the optimal path and exhibits more instability.

4.3.2   10*10 Grid Map Scenario
In Experiment Scenario 2, similar to Experiment Scenario 1, each algorithm was trained 100 times, and the most frequent path result among the trials was selected as the best path for that algorithm. The best paths obtained by the three algorithms in this experiment are illustrated in Fig. 6, where the red dashed lines represent the optimal paths determined by each algorithm.

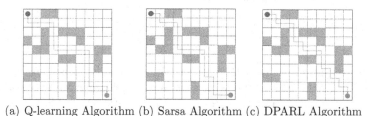

(a) Q-learning Algorithm (b) Sarsa Algorithm (c) DPARL Algorithm

Fig. 6. Three Different Algorithms' Optimal Paths (Color figure online)

From Fig. 6, it is evident that all three algorithms are capable of providing optimal paths. However, Q-learning algorithm demonstrates fewer turns in comparison to Sarsa algorithm, while the DPARL algorithm exhibits more turns. In addition to this, the paper also compares the performance of the three algorithms by calculating the average runtime, average step count, time taken for the first detection of the optimal path, and the number of iterations for the first detection of the optimal path over 100 runs. The results are summarized in Table 5.

Table 5. Performance Comparison of Three Algorithms (Averaged Over 100 Runs)

Algorithm	Q-learning	Sarsa	DPARL
Average Runtime (ms)	7707.2	517.5	490.7
Average Step Count(steps)	597.4	36.9	32.9
Time for 1st Optimal Path(ms)	7299.1	201.3	161.5
Iterations for 1st Path	135.4	80.3	66.5

From Table 5, it is evident that among the three algorithms, DPARL algorithm has the shortest runtime, while Q-learning's runtime is 14 times that of the former two. Looking at the average number of steps, both DPARL and Sarsa algorithms exhibit similar values, significantly fewer than Q-learning algorithm. Regarding the time taken for the first detection of the optimal path, DPARL algorithm performs the best and Q-learning has the longest time. In terms of the number of iterations required for the first detection of the optimal path, DPARL algorithm outperforms the others, while Q-learning requires the most iterations. This indicates that DPARL algorithm is capable of finding the optimal path with fewer iterations and in less time, demonstrating good stability, and resulting in fewer occurrences of ineffective paths.

In Scene 2, the path planning was performed using three different algorithms. The reward curves for each algorithm in a single training run are shown in Fig. 7. The horizontal axis represents the current iteration episodes, while the vertical axis represents the total reward obtained in each episode.

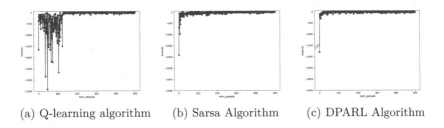

(a) Q-learning algorithm   (b) Sarsa Algorithm   (c) DPARL Algorithm

Fig. 7. The rewards curves for the three different algorithms

From Fig. 7, it is evident that when using the DPARL algorithm, the intelligent agent is able to reach the optimal path earlier and more frequently. As the complexity of the scene increases, the DPARL algorithm exhibits strong stability, avoiding the occurrence of overly long paths. This algorithm appears to be well-suited for path planning in intricate maps.

5   Conclusions

In complex environments lacking prior environmental information, traditional path planning algorithms suffer from drawbacks such as high computational complexity, low efficiency, and unstable results. To address this challenge, this paper proposes a path planning strategy called DPARL. This strategy increases the random search probability in the early stage and decreases it in the later stage of the program, enhancing the interaction efficiency between the agent and the environment, and improving the performance of the existing algorithm. The paper compares the DPARL with the Q-learning algorithm and the Sarsa algorithm. The DPARL demonstrates more comprehensive search, better convergence, and the ability to find optimal paths with fewer iterations and less time. Additionally, it maintains high stability even in complex maps.

In future work, further enhancement of the adaptability of the adaptive algorithm is needed. Additionally, applying parameter adaptive path planning algorithm to high-precision modeling environments is also a potential research area.

References

1. Chen, H., Ji, Y., Niu, L.: Reinforcement learning path planning algorithm based on obstacle area expansion strategy. Intell. Serv. Robot. **13**(2), 289–297 (2020)
2. Devo, A., Costante, G., Valigi, P.: Deep reinforcement learning for instruction following visual navigation in 3D maze-like environments. IEEE Rob. Autom. Lett. **5**(2), 1175–1182 (2020)
3. Dijkstra, E.W.: A note on two problems in connexion with graphs. In: Edsger Wybe Dijkstra: His Life, Work, and Legacy, pp. 287–290 (2022)
4. Hart, P.E., Nilsson, N.J., Raphael, B.: A formal basis for the heuristic determination of minimum cost paths. IEEE Trans. Syst. Sci. Cybern. **4**(2), 100–107 (1968)
5. Jiang, L., Huang, H., Ding, Z.: Path planning for intelligent robots based on deep q-learning with experience replay and heuristic knowledge. IEEE/CAA J. Automatica Sinica **7**(4), 1179–1189 (2019)
6. Patle, B., Pandey, A., Parhi, D., Jagadeesh, A., et al.: A review: on path planning strategies for navigation of mobile robot. Defence Technol. **15**(4), 582–606 (2019)
7. Polydoros, A.S., Nalpantidis, L.: Survey of model-based reinforcement learning: applications on robotics. J. Intell. Rob. Syst. **86**(2), 153–173 (2017)
8. Santiago, R.M.C., De Ocampo, A.L., Ubando, A.T., Bandala, A.A., Dadios, E.P.: Path planning for mobile robots using genetic algorithm and probabilistic roadmap. In: 2017IEEE 9th International Conference on Humanoid, Nanotechnology, Information Technology, Communication and Control, Environment and Management (HNICEM), pp. 1–5. IEEE (2017)
9. Sutton, R.S., Barto, A.G.: Reinforcement Learning: An Introduction. MIT press, Cambridge (2018)
10. Szepesvári, C.: Algorithms for Reinforcement Learning. Springer, Heidelberg (2022). https://doi.org/10.1007/978-3-031-01551-9
11. Wei, J., De Hua, Z., Shuangbao, M., Gaocheng, Y., Wei, C.: Dynamic walking characteristics and control of four-wheel mobile robot on ultra-high voltage multi-split transmission line. Trans. Inst. Meas. Control. **44**(6), 1309–1322 (2022)

12. Yang, Y., Juntao, L., Lingling, P.: Multi-robot path planning based on a deep reinforcement learning DQN algorithm. CAAI Trans. Intell. Technol. **5**(3), 177–183 (2020)
13. Zhang, H.Y., Lin, W.M., Chen, A.X.: Path planning for the mobile robot: a review. Symmetry **10**(10), 450 (2018)

On the Mating Between a Polygonal Curve and a Convex Polygon

Jin-Yi Liu[✉]

School of AI and Software, Liaoning Petrochemical University, Fushun 113001,
Liaoning, People's Republic of China
j_y_liu@sina.com

Abstract. Given a simple polygonal curve with m edges and a convex polygon with n edges lying respectively in two distinct parallel planes, we consider both the decision version and the optimization version of their mating problem. The decision version asks for whether a simple polyhedron can be constructed by a triangle connection between them without new vertices being inserted, and the optimization version is to obtain an optimal polyhedron if possible. This restricted curve-polygon mating problem is a natural variant of the previously studied polygon-polygon mating problem, whose computational complexity is still open even in the special case with one polygon being convex. In this paper, we first present an $O(mn^4)$-time dynamic-programming algorithm for both the decision version and the optimization version, and then present an $O(mn + n^2)$-time greedy algorithm for the decision version only. Additionally, we show that whether a polygonal curve is potentially matable with some convex polygon can be decided in linear time.

Keywords: Mating · Polygonal curve · Convex polygon · Polyhedron

1 Introduction

Motivated by exploring the fundamental issues raised in the reconstruction of three-dimensional objects from parallel slices, an extensively studied topic in the area of medical imaging, O'Rourke [8] first introduced the terminology "mating" and proposed the polygon-polygon mating problem with his students [5]. We now recite its definition as follows. Let A and B be two polygons, assumed lying respectively in two distinct parallel planes. A mating of A and B is a triangle connection between A and B which forms a simple polyhedron, and a triangle connection is an assignment that maps each edge of A to a single vertex of B and maps each edge of B to a single vertex of A. Notably, the double-pyramid connection is not a mating of two polygons, and Steiner vertices are not allowed to be inserted if not specially mentioned.

Given two simple polygons, the existence of a mating had ever been taken for granted for years until Gitlin et al. [5] presented a nonmatable pair of polygons, where one polygon is a triangle and the other is a labyrinthine polygon with 63

© The Author(s), under exclusive license to Springer Nature Switzerland AG 2024
W. Wu and J. Guo (Eds.): COCOA 2023, LNCS 14462, pp. 240–252, 2024.
https://doi.org/10.1007/978-3-031-49614-1_18

vertices. Barequet et al. [1] simplified the labyrinthine polygon to 45 vertices and additionally gave an unbounded family of nonmatable pairs of polygons. These counterexamples immediately aroused the question: given two polygons, is there a mating between them? This is just the decision version of the polygon-polygon mating problem (called *matability* in [1]). About this problem, the following facts have been made clear: (1) the matability is invariant to a polygon's translation and uniform scaling [8]; (2) the cycling constraint and some shadow conditions are necessary to guarantee non-self-intersection [5]; (3) there are several classes of polygon-polygon pairs (e.g., two star-shaped polygons) that are sufficiently matable [1]. Besides, a recent work by Biedl et al. [2] shows that if the two polygons have the same number of vertices and the mating is limited to "banded" connection, then their matability can be decided in polynomial time. However, since no conditions which are both necessary and sufficient (polynomial-time checkable) have been known, whether this problem is tractable is still open. It is true even in the special case with one polygon being convex (by a personal communication with Gill Barequet).

The curve-polygon mating problem is a natural variant of the polygon-polygon mating problem. Here, the input is a simple polygonal curve P with m edges and a simple polygon Q with n edges, assumed lying on two parallel planes. A triangle connection between them is an assignment that maps each edge of Q to a single vertex of P and maps each edge of P, however, to two distinct vertices of Q. A mating of them is a triangle connection that forms a mountain-like simple polyhedron with exactly $m + n + 1$ vertices and $2m + n$ connecting triangles. Apparently this definition obeys Euler's formula for polyhedra too. While it seems that the curve-polygon mating problem can hardly be exploited in the area of medical imaging, it may have potential applications in the area of constructing objects from sketches [10] or the area of planar shape morphing [6].

Although the curve-polygon mating problem possesses similar properties with the polygon-polygon mating problem, it cannot be reduced to the latter, neither vice versus. But from the fact that there is a very simple nonmatable curve-polygon pair (Fig. 1(a)), we can presume that it may be easier to solve than the latter. But it is not easy enough, since it is almost unimaginable that the so curled four-edge chain in Fig. 1(b) is matable with a convex quadrilateral. To the best of the author's knowledge, there are only two sufficient conditions for the trivial cases of the problem: a point or a line segment is matable with any simple polygon [5].

In this paper, restricting the polygon to being convex, we try to solve in polynomial time both the decision version and the optimization version of the curve-polygon mating problem. The optimization version of the polygon-polygon merging (not mating) problem has ever been considered by many authors under the goal to obtain a good reconstructed object, where several optimization criteria have been employed, including Maximum Volume [7], Minimum Surface Area [4], and Minimum Spanning Length [3]. Generally, all the three optimization objectives can be accomplished in cubic time ($O(m^2n)$ or $O(mn^2)$) by a dynamic-programming schema. However, all the three optimization criteria, as well as others (e.g., the angle criterion in [9]), are no guarantee of non-self-intersection. Unlike these previous approaches, the pursuit of this paper is a possible validity-guaranteed optimal object.

The results and organization of the paper are as follows. After giving some preliminary notations and definitions in Sect. 2, we in Sect. 3 derive a necessary and sufficient condition for the existence of a mating between a polygonal curve and a convex polygon. Interestingly, this condition is unrelated to the edge lengths of the input curve. Based on this condition, we present a series of polynomial-time algorithms in Sect. 4. We first present an $O(mn^4)$-time dynamic-programming algorithm for both the decision version and the optimization version, where the optimization criteria can be any of the above three. Then we present an $O(mn + n^2)$-time greedy algorithm for the decision version only. Also in this section, we show that whether a given polygonal curve is potentially matable with some convex polygon can be decided in linear time, and if yes, we present a linear-time algorithm for generating such a convex polygon. At last in Sect. 5, we conclude the paper and raise some questions that deserve research in the future.

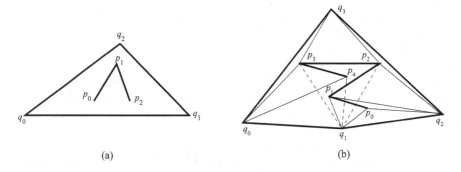

Fig. 1. (a) A nonmatable curve-polygon pair. (b) A matable curve-polygon pair and their one mating viewed orthogonally.

2   Preliminaries

In the rest of the paper, we always use P to denote the given m-edge polygonal curve, described by the vertex sequence $(p_0, p_1, \ldots, p_m)$, and use Q to denote the given n-edge (maybe non-strict) convex polygon, described by the vertex cycle $(q_0, q_1, \ldots, q_{n-1})$. W.l.o.g., we assume that Q is in the xoy plane with the counterclockwise cycling direction, and P is in the plane $z = h$ with h being a given positive height. (In fact, if we are only concerned with the decision version or the optimization version under the Maximum Volume criterion, the actual value of h is not needed.) That is to say, the inputs of our problem are only x-

and y-coordinates of $m+n+1$ vertices and the possible h. For the vertices of Q and sub-chains of Q, we follow the convention that modulo operations on the indices are omitted. For example, when we use q_{i-1} and q_{i+1} for some $0 \le i < n$, their indices actually refer to $(n+i-1) \bmod n$ and $(i+1) \bmod n$ respectively.

For $0 \le k \le \ell \le m$, we use $P_{k\ell}$ to denote the sub-curve of P formed by the vertex sequence $(p_k, p_{k+1}, \ldots, p_\ell)$. For $0 \le i \ne j < n$, we use Q_{ij} to denote the sub-chain of Q formed by the vertex sequence $(q_i, q_{i+1}, \ldots, q_{j-1}, q_j)$. According to this notation, $Q_{i+1,i}$ includes all the edges of Q except $\overline{q_i q_{i+1}}$, and $Q_{i,i+1}$ is just the edge $\overline{q_i q_{i+1}}$. We especially define Q_{ii} to be empty. For two sub-chains of Q, Q_{ij} and $Q_{i'j'}$, if the vertex sequence of $Q_{i'j'}$ is a substring of the vertex sequence of Q_{ij}, we use $Q_{i'j'} \subseteq Q_{ij}$ to denote their relation.

Throughout the paper, we use vertex-ordered line segments, e.g., $\overline{ab}$ represents the closed line segment starting from point a and ending at point b. We define $\overrightarrow{ab} := b - a$ to be the (free) vector directed from point a to point b. In the whole paper we only deal with planar vectors. We follow the normal notations for the cross product and the dot product of two planar vectors. To characterize the relations among planar vectors, we define several other notations for them, which perhaps seem "abnormal." Suppose that $\mathbf{u}$ and $\mathbf{v}$ are two non-degenerate planar vectors. First we use $\mathbf{u} \stackrel{\rightarrow}{=} (\ne)\mathbf{v}$ to denote whether (or not) $\mathbf{u}$ has the same direction with $\mathbf{v}$. Then we define a closed vector range $[\mathbf{u},\mathbf{v}]$ as follows:

$$[\mathbf{u}, \mathbf{v}] := \begin{cases} \{\alpha\mathbf{u} + \beta\mathbf{v} : \alpha \ge 0, \beta \ge 0, \alpha + \beta > 0\} & \text{if } \mathbf{u} \stackrel{\rightarrow}{\ne} -\mathbf{v} \\ \{\mathbf{w} : \mathbf{u} \times \mathbf{w} \ge 0, \mathbf{w} \ne \mathbf{0}\} & \text{if } \mathbf{u} \stackrel{\rightarrow}{=} -\mathbf{v}. \end{cases}$$

Note that when $\mathbf{u} \stackrel{\rightarrow}{\ne} -\mathbf{v}$, we have $[\mathbf{u}, \mathbf{v}] = [\mathbf{v}, \mathbf{u}]$. This notation $[\mathbf{u},\mathbf{v}]$ allows us to simply represent the vector relations induced by sub-chains of Q.

Observation 1. *Let $Q_{i''j''} \subseteq Q_{i'j'} \subseteq Q_{ij}$ be three nested sub-chains of Q. Then by the convexity of Q, we have $[\overrightarrow{q_j q_{i''}}, \overrightarrow{q_i q_{j''}}] \subseteq [\overrightarrow{q_{j'} q_{i''}}, \overrightarrow{q_{i'} q_{j''}}], [\overrightarrow{q_j q_{i'}}, \overrightarrow{q_i q_{j'}}].$*

For a connecting triangle between P and Q, we also use vertex-ordered denotations such that it has a unique normal, and say that it is *positive* if its normal vector points outwards from the constructed object. We now define a mating of P and Q to be *positive* if all the connecting triangles are positive. From the knowledge of a simple polyhedron, we can immediately claim that there is a mating of P and Q if and only if there is positive mating of P and Q. Henceforth, in the rest of the paper, we shall only pursue a positive mating for both the decision version and the optimization version of the problem.

Now, let us consider the two connecting triangles sharing an edge of P in a mating. We view them integrally as a patch of surface, and call it a *ridging triangle-pair*. We use $\Diamond p_k p_{k+1} q_i q_j$ to denote the ridging triangle-pair sharing the edge $\overline{p_k p_{k+1}}$, and use the convention that its two triangles are $\triangle p_k p_{k+1} q_i$ and $\triangle p_{k+1} p_k q_j$. This convention guarantees that $\triangle p_k p_{k+1} q_i$ and $\triangle p_{k+1} p_k q_j$ must be simultaneously positive or simultaneously negative. Based on the assumption on P and Q, we immediately have a rule to decide whether a ridging triangle-pair is positive: $\Diamond p_k p_{k+1} q_i q_j$ is positive iff $\overrightarrow{q_i q_j} \times \overrightarrow{p_k p_{k+1}} > 0$.

Next, let us explore the triangle-fan in a mating determined by some p_k of P and some Q_{ij} of Q. If the triangles in the fan are all positive, they must be in the form of $\triangle p_k q_i q_{i+1}$, $\triangle p_k q_{i+1} q_{i+2}$, ..., $\triangle p_k q_{j-1} q_j$, by the direction of Q. For convenience, we use $\mathcal{F}(p_k, Q_{ij})$ to denote this triangle-fan and call it a *hilling triangle-fan*. Unlike a ridging triangle-pair, the definition of $\mathcal{F}(p_k, Q_{ij})$ guarantees that it must be positive.

The following lemma, which characterizes two neighboring ridging triangle-pairs and induced hilling triangle-fans, can be viewed as a fitting of the cycling constraint (Theorem 1 of [4] or Lemma 6 of [5]) to the curve-polygon mating problem. But we give an independent proof.

Lemma 1. *Let $\diamondsuit p_{k-1} p_k q_{i'} q_{j'}$ and $\diamondsuit p_k p_{k+1} q_i q_j$ be two positive ridging triangle-pairs in a positive mating of P and Q. Then $Q_{i'j'} \subseteq Q_{ij}$, and $\mathcal{F}(p_k, Q_{ii'})$ as well as $\mathcal{F}(p_k, Q_{j'j})$ is a part of the mating.*

Proof. Let us first consider $\triangle p_{k-1} p_k q_{i'}$ and $\triangle p_k p_{k+1} q_i$. If $q_i \neq q_{i'}$, there exists a hole $\mathcal{H}$ on the surface of the constructed object which takes as boundary $\overline{p_k q_i}$, $\overline{p_k q_{i'}}$, and a sub-chain of Q between q_i and $q_{i'}$. This hole $\mathcal{H}$ must be filled with a positive hilling triangle-fan which is either $\mathcal{F}(p_k, Q_{ii'})$ or $\mathcal{F}(p_k, Q_{i'i})$. But in fact, the latter is impossible since $\triangle p_k q_{i-1} q_i$ belongs to it and $\triangle p_k q_{i-1} q_i$ is not consistent with the known $\triangle p_k p_{k+1} q_i$ (for their common edge is not in reversed orders). On the contrary, $\mathcal{F}(p_k, Q_{ii'})$ can appropriately fill the hole $\mathcal{H}$. Thus, we have the conclusion that $\mathcal{F}(p_k, Q_{ii'})$ is a part of the mating.

Similarly, we can conclude that $\mathcal{F}(p_k, Q_{j'j})$ is also a part of the mating. These two conclusions also imply that $q_{j'}, q_j \notin Q_{ii'}$ and $q_{i'}, q_i \notin Q_{j'j}$, so there is only one possibility: $Q_{i'j'} \subseteq Q_{ij}$.   □

Applying induction on k from Lemma 1, we immediately derive a necessary condition for the matability of P and Q: If there is a positive mating of P and Q, then there must be a nested sequence of Q's sub-chains, $Q_{i_1 j_1} \subseteq Q_{i_2 j_2} \subseteq \cdots \subseteq Q_{i_m j_m}$, such that each $\diamondsuit p_{k-1} p_k q_{i_k} q_{j_k}$ is positive for $k = 1, \ldots, m$. But this condition may not be sufficient because the induced ridging triangle-pairs and hilling triangle-fans may be intersecting.

Here, following [1,5], the notion "intersecting" (with its verb or noun) always excludes "normally touching," and it is concise to describe it with the word *stab*. Given a line segment $\overline{ab}$ and a triangle $\triangle$ in the space, we say that $\overline{ab}$ stabs $\triangle$ if they share exactly one common point c such that $c \notin \{a, b\}$. Then given two connecting triangles $\triangle$ and $\triangle'$ between P and Q, we say they are intersecting if and only if one connecting edge of $\triangle$ stabs $\triangle'$, or one connecting edge of $\triangle'$ stabs $\triangle$. About the intersection of two P-based triangles, we have the following observation which can be easily verified.

Observation 2. *Let $\overline{p_{k-1} p_k}$ and $\overline{p_{\ell-1} p_\ell}$ be two distinct edges of P, and q_i and q_j be two distinct vertices of Q. Then $\triangle p_{k-1} p_k q_i$ and $\triangle p_{\ell-1} p_\ell q_j$ are intersecting iff there are a point $p' \in \overline{p_{k-1} p_k}$ and a point $p \in \overline{p_{\ell-1} p_\ell}$ such that $\overrightarrow{p'p} \rightleftarrows \overrightarrow{q_j q_i}$.*

For considering a part of a mating, we need a notation for a *sub-mating*. For $1 \leq k \leq m$, a sub-mating $\mathcal{M}(P_{0,k}, Q_{ij})$ is a part of a positive mating of P and Q, which is induced by $P_{0,k}$ and Q_{ij}. That is, if it exists, $\mathcal{M}(P_{0,k}, Q_{ij})$ is a simple polygonal surface which takes $\overline{q_j p_k}$, $\overline{p_k p_i}$, and Q_{ij} as its closed boundary. Then we give a property of a sub-mating which is useful in the next section.

Lemma 2. *Let $\mathcal{M}(P_{0,k}, Q_{ij})$ be a sub-mating in a positive mating of P and Q, and let $\overline{q_r p_\ell}$ be a connecting edge of it. Then $\overline{q_r p_\ell}$ cannot stab $\triangle p_k q_j q_i$ from the negative side to the positive side.*

Proof. First we point out that $\overline{q_r p_\ell}$ cannot stab $\triangle p_k q_j q_i$ at its boundary since $\overline{q_r p_\ell}$ must be shared by two connecting triangles. Now suppose that $\overline{q_r p_\ell}$ stabs $\triangle p_k q_j q_i$ at an interior point s of $\triangle p_k q_j q_i$ from its negative side to its positive side. Then, the intersection of $\mathcal{M}(P_{0,k}, Q_{ij})$ and the plane $\Pi_{\triangle p_k q_j q_i}$ must include a continuous polygonal curve C that starts from s and ends at an intersection point $p = P_{\ell,k-1} \cap \Pi_{\triangle p_k q_j q_i}$. The existence of p is based on the fact that p_{k-1} is on the negative side of $\triangle p_k q_j q_i$ (by the positiveness of $\Diamond p_{k-1} p_k q_i q_j$) and p_ℓ is on the positive side of $\triangle p_k q_j q_i$, and the continuousness of C is enabled by the fact that $\mathcal{M}(P_{0,k}, Q_{ij})$ is made up of connecting triangles. Now, considering that on $\Pi_{\triangle p_k q_j q_i}$, s is inside $\triangle p_k q_j q_i$ and p is outside $\triangle p_k q_j q_i$, and also considering that C cannot pass through p_k (since p cannot belongs to $\overline{p_{k-1} p_k}$), we have that either $\overline{p_k q_i}$ or $\overline{p_k q_j}$ must strictly intersect C, i.e., either $\overline{p_k q_i}$ or $\overline{p_k q_j}$ must stab a connecting triangle of $\mathcal{M}(P_{0,k}, Q_{ij})$, leading to a contradiction. $\square$

Remark. In the arguments for Lemmas 1 and 2, the convexity of Q is not exploited, thus they hold for general curve-polygon pairs. But if Q is a general simple polygon, the decision whether a ridging triangle-pair $\Diamond p_k p_{k+1} q_i q_j$ is positive needs additional conditions beyond $\overrightarrow{q_i q_j} \times \overrightarrow{p_k p_{k+1}} > 0$.

At the end of this section, we mention that although the property that the matability is invariant to translation and uniform scaling was proved for polygon-polygon pairs in [8], it also holds for curve-polygon pairs (no mater whether Q is convex).

3   A Necessary and Sufficient Condition

In this section, we will derive a necessary and sufficient condition for the matability of P and Q by exploring the possible intersections among the ridging triangle pairs and hilling triangle-fans induced by a nested subsequence of Q's sub-chains. We start with exploring the two hilling triangle-fans sandwiched between two neighboring ridging triangle-pairs.

Lemma 3. *Suppose that there are $Q_{i'j'} \subseteq Q_{ij}$ with $\Diamond p_{k-1} p_k q_{i'} q_{j'}$ and $\Diamond p_k p_{k+1} q_i q_j$ being both positive. Then $\mathcal{F}(p_k, Q_{ii'})$ and $\mathcal{F}(p_k, Q_{j'j})$ are non-intersecting with $\Diamond p_{k-1} p_k q_{i'} q_{j'}$ or $\Diamond p_k p_{k+1} q_i q_j$.*

Proof. By the convexity of Q and the positiveness of the ridging triangle-pairs, these two triangle-fans are separated from $\Diamond p_k p_{k+1} q_i q_j$ by the plane $\Pi_{\triangle p_k q_i q_j}$, and separated from $\Diamond p_{k-1} p_k q_{i'} q_{j'}$ by the plane $\Pi_{\triangle p_k q_{i'} q_{j'}}$, except the common edges. $\square$

How about two neighboring ridging triangle-pairs themselves? At first glance, $\Diamond p_{k-1}p_kq_{i'}q_{j'}$ and $\Diamond p_kp_{k+1}q_iq_j$ seem to be non-intersecting under the condition that $Q_{i'j'} \subseteq Q_{ij}$ and they are both positive, but they can really intersect. Although their intersection can be easily checked with spacial geometry, the following planar checking conditions are crucial for deriving our algorithms.

Lemma 4. *Suppose that there are $Q_{i'j'} \subseteq Q_{ij}$ with $\Diamond p_{k-1}p_kq_{i'}q_{j'}$ and $\Diamond p_kp_{k+1}$ q_iq_j being both positive. Then we have two equivalent conditions for the relations between $\Diamond p_{k-1}p_kq_{i'}q_{j'}$ and $\Diamond p_kp_{k+1}q_iq_j$: they are*

(a) intersecting iff $[\overrightarrow{q_jq_{i'}}, \overrightarrow{q_iq_{j'}}] \subset [\overrightarrow{p_{k-1}p_k}, \overrightarrow{p_kp_{k+1}}]$, and
(b) non-intersecting iff $[\overrightarrow{q_jq_{i'}}, \overrightarrow{q_iq_{j'}}] \cap [\overrightarrow{p_{k-1}p_k}, \overrightarrow{p_kp_{k+1}}] = \emptyset$.

Proof. First, we show that it is impossible that $\overrightarrow{p_{k-1}p_k} \in [\overrightarrow{q_jq_{i'}}, \overrightarrow{q_iq_{j'}}]$ or $\overrightarrow{p_kp_{k+1}} \in [\overrightarrow{q_jq_{i'}}, \overrightarrow{q_iq_{j'}}]$ under the given conditions. By the convexity and direction of Q, we have $\overrightarrow{q_jq_{i'}} \times \overrightarrow{q_iq_{j'}} \geq 0$, $\overrightarrow{q_jq_{i'}} \times \overrightarrow{q_iq_j} \geq 0$, and $\overrightarrow{q_iq_{j'}} \times \overrightarrow{q_iq_j} \geq 0$. So if $\overrightarrow{p_kp_{k+1}} \in [\overrightarrow{q_jq_{i'}}, \overrightarrow{q_iq_{j'}}]$, we immediately have $\overrightarrow{p_kp_{k+1}} \times \overrightarrow{q_iq_j} \geq 0$, which is a contradiction to the positiveness of $\Diamond p_kp_{k+1}q_iq_j$. Similarly we can argue that $\overrightarrow{p_{k-1}p_k} \notin [\overrightarrow{q_jq_{i'}}, \overrightarrow{q_iq_{j'}}]$. By now, we have only two possibilities for the four vectors $\overrightarrow{q_jq_{i'}}$, $\overrightarrow{q_iq_{j'}}$, $\overrightarrow{p_{k-1}p_k}$, and $\overrightarrow{p_kp_{k+1}}$, which are respectively given in (a) and (b).

For the "if" part of (a), we note that for any $p' \in \overline{p_{k-1}p_k}$ and any $p \in \overline{p_kp_{k+1}}$ with $p' \neq p$, we have $\overrightarrow{p'p} \in [\overrightarrow{p_{k-1}p_k}, \overrightarrow{p_kp_{k+1}}]$. So, if $[\overrightarrow{q_jq_{i'}}, \overrightarrow{q_iq_{j'}}] \subset [\overrightarrow{p_{k-1}p_k}, \overrightarrow{p_kp_{k+1}}]$, there must be some $p' \in \overline{p_{k-1}p_k}$ and some $p \in \overline{p_kp_{k+1}}$ such that $\overrightarrow{p'p} \rightleftharpoons \overrightarrow{q_jq_{i'}}$, leading to the intersection between $\triangle p_{k-1}p_kq_{i'}$ and $\triangle p_kp_{k+1}q_j$, by Obs. 2.

For the "if" part of (b), by the convexity of Q, we note that $\overrightarrow{q_iq_{i'}}, \overrightarrow{q_jq_{j'}} \in [\overrightarrow{q_jq_{i'}}, \overrightarrow{q_iq_{j'}}]$ if they are non-degenerate. Hence if $[\overrightarrow{q_jq_{i'}}, \overrightarrow{q_iq_{j'}}] \cap [\overrightarrow{p_{k-1}p_k}, \overrightarrow{p_kp_{k+1}}] = \emptyset$, there are no such $p' \in \overline{p_{k-1}p_k}$ and $p \in \overline{p_kp_{k+1}}$ that $\overrightarrow{p'p} \rightleftharpoons \overrightarrow{q_jq_{i'}}, \overrightarrow{q_iq_{j'}}, \overrightarrow{q_iq_{i'}}$, or $\overrightarrow{q_jq_{j'}}$. That is, by Obs. 2, there are no intersections between $\triangle p_{k-1}p_kq_{i'}$ and $\triangle p_kp_{k+1}q_j$, between $\triangle p_{k-1}p_kq_{j'}$ and $\triangle p_kp_{k+1}q_i$, between $\triangle p_{k-1}p_kq_{i'}$ and $\triangle p_kp_{k+1}q_i$, or between $\triangle p_{k-1}p_kq_{j'}$ and $\triangle p_kp_{k+1}q_j$; or simply say, there is no intersection between $\Diamond p_{k-1}p_kq_{i'}q_{j'}$ and $\Diamond p_kp_{k+1}q_iq_j$.

Since there are only two possibilities, we have no need to argue the "only if" parts for (a) and (b). □

About two non-neighboring ridging triangle-pairs, we have an important necessary condition: Two non-neighboring ridging triangle-pairs are non-intersecting if their in-between neighboring ones are all non-intersecting.

Lemma 5. *Let $P_{k-1,\ell}$ be a sub-curve of P with $1 < k+1 < \ell \leq m$. Suppose there are $Q_{i_kj_k} \subseteq Q_{i_{k+1}j_{k+1}} \subseteq \cdots \subseteq Q_{i_\ell j_\ell}$ such that each $\Diamond p_{r-1}p_rq_{i_r}q_{j_r}$ is positive for $r = k, \ldots, \ell$, and each neighboring pair $\Diamond p_{r-1}p_rq_{i_r}q_{j_r}$ and $\Diamond p_rp_{r+1}q_{i_{r+1}}q_{j_{r+1}}$ are non-intersecting for $r = k, \ldots, \ell-1$. Then $\Diamond p_{k-1}p_kq_{i_k}q_{j_k}$ and $\Diamond p_{\ell-1}p_\ell q_{i_\ell}q_{j_\ell}$ are non-intersecting.*

Proof. Suppose for the aim of contradiction that $\Diamond p_{k-1}p_kq_{i_k}q_{j_k}$ and $\Diamond p_{\ell-1}$ $p_\ell q_{i_\ell}q_{j_\ell}$ are intersecting. Then by Obs. 2, there must be a point $s \in \overline{p_{k-1}p_k}$ and a point $t \in \overline{p_{\ell-1}p_\ell}$ such that $\overrightarrow{st} \rightleftharpoons \overrightarrow{q_{j_\ell}q_{i_k}}, \overrightarrow{q_{i_\ell}q_{j_k}}, \overrightarrow{q_{i_\ell}q_{i_k}}$, or $\overrightarrow{q_{j_\ell}q_{j_k}}$. In any of the four cases, we have that $\overrightarrow{st} \in [\overrightarrow{q_{j_\ell} q_{i_k}}, \overrightarrow{q_{i_\ell} q_{j_k}}]$. Let p_r be a farthest vertex to the line through $\overrightarrow{st}$ among the vertices of $P_{k,\ell-1}$. (If there are multiple farthest vertices, p_r can be any one of them.) We now claim that $\Diamond p_{r-1} p_r q_{i_r} q_{j_r}$ and $\Diamond p_r p_{r+1} q_{i_{r+1}} q_{j_{r+1}}$ must be intersecting, which is a contradiction. The reason of the claim is as follows. Let ζ be the line passing through p_r and parallel to $\overrightarrow{st}$. We use the following way to choose two different points p' and p respectively on $\overline{p_{r-1} p_r}$ and $\overline{p_r p_{r+1}}$: if p_{r-1} is on ζ, we choose $p' := p_{r-1}$ and $p := p_r$; if p_{r+1} is on ζ, we choose $p' := p_r$ and $p := p_{r+1}$; otherwise, since p_r is a farthest vertex to the line through $\overrightarrow{st}$, there must be a line ξ which is parallel to ζ and intersects $\overline{p_{r-1} p_r}$ and $\overline{p_r p_{r+1}}$ at exactly two points, then we choose $p' := \xi \cap \overline{p_{r-1} p_r}$ and $p := \xi \cap \overline{p_r p_{r+1}}$. In any of the three cases, since P is simple, we have $\overrightarrow{p'p} \stackrel{\circ}{=} \overrightarrow{st}$. Considering that $\overrightarrow{st} \in [\overrightarrow{q_{j_\ell} q_{i_k}}, \overrightarrow{q_{i_\ell} q_{j_k}}] \subseteq [\overrightarrow{q_{j_{r+1}} q_{i_r}}, \overrightarrow{q_{i_{r+1}} q_{j_r}}]$ by Obs. 1, we have $\overrightarrow{p'p} \in [\overrightarrow{q_{j_{r+1}} q_{i_r}}, \overrightarrow{q_{i_{r+1}} q_{j_r}}]$. Since we additionally have $\overrightarrow{p'p} \in [\overrightarrow{p_{r-1} p_r}, \overrightarrow{p_r p_{r+1}}]$, we now have $[\overrightarrow{q_{j_{r+1}} q_{i_r}}, \overrightarrow{q_{i_{r+1}} q_{j_r}}] \cap [\overrightarrow{p_{r-1} p_r}, \overrightarrow{p_r p_{r+1}}] \neq \emptyset$, implying that $\Diamond p_{r-1} p_r q_{i_r} q_{j_r}$ and $\Diamond p_r p_{r+1} q_{i_{r+1}} q_{j_{r+1}}$ are intersecting, by condition (b) of Lemma 4. □

Now it is the time to give out the whole sufficient condition, and then the whole necessary and sufficient condition. We say that a nested sequence of Q's sub-chains $Q_{i_1 j_1} \subseteq Q_{i_2 j_2} \subseteq \cdots \subseteq Q_{i_m j_m}$ is *viable* if each $\Diamond p_{k-1} p_k q_{i_k} q_{j_k}$ is positive for $k = 1, \ldots, m$, and each neighboring pair $\Diamond p_{k-1} p_k q_{i_k} q_{j_k}$ and $\Diamond p_k p_{k+1} q_{i_{k+1}} q_{j_{k+1}}$ are non-intersecting for $k = 1, \ldots, m-1$.

Lemma 6. *Suppose that there is a viable nested sequence of Q's sub-chains $Q_{i_1 j_1} \subseteq Q_{i_2 j_2} \subseteq \cdots \subseteq Q_{i_m j_m}$. Then all the induced ridging triangle-pairs and hilling triangle-fans are non-intersecting.*

Proof. The proof is by induction. First we note that the terminal $\mathcal{F}(p_0, Q_{i_1 j_1})$ doesn't intersect $\Diamond p_0 p_1 q_{i_1} q_{j_1}$.

Now suppose that for some $1 \leq k < m$, there is no intersection among all the hilling triangle-fans and ridging triangle-pairs induced by the subsequence $Q_{i_1 j_1} \subseteq Q_{i_2 j_2} \subseteq \cdots \subseteq Q_{i_k j_k}$, i.e., they form a $\mathcal{M}(P_{0,k}, Q_{i_k j_k})$. With the existence of Lemma 3 and 5, what we need to argue in the inductive step is only that neither $\mathcal{F}(p_k, Q_{i_{k+1} i_k})$ nor $\mathcal{F}(p_k, Q_{j_k j_{k+1}})$ intersects $\mathcal{M}(P_{0,k}, Q_{i_k j_k})$. In fact, let $\overline{q_r p_\ell}$ be a connecting edge of $\mathcal{M}(P_{0,k}, Q_{i_k j_k})$. Then by the convexity of Q, if $\overline{q_r p_\ell}$ stabs any triangle of $\mathcal{F}(p_k, Q_{i_{k+1} i_k})$ or $\mathcal{F}(p_k, Q_{j_k j_{k+1}})$, it must first stabs $\triangle p_k q_{i_k} q_{j_k}$ since q_r cannot lie in the positive side of any triangle of $\mathcal{F}(p_k, Q_{i_{k+1} i_k})$ or $\mathcal{F}(p_k, Q_{j_k j_{k+1}})$. But this is impossible by Lemma 2.

At last, considering that the terminal $\mathcal{F}(p_m, Q_{j_m i_m})$ and $\mathcal{M}(P_{0,m}, Q_{i_m j_m})$ are also non-intersecting, we accomplish the proof. □

Theorem 1. *There is a (positive) mating of P and Q if and only if there is a viable nested sequence of Q's sub-chains $Q_{i_1 j_1} \subseteq Q_{i_2 j_2} \subseteq \cdots \subseteq Q_{i_m j_m}$.*

Corollary 1. *The matability P and Q is invariant to edge length changing to P, as long as the edge directions and the simplicity of P are preserved.*

4 The Algorithms

4.1 The Dynamic-Programming Algorithm

Based on Theorem 1, we can easily derive a dynamic-programming algorithm to solve the optimization version and the decision version at the same time. The algorithm either returns an optimal positive mating of P and Q, or decide its non-existence. The optimization criteria can be any of Maximum Volume, Minimum Area, and Minimum Spanning Length. For convenience, we unitively treat all the three optimization criteria as minimization criteria. This is feasible because maximizing the volume is equivalent to minimizing its negation.

We first describe the dynamic-programming algorithm in a straightforward way, and then improve on it. In our description, we use $|\mathcal{S}|$ to denote the weight (volume, area, or length) induced by a connecting patch $\mathcal{S}$.

To fit to the dynamic-programming paradigm, we directly use a 3-dimensional table $A[1..m, 0..n-1, 0..n-1]$, where for $1 \leq k \leq m$ and $0 \leq i \neq j < n$, $A[k, i, j]$ is defined as the minimum of the weights induced by all possible $\mathcal{M}(P_{0,k}, Q_{ij})$'s. The entries of table A can be computed by the recursive formula

$$A[k+1, i, j] = \begin{cases} \infty & \text{if } \Diamond p_k p_{k+1} q_i q_j \text{ not positive,} \\ |\mathcal{F}(p_0, Q_{ij})| + |\Diamond p_0 p_1 q_i q_j| & \text{if } k = 0 \text{ and } \Diamond p_0 p_1 q_i q_j \text{ positive,} \\ \min_{Q_{i'j'} \subseteq Q_{ij}} w(k, i, j, i', j') & \text{if } k > 0 \text{ and } \Diamond p_k p_{k+1} q_i q_j \text{ positive,} \end{cases}$$

where $w(k, i, j, i', j') := \infty$ if $\Diamond p_k p_{k+1} q_i q_j$ intersects $\Diamond p_{k-1} p_k q_{i'} q_{j'}$, and otherwise

$$w(k, i, j, i', j') := A[k, i', j'] + |\mathcal{F}(p_k, Q_{ii'})| + |\mathcal{F}(p_k, Q_{j'j})| + |\Diamond p_k p_{k+1} q_i q_j|.$$

After all the entries of the table A are computed (with memoization or in a bottom-up way), we need another step to compute $\min_{0 \leq i \neq j < n}(A[m, i, j] + |\mathcal{F}(p_m, Q_{ji})|)$, which is the final optimal value. If the final optimal value is not ∞, there must exist an optimal mating, and we can reconstruct it with a standard reconstruction procedure, by using another table B also with $m \times n \times n$ entries, maintained in the process of evaluating table A.

By now, what we have described is an $O(mn^5)$-time and $O(mn^2)$-space algorithm. The time bound is determined by the process of evaluating table A. There are $O(mn^2)$ entries in table A. For each entry (except the one in the first layer), we have to access $O(n^2)$ possible $Q_{i'j'}$'s, and for each $Q_{i'j'}$, we need two $O(n)$-time innermost loops to compute $|\mathcal{F}(p_k, Q_{ii'})|$ and $|\mathcal{F}(p_k, Q_{j'j})|$.

We observe two tricks to improve the current straightforward algorithm. The first trick is based on the fact that all the entries in the $(k+1)$-th layer of table A is only dependent on the entries on the k-th layer. Thus we can use two $n \times n$ table to substitute for the 3-dimensional table A. However, this trick doesn't affect the asymptotical space bound because the 3-dimensional table B cannot be contracted. The second trick is to remove the two innermost loops for computing $|\mathcal{F}(p_k, Q_{ii'})|$ and $|\mathcal{F}(p_k, Q_{j'j})|$, by using preprocessing at the beginning. Precisely, we employ another 3-dimensional table $W[1..m, 0..n-1, 0..n-1]$ with $W[k, i, j]$

defined as $|\mathcal{F}(p_k, Q_{ij})|$. In the preprocessing step, we evaluate each $W[k, i, j]$ for $1 \leq k \leq m$ and $0 \leq i \neq j < n$, in at most $O(mn^3)$ time. Then, in the process of evaluating table A, for each $Q_{i'j'} \subseteq Q_{i,j}$ only constant time is needed because we can now compute $|\mathcal{F}(p_k, Q_{ii'})| + |\mathcal{F}(p_k, Q_{j'j})|$ by

$$|\mathcal{F}(p_k, Q_{ii'})| + |\mathcal{F}(p_k, Q_{j'j})| = W[k, i, j] - W[k, i', j'].$$

(If the optimization criterion is Minimum Spanning Length, the formulae in this subsection needs some slight revision to avoid duplicate or omitted counting.) Therefore, the second trick decrease the time bound to $O(mn^4)$.

Theorem 2. *In $O(mn^4)$ time and in $O(mn^2)$ space, we can obtain an optimal mating of P and Q if it exists, or decide its non-existence.*

4.2   The Greedy Algorithm

If we are only concerned with the decision version, the mating problem can be solved more efficiently with a greedy approach. For $0 \leq i_1 < n$, our greedy algorithm calls Algorithm 1 repeatedly, until finding a viable nested sequence of Q's sub-chains $Q_{i_1 j_1} \subseteq Q_{i_2 j_2} \subseteq \cdots \subseteq Q_{i_m j_m}$ with $Q_{i_1 j_1} = Q_{i_1, i_1+1}$, with either $i_{k+1} = i_k$ or $j_{k+1} = j_k$ or both for $1 \leq k < m$, and with each $Q_{i_k j_k}$ as short as possible. All the three properties are realized in Algorithm 1.

The time analysis of our greedy algorithm is easy: Algorithm 1 is called at most n times, and in Algorithm 1, with the increase of k from 1 to m, the sub-chains $Q_{i_k j_k}$ never decreases; so the total time cost is $O((m+n)n)$. Its correctness is guaranteed by the following three claims, whose proofs are omitted here.

Claim 1. Suppose that there are $Q_{i'j'} \subseteq Q_{ij}$ with either $i = i'$ or $j = j'$ or both, and with $\Diamond p_{k-1} p_k q_i q_{j'}$ and $\Diamond p_k p_{k+1} q_i q_j$ being both positive. Then $\Diamond p_{k-1} p_k q_{i'} q_{j'}$ and $\Diamond p_k p_{k+1} q_i q_j$ are non-intersecting.

Claim 2. Suppose that there are $Q_{i'j'} \subseteq Q_{ij}$ with $\Diamond p_{k-1} p_k q_{i'} q_{j'}$ and $\Diamond p_k p_{k+1} q_i q_j$ being both positive and non-intersecting. Then, if $\overrightarrow{p_{k-1} p_k} \times \overrightarrow{p_k p_{k+1}} \geq 0$, we have either $\Diamond p_k p_{k+1} q_{i'} q_{j'}$ or $\Diamond p_k p_{k+1} q_{i'} q_j$ or both are positive (and non-intersecting with $\Diamond p_{k-1} p_k q_i q_{j'}$), and otherwise, we have either $\Diamond p_k p_{k+1} q_{i'} q_{j'}$ or $\Diamond p_k p_{k+1} q_i q_j$ or both are positive (and non-intersecting with $\Diamond p_{k-1} p_k q_{i'} q_{j'}$)

Claim 3. Suppose that there are $Q_{i''j''} \subseteq Q_{i'j'} \subseteq Q_{ij}$ such that $\Diamond p_{k-1} p_k q_{i''} q_{j''}$, $\Diamond p_k p_{k+1} q_{i'} q_{j'}$, and $\Diamond p_k p_{k+1} q_i q_j$ are all positive. Then, $\Diamond p_{k-1} p_k q_{i''} q_{j''}$ and $\Diamond p_k p_{k+1} q_{i'} q_{j'}$ are non-intersecting iff $\Diamond p_{k-1} p_k q_{i''} q_{j''}$ and $\Diamond p_k p_{k+1} q_i q_j$ are non-intersecting.

Claim 1 guarantees that there is no need to do intersection check for two neighboring ridging triangle-pairs obtained in Algorithm 1. Claim 2 guarantees that each $Q_{i_k j_k}$ obtained in Algorithm 1 is as short as possible. At last, Claim 3 guarantees that there is any viable nested sequence of Q's sub-chains, if and only if there is the one found by our greedy algorithm.

Theorem 3. *Whether there is a mating of P and Q can be decided in $O(mn + n^2)$ time and $O(m + n)$ space.*

Algorithm 1. Decide whether there is a positive mating starting with Q_{i_1,i_1+1}

Input: P, Q, and an index i_1 with $0 \leq i_1 < n$.
Output: The indices for viable $Q_{i_1,i_1+1} \subseteq Q_{i_2 j_2} \subseteq \cdots \subseteq Q_{i_m j_m}$ if yes, or false otherwise.

1: $q_{j_1} := q_{i_1+1}$
2: **if** $\Diamond p_0 p_1 q_{i_1} q_{j_1}$ is not positive **then**
3:     **return** false
4: **for** $k := 1$ to $m - 1$ **do**
5:     **if** $\overrightarrow{p_{k-1} p_k} \times \overrightarrow{p_k p_{k+1}} \geq 0$ **then**
6:         let q_b be the first vertex satisfying that $\Diamond p_k p_{k+1} q_{i_k} q_b$ is positive, when scanning $Q_{j_k i_k}$ in the counterclockwise order from q_{j_k} to q_{i_k}
7:         **if** q_b exists **then**
8:             $q_{i_{k+1}} := q_{i_k}, q_{j_{k+1}} := q_b$
9:         **else**
10:             **return** false
11:     **else**
12:         let q_a be the first vertex that satisfying $\Diamond p_k p_{k+1} q_a q_{j_k}$ is positive, when scanning $Q_{j_k i_k}$ in the clockwise order from q_{i_k} to q_{j_k}
13:         **if** q_a exsits **then**
14:             $q_{i_{k+1}} := q_a, q_{j_{k+1}} := q_{j_k}$
15:         **else**
16:             **return** false
17: **return** the indices for obtained $Q_{i_1 j_1} \subseteq Q_{i_2 j_2} \subseteq \cdots \subseteq Q_{i_m j_m}$

4.3 Characterizing Polygonal Curves Matable with Some Convex Polygon

Intuitively, if P is very spiralling, it cannot be matable with any convex polygon. This subsection confirms this intuition by providing algorithms. Let $\mathbf{U}$ be the universe of all non-degenerate planar vectors.

Lemma 7. *P is matable with some Q iff $\bigcup_{k=1}^{m-1} [\overrightarrow{p_{k-1} p_k}, \overrightarrow{p_k p_{k+1}}] \neq \mathbf{U}$.*

Proof. Suppose that there is a mating between P and some Q. Then by Theorem 1 and condition (b) of Lemma 4, there must be viable $Q_{i_1 j_1} \subseteq Q_{i_2 j_2} \subseteq \cdots \subseteq Q_{i_m j_m}$ satisfying that $[\overrightarrow{q_{j_{k+1}} q_{i_k}}, \overrightarrow{q_{i_{k+1}} q_{j_k}}] \cap [\overrightarrow{p_{k-1} p_k}, \overrightarrow{p_k p_{k+1}}] = \emptyset$ for each $1 \leq k < m$. Since $[\overrightarrow{q_{j_m} q_{i_1}}, \overrightarrow{q_{i_m} q_{j_1}}] \subseteq [\overrightarrow{q_{j_{k+1}} q_{i_k}}, \overrightarrow{q_{i_{k+1}} q_{j_k}}]$ for each $1 \leq k < m$, we have $[\overrightarrow{q_{j_m} q_{i_1}}, \overrightarrow{q_{i_m} q_{j_1}}] \cap [\overrightarrow{p_{k-1} p_k}, \overrightarrow{p_k p_{k+1}}] = \emptyset$ for each $1 \leq k < m$. That is, we have $\bigcup_{k=1}^{m-1} [\overrightarrow{p_{k-1} p_k}, \overrightarrow{p_k p_{k+1}}] \neq \mathbf{U}$, because $[\overrightarrow{q_{j_m} q_{i_1}}, \overrightarrow{q_{i_m} q_{j_1}}]$ cannot be empty.

We now prove the "if" part by showing that if $\bigcup_{k=1}^{m-1} [\overrightarrow{p_{k-1} p_k}, \overrightarrow{p_k p_{k+1}}] \neq \mathbf{U}$, we can generate a convex polygon Q which is matable with P. First we note that $\bigcup_{k=1}^{m-1} [\overrightarrow{p_{k-1} p_k}, \overrightarrow{p_k p_{k+1}}]$ is a closed vector range because $[\overrightarrow{p_{k-1} p_k}, \overrightarrow{p_k p_{k+1}}] \cap [\overrightarrow{p_k p_{k+1}}, \overrightarrow{p_{k+1} p_{k+2}}] \neq \emptyset$ for $0 < k < m-1$. Thus $\mathbf{U} \setminus \bigcup_{k=1}^{m-1} [\overrightarrow{p_{k-1} p_k}, \overrightarrow{p_k p_{k+1}}]$ is an open vector range, and we denote it by $\overline{\mathbf{R}}$. Once $\overline{\mathbf{R}}$ is known, we can construct Q as follows. Set Q to be comprised of $2m + 2$ vertices, all on the boundary of a "slim" rectangle (maybe not axis-aligned). The four corners of the rectangle are q_0, q_m, q_{m+1}, and q_{2m+1} with $Q_{m,m+1} \subseteq Q_{0,2m+1}$, and the other vertices are uniformly aligned on $\overline{q_0 q_m}$ and $\overline{q_{m+1} q_{2m}}$, such that for $0 \leq i \leq m$, each $\overline{q_i q_{2m+1-i}}$ is perpendicular to $\overline{q_0 q_m}$ and $\overline{q_{m+1} q_{2m+1}}$. The size and direction of the rectangle are so arranged that each $[\overrightarrow{q_{2m+1-i} q_{i+1}}, \overrightarrow{q_i q_{2m-i}}] \subset \overline{\mathbf{R}}$ for $0 \leq i < m$. An example of the given P and the constructed Q are shown in Fig. 2. Note that our rule to construct Q guarantees that all $\overline{q_i q_{2m-i}}$ are parallel for $0 \leq i < m$, and all $\overline{q_i q_{2m+2-i}}$ are parallel for $1 \leq i \leq m$.

We now give a claim about the constructed Q: For any $1 \leq k \leq m$ and any $0 \leq i < m$, one of $\Diamond p_{k-1} p_k q_i q_{2m-i}$ and $\Diamond p_{k-1} p_k q_{i+1} q_{2m+1-i}$ must be positive. The reason of this claim is as follows. From $[\overrightarrow{q_{2m+1-i} q_{i+1}}, \overrightarrow{q_i q_{2m-i}}] \subset \overline{\mathbf{R}}$, we have $\overrightarrow{p_{k-1} p_k} \notin [\overrightarrow{q_{2m+1-i} q_{i+1}}, \overrightarrow{q_i q_{2m-i}}]$. Then making a deduction on these three vectors, we immediately have either $\overrightarrow{q_i q_{2m-i}} \times \overrightarrow{p_{k-1} p_k} > 0$ or $\overrightarrow{q_{i+1} q_{2m+1-i}} \times \overrightarrow{p_{k-1} p_k} > 0$ or both.

Based on the above claim, we now show that Q is matable with P by obtaining a viable nested sequence $Q_{i_1 j_1} \subseteq Q_{i_2 j_2} \subseteq \cdots \subseteq Q_{i_m j_m}$. At the beginning, we choose $Q_{i_1 j_1}$ to be either $Q_{m,m+2}$ or $Q_{m-1,m+1}$ according to which is positive. Now suppose that $Q_{i_k j_k}$ has been chosen for some $1 \leq k < m$, and it is either $Q_{i,2m-i}$ or $Q_{i+1,2m+1-i}$ for some i. Then we choose $Q_{i_{k+1} j_{k+1}}$ as

$$Q_{i_{k+1} j_{k+1}} := \begin{cases} Q_{i,2m-i} & \text{if } Q_{i_k j_k} = Q_{i,2m-i}, \Diamond p_k p_{k+1} q_i q_{2m-i} \text{ positive,} \\ Q_{i,2m+2-i} & \text{if } Q_{i_k j_k} = Q_{i,2m-i}, \Diamond p_k p_{k+1} q_i q_{2m+2-i} \text{ positive,} \\ Q_{i+1,2m+1-i} & \text{if } Q_{i_k j_k} = Q_{i+1,2m+1-i}, \Diamond p_k p_{k+1} q_{i+1} q_{2m+1-i} \text{ positive,} \\ Q_{i-1,2m+1-i} & \text{if } Q_{i_k j_k} = Q_{i+1,2m+1-i}, \Diamond p_k p_{k+1} q_{i-1} q_{2m+1-i} \text{ positive.} \end{cases}$$

If two cases are satisfied in the above formula, we can choose any one. Anyway, $Q_{i_{k+1} j_{k+1}}$ is either unchanged or moved by one block, so the i in the above formula is guaranteed in the range $[m-k, m-1]$. In the end, we obtain a sequence $Q_{i_1 j_1} \subseteq Q_{i_2 j_2} \subseteq \cdots \subseteq Q_{i_m j_m}$ satisfying that either $i_{k+1} = i_k$ or $j_{k+1} = j_k$ or both for each $1 \leq k < m$, and $\Diamond p_{k-1} p_k q_{i_k} q_{j_k}$ is positive for each $1 \leq k \leq m$. According to Claim 1, such an obtained sequence $Q_{i_1 j_1} \subseteq Q_{i_2 j_2} \subseteq \cdots \subseteq Q_{i_m j_m}$ must be viable. $\square$

Theorem 4. *Given P, in $O(m)$ time we can decide whether it is matable with some convex polygon, and if yes, we can generate such a convex polygon also in $O(m)$ time.*

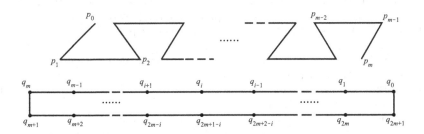

Fig. 2. An example of the given P and the constructed Q, such that they are matable.

5   Conclusion

Up to the polynomial-time sense, this paper has solved a restricted curve-polygon mating problem by providing a series of algorithms. While the arguments are somewhat long, each algorithm can be implemented very simply, say, with at most one hundred lines of programming code. We raise the following questions which deserve further research:

1. Can the optimization algorithm and the decision algorithm for P and Q presented in this paper be sped up?
2. Can the general curve-polygon mating problem be solved in polynomial time?
3. How to decide the matability of P and Q if Steiner vertices are allowed to be added on the edges of Q? From Theorem 1, we know that adding Steiner vertices on P does not help the matability at all. But on Q, it may help (consider Fig. 1(a) with one vertex added on the edge $\overline{q_0 q_1}$), or may not help (consider the case where Q is a triangle and P is a three-edge curve with each edge parallel to an edge of Q).

Besides the polygon-polygon and curve-polygon mating problems mentioned in this paper, the curve-curve mating problem (also in the sense to construct a simple polyhedron) is also well-defined and has potential applications too. To the knowledge of the author, no previous study exists on this mating problem.

References

1. Barequet, G., Steiner, A.: On the matability of polygons. Int. J. Comput. Geom. Appl. **18**, 469–506 (2008)
2. Biedl, T., Bulatovic, P., Irvine, V. Lubiw, A., Merkel, O., Naredla, A.M.: Reconstructing a polyhedron between polygons in parallel slices. In: 31st Canadian Conference on Computational Geometry, pp. 139–145. Edmonton, Alberta (2019)
3. Christiansen, H.N., Sederberg, T.W.: Conversion of complex contour line definitions into polygonal element mosaics. Comput. Graphics **13**, 187–192 (1978)
4. Fuchs, H., Kedem, Z.M., Uselton, S.P.: Optimal surface reconstruction from planar contours. Commun. ACM **20**, 693–702 (1977)
5. Gitlin, C., O'Rourke, J., Subramanian, V.: On reconstructing polyhedra from parallel slices. Int. J. Comput. Geom. Appl. **6**, 103–122 (1996)
6. Guibas, L., Hershberger, J., Suri, S.: Morphing simple polygons. Discrete Comput. Geom. **24**, 1–34 (2000)
7. Keppel, E.: Approximating complex surface by triangulation of contour lines. IBM J. Res. Dev. **19**, 2–11 (1975)
8. O'Rourke, J.: On the scaling heuristic for reconstruction from slices. Graph. Model Image Process. **56**, 420–423 (1994)
9. Welzl, E., Wolfers, B.: Surface reconstruction between simple polygons via angle criteria. In: Lengauer, T. (ed.) ESA 1993. LNCS, vol. 726, pp. 397–408. Springer, Heidelberg (1993). https://doi.org/10.1007/3-540-57273-2_75
10. Zou, M., Ju, T., Carr, N.: An algorithm for triangulating multiple 3D polygons. Comput. Graph. Forum **32**, 157–166 (2013)

A Faster Parameterized Algorithm for Bipartite 1-Sided Vertex Explosion

Yunlong Liu[1]($\boxtimes$)(iD), Guang Xiao[1], Ao Liu[2], Di Wu[3], and Jingui Huang[1]($\boxtimes$)(iD)

[1] College of Information Science and Engineering, Hunan Provincial Key Laboratory of Intelligent Computing and Language Information Processing, Hunan Normal University, Changsha 410081, People's Republic of China
{ylliu,xiaoguang,hjg}@hunnu.edu.cn
[2] Xiangtan University, Xiangtan 411101, People's Republic of China
hnsdxjp@163.com
[3] School of Computer Science and Engineering, Central South University, Changsha 410083, People's Republic of China
csuwudi@csu.edu.cn

Abstract. Given a bipartite graph $G = (T \cup B, E)$, the problem BIPAR-TITE 1-SIDED VERTEX EXPLOSION is to decide whether there exists a planar 2-layer embedding of G after exploding at most k vertices of B. For this problem, which is known to be **NP**-complete, parameterized algorithms have received increasing attention more recently. In this paper, we focus on the problem parameterized by the number k of allowed exploded vertices of B and develop a faster algorithm for it. More specifically, we show that this parameterized problem admits a kernel of at most $10.5k$ vertices, and present a fixed-parameter tractable algorithm running in time $\mathcal{O}(2.31^{k} \cdot m)$, where m is the number of edges of G.

Keywords: Vertex explosion · Planar embedding · Kernel · Branching

1 Introduction

Bipartite graphs are mostly used in modeling the relationships between two disjoint sets of entities. Typical examples include the 2-layer networks between two communities [15], the tangle-gram layouts for comparing phylogenetic trees [16], the representation for the relationship between human anatomical structures and cell types [14], and so on. Drawing a bipartite graph G in a visualizable and understandable way is then a primary step in these applications.

A *2-layer drawing* of a bipartite graph $G = (T \cup B, E)$ is a drawing which maps the vertices in T to points on a line and those in B to points on another parallel line, and maps edges in E to straight-line segments connecting their

This research was supported in part by the National Natural Science Foundation of China under Grant (No.61572190), Hunan Provincial Science and Technology Program (No.2018TP1018), and Changsha Municipal Natural Science Foundation (Grant No. kq2202247).

W. Wu and J. Guo (Eds.): COCOA 2023, LNCS 14462, pp. 253–266, 2024.
https://doi.org/10.1007/978-3-031-49614-1_19

respective end points [8]. In particular, a 2-layer drawing $\mathcal{D}$ of a bipartite graph G is called a *planar 2-layer embedding* of G if there are no edge-crossings in $\mathcal{D}$.

Deciding whether a given bipartite graph admits a planar 2-layer embedding can be done in linear time [7]. When a bipartite graph doesn't admit any planar 2-layer embedding, an alternative approach is to explode some of its vertices such that the resulting graph does. Given a vertex v in a graph G, a *vertex explosion* operation on v means replacing v by $\deg(v)$ vertices of degree 1, each incident to exactly one edge that was originally incident to v, where $\deg(v)$ denotes the degree of v in G [3]. Motivated by the fact that the two vertex sets of a bipartite graph play different roles in some applications, the problem BIPARTITE 1-SIDED VERTEX EXPLOSION, where vertex explosion is confined to only one vertex set, has been formulated in recent years [1,2,14].

More formally, the BIPARTITE 1-SIDED VERTEX EXPLOSION problem asks, for a given bipartite graph $G = (T \cup B, E)$, whether there is a planar 2-layer embedding of G after exploding at most k vertices of B (abbreviated as **BSVE**). Figure 1 gives a planar 2-layer embedding of a bipartite graph after exploding 2 vertices. The BSVE problem is known to be **NP**-complete [2,5].

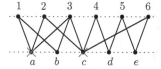 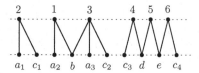

Fig. 1. A bipartite graph G (left) and a planer 2-layer embedding of G after exploding 2 vertices a and c (right).

Parameterized algorithms have recently been an efficient approach to deal with many NP-hard problems in graph drawing [4,12,13]. In particular, parameterized algorithms for the BSVE problem have received attention more recently [1,3]. When parameterized by the number of allowed exploded vertices, its parameterized version can be formally defined as follows [1].

BIPARTITE 1-SIDED VERTEX EXPLOSION (p-**BSVE**)
Input: A bipartite graph $G = (T \cup B, E)$, a positive integer k;
Parameter: k;
Task: Decide whether there is a planar 2-layer embedding of G after exploding at most k vertices of B?

For the p-BSVE problem, Ahmed et al. [1] showed a kernel of at most $\mathcal{O}(k^6)$ vertices, which results in a parameterized algorithm with running time $2^{O(k^6)} \cdot m$, where m is the number of edges of G. More recently, Baumann et al. [3] studied this problem in a more general setting (named as POVE), i.e., for an undirected graph $G = (V, E)$, a set $S \subseteq V$, and a positive integer k, decide whether there exists a set $W \subseteq S$ with $|W| \leq k$ such that the graph resulting from exploding all vertices in W admits a 2-layer drawing without crossings. They showed a kernel of at most $16k^2 + 16k$ vertices and a branching algorithm of running time $\mathcal{O}(4^k \cdot m)$ for the POVE problem.

Improving the running time of the algorithm for the p-BSVE problem was posed as an open question [1]. Baumann et al.'s algorithm [3] was proposed with respect to a more general setting. Thus, we further study the p-BSVE problem and aim to propose a faster algorithm by exploiting the special properties of bipartite graphs. In this paper, we present an algorithm running in time $\mathcal{O}(2.31^k \cdot m)$, where m is the number of edges of G. Our algorithm consists of the following two parts:

1. We show a kernel of at most $10.5k$ vertices. Our work on kernelization includes making a more refined analysis on some reduction rules proposed by Ahmed et al. [1], employing the strategy used by Baumann et al. [3] to shorten each long path to constant size, and introducing a new reduction rule to eliminate a type of special substructures. The kernelization procedure is not only the basis of our algorithm, but also has much independent interest.
2. We present a series of efficient branching rules. In particular, we propose a novel approach to derive these rules, in which we first reduce the instance exhaustively such that multiple forbidden substructures stay close to each other, and then exploit the relationship among them.

2   Terminology and Notations

We consider only undirected simple bipartite graphs without isolated vertices. Let $G = (T \cup B, E)$ be a bipartite graph, where T is the set of top vertices and B is the set of bottom vertices. We also denote by $V(G)$ and $E(G)$ the set of its vertices and edges, respectively. For a vertex $v \in V(G)$, we use $N(v)$ to denote the set of neighbors of v, $N^2(v)$ to denote the set of vertices, excluding $\{v\} \cup N(v)$, that are adjacent to a vertex in $N(v)$, and $deg(v)$ to denote the degree of v. For a subset $S \subset V(G)$, we denote by $N(S) = \bigcup_{v \in S} N(v) \setminus S$. A vertex $u \in N(v)$ is called a *pendant neighbor* of v if u has degree 1. For a vertex $v \in V(G)$, we let $\deg^*(v) = |\{u \mid u \in N(v) \text{ and } \deg(u) > 1\}|$ denote the degree of v ignoring its pendent neighbors. If $\deg^*(v) = d$, we refer to v as a vertex of *degree* d. An edge e connecting two vertices u and v is denoted by uv. A path P connecting two endpoints u and v is denoted by P_{uv}. Moreover, we call a bipartite graph G as an *empty graph* if $V(G) = \emptyset$.

Given two sets A and B, we use $A \setminus B$ to denote the set of all elements that belong to A but not to B, $|A|$ to denote the number of elements in A. Given a graph G and any vertex $v \in V(G)$, we use $G - v$ to denote the subgraph of G obtained by removing the vertex v together with all the edges incident to it from G.

We use a series of reduction rules and branching rules for the presentation of our search tree algorithm. For an instance (G, k) of the p-BSVE problem, we say that a reduction rule $\mathcal{R}_1$ is *safe* if $\mathcal{R}_1$ reduces (G, k) to (G', k') such that (G, k) is a yes-instance if and only if (G', k') is. Given an instance (G, k), a branching rule $\mathcal{R}_2$ creates $t \geq 2$ subinstances $(G'_1, k'_1), \ldots, (G'_t, k'_t)$. We also say that the rule $\mathcal{R}_2$ is *safe* if (G, k) is a yes-instance if and only if (G'_i, k'_i) is for some $1 \leq i \leq t$. We also use the *branching number* to describe the base of the (exponential) search tree size, which can be computed using standard branching analysis tools [6].

Because of the space limit, most of the proofs are deferred to a full version.

3   Linear Kernel

For the p-BSVE problem, Ahmed et al. [1] gave a kernel of at most $\mathcal{O}(k^6)$ vertices. Baumann et al. [3] obtained a kernel of at most $16k^2 + 16k$ vertices for a more general version of it. To obtain a smaller kernel, we will follow the work done by Ahmed et al. [1] and conduct a further investigation, which includes making a more refined analysis on the number of vertices in the reduced instance, employing the technique used by Baumann et al. [3] to shorten the long paths, and introducing a new reduction rule to eliminate a type of special substructures.

We start with restating some reduction rules introduced by Ahmed et al. [1]. For a given instance $(G_0 = (T_0 \cup B_0, E_0), k_0)$, let $B_{0,tr} = \{v \mid v \in B_0$ and $\deg^*(v) \geq 3\}$. Obviously, each vertex in $B_{0,tr}$ must be exploded.

Reduction Rule 1 ([1]). Remove the vertices in $B_{0,tr}$ from graph G_0 and let $k_1 = k_0 - |B_{0,tr}|$.

Denote by $G_1 = (T_1 \cup B_1, E_1)$ the graph obtained from G_0 by executing Reduction Rule 1.

Reduction Rule 2 ([1]). Let $v \in T_1 \cup B_1$ be a vertex with $\deg(v) = 1$ and let u be the unique neighbor of v in G_1. If $\deg(u) \geq 3$, then remove v from G_1. The new instance is $(G_1 - v, k_1)$.

After executing Reduction Rule 2 exhaustively, we denote by $(G_2 = (T_2 \cup B_2, E_2), k_2)$ the resulting instance, where $k_2 = k_1$. Note that for each vertex $u \in B_2$, the degree of u is at most 2.

For the graph G_2, Ahmed et al. [1] showed a degree bound on the vertices in T_2. Namely, for each vertex $v \in T_2$, it holds that $\deg(v) \leq k_2 + 2$ if there exists a planar 2-layer drawing of G_2 with at most k_2 exploded vertices [1]. Let $T_{2,deg(v) \geq 3}$ be the set of all vertices of degree at least three in T_2. Ahmed et al. [1] also obtained a bound on the number of vertices in $T_{2,deg(v) \geq 3}$.

Lemma 1. *([1]) It holds that $|T_{2,deg(v) \geq 3}| \leq 2k_2$ if there exists a planar 2-layer drawing of G_2 with at most k_2 exploded vertices.*

Based on the degree bound on the vertices in T_2 and Lemma 1, Ahmed et al. [1] claimed that $|N(T_{2,deg(v) \geq 3})| \leq 2k_2 \cdot (k_2 + 2)$ if (G_2, k_2) is a yes-instance. Herein, by making an analysis from a different perspective, we obtain an improved upper bound for it.

Lemma 2. *In graph G_2, the set $N(T_{2,deg(v)\geq 3})$ has at most $5k_2$ vertices if there exists a planar 2-layer embedding of G_2 with at most k_2 exploded vertices.*

Proof. Assume that $G_2 = (T_2 \cup B_2, E_2)$ admits a planar 2-layer embedding $\mathcal{E}$ after exploding at most k_2 vertices of B_2. Let v be an arbitrary vertex in $T_{2,deg(v)\geq 3}$. According to Reduction Rule 2, vertex v has not any pendent neighbors in B_2, therefore, to obtain a planar 2-layer embedding of G_2, all but at most two neighbors of v must be exploded. In other words, vertex v has at most 2 neighbors that are not exploded in $\mathcal{E}$. Therefore, by Lemma 1, $N(T_{2,deg(v)\geq 3})$ contains at most $2k_2 \times 2 = 4k_2$ vertices that are not exploded in $\mathcal{E}$. Suppose for contradiction that $N(T_{2,deg(v)\geq 3})$ contains at least $5k_2 + 1$ vertices. Then, at least $(5k_2 + 1) - 4k_2 = k_2 + 1$ vertices in $N(T_{2,deg(v)\geq 3})$ must be exploded, which contradicts the assumption that G_2 has at most k_2 exploded vertices.   □

Let C denote the set of vertices in $T_{2,deg(v)\geq 3} \bigcup N(T_{2,deg(v)\geq 3})$. The subgraph induced by the vertices in C is called the *core* of G_2 [1]. Observe that some vertices in $(T_2 \cup B_2) \setminus C$ induce some isolated cycles or paths (i.e., these cycles or paths are not connected to the core of G_2). Let $\mathcal{I}_1$ (resp. $\mathcal{I}_2$) be the set of isolated cycles (resp. paths) induced by some vertices in $(T_2 \cup B_2) \setminus C$. Each cycle in $\mathcal{I}_1$ and each path in $\mathcal{I}_2$ can be handled separately.

Reduction Rule 3.

3.1 **For** each cycle $l \in \mathcal{I}_1$ **do** : remove l; take one vertex $v \in B_2$ on l as an exploded vertex; set $k_2 = k_2 - 1$ ([1]).

3.2 For each path in $\mathcal{I}_2$, we remove it separately.

Denote by $(G_3 = (T_3 \cup B_3, E_3), k_3)$ the instance resulted from (G_2, k_2) by executing Reduction Rule 3 exhaustively. Let $\mathcal{P}$ be the set of paths induced by some vertices in $V(G_3) \setminus T_{3,deg(v)\geq 3}$ and connected to the core of G_3. Note that the core of G_3 is also that of G_2. Now, we deal with the long paths in $\mathcal{P}$. In dealing with each such path $p \in \mathcal{P}$, we employ the technique used by Baumann et al. [3], i.e., shortening each path to constant size. Since the input graph in the p-BSVE problem is a bipartite graph, our rule is more straight-forward.

We divide the paths in $\mathcal{P}$ into two classes. The first class, denoted by $\mathcal{P}_1$, consists of the paths whose two endpoints lie in $N(T_{3,deg(v)\geq 3})$. The second class, denoted by $\mathcal{P}_2$, consists of those in $\mathcal{P} \setminus \mathcal{P}_1$. Let $P_{uv} \in \mathcal{P}$ be a path connecting endpoints u and v.

Reduction Rule 4 illustrated in Fig. 2.

4.1 If $P_{uv} \in \mathcal{P}_1$ has length at least 4, then shorten P_{uv} until P_{uv} has length 2, i.e., iteratively remove one of the middle vertices of P_{uv} and identify its two neighbors.

4.2 If $P_{uv} \in \mathcal{P}_2$ has length at least 2, then shorten P_{uv} until it has length 1, i.e., for the path P_{uv}, remove all vertices but u and w, where w is the first neighbor of u on P_{uv}.

Denote by $(G_4 = (T_4 \cup B_4, E_4), k_4)$ the instance resulted from (G_3, k_3) by executing Reduction Rule 4 exhaustively, where $k_4 = k_3$. For the instance (G_4, k_4), we have the following lemma.

Fig. 2. A path $P_{uv} \in \mathcal{P}_1$ of length 8 is reduced to one of length 2 (left) and another path $P_{uv} \in \mathcal{P}_2$ of length 8 is reduced to one of length 1 (right).

Lemma 3. (G_4, k_4) *is a yes-instance of the p-BSVE problem if and only if* (G_3, k_3) *is a yes-instance of the p-BSVE problem.*

We now introduce a new reduction rule for eliminating a type of special substructures that may be contained in graph G_4.

Reduction Rule 5. Let $r \in T_{4,deg(v) \geq 3}$ have h neighbors of degree 2 and let $S = \{t \mid t \in N^2(r) \text{ and } 1 \leq \deg(t) - |N(r) \cap N(t)| \leq 2\}$. If there exist at least two vertices of degree 2 in $N(r)$, say s_1 and s_2, such that $\deg^*(s_1) = 1$ and $\deg^*(s_2) = 1$ respectively, then put the vertices in $N(r) \setminus \{s_1, s_2\}$ into the solution set, set $k_4 = k_4 - (h - 2)$, and delete all vertices in $\{r\} \cup N(r)$. Furthermore, for each vertex $v \in S$, if a new isolated path or cycle (resp. an unisolated path) passing through v still occurs in the current graph, then deal with it by using Reduction Rule 3 (resp. Reduction Rule 4). See Fig. 3 for an illustration.

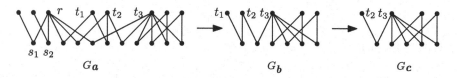

Fig. 3. An example for Reduction Rule 5. Here, graph G_b is obtained from graph G_a by deleting all vertices in $\{r\} \cup N(r)$; graph G_c is obtained from graph G_b by shortening the path $P_{t_1 t_3}$.

Lemma 4. *Reduction Rule 5 is safe.*

Denote by $(G_5 = (T_5 \cup B_5, E_5), k_5)$ the instance resulted from (G_4, k_4) by executing Reduction Rule 5 exhaustively. Based on Lemma 4, we obtain a linear kernel for the p-BSVE problem.

Theorem 1. *The p-BSVE problem admits a kernel with at most $10.5k_0$ vertices.*

Proof. Let $(G_5 = (T_5 \cup B_5, E_5), k_5)$ be a reduced instance of the p-BSVE problem. Denote by $T_{5,deg(v) \geq 3}$ the set of all vertices of degree at least 3 in T_5. By Lemma 1 and 2, the set C in G_5, i.e., $T_{5,deg(v) \geq 3} \bigcup N(T_{5,deg(v) \geq 3})$, contains at most $2k_5 + 5k_5 = 7k_5$ vertices. Next, we bound the number of vertices in $V(G_5) \setminus C$ by estimating the number of paths in $\mathcal{P}_1 \cup \mathcal{P}_2$ induced by some vertices in $V(G_5) \setminus$

$T_{5,deg(v)\geq 3}$. By Lemma 2, it holds that $|N(T_{5,deg(v)\geq 3})| \leq 5k_5$. Observe that for each path in $\mathcal{P}_2$, only one endpoint lies in $N(T_{5,deg(v)\geq 3})$. Instead, for each path in $\mathcal{P}_1$, both endpoints lie in $N(T_{5,deg(v)\geq 3})$. Thus, $|\mathcal{P}_1 \cup \mathcal{P}_2|$ achieves its maximum value only if $|\mathcal{P}_2|$ achieves its maximum value. Furthermore, since G_5 has been processed by Reduction Rules 5, each vertex in $T_{5,deg(v)\geq 3}$ has at most one neighbor serving as an endpoint of a path in $\mathcal{P}_2$. Thus, the maximum value for $|\mathcal{P}_2|$ is $2k_5$. When $|\mathcal{P}_2| = 2k_5$, it holds that $|\mathcal{P}_1| = (5k_5 - 2k_5)/2 = 1.5k_5$. Moreover, after being reduced by Reduction Rule 4, each path in $\mathcal{P}_1 \cup \mathcal{P}_2$ contains only one vertex in $V(G_5) \setminus C$. It follows that $|V(G_5) \setminus C| \leq (2k_5 + 1.5k_5) \times 1 = 3.5k_5$, which means that $|V(G_5)| \leq 7k_5 + 3.5k_5 = 10.5k_5 \leq 10.5k_0$. □

4 Efficient Branching Rules

Let $(G = (T \cup B, E), k)$ be an instance reduced by the reduction rules in Sect. 3 exhaustively. In this section, we present a processing rule and several efficient branching rules on graph G. To facilitate the description of these rules, we introduce a special subgraph named W_2 as follows.

Formally, a W_2 *substructure* is a bipartite subgraph consisting of a *root* vertex r adjacent to three distinct vertices of degree 2. See Fig. 4 (a) for an example. We use $r(s_1, s_2, s_3)$ to denote the W_2 substructure rooted by r that has three neighbors s_1, s_2 and s_3. Given a bipartite graph $(T \cup B, E)$ containing a W_2 substructure $r(s_1, s_2, s_3)$, we from now on assume that $r \in T$. The W_2 substructure is a variant of N_2 substructure introduced by Baumann et al. [3] with respect to the bipartite graphs. It can be easily inferred from [3] that a bipartite graph G admits a planar 2-layer embedding if and only if G is acyclic and contains no W_2 substructure. Hence, the functions of our processing rule and branching rules are to eliminate the cycles and the W_2 substructures in G.

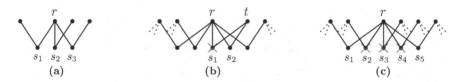

Fig. 4. A subgraph W_2 (a), a situation for Processing Rule 1 (b), and a situation for Branching Rule 2 (c).

Considering the description of our rules in graphs without any cycle C_4 becomes more concise, we first present a rule to eliminate the C_4 in G. We still denote by $T_{deg(v)\geq 3}$ the set of all vertices of degree at least three in T.

Processing Rule 1. Let $r, s_1, t,$ and s_2 induce a cycle C_4 in G, where $r \in T_{deg(v)\geq 3}$. Then, eliminating this C_4 can be done by arbitrarily exploding one vertex in $\{s_1, s_2\}$, and the parameter k is decreased by 1. See Fig. 4 (b) for an illustration.

Lemma 5. *Processing Rule 1 is safe. Moreover, one can find a C_4 in G (if exists) in time $\mathcal{O}(m)$, where m is the number of edges of G.*

Let $r \in T_{deg(v) \geq 3}$. Assume that r has h (for $h \geq 3$) neighbors of degree 2. Since only at most 2 neighbors of r can be safely drawn in any planar 2-layer embedding of G, $h - 2$ neighbors of r must be simultaneously exploded. In the following, we distinguish three cases to deal with the neighbors of r based on the degree h.

Case 1: $h \geq 5$, namely, r has at least 5 neighbors of degree 2.

Branching Rule 2. Let r have h (for $h \geq 5$) neighbors of degree 2. Then, eliminating the subgraph induced by vertices in $\{r\} \cup N(r) \cup N^2(r)$ can be done by $\binom{h}{h-2}$ ways, i.e., enumerating all possible ways of exploding $h - 2$ vertices among the h neighbors, and the parameter k is decreased by $h - 2$ in each case. Figure 4 (c) gives one way for exploding 3 neighbors of a vertex with degree 5.

Lemma 6. *Branching Rule 2 is safe, and the corresponding branching number is at most 2.16.*

Case 2: $h = 4$, namely, r has exactly 4 neighbors of degree 2.

In this case, if we directly deal with them using the same approach as Branching Rule 2, then the branching number is at least 2.44. To improve the branching number, we now propose a novel branching rule for them.

For ease of presentation, we from now on use the notation *exploding (v_1, v_2)* to denote exploding both v_1 and v_2 at the same time.

Branching Rule 3. Let r have exactly 4 neighbors of degree 2, say s_1, s_2, s_3, and s_4. Then, eliminating the subgraph induced by vertices in $\{r\} \cup N(r) \cup N^2(r)$ can be done by exploding s_1, exploding (s_2, s_3), exploding (s_2, s_4), or exploding (s_3, s_4), decrease parameter 1 for the first case and 2 for other three cases. See Fig. 5 for an illustration.

$$s_1\ s_2\ s_3\ s_4 \qquad s_1\ s_2\ s_3\ s_4 \qquad s_1\ s_2\ s_3\ s_4 \qquad s_1\ s_2\ s_3\ s_4$$
$$\text{(a)} \qquad\qquad \text{(b)} \qquad\qquad \text{(c)} \qquad\qquad \text{(d)}$$

Fig. 5. Four branches in Branching Rule 3.

Note that in the resulting graph G' obtained from G by exploding s_1, the vertex r still has *three* neighbors of degree 2, which is forbidden in any planar 2-layer embedding of G'. The task of exploding another neighbor of r is moved to case 3, which contains the special rules dealing with the vertices of degree* 3.

Lemma 7. *Branching Rule 3 is safe, and the corresponding branching number is at most 2.31.*

Case 3: $h = 3$, namely, r has exactly three neighbors of degree 2.

If we use s_1, s_2 and s_3 to denote the neighbors of r, then the subgraph induced by vertices in $\{r\} \cup N(r) \cup N^2(r)$ is exactly a W_2 substructure and can be denoted by $r(s_1, s_2, s_3)$. Given a W_2 substructure $r(s_1, s_2, s_3)$, there are 3 basic ways to eliminate it, namely, exploding s_1, exploding s_2, and exploding s_3. To derive more efficient branching rules, we present a novel strategy that is quite distinct from some popular approaches in [9–11].

We deal with the instance by using a series of reduction rules and simple branching rules exhaustively such that the resulting graph becomes a special combinatorial structure, where distinct W_2 substructures stay close to each other. When eliminating multiple W_2 substructures, we can find some redundant solutions that can be safely discarded, which leads to some more efficient branching rules.

In the following, we first introduce a simple branching rule to deal with a type of substructures containing a single W_2.

Branching Rule 4.1. Let $r(s_1, s_2, s_3)$ be a W_2 substructure in G such that only s_2 has one pendant neighbor t_2. Then, eliminating this W_2 substructure can be done by two branches: exploding s_1 and exploding s_3. The parameter k is decreased by 1 in each case.

Lemma 8. *Branching Rule 4.1 is safe, and the branching number is 2.*

Next, we introduce two branching rules to eliminate multiple W_2 substructures simultaneously.

Given two W_2 substructures $r_1(s_1, s_2, s_3)$ and $r_2(p_1, p_2, p_3)$, we define two structure relationships between them. (1) Two W_2 substructures $r_1(s_1, s_2, s_3)$ and $r_2(p_1, p_2, p_3)$ are called *jointed* with each other if $\{s_1, s_2, s_3\} \cap \{p_1, p_2, p_3\} \neq \emptyset$. See Fig. 6 (a) for an example. (2) Two W_2 substructures $r_1(s_1, s_2, s_3)$ and $r_2(p_1, p_2, p_3)$ are called *adjacent* to each other if there exists at least one vertex $s \in \{s_1, s_2, s_3\}$ and at least one vertex $p \in \{p_1, p_2, p_3\}$ such that $N(s) \setminus \{r_1\} = N(p) \setminus \{r_2\}$. Specifically, $r_1(s_1, s_2, s_3)$ and $r_2(p_1, p_2, p_3)$ are called adjacent by (s_i, p_j) if $N(s_i) \setminus \{r_1\} = N(p_j) \setminus \{r_2\}$, where $1 \leq i, j \leq 3$. Figure 6 (b) gives two W_2 substructures adjacent by (s_3, p_1). Note that it is possible that two W_2 substructures are both jointed and adjacent. Figure 6 (c) gives an example.

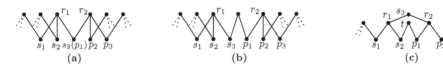

Fig. 6. (a) Two W_2 substructures are jointed with each other; (b) two W_2 substructures are adjacent to each other; and (c) two W_2 substructures are both jointed and adjacent.

Given two adjacent W_2 substructures, we have the following observation.

Lemma 9. *Let $r_1(s_1, s_2, s_3)$ and $r_2(p_1, p_2, p_3)$ be two W_2 substructures in G such that they are adjacent by (s_3, p_1). If there exists a planar 2-layer embedding $\mathcal{E}_1$ of G_1 obtained from G by exploding (s_3, p_1), then there exists a planar 2-layer embedding $\mathcal{E}_2$ of G_2 obtained from G by exploding (s_3, p_2) and a planar 2-layer embedding $\mathcal{E}_3$ of G_3 obtained from G by exploding (s_3, p_3).*

Based on Lemma 9, we now introduce an efficient branching rule for the structure which consists of two adjacent W_2 substructures.

Branching Rule 4.2. Let $r_1(s_1, s_2, s_3)$ and $r_2(p_1, p_2, p_3)$ be adjacent by (s_3, p_1).

(1) If $|N(r_1) \cup N(r_2)| = 5$, then eliminating both $r_1(s_1, s_2, s_3)$ and $r_2(p_1, p_2, p_3)$ can be determinately done by exploding the vertex in $N(r_1) \cap N(r_2)$. See Fig. 6 (c) for an example, in which s_3 is exploded.

(2) Otherwise, namely, $|N(r_1) \cup N(r_2)| = 6$, then eliminating both $r_1(s_1, s_2, s_3)$ and $r_2(p_1, p_2, p_3)$ can be done by four branches. See Fig. 6 (b) for an example, in which the branches consist of exploding (s_1, p_2), exploding (s_1, p_3), exploding (s_2, p_2), and exploding (s_2, p_3)[1].

Lemma 10. *Branching Rule 4.2 is safe, and the corresponding branching number is at most 2.*

An instance is called an *exhaustively processed instance* if it has been processed by Reduction Rules 1–5, Processing Rule 1, and Branching Rules 2, 3, 4.1, 4.2 exhaustively. In an exhaustively processed instance, the W_2 substructures stay close to each other. More precisely, given a W_2 substructure, it is jointed with three W_2 substructures at the same time (we will later show it in Lemma 12).

Based on Branching Rule 4.2, we can eliminate *four* W_2 substructures simultaneously in an exhaustively processed instance, which leads to another efficient branching rule. Let $r_1(s_1, s_2, s_3)$ be a W_2 substructure. Assume that it is jointed with $r_2(p_1, p_2, p_3)$, $r_3(q_1, q_2, q_3)$, and $r_4(t_1, t_2, t_3)$ at the same time. Based on the number of vertices in $N(r_2) \cup N(r_3) \cup N(r_4) \setminus N(r_1)$, we distinguish four combinatorial structures among them. See Fig. 7 for an illustration.

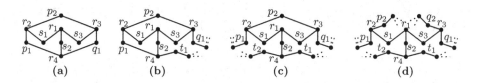

(a)   (b)   (c)   (d)

Fig. 7. Four combinatorial structures in which a W_2 substructure is jointed with other three W_2 substructures at the same time.

Branching Rule 4.3. Let $r_1(s_1, s_2, s_3)$ be jointed with three W_2 substructures simultaneously, say $r_2(p_1, p_2, p_3)$, $r_3(q_1, q_2, q_3)$, and $r_4(t_1, t_2, t_3)$.

---

[1] For brevity, we only present this general rule, although there are some more refined rules for special subcases including $|N^2(r_1) \cap N^2(r_2)| = i$, for $i = 2$ or $i = 3$.

(1) If $|N(r_2) \cup N(r_3) \cup N(r_4) \setminus N(r_1)| = 3$, then eliminating these W_2 substructures can be determinately done by exploding three vertices. See Fig. 7 (a) for an example, in which exploding (s_1, p_1, q_1) is done.

(2) If $|N(r_2) \cup N(r_3) \cup N(r_4) \setminus N(r_1)| = 4$, then eliminating these W_2 substructures can be determinately done by exploding two vertices. See Fig. 7 (b) for an example, in which exploding (s_2, p_2) is done.

(3) If $|N(r_2) \cup N(r_3) \cup N(r_4) \setminus N(r_1)| = 5$, then eliminating these W_2 substructures can be done by 9 branches. See Fig. 7 (c) for an example, in which the branches consist of exploding (s_2, p_2), exploding (s_1, p_2, t_1), exploding (s_1, p_2, t_2), exploding (s_1, q_1, t_1), exploding (s_1, q_1, t_2), exploding (s_3, p_1, t_1), exploding (s_3, p_1, t_2), exploding (s_3, p_2, t_1), and exploding (s_3, p_2, t_2).

(4) Otherwise, namely, $|N(r_2) \cup N(r_3) \cup N(r_4) \setminus N(r_1)| = 6$, then eliminating these W_2 substructures can be done by 12 branches. See Fig. 7 (d) for an example, in which the branches consist of exploding (s_1, q_1, t_1), exploding (s_1, q_1, t_2), exploding (s_1, q_2, t_1), exploding (s_1, q_2, t_2), exploding (s_2, p_1, q_1), exploding (s_2, p_1, q_2), exploding (s_2, p_2, q_1), exploding (s_2, p_2, q_2), exploding (s_3, p_1, t_1), exploding (s_3, p_1, t_2), exploding (s_3, p_2, t_1), and exploding (s_3, p_2, t_2).

Lemma 11. *Branching Rule 4.3 is safe, and the branching number is bounded by 2.29.*

5   A Whole Parameterized Algorithm

On the algorithms for the p-BSVE problem, Ahmed et al. [1] directly used a brute-force searching on the kernel, which runs in time $2^{\mathcal{O}(k^6)} \cdot m$, where m is the number of edges of the input graph. More recently, in studying a general version of this problem, Baumann et al. [3] employed a basic branch-and-bound process to eliminate a kind of forbidden substructures, which runs in time $\mathcal{O}(4^k \cdot m)$.

Now, we combine the kernelization procedure consisting of all reduction rules in Sect. 3 and the branch-and-bound process consisting of the processing rule and all branching rules in Sect. 4, which leads to an algorithm of running time $\mathcal{O}(2.31^k \cdot m)$. Figure 8 describes the main steps of our algorithm. Herein, the reduction rules include all rules in Sect. 3, which are executed in the order they appear; Rules 1–4.3 stand for the corresponding rules in Sect. 4.

To guarantee step 3.6 would go smoothly, we give the following lemma.

Lemma 12. *Let $r_1(s_1, s_2, s_3)$ be a W_2 substructure in an exhaustively processed instance $(G = (T \cup B, E), k)$. Then, for each i ($1 \le i \le 3$), one can always find another W_2 substructure X_i such that $r_1(s_1, s_2, s_3)$ and X_i are jointed with each other. Moreover, all root vertices on $X_1, X_2,$ and X_3 are distinct.*

Proof. We first show that one can find another W_2 substructure X_1 from s_1 such that $r_1(s_1, s_2, s_3)$ and X_1 are jointed with each other. Since $s_1 \in B$, s_1 has degree 2. Assume that $N(s_1) \setminus \{r_1\} = \{u\}$, where $u \in T$. If u has degree 1, then it has been processed by Reduction Rule 5 or Branching Rule 4.1. If the degree of u is

---

Algorithm EXPV(G, k)

INPUT: a bipartite graph $G = (T \cup B, E)$ and a positive integer k;

OUTPUT: "Yes" if there is a planar 2-layer embedding of G after exploding at
    most k vertices of B; or report "No".

/** assume G is reduced by the reduction rules exhaustively.

1.   **if** $(k < 0)$ or $(k = 0$ but G is not an empty graph) **then** return "No";
2.   **if** $(k \geq 0$ and G is an empty graph) **then** return "Yes";
3.   **switch**
3.1 Case 1: apply Rule 1, and recursively work on the reduced instance;
3.2 Case 2: apply Rule 2, and recursively work on the reduced instances;
3.3 Case 3: apply Rule 3, and recursively work on the reduced instances;
3.4 Case 4.1: apply Rule 4.1, and recursively work on the reduced instances;
3.5 Case 4.2: apply Rule 4.2, and recursively work on the reduced instances;
3.6 Case 4.3: apply Rule 4.3, and recursively work on the reduced instances.

Fig. 8. A parameterized algorithm for the p-BSVE problem.

more than 3, then it has been processed by Branching Rule 2 or 3. In the following, we argue that u does not have degree 2. Assume towards a contradiction that u has degree 2. Let $N(u) \setminus \{s_1\} = \{w\}$, where $w \in B$. Note that w has degree 2. If $w \in N(r_1) \setminus \{s_1\}$, then u has been processed by Processing Rule 1. Otherwise, namely, $w \notin N(r_1) \setminus \{s_1\}$, let $N(w) \setminus \{u\} = \{v\}$. If the degree of v is more than 3, then it has also been processed by Branching Rule 2 or 3. Since G has been exhaustively reduced by Reduction Rule 4, vertex v does not have degree 2 or 1. If v has degree 3, then $\{v\} \cup N(v) \cup N^2(v)$ induce a W_2 substructure X such that X and $r_1(s_1, s_2, s_3)$ are adjacent to each other, which leads to the executing of Branching Rule 4.2. In a word, vertex u having degree 2 contradicts the assumption that G is an exhaustively processed graph. Thus, vertex u must have degree 3 and can be viewed as the root of another W_2 substructure X_1, say $u(p_1, p_2, p_3)$. Moreover, $\{s_1, s_2, s_3\} \cap \{p_1, p_2, p_3\} = \{s_1\}$, namely, $r_1(s_1, s_2, s_3)$ and X_1 are jointed with each other. Along the same line, one can also find a W_2 substructure X_2 (resp. X_3) from s_2 (resp. s_3) such that $r_1(s_1, s_2, s_3)$ and X_2 (resp. X_3) are jointed with each other. Moreover, since Processing Rule 1 has been executed, all root vertices on X_1, X_2, and X_3 are distinct.   □

As shown in Fig. 8, the instance at the beginning of step 3.6 has been processed exhaustively. By Lemma 12, Branching Rule 4.3 will be executed smoothly, which leads to the following conclusion.

Theorem 2. *The algorithm EXPV*(G, k) *solves the p-BSVE problem in time* $\mathcal{O}(2.31^k \cdot m)$, *where m is the number of edges of G.*

6   Conclusion and Further Work

We show a kernel of at most $10.5k$ vertices and a parameterized algorithm of running time $\mathcal{O}(2.31^k \cdot m)$ for the BIPARTITE 1-SIDED VERTEX EXPLODING problem with respect to the number of exploded vertices, where m is the number of edges of the input graph. Our algorithm consists of a kernelization procedure and a branch-and-bound process, which work in an alternating fashion.

Some problems are interesting and deserve further research. (1) Our algorithm may be further improved by exploiting more refined branching rules. (2) We believe that our strategy for deriving branching rules can be applied to other graph drawing problems.

Acknowledgements. The authors thank the anonymous referees for their valuable comments and suggestions.

References

1. Ahmed, R., Kobourov, S., Kryven, M.: An FPT algorithm for bipartite vertex splitting. In: Angelini P., Hanxleden R. von (eds.) Graph Drawing and Network Visualization. GD 2022. LNCS, vol. 13764. Springer, Cham (2023). https://doi.org/10.1007/978-3-031-22203-0_19
2. Ahmed, R., et al.: Splitting vertices in 2-layer graph drawings. IEEE Comput. Graph. **43**(3), 24–35 (2023)
3. Baumann, J., Pfretzschner, M., Rutter, I.: Parameterized complexity of vertex splitting to pathwidth at most 1. In: Paulusma, D., Ries, B. (eds.) Graph-Theoretic Concepts in Computer Science. WG 2023. LNCS, vol. 14093. Springer, Cham (2023). https://doi.org/10.1007/978-3-031-43380-1_3
4. Bhore, S., Ganian, R., Montecchiani, F., Nöllenburg, M.: Parameterized algorithms for queue layouts. J. Graph Algorithms Appl. **26**(3), 335–352 (2022)
5. Chaudhary, A., Chen, D.Z., Hu, X.S., Niemier, M.T., Ravichandran, R., Whitton, K.: Fabricatable interconnect and molecular QCA circuits. IEEE Trans. Comput. Aided Des. Integr. Circuits Syst. **26**(11), 1978–1991 (2007)
6. Cygan, M., et al.: Parameterized Algorithms. Springer, Cham (2016). https://doi.org/10.1007/978-3-319-21275-3
7. Eades, P., McKay, B.D., Wormald, N.C.: On an edge crossing problem. In: ACSC 1986, pp. 327–334 (1986)
8. Eades, P., Wormald, N.C.: Edge crossings in drawings of bipartite graphs. Algorithmica **11**(4), 379–403 (1994)
9. Gramm, J., Guo, J., Hüffner, F., Niedermeier, R.: Automated generation of search tree algorithms for hard graph modification problems. Algorithmica **39**(4), 321–347 (2004)
10. Liu, Y., Wang, J., Guo, J., Chen, J.: Complexity and parameterized algorithms for cograph editing. Theoret. Comput. Sci. **461**, 45–54 (2012)
11. Liu, Y., Wang, J., You, J., Chen, J., Cao, Y.: Edge deletion problems: branching facilitated by modular decomposition. Theoret. Comput. Sci. **573**, 63–70 (2015)
12. Liu, Y., Chen, J., Huang, J., Wang, J.: On parameterized algorithms for fixed-order book thickness with respect to the pathwidth of the vertex ordering. Theoret. Comput. Sci. **873**, 16–24 (2021)
13. Liu, Y., Chen, J., Huang, J.: On book thickness parameterized by the vertex cover number. Sci. Chin. Inf. Sci. **65**(4), 1–2 (2022). https://doi.org/10.1007/s11432-021-3405-x
14. Paul, H., Börner, K., Herr II, B.W., Quardokus, E.M.: ASCT+B REPORTER. https://hubmapconsortium.github.io/ccf-asct-reporter/. Accessed 06 June 2022

15. Pezzotti, N., Fekete, J.D., Höllt, T., Lelieveldt, B.P.F., Eisemann, E., Vilanova, A.: Multiscale visualization and exploration of large bipartite graphs. Comput. Graph. Forum **37**(3), 549–560 (2018)
16. Scornavacca, C., Zickmann, F., Huson, D.H.: Tanglegrams for rooted phylogenetic trees and networks. Bioinformatics **27**(13), i248–i256 (2011)

Multi-winner Approval Voting with Grouped Voters

Yinghui Wen[1]([⊠]), Chunjiao Song[1], Aizhong Zhou[2], and Jiong Guo[1]([⊠])

[1] Shandong University, Qingdao, Shandong, China
{yhwen,chjsong}@mail.sdu.edu.cn, jguo@sdu.edu.cn
[2] Ocean University of China, Qingdao, Shandong, China
zhouaizhong@ouc.edu.cn

Abstract. We consider the general case of approval-based committee elections, where some attributes divide the voters into diverse groups which vary in size. This scenario occurs in applications like the presidential election, where voters come from different parties, or the student board election at a university with students from different schools. However, all existing committee election rules either are derived for the single-group case, or neglect the welfare of groups with few votes. Therefore, new voting rules are needed for this setting. In this paper, We propose two natural axioms for this setting, namely, small group benefited representation (SGBR) and large group benefited representation (LGBR). SGBR requires that if the committee size exceeds the number of groups, at least one candidate approved by each group is in the winning committee. LGBR requires that the winning committee must have at least as many candidates approved by a large group as by a small group. Based on the axioms, we propose three models and investigate parameterized complexity of the models with respect to various parameters. We show that all models are fixed-parameter tractable (FPT) when parameterized by the number n of votes, whereas they become fixed-parameter intractable when parameterized by the size k of the committee or d of the satisfaction bound.

Keywords: Parameterized Complexity · Voting Problems · Computational Social Choice

1 Introduction

Voting problems, which form a core topic in the field of artificial intelligence and computational social choice [8,13], have great significance in serving as a tool to aggregate conflicting preferences [8,13] and receive a considerable amount of attention [28]. The axiomatic properties as well as algorithmic and computational aspects of voting problems have been extensively studied [2,9], where voters express their preferences over all candidates and the goal is to compute winners according to voting rules. Herein, the most widely used and straightforward voting rules are based on *Approval Voting (AV)*, which is originally defined for dichotomous votes, where each vote assigns an approval to each of her favorite

© The Author(s), under exclusive license to Springer Nature Switzerland AG 2024
W. Wu and J. Guo (Eds.): COCOA 2023, LNCS 14462, pp. 267–278, 2024.
https://doi.org/10.1007/978-3-031-49614-1_20

candidates and all other candidates receive disapproval. The winner set consists of those candidates who receive the most approvals.

AV has many desirable properties in single-winner case including simplicity, monotonicity and robustness against manipulation [5,17]. But it becomes less favorable for the case of multiple winners and the most significant drawback is the lack of egalitarian [20]. Attempting to address fairness when using AV for multi-winner voting, some variants of AV have been introduced in the literature [20]. Among them are *Proportional Approval Voting (PAV)* [27], *Satisfaction Approval Voting (SAV)* [7], *Approval Chamberlin-Courant Voting (CCAV)* [12,27], and *Minimax approval voting (MAV)* [6]. The AV, SAV, PAV, and CCAV use a score to represent each vote's satisfaction with respect to the committee, and MAV use a score to represent each vote's dissatisfaction with respect to the committee. The goal is to select a committee, which maximizes the sum of all votes' satisfaction scores for AV, SAV, PAV, and CCAV, or minimizes the maximum of all votes' dissatisfaction scores for MAV. Among the rules, only AV and SAV are polynomial-time solvable the others are NP-hard [2,22,25].

Most of previous researches of committee elections consider each voter as an individual, who is independent of other voters and there is no relation among the voters. However, we often have the scenario in real-world applications, where every voter belongs to a group due to some attributes, which can be formulated as committee election problems with vote attributes. Below we describe several scenarios.

Student Board Election. The first example is the student board election of a university. Here, students are voters and the schools or departments they belong to define an attribute of the voters.

International Sports Election. The second one is people from different countries try to elect a sort of place to hold an international sports event, where each voter has a natural attribute that the country he comes from.

Favorite Singers and Best Film. In some TV shows, like The Singer, audiences are asked to vote for TOP3 favorite singers and the audiences from different ages may like different kind of singers. Similarly, for the election of the best film of the year with votes being given by different film websites, voters from the same website may have the same taste.

Under these scenarios, voters are partitioned into different groups according to these attributes. The groups might admit significantly different sizes. In the first example, a medical school normally has thousands of students, while less than a hundred students are enrolled in a school of sport sciences. Obviously, the rules for committee elections without voter attributes are not appropriate for this application. For example, if PAV is applied, then the opinion of the sport sciences students might be completely ignored.

Some researches considered the case where the candidates are defined by some attributes and the goal is to select a committee that for each attribute offers a certain representation [4,10,11,18,21]. While, there are few researches consider the case of committee elections, where the voters are associated with some attributes. In addition, a lot of axioms studied these years only focus on the welfare of large groups, the welfare of a group with few voters always be neglected [1,16]. For instance, *Justified Representation (JR)* is introduced to make sure that if a large enough group of voters exhibits agreement by supporting the same candidate, then at least one voter in this group has an approved candidate in the winning committee.

Therefore, new voting rules are needed for this setting, which takes the following consideration into account: (1) Even small groups have the right to be represented in the committee. (2) Candidates approved by a large group have no less chance to be members of the committee than that approved by a small group. We formally define the two axioms in Sect. 2, namely, *Small group benefited representation (SGBR)* and *Large group benefited representation (LGBR)*. Based on the above two goals, we propose three models, namely, *Group Representative-σ-π (GR-σ-π)*, *Group Average-π (GA-π)*, and *Group Egalitarian-π (GE-π)*.

GR-σ-π. With $\pi \in$ {AV, SAV, PAV, CCAV, MAV}, this voting rule can be thought as having two rounds. To be more specific, we first use an internal election rule σ to select or construct a set of votes for each group based on the votes in the group, which will serve as that group's representation, called the representative votes of the group. Then, based on the representative votes of all groups, we use a voting rule π to select a winning committee.

GA-π. With $\pi \in$ {SAV, PAV, CCAV, MAV}, this rule uses the average satisfaction of the votes in a group as the satisfaction of that group. More precisely, use the average π-score of the votes in a group as the π-score of that group. Then, find a committee maximizes the minimum score of all groups for $\pi \in$ {SAV, PAV, CCAV}, or a committee minimizes the maximum score of all groups.

GE-π. Despite having rules that satisfy at least one of the LGBR and SGBR as we will show in Sect. 3, GR-σ-π and GA-π both have shortcomings. In GR-σ-π, the satisfaction of minorities in a group is ignored; more specifically, a group member whose approved candidates differ from the majority's of the group may not be taken into account. In GA-π with $\pi \in$ {PAV, CCAV}, if a group contains some individuals whose choices for candidates differ from those of the majority, the group's score is likely to be lower. Here, we design two voting rules to overcome these drawbacks, called Group Egalitarian-PAV (GE-PAV) and Group Egalitarian-CCAV (GE-CCAV), such that a minority's opinion does not lead to a decrease in the group's score, and furthermore, an increase in the group's score if the candidates approved by minorities are members of the winning committee.

Related Work. The goal of selecting a subset of candidates with different attributes under fairness constraints has recently been the focus of a lot of research [4,10,11,18,21]. Fairness constraints are typically captured by absolute upper bounds and/or lower bounds on the number of selected candidates in specific attributes, or proportional representative of selected candidates in specific attribute. In contrast to our setup where the groups of voters are provided in the input, Talmon [26] and Faliszewski and Talmon [15] studied the question of how to partition the votes into disjoint groups. A vertex-labeled graph with each vertex representing a vote is provided as input; the task is to divide the graph into disjoint groups and assign a member of the committee to each group, so that each vote is represented by one of her preferred alternatives. The axioms that concern the fairness of groups of votes also been studied these years, namely, Justified Representation (JR) [1], Extended Justified Representation (EJR) [1] and Proportional Justified Representation (PJR) [16]. They concentrate on the scenario where a number of voters supporting the same candidates form a group, and at least a certain number of voters in this group has an approved candidate in the winning committee. While, the welfare of a group with few voters might be neglected. For instance, PAV satisfies all the axioms but fails to fulfil the SGBR, the small group benefited representation.

2   Preliminaries

In this section, we introduce the definitions and notations used in our models for the committee elections with grouped voters. A committee election with grouped voters can be denoted as $E = (C, V, \zeta)$, where $C = \{c_1, c_2, \ldots, c_m\}$ is the set of the candidates, $V = \{v_1, v_2, \ldots, v_n\}$ is a list of voters represented by their votes, and $\zeta = \{G_1, G_2, \ldots, G_\ell\}$ denotes the set of groups with $G_i \bigcap G_j = \emptyset$ and $\bigcup_{i=1}^{\ell} G_i = V$. In this paper, we interchangeably use the terms vote and voter. The number of the votes in a certain group $G_i \in \zeta$ is denoted as $|G_i| = |\{v \mid v \in G_i\}|$. We focus on approval votes, where an approval vote $v_i \in V$ can be considered as a $\{0, 1\}$-vector of length m. The x-th position of v_i is denoted as $v_i[x]$ with $v_i[x] \in \{0, 1\}$, where $v_i[x] = 1$ (or 0) means that the candidate c_x is approved (or disapproved) by v_i. Given a vote $v \in V$ and a subset of candidates $C' \subseteq C$, we let $v \cap C'$ denote the set of candidates approved by v in C'. Let k be a non-negative integer. A k-committee is a k-size subset of candidates. A k-committee selection rule maps each election (C, V) and every non-negative integer k with $k \leq |C|$ to a collection of k-committees of C with the winning k-committees of (C, V) having the optimal scores under this rule.

2.1   Approval Voting Rules

We first introduce some important approval-based multi-winner voting rules, namely, AV, SAV, PAV, CCAV, and MAV. With respect to each rule, each k-subset of C receives a score and the winning k-committees are those with the desired score.

Under AV, the score of a candidate $c \in C$, denoted as $AV(c)$, is the number of votes approving c. Given a subset $C' \subseteq C$, $AV(C') = \sum_{c \in C'} AV(c)$. Under the SAV, PAV, CCAV, and MAV, given a subset $C' \subseteq C$ and a vote $v \in V$, the scores with respect to C' and v are set as follows. $SAV(v, C') = \frac{|v \cap C'|}{|v|}$. $PAV(v, W) = 1 + \frac{1}{2} + \cdots + \frac{1}{|v \cap C'|}$. If $v \cap C' \neq \emptyset$, $CCAV(v, C') = 1$; otherwise,

$\mathrm{CCAV}(v, C') = 0$. $\mathrm{SAV}(v, W) = \mathcal{H}(v, C')$ with $\mathcal{H}$ being the Hamming distance between v and C', that is, $\mathcal{H}(v, C') = |C_v \setminus C'| + |C' \setminus C_v|$ with C_v being the candidates approved by v.

Given a vote set $V' \in V$ and a candidate set $C' \in C$, the score with respect to V' and C' is set as $\pi(V', C') = \sum_{v \in V'} \pi(v, C')$ with π being SAV, PAV, and CCAV. For MAV, we have $\mathrm{MAV}(V', C') = \max_{v \in V'} \mathrm{MAV}(v, C')$.

The π-*Winner Determination* (π-WD) problem is defined as: **Input:** given an election $E = (C, V)$, a committee size k and a rational number d. **Question:** is there a committee $W \subseteq C$ with $|W| = k$ satisfying that, $\mathrm{AV}(W) \geq d$ with $\pi = \mathrm{AV}$, or $\pi(V, W) \geq d$ with $\pi \in \{\mathrm{SAV}, \mathrm{PAV}, \mathrm{CCAV}\}$, or $\pi(V, W) \leq d$ with $\pi = \mathrm{MAV}$.

2.2 Axioms

We introduce two axioms for multi-winner approval voting with grouped votes, namely, large group benefited representation and small group benefited representation. The first axiom captures the intuition that a large group deserves no fewer representatives than a small group. The second axiom considers that even a small group still deserve at least one representative if the committee size is no less than the number of groups.

Definition 1. Let $E = (C, V, \zeta)$ be an election, k be an integer, W be a committee with $|W| = k$:

1. W provides **large group benefited representation (LGBR)** for E if there do not exist two groups $G_i, G_j \in \zeta$ with $|G_i| > |G_j|$, such that $|\bigcup_{v \in G_i} v \cap W| < |\bigcup_{v' \in G_j} v' \cap W|$.
2. W provides **small group benefited representation (SGBR)** for E if for each group $G_i \in \zeta$, we have $(\bigcup_{v \in G_i} v) \cap W \neq \emptyset$.

We say that an approval-based voting rule satisfies LGBR (SGBR) if for each election E and target committee size k it outputs a committee providing LGBR (SGBR).

2.3 The Models

Given an election $E = (C, V, \zeta)$ and an integer k of the target committee size, we define the score of a group $G_i \in \zeta$ with respect to a k-size subset of C, denoted as C', as follows.

Group Representative-σ-π (GR-σ-π). We set GR-σ-$\pi(G_i, C') = \pi(V^i, C')$ with V^i being the set of representative votes of G_i.

Group Average-π (GA-π). We use the average π-score of the votes in a group as the π-score of that group, that is, GA-$\pi(G_i, C') = \frac{\sum_{v \in G_i} \pi(v, C')}{|G_i|}$.

Group Egalitarian-π (GE-π). GE-$\mathrm{PAV}(G_i, W) = 1 + \frac{1}{2} + \frac{1}{3} + \cdots + \frac{1}{\sum_{v \in G_i} |v \cap W|}$. In GE-CCAV, the input contains a set of integer $\{t_1, \cdots, t_\ell\}$.
GE-$\mathrm{CCAV}(G_i, W) = 1 + \frac{1}{2} + \frac{1}{3} + \cdots + \frac{1}{|\{v | v \in G_i, |v \cap W| \geq t_i\}|}$.

We now have all tools to define the problem of this paper, called *Winner Determination* for τ (τ-WD), where $\tau \in$ {GR-σ-π_1, GE-π_2} with $\pi_1 =$ {AV, SAV, PAV, CCAV, MAV} and $\pi_2 =$ {PAV, CCAV}.

Winner Determination for τ (τ-WD)
Input: An election $E = (C, V, \zeta)$, a positive integer k, and a positive rational number d, and a set of integer $\{t_1, \cdots, t_{|\zeta|}\}$ for GE-CCAV.
Question: Is there a k-size subset $W \subseteq C$ satisfying $\max_{G \in \zeta} \tau(G, W) \leq d$ for π being MAV, or $\sum_{G \in \zeta} \tau(G, W) \geq d$ for π being others?

GA-π_3-WD with $\pi_3 =$ {SAV, PAV, CCAV, MAV} can be defined similarly by replacing the question as: Is there a k-size subset $W \subseteq C$ satisfying that (1) for π being MAV, $\max_{G \in \zeta} \tau(G, W) \leq d$, or (2) $\min_{G \in \zeta} \tau(G, W) \geq d$ for π being others?

In this paper, we consider the following parameters: $m = |C|$, $n = |V|$, $\ell = |\zeta|$, k, the maximal size of groups $\max_i |G_i|$, and d (called the total satisfaction bound).

3   Large/Small Group Benefited Representation

GR-σ-π. We first show that, no matter what σ is, it might be a bad idea to allow each group to have multiple representative votes.

Theorem 1. *(*) (1) If there exists a group having more than one representative vote, then GR-σ-π does not satisfy SGBR even with $\ell = 2$, where $\pi \in \{AV, SAV, PAV, CCAV, MAV\}$. (2) If groups can have more than one representative vote, then GR-σ-π does not satisfy LGBR even with $\ell = 3$, where $\pi \in \{AV, SAV, PAV, CCAV, MAV\}$.*

The theorem above implies that if we use π to handle the grouped voters case without making any changes, then π does not satisfy SGBR and LGBR with $\pi \in$ {AV, SAV, PAV, CCAV, MAV}. Here, we only need to let all votes in a group be the representative votes of the group, which can be seen as an internal election rule σ.

Corollary 1. *AV, SAV, PAV, CCAV, and MAV do not satisfy both SGBR and LGBR.*

We then look into the possibility that each group only has one representative vote. In this situation, some voting rules can fulfil SGBR even while none of them satisfies LGBR.

Theorem 2. *(*) If each group has exactly one representative vote, then (1) GR-σ-π does not satisfy SGBR even with $\ell = 3$, where $\pi \in \{AV, SAV\}$. (2) GR-σ-π satisfies SGBR with $\pi \in \{PAV, CCAV\}$. (3) if $|v^{G_i}| = |v^{G_j}|$ with $G_i, G_j \in \zeta$, GR-σ-MAV satisfies SGBR. (4) if $|v^{G_i}| > |v^{G_j}|$ with $|G_i| > |G_j|$ for all $G_i, G_j \in \zeta$, GR-σ-MAV does not satisfy SGBR even with $\ell = 2$.*

It is easy to show that GR-σ-π does not satisfy LGBR, since each group have exactly one representative vote. As a result, the voting rule cannot make use of the information about the sizes of the groups. Indeed, GR-σ-π fails to fulfill LGBR even if we allow $|v^{G_i}| > |v^{G_j}|$ if $|G_i| > |G_j|$ with $\pi \in \{$AV, SAV, PAV, CCAV$\}$., where $G_i, G_j \in \zeta$, v^{G_i}, v^{G_j} are the representative votes of G_i, G_j, and $|v^{G_i}|(|v^{G_j}|)$ is the number of candidates approved by $v^{G_i}(v^{G_j})$.

Theorem 3. (*) *If each group has exactly one representative vote, then (1) GR-σ-π does not satisfy LGBR even with $\ell = 3$, where $\pi \in \{AV, SAV, PAV, CCAV\}$. (2) if $|v^{G_i}| = |v^{G_j}|$ with $G_i, G_j \in \zeta$, GR-σ-MAV does not satisfy LGBR even with $\ell = 2$. (3) if $|v^{G_i}| > |v^{G_j}|$ with $|G_i| > |G_j|$ for all $G_i, G_j \in \zeta$, GR-σ-MAV satisfies LGBR.*

It can be seen that even if there is only one representative vote per group, most of the voting rules studied in this subsection fail to satisfy both LGBR and SGBR. Finding a voting rule that fulfills both axioms in this setting is therefore important. To address this issue, we define a voting rule Group-based Generalized Approval Voting (GGAV) as follows, which can be seen as generalization of GAV [19]. For each group $G_i \in \zeta$, there is a score-vector $w_i = \{a_i^1, \cdots, a_i^m\}$. The score of a subset of C, denoted as W, with respect to G_i and ζ are defined as $\text{GGAV}(G_i, W) = \sum_{j=1}^{|v^{G_i} \cap W|} a_i^j$, and $\text{GGAV}(\zeta, W) = \sum_{G_i \in \zeta} \text{GGAV}(G_i, W)$. By carefully designing each vector, we can make GGAV satisfy LGBR and SGBR. Given a set of groups $\zeta = \{G_1, \cdots, G_\ell\}$ with $|G_p| \geq |G_q|$ if $p > q$, we set a_i^j with $1 \leq i \leq \ell$ as follows. (1) $a_i^k = \ell - i + 1$; (2) $a_i^j = 0$, for $k < j \leq m$. (3) $a_i^j = \sum_{j+1 < \beta \leq k} \sum_{1 \leq \alpha \leq \ell} a_\alpha^\beta + \sum_{i+1 < \gamma \leq \ell} a_\gamma^j$, for $1 \leq j < k$. In other word, if we denote a_i^j as $b_{(k-j) \times k + (n-i+1)}$, then $b_p = \sum_{1 \leq q < p} b_q$ with $p > \ell$. By doing this, We say the score-vectors are set to grouped setting. It is easy to see that, GR-σ-PAV is a special case of GR-σ-GGAV by setting w_i as $w_i = \{1, \frac{1}{2}, \frac{1}{3}, \cdots, \frac{1}{m}\}$ for each group $G_i \in \zeta$. Therefore, we do not study the parameterized complexity of GGAV in Sect. 4, since all the hardness results of GR-σ-PAV hold for GGAV.

Theorem 4. (*) *If each group has exactly one representative vote, GR-σ-GGAV satisfies both LGBR and SGBR with score-vectors being set to grouped setting.*

Theorem 5. (*) (1) *GA-π satisfies SGBR with $\pi \in \{SAV, PAV, CCAV\}$. (2) GA-MAV does not satisfy SGBR. (3) GA-π does not satisfy LGBR even with $\ell = 2$, where $\pi \in \{SAV, PAV, CCAV, MAV\}$.*

Unfortunately, even GE-PAV and GE-CCAV can overcome the shortcomings of GR-σ-Π and GA-Π, neither of them satisfies the LGBR and SGBR.

Theorem 6. (*) *GE-PAV and GE-CCAV do not satisfy the LGBR and SGBR.*

4 Parameterized Complexity

In this section, we demonstrate the parameterized complexity results of GR-σ-π-WD, GA-π-WD, and GE-π-WD with parameter being $n, m, k, \ell, \max_i |G_i|$, and d. With σ being AV, SAV, or t-Count, it is obvious that GR-σ-AV-WD, GR-σ-SAV-WD, and GA-SAV-WD can be solved in polynomial time. All other problems are NP-hard since PAV-WD, CCAV-WD, and MAV-WD are NP-hard even in the non-grouped setting. Unsurprisingly, parameterized complexity of GR-σ-π-WD with $\pi \in \{$PAV, CCAV, MAV$\}$ is quite similar. Specifically, for a certain parameter, if GR-σ-PAV is W-hard, then GR-σ-CCAV and GR-σ-MAV are also W-hard; likewise, if GR-σ-PAV is FPT, GR-σ-CCAV and GR-σ-MAV are also FPT. To GA-π-WD, the same thing took place. Therefore, instead of displaying the parameterized complexity of all models of GR-σ-π-WD and GA-π-WD, we choose GR-(t-Count)-PAV-WD and GA-MAV-WD to serve as exemplars of R-σ-π-WD and GA-π-WD and examine their parameterized complexity. In addition, we also investigate the parameterized complexity of GE-PAV-WD and GE-CCAV-WD. See Table 1 for the summary of the results.

Table 1. Summary of the results. The results with m as parameter are trivial (by trying all size-k subsets of candidates in $O^*(2^m)$ time). Here, $m = |C|$, $n = |V|$, $\ell = |\zeta|$, k is the size of committee, $\max_i |G_i|$ is the maximal size of groups, and d is the total satisfaction bound.

	GE-CCAV	GE-PAV	GR-(t-Count)-PAV	GA-MAV
m	FPT	FPT	FPT	FPT
n	FPT	FPT	FPT	FPT
ℓ	Para-NP-hard	open	FPT	open
$\max_i |G_i|$	W[2]-hard	Para-NP-hard	Para-NP-hard	Para-NP-hard
k	W[2]-hard	W[2]-hard	W[1]-hard	W[2]-hard
d	W[1]-hard	open	open	W[2]-hard

Theorem 7. (*) (1) GE-CCAV-WD is FPT with respect to n.
(2) Even with only one group, GE-CCAV-WD is NP-hard and W[2]-hard with k as parameter.
(3) GE-CCAV-WD is NP-hard even when $\max_i |G_i| = 1$.
(4) GE-CCAV-WD is W[1]-hard with d as parameter.
(5) GE-PAV-WD is FPT with respect to n.
(6) GE-PAV-WD and GR-(t-Count)-PAV-WD are W[1]-hard with k as parameter.
(7) GR-(t-Count)-PAV-WD is FPT with respect to ℓ or n. (8) GA-MAV-WD is NP-hard even if $\max_i |G_i| = 1$.

Theorem 8. GA-MAV-WD is FPT with n as parameter.

Proof. In the following, we use the tool of integer linear program (ILP) to prove the theorem.

We can think of all votes as an $n \times m$ matrix M with binary values. From the column perspective, there are m columns which can be considered as a collection of n-dimensional vectors. We call two columns identical, if both columns contain

the same value at each position. The set of pairwise identical columns is called
a *column type*. Clearly, there are at most 2^n different column types. Moreover,
let T denote the set of different column types, and for each type $t \in T$, let n_t
denote the number of columns of type t in the input. Additionally, let $\Sigma = \{0, 1\}$.
The ILP can be formulated as follows. It contains 2×2^n variables $x_{t,\varphi}$, where
t denotes a column type and $\varphi \in \Sigma$. The value of $x_{t,\varphi}$ denotes the number of
columns of type t whose corresponding positions in the winning committee W
are set to be φ. We use $\varphi_{t,i,j}$ to denote the value of the vote v_i^j at the positions
corresponding to columns of type t. Considering that the goal of GA-MAV-WD
is to minimize the maximum score among all groups, we aim to minimize

$$\max_{1 \leq i \leq \ell} \frac{\sum_{1 \leq j \leq |G_i|} \sum_{t \in T} \sum_{\varphi \in (\Sigma \setminus \{\varphi_{t,i,j}\})} x_{t,\varphi}}{|G_i|},$$

where ℓ denotes the number of groups and the number of votes in a certain
group G_i is represented by $|G_i|$. Notice that the above objective function can be
replaced by the following constraint by introducing the maximum distance d:

$$\sum_{1 \leq j \leq |G_r|} \sum_{t \in T} \sum_{\varphi \in (\Sigma \setminus \{\varphi_{t,i,j}\})} x_{t,\varphi} \leq d \times |G_i|, \ \forall 1 \leq i \leq \ell$$

Doing so, we arrive at an ILP without objective function. In addition, we add
the following constraints:

$$\sum_{\varphi \in \Sigma} x_{t,\varphi} = n_t, \ \forall t \in T,$$

which means that each column is assigned a value of 0 or 1 in the corresponding position of W, determining whether the corresponding candidate is
selected in W or not. All variables $x_{t,\varphi}$ must be non-negative integers, that
is, $x_{t,\varphi} \in \{0, 1, 2, \ldots, n_t\}, \forall t \in T$ and $\forall \varphi \in \{0, 1\}$. We need an equation constraint to make sure that the winning committee contains exactly k candidates:
$\sum_{t \in T} x_{t,1} = k$.

If there is a solution for the above ILP instance, then we can construct a size-k
committee W by adding $x_{t,1}$ many candidates whose corresponding column type
is t to W. Thus, we can give an ILP formulation for GA-MAV-WD, where the
number of variables depends solely on the parameter value n, the total number
of the votes. It is easy to verify that the above ILP has a solution if and only if
the GA-MAV-WD instance has a solution. The theorem follows from the result
of Lenstra's [23].   □

Theorem 9. *GA-MAV-WD is W[2]-hard with respect to parameters k and d.*

Proof. We prove the theorem by reducing Dominating Set to GA-MAV-WD. The
Dominating Set problem is defined as below. **Input:** A non-negative integer k'
and an undirected graph $G' = (V', E')$ with $|V'| = n'$ and $|E'| = m'$. **Question:**
Is there a subset $S \subseteq V'$ with k' vertices such that every vertex $v \in V'$ is
contained in S or has at least one neighbor in S?

Dominating Set is W[2]-hard with respect to k' [14]. Without loss of generality, we assume each vertex in G' has a degree at least k', since we can add a k'-clique to G' if there is a vertex $v' \in G'$, whose degree is less than k', then make v' be adjacent to each vertex of the added k'-clique. Given an instance (G', k') of Dominating Set, we construct an instance $F((C, V, \zeta), k, d)$ as follows. We create a candidate for each vertex in the graph G'. We call these candidates as "real candidates", denoted as $c_1, c_2, \ldots, c_m$ with $m = n'$. In addition, we construct $3n'$ "dummy candidate" denoted as: $c'_1, c'_2, \ldots, c'_{3n'}$. There are $4n'$ candidates in total. For each vertex $v'_i \in V'$, we construct a group G_i, where the votes one-to-one correspond to the edges incident to v_i. In addition, we add a "special" vote to each group. Therefore, the total number $|G_i|$ of votes in group G_i is $deg(v'_i) + 1$, where $deg(v'_i)$ denotes the degree of vertex v'_i, that is, the number of edges incident to v'_i. Observe that each edge in G' corresponds to two votes, because it has two endpoints. The total number n of votes is $2m' + n'$. For a vote v_j in group G_i constructed for the edge $e_j = \{v'_r, v'_s\}$, it solely approves the two real candidates c_r and c_s, who correspond to the endpoints v'_r and v'_s, but disapproves of all other candidates. We can consider this vote as a vector with two positions of value 1 and $4n' - 2$ positions of value 0. For the special vote in group G_i, it only approves three dummy candidates, $c'_{(i-1) \times 3+1}, c'_{(i-1) \times 3+2}$ and $c'_{(i-1) \times 3+3}$ but disapproves all other candidates. In fact, each dummy candidate is approved only once, because each special vote approves three distinct dummy candidates. By doing so, adding a dummy candidate to the committee is never better than adding a real candidate. We denote a committee as a $\{0, 1\}^{4n'}$ vector which has exactly k' 1's in the real candidate part, so that the Hamming distance of each special vote to the winning committee is $k' + 3$. Then, let $k = k'$ and $d = k' + 2$. We show the equivalence between the instances in the following. Here we omit the "$\Rightarrow$".

"$\Leftarrow$": Assume that there is a solution of GA-MAV-WD, which means that there is a winning committee W of size k' satisfying GA-MAV$(G_i, W) \leq d = k' + 2$ with $1 \leq i \leq n'$. Let $W = W_1 \bigcup W_2$, where W_1 denotes the set of real candidates in W and W_2 denotes the set of dummy candidates in W. We consider the following cases according to whether $|W_2| = 0$ or not.

Case 1: $|W_2| = 0$: Then, $|W_1| = k'$. That is, all candidates in W are real candidates. Let S be the set of the vertices corresponding to the candidates in W. Since the score of each group is at most $d = k' + 2$ and each vote in the group approves either two real candidates or three dummy candidates, there is at least one vote in this group G_i whose Hamming distance to W is at most k', meaning that this vote approves at least one candidate in W and thus there is at least one neighbor of the corresponding vertex v'_i in S. This implies that the set S forms a dominating set.

Case 2: $|W_2| \neq 0$: Let $|W_2| = j$ and $|W_1| = k' - j$. By the construction, each dummy candidate is approved exactly once by a special vote in a group. Thus, adding a dummy candidate to W can only decrease the score of one group. Then we can replace this dummy candidate by a real candidate, which is approved by other votes in this group but not in W. With the degree of each vertex in G'

being at least k', we can conclude that such a real candidate exists. By doing so, we decrease $|W_2|$ by one without increasing the score of any group. Repeating this replacing operation for all dummy candidates in W_2, we arrive at another solution W, containing only real candidates, and Case 1 applies. In summary, a solution of GA-MAV-WD implies a dominating set in G'. $\square$

5   Concluding Remarks

In this paper, we propose three models to deal with the case of approval-based committee elections with grouped voters. At the same time, we propose two axioms named Large group benefited representation and small group benefited representation, and investigate whether the proposed models satisfy the two axioms. We show that all models can hardly satisfy both axioms except the GGAV with the score-vectors being set to grouped setting. We show that all models are fixed-parameter tractable (FPT) when parameterized by the number n of votes, whereas they become fixed-parameter intractable when parameterized by the size k of the committee or d of the satisfaction bound.

We left four questions in Table 1 open, GE-PAV-WD, GR-(t-Count)-PAV-WD and GCMAV-WD with respect to d, GE-PAV-WD and GA-MAV-WD with the parameterization by ℓ. Thus, one future research goal is to resolve the parameterized complexity for them.

Inspired by the work of Baumeister and Dennisen [3], we propose the direction for future research to extend the voting models to other forms of votes, such as trichotomous votes, complete linear orders, and partial linear orders. Another task worthy of detailed study is the problem of coalitional manipulation in the case of committee elections with grouped voters [24].

References

1. Aziz, H., Brill, M., Conitzer, V., Elkind, E., Freeman, R., Walsh, T.: Justified representation in approval-based committee voting. Soc. Choice Welfare **48**(2), 461–485 (2017)
2. Aziz, H., Gaspers, S., Gudmundsson, J., Mackenzie, S., Mattei, N., Walsh, T.: Computational aspects of multi-winner approval voting. In: AAMAS 2015, pp. 107–115 (2015)
3. Baumeister, D., Dennisen, S., Rey, L.: Winner determination and manipulation in minisum and minimax committee elections. In: ADT 2015. vol. 9346, pp. 469–485 (2015)
4. Bei, X., Liu, S., Poon, C.K., Wang, H.: Candidate selections with proportional fairness constraints. Auton. Agent. Multi-Agent Syst. **36**(1), 5 (2022)
5. Brams, S.: Mathematics and democracy: designing better voting and fair-division procedures. Math. Comput. Modell. **48**(9–10), 1666–1670 (2008)
6. Brams, S., Kilgour, D.M., Sanver, M.R.: A minimax procedure for electing committees. Public Choice **132**, 401–420 (2007)
7. Brams, S.J., Kilgour, D.M.: Satisfaction approval voting. In: Fara, R., Leech, D., Salles, M. (eds.) Voting Power and Procedures. SCW, pp. 323–346. Springer, Cham (2014). https://doi.org/10.1007/978-3-319-05158-1_18

8. Brandt, F., Conitzer, V., Endriss, U., Lang, J., Procaccia, A.D. (eds.): Handbook of Computational Social Choice. Cambridge University Press, Cambridge (2016)
9. Bredereck, R., Chen, J., Faliszewski, P., Guo, J., Niedermeier, R., Woeginger, G.J.: Parameterized algorithmics for computational social choice: Nine research challenges. Tsinghua Sci. Technol. **19**(4), 358–373 (2014)
10. Bredereck, R., Faliszewski, P., Igarashi, A., Lackner, M., Skowron, P.: Multiwinner elections with diversity constraints. In: Proceedings of the 32nd AAAI Conference on Artificial Intelligence, pp. 933–940 (2018)
11. Celis, L.E., Huang, L., Vishnoi, N.K.: Multiwinner voting with fairness constraints. In: Proceedings of the 27th International Joint Conference on Artificial Intelligence, pp. 144–151 (2018)
12. Chamberlin, J., Courant, P.: Representative deliberations and representative decisions: proportional representation and the Borda rule. Am. Polit. Sci. Rev. **77**(3), 718–733 (1983)
13. Conitzer, V.: Making decisions based on the preferences of multiple agents. Commun. ACM **53**(3), 84–94 (2010)
14. Downey, R., Fellows, M.: Parameterized Complexity. Springer Science & Business Media (2012)
15. Faliszewski, P., Talmon, N.: Between proportionality and diversity: balancing district sizes under the Chamberlin-courant rule. In: Proceedings of the 17th International Conference on Autonomous Agents and Multi-Agent Systems, pp. 14–22 (2018)
16. Fernández, L., et al.: Proportional justified representation. In: Proceedings of the 31st AAAI Conference on Artificial Intelligence, pp. 670–676 (2017)
17. Fishburn, P.: Axioms for approval voting: direct proof. J. Econ. Theory **19**(1), 180–185 (1978)
18. Ianovski, E.: Electing a committee with dominance constraints. Ann. Oper. Res. **318**(2), 985–1000 (2022)
19. Kilgour, D.M., Marshall, E.: Approval balloting for fixed-size committees. Electoral systems: paradoxes, assumptions, and procedures, pp. 305–326 (2012)
20. Kilgour, M.: Approval balloting for multi-winner elections. In: Handbook on Approval Voting, pp. 105–124. Springer (2010). https://doi.org/10.1007/978-3-642-02839-7_6
21. Lang, J., Skowron, P.: Multi-attribute proportional representation. Artif. Intell. **263**, 74–106 (2018)
22. LeGrand, R., Markakis, E., Mehta, A.: Some results on approximating the minimax solution in approval voting. In: AAMAS 2007, p. 198 (2007)
23. Lenstra, H.: Integer programming with a fixed number of variables. Math. Oper. Res. **8**(4), 538–548 (1983)
24. Obraztsova, S., Zick, Y., Elkind, E.: On manipulation in multi-winner elections based on scoring rules. In: AAMAS 2013, pp. 359–366 (2013)
25. Procaccia, A.D., Rosenschein, J.S., Zohar, A.: On the complexity of achieving proportional representation. Soc. Choice Welfare **30**(3), 353–362 (2008)
26. Talmon, N.: Structured proportional representation. Theoret. Comput. Sci. **708**, 58–74 (2018)
27. Thiele, T.N.: Om flerfoldsvalg. Oversigt over det Kongelige Danske Videnskabernes Selskabs Forhandlinger **1895**, 415–441 (1895)
28. Zwicker, W.S.: Introduction to the theory of voting. In: Handbook of Computational Social Choice, pp. 23–56. Cambridge University Press (2016)

EFX Allocation to Chores over Small Graph

Huahua Miao[1,2], Sijia Dai[1,2], Yicheng Xu[1,2], and Yong Zhang[1,2(✉)]

[1] Shenzhen Institute of Advanced Technology, Chinese Academy of Sciences, Shenzhen, People's Republic of China
{hh.miu,sj.dai,yc.xu,zhangyong}@siat.ac.cn
[2] University of Chinese Academy of Sciences, Beijing, People's Republic of China

Abstract. When allocating indivisible items among agents, achieving envy-free (EF) allocation is not always feasible. Hence a specific area of interest lies in determining whether envy-freeness up to any item (EFX) allocation is feasible for indivisible items. The existence of EFX allocations poses a significant open problem in the field of fair division, even when considering additive valuations. However, while there is a wealth of research on the allocation of goods, relatively little is known about the allocation of chores. Notably, for instances involving bi-valued valuations, existence results have only been established for cases involving three agents. Therefore, we study a natural relaxation of these two fairness constraints, where agents are located on the node of a linear graph and the envy is only possible between adjacent agents. Our main contribution lies in determining the impact of the number of special agents and the presence of arrows in scenarios involving four agents, finding the algorithm that guarantees an EFX allocation when allocating m indivisible bi-valued chores among four linearly structured agents.

Keywords: Fair allocation · EFX · Chores

1 Introduction

The study of fair allocation problem originated with the formal introduction of the cake-cutting problem by *Banach, Knaster and Steinhaus*. Over time, the focus has expanded to include the allocation of a set M of m items to a group N of n agents, where each agent may have a distinct valuation function for the items. In cases where the valuation functions yield positive values, the items are considered goods such as resources, while negative values correspond to chores, such as tasks and housework.

Several notions of fairness have been proposed and analyzed in the last two decades. Among these, envy-free (EF) stands out as the most compelling criterion [6]. An envy-free allocation of goods ensures that no agent prefers the

Supported by NSFC (Grant No.12071460, 12371321) and the Shenzhen Science and Technology Program (Grant No.CJGJZD20210408092806017).

W. Wu and J. Guo (Eds.): COCOA 2023, LNCS 14462, pp. 279–291, 2024.
https://doi.org/10.1007/978-3-031-49614-1_21

set of goods allocated to any other agent over their own. Unfortunately, when dealing with indivisible goods, the existence of an envy-free allocation is not guaranteed [8]. Consequently, researchers have explored natural relaxations of envy-free, such as envy-free up to one good (EF1) [15] and envy-free up to any good (EFX) [9].

It is known that EF1 allocations are guaranteed to exist and can be found in polynomial time for goods [15], chores, and the mixture of the two types [4]. For the case of goods, *Plaut and Roughgarden* [18] have shown that EFX conditions exist in some special cases: (1) identical (combinatorial) valuations, (2) IDO (identical ordering) additive valuations, and (3) $n = 2$. *Chaudhury et al.* [10], *Amanatidis et al.* [1] and *Mahara et al.* [16] further extended the existence of EFX allocations to the cases when (4) $n = 3$, (5) bi-valued valuations, (6) binary valuations under the assumption of budget constraint [11], and (7) when all agents have one of two (general) valuation functions.

Compared to the allocation of goods, there is limited knowledge concerning an EFX allocation for chores. Existing research indicates that an EFX allocation is guaranteed to exist for IDO instances [14] and instances with leveled preferences [12]. Under bi-valued instances, *Zhou and Wu* [20] demonstrate the existence of a polynomial time algorithm for computing an EFX allocation under bi-valued instances when there are three agents.

A few other works have explored fairness concepts using non-complete graphs in the context of resource allocation. In this line of research [5,7,13,19], graphs are used to characterize feasible allocations, emphasizing that the resources allocated to each agent should be connected. Another aspect of this research involves placing agents on the vertices of an undirected graph G, where agents can only view the allocations of their adjacent agents in the graph. For example, envy-free housing allocation over a graph has been investigated, where agents receive one good each and must not envy their neighbors [3]. *Aziz et al.* [2] focus on finding allocations that are epistemically envy-free, meaning that no vertex envies its neighbors. Moreover, they ensure that for any vertex x, there exists an allocation of the remaining goods to the other agents such that x does not envy any other agent. Recently, *Payan et al.* [17] show that when G is represented as a three-edge path, a G-EFX allocation exists for agents with goods.

Our Contributions. In this paper, our main focus is to investigate linear-graph-based relaxations of EFX allocation. We make two key contributions: First, we demonstrate that when there are three arrows connecting four agents in the graph structure, it is possible to compute an EFX allocation in polynomial time. Second, we establish that even in scenarios where there are no three arrows connecting four agents, an EFX allocation can still be computed. These results provide a practical algorithmic approach for achieving EFX allocation in such cases.

2   Preliminaries

We consider how to fairly allocate a set of m indivisible chores M to a group of n agents N, and agents are linear structured with bi-valued cost functions.

A bundle is defined as a subset of items denoted as $X \subseteq M$. An allocation is represented by an n-partition $\mathbf{X} = (X_1, \cdots, X_n)$ of the items, where $X_i \cap X_j = \emptyset$ for all $i \neq j$ and $\cup_{i \in N} X_i = M$. In the allocation $\mathbf{X}$, each agent $i \in N$ receives bundle X_i. Each agent $i \in N$ has an additive cost function $c_i : 2^M \to \mathbb{R}^+ \cup \{0\}$. Specifically,, for any $i \in N$ and $X \subseteq M$, $c_i(S) = \sum_{e \in X} c_i(\{e\})$.

To enhance readability, we write $c_i(e)$ instead of $c_i(\{e\})$, It is assumed that all cost functions are normalized i.e., for any $i \in N, c_i(M) = 1$. For a bundle X_i and chore e, we will write $X_i + e$ or $X_i - e$ to denote $X_i \cup \{e\}$ or $X_i \backslash \{e\}$, respectively.

Definition 1. (Linear fair allocation problem(LFA)) *An instance of the LFA is a triple tuple* $\mathcal{I} = (G, \mathcal{N}, \mathcal{C})$ *where*
- *$G = (V, E)$ is a path graph. V represents the set of vertices, symbolizing individual agents, while E represents the set of edges, which denotes the presence of envy among these agents,*
- *$\mathcal{N} = 1, \cdots, n$ is a set of agents,*
- *$\mathcal{C}$ is an n-tuple of additive cost function c_i.*

As Fig. 1 shows, we refer to elements of V as agents.

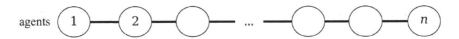

Fig. 1. linear structured agents.

We primarily focus on studying the existence of EFX allocation in bi-valued instances. In a bi-valued instance, for any $i \in N$ and $g \in M$, there exist constants $a, b \geq 0, c_i(e) \in \{a, b\}$. Equivalently, for any $a \neq b$, we can scale the cost function so that $c_i(e) \in \{\epsilon, 1\}$, where $\epsilon \in [0, 1)$.

Due to the particularity of bi-valued instances, we only need to consider two cost values. In the process of allocating chores, the classical Round-Robin algorithm will the agent who finishes picking items later jealous of the agents who finish picking items before her, different valuations of the same item by different agents become the key to solve the problem. When an agent's jealousy of another agent satisfies EF1, if there is still an unallocated sunken item, the envied agent will estimate its cost smaller, while the envied agent will estimate its cost larger. Then assigning this item to the envy-generating agent would effectively implement EFX allocation. Of course this is the simplest case, more complex cases cannot be solved so easily, and to cope with more cases we give the following definition

Definition 2. (Special chore). *We remark chore* $e \in M$ *a special chore to agent i, if there exists a set of neighbors $i - 1, i, i + 1 \in V$, satisfying $c_{i-1}(e) = 1, c_i(e) = \epsilon$, $c_{i+1}(e) = 1$, and we use e_i to represent this chore.*

If this chore satisfies $c_{i-1}(e) = \epsilon, c_i(e) = \epsilon$, $c_{i-1}(e) = 1$, we will use $e_{i,i+1}$ to represent this special chore, respectively. (When agent i is at an endpoint, only one-sided neighbor is considered)

Definition 3. (Special agent). *We remark agent $i \in V$ a special agent if she possesses at least one special chore, and if there exists a chore e, a pair of neighbors $(i,j) \in V$ satisfying $c_i(g) = \epsilon$, $c_j(e) = 1$, we say i is a special agent to j.*

Definition 4. (Envy-free(EF)) *An allocation $\mathbf{X} = (X_1, X_2, \cdots, X_n)$ is EF for chores if for every $(i,j) \in V$, $c_i(X_i) \leq c_i(X_j)$.*

Definition 5. *An allocation $\mathbf{X} = (X_1, X_2, \cdots, X_n)$ is*

a) *Envy-free up to one item(**EF1**) if for every $(i,j) \in V$, there exists a chore $g \in X_j$, such that $c_i(X_i \backslash \{g\}) \leq c_i(X_j)$.*
b) *Envy-free up to any item(**EFX**) if for every $(i,j) \in V$ and every chore $e \in X_j$, it holds that $c_i(X_i \backslash \{e\}) \leq c_i(X_j)$.*

We say the agent i strongly envies an agent j if there exists some chore $g \in X_j$ such that $c_i(X_i \backslash \{e\}) > c_i(X_j)$. When the agents are arranged on the graph G, our goal is to output an allocation X of the set of chores M among the agents N such that there is no edge $(i,j) \in E$ with agent i strongly envious of agent j.

The algorithm we use to compute an EFX allocation is based on the Round-Robin algorithm and Divide-and-Choose algorithm. Initially, all chores start off unallocated. In our proof, unless otherwise specified, it is assumed that the special chores corresponding to different special agents will not be the same.

3   Warm-Up: EFX Allocation for Three Agents and Star Structured Agents

Consider the case where three agents are structured in a non-complete graph, specifically in a linear structure. In this scenario, a straightforward approach can be employed. The central agent initially divides the items into three bundles that satisfy her EFX condition(i.e., no matter which package the central agent gets, her jealousy of the other two bundles is EFX satisfied). Subsequently, the remaining two agents take turns selecting their preferred bundle.

Similarly, when the agents are structured in a star pattern, with one central agent connected to multiple peripheral agents, envy can only occur from the peripheral agents towards the central agent. The same algorithm is applicable in this case. The central agent divides the items into n bundles that meet her EFX condition, and the peripheral agents then select their preferred bundles randomly.

Therefore, for both linear structured agents (with 3 agents) and star structured agents (with n agents), an EFX allocation can be computed for both goods and chores, regardless of the specific valuation functions.

Proposition 1. *For linear structured agents consisting of either 3 agents or n agents arranged in a star structured, an EFX allocation can be computed for both goods and chores, regardless of the specific valuation functions.*

4 An EFX Allocation for LFA

In the following, we present an algorithm that computes an EFX allocation for LFA when $n = 4$.

We use the Fig. 2 in the following as an example to illustrate the existence of a special agent. Case (a) indicates that i is a special agent to j, and j is not a special agent to i. Case (b) indicates that i and j have identical valuation functions. Case (c) indicates that i may be a special agent to j, and j may be a special agent to i (means that it is not important whether i is a special agent for j or j is a special agent for i). It should be noted that if an agent is a special agent (represented by a shadow), there must be an arrow pointing from that agent.

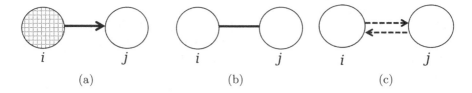

Fig. 2. The presence or absence of special agents.

The Round-Robin algorithm takes a set of items S' and an ordering $\mathcal{L}$ of the agents as input. The algorithm proceeds by allowing agents to choose their favorite chore (the one with the minimum cost) one by one, following the specified order until all items are allocated. It should be noted that unlike allocating goods, our order $\mathcal{L}$ in allocating chores refers to the ending order, that is, if $\mathcal{L}$:(1,2,3), it means that agent 3 takes the last item in S' as the last person to take it. We call the output allocation **X** a Round-Robin allocation.

In this algorithm, the order of agents $\mathcal{L}$ is determined by the order of arrows. When there are $(n-1)$ arrows connecting n agents, the agent with an out-degree of 0 is randomly placed at the end. Then, the sorting proceeds from back to front according to the order indicated by the arrows (when there exists an agent with an out-degree of 2, the agents pointed to by the arrows are randomly arranged). For example, in the case of four agents, if agent 1 has arrows pointing to agent 2, agent 2 and agent 4 both have an arrow pointing to agent 3, the order could be 1,2,4,3 or 1,4,2,3. To describe the allocation process, we index each round using integers $1, 2, \cdots$, and the last round agent i received a chore is denoted by r_i.

Lemma 1. *Let $\mathcal{L}$ be any ordering of N and $X_0 = (\emptyset, \cdots, \emptyset)$. Then Algorithm 1 with input $(N, S', X, \mathcal{L})$ produces an EF1 allocation in polynomial time.* x

---

Algorithm 1: Round-Robin

Input: set of agents: N, set of unallocated chores: S', partial allocation: $\mathbf{X}$, an ordering of N: $\mathcal{L}$

1   $X_i = \emptyset$ for all i,and $k = 1$;
2   **while** $S' \neq \emptyset$ **do**
3      $g = argmin_{e \in S'} c_{\mathcal{L}[k]}(e)$;
4      $X_{\mathcal{L}[k]} = X_{\mathcal{L}[i]} \cup \{e\}$;
5      $S' = S' \backslash \{g\}$;
6      $k = k + 1 \bmod n$;

Output: $\mathbf{X}$

Proof. Consider two agents $i \neq j$. Without loss of generality, assume that $r_i < r_j$, for every round r in which agent i receives a chore e, there exists a round r' satisfying $r < r' < r + n$, in which agent j received an chore e'.

Now, since e has the minimum cost among the agents in round r, we have $c_i(e) \leq c_i(e')$ which leads to $c_i(X_i) \leq c_i(X_j)$.

Similarly, for agent j, excluding the last chore e'' agent j received, we have $c_j(X_j - e'') \leq c_j(X_i)$.

Therefore, the algorithm produces an EF1 allocation. Furthermore, the algorithm runs in polynomial time as it iterates through the set of unallocated chores, and each iteration can be performed in constant time. □

Lemma 2. *Given the Round-Robin allocation on items set M' and agents i and j that are neighbors to each other, if i is a special agent to j, then in the allocation $(\cdots, X_i + e, X_j, \cdots)$, agent i is EFX towards agent j, while agent j is EF towards agent i.*

Proof. From Lemma 1, we have $c_i(X_i) = c_i(X_i + e'') - \epsilon \leq c_i(X_j)$, which leads to agent i being EFX towards agent j. As for agent j, we have $c_j(X_i + e'') = c_j(X_i) + 1 \geq c_j(X_j)$, since agent j does not envy agent i by more than one chore. Therefore, agent j is EF towards agent i. □

Algorithm 2: New Divide-and-Choose

Input: set of agents: N, set of unallocated chores: S', partial allocation: $\mathbf{X}$, an ordering of N: $\mathcal{L}$

1   Initialize: $|N| = n, X_i = \emptyset$ for all i,and $k = 1$;
2   let $\mathcal{L}[n]$ divides S' into n bundles $\mathbf{P} = (P_1, \cdots, P_n)$ which satisfy her EFX;
3   **while** $S' \neq \emptyset$ **do**
4      $X_{\mathcal{L}[k]} = argmin\{c_{\mathcal{L}[k]}(P_1), \cdots, c_{\mathcal{L}[k]}(P_n)\}$;
5      $P' = P' \backslash X_{\mathcal{L}[k]}$;
6      $k = k + 1 \bmod n$;

Output: $\mathbf{X}$

The New Divide-and-Choose algorithm takes a set of unallocated chores S' and an ordering $\mathcal{L}$ of the agents as input. The algorithm proceeds by allowing the last agent in $\mathcal{L}$ to divide S' into n bundles, then has each agent choose their favorite bundle by ordering $\mathcal{L}$. The meaning of n bundles **P** satisfying the EFX of the agent $\mathcal{L}[n]$ is that no matter which bundle agent $\mathcal{L}[n]$ receives, she will not develop strong envy towards the agent who receives other bundles.

Based on the aforementioned discussion, we now delve into the analysis of the number of special agents involved.

4.1   At Least Three Special Agents

Let us first examine the scenario where there are at least three special agents. Without loss of generality, we assume that agent 1 and agent 2 are the special agents, and at least one of agent 3 or agent 4 also holds this designation.

Lemma 3. *For the case where there are three solid arrows connecting the four agents, an EFX allocation can be efficiently computed using the Round-Robin algorithm in polynomial time.*

Proof. If there are solid arrows between any two agents, Algorithm 1 can be employed to compute an EFX allocation. These arrows can be categorized into two distinct scenarios. The first scenario arises when all three arrows point in the same direction (see Fig. 3) situation, we designate the agent without any in-degree as agent 1.

The second scenario occurs when one of the arrows exhibits the opposite direction compared to the other two arrows. In this case we can always find an agent with an in-degree of 2, we label this agent as agent 2.   □

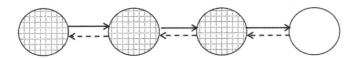

Fig. 3. 3 arrows point in the same direction.

Lemma 4. *When we can not find the three solid arrows, an EFX allocation can be computed.*

Proof. If we fail to find the three arrows that connect the four agents, it indicates that either agent 2 and agent 3 or agent 3 and agent 4 possess identical valuation functions. Base on the location of identical agents, we can use Algorithm 3 to compute an EFX allocation.

In Algorithm 3, agents with zero in-degree and out-degree determine the order of $\mathcal{L}$. When such an agent is at an endpoint, they and their neighbors are randomly sorted at the beginning of $\mathcal{L}$. Agents from the other endpoint follow, and the last agent is placed at the end of $\mathcal{L}$. Alternatively, Algorithm 2 is employed.

Algorithm 3: Double Divide-and-Choose

Input: N, S',**X**, $\mathcal{L}$, $\mathcal{L}'$

1 Initialize: $|N| = 4$,$X_i = \emptyset$ for all i, $k = 1$;

2 **if** *one of the agents at the endpoints, named agent q, has zero outdegree and zero indegree (call her neighbor as agent p)* **then**

3 $\quad$ let $\mathcal{L}[4]$ divides S' into 4 bundles $\mathbf{P} = (P_1, \cdots, P_4)$ which satisfy her EFX;

4 $\quad$ **while** $S' \neq \emptyset$ **do**

5 $\quad\quad$ $X_{\mathcal{L}[k]} = argmin\{c_{\mathcal{L}[k]}(P_1), \cdots, c_{\mathcal{L}[k]}(P_4)\}$;

6 $\quad\quad$ $P' = P' \setminus X_{\mathcal{L}[k]}$;

7 $\quad\quad$ $k = k + 1 \bmod n$;

8 $\quad$ $S' = X_p$;

9 $\quad$ $X_p = \emptyset$;

10 $\quad$ let agent p divides S' into 2 bundles $P' = (P'_1, P'_2)$ which satisfy her EFX;

11 $\quad$ **if** $S' \neq \emptyset$ **then**

12 $\quad\quad$ $X_p = argmax\{c_{\mathcal{L}[4]}(P'_1), c_{\mathcal{L}[4]}(P'_2)\}$;

13 $\quad\quad$ $X_q = P' \setminus X_p$;

14 **else**

15 $\quad$ Computing Divide-and Choose algorithm with $\mathcal{L}' : (1, 4, 3, 2)$;

Output: (X_1, X_2, X_3, X_4)

- When agent 2 and agent 3 have the identical valuation functions, then use algorithm 2 with $\mathcal{L}$:(1,4,2,3). We have

$$c_1(X_1) \leq c_2(X_2), c_4(X_4) \leq c_3(X_3)$$

$$c_2(X_2) - \epsilon \leq c_2(X_i), c_3(X_3) - \epsilon \leq c_3(X_i)$$

and our results absolutely can meet the EFX condition.
- Considering agent 3 and agent 4 have the identical valuation functions, and then use Algorithm 3 with $\mathcal{L}$:(1,3,3,2), $p = 3, q = 4$. We call the bundle obtained by agent 3 in the first round X'_3 and X''_3, cause agent 3 and agent 4 have identical valuation functions. Hence,

$$c_3(X'_3) \leq c_3(X_2), \quad c_3(X''_3) \leq c_3(X_2).$$

For agent 1, $c_1(X_1) \leq c_1(X_2)$, for agent 4, and for every $e \in M$, $c_4(X_4 - g) \leq c_3(X_3)$.

For agent 2,

$$c_2(X_2 - e) \leq c_2(X'_3) \tag{1}$$

$$c_2(X_2 - e) \leq c_2(X''_3) \tag{2}$$

adding inequality 1 and 2 yields the following inequality

$$c_2(X_2) - c_2(e) \leq c_2(X'_3 + X''_3)/2$$
$$\leq \max\{c_2(X'_3), c_2(X''_3)\}$$

Recall that $X_3 = \max\{c_2(P_1), c_2(P_2)\}$, and $X_3 \cup X_4 = X_3' \cup X_3''$. We have

$$c_2(X_2) - c_2(e) \leq c_2(X_3 + X_4)/2$$
$$\leq \max\{c_2(X_3), c_2(X_4)\}$$
$$= c_2(X_3)$$

then agent 2 is EFX towards agent 3.
For agent 3, $c_3(X_i) = c_4(X_i)$,

$$c_3(X_3 + X_4) = c_3(X_3) + c_4(X_4)$$
$$\leq 2 \cdot c_3(X_2)$$

- When $c_3(X_3) \leq c_3(X_4)$, we have $c_3(X_3) \leq c_3(X_3 + X_4)/2 \leq c_3(X_2)$.
- When $c_3(X_4) \leq c_3(X_3)$, then $c_3(X_3 - e) \leq c_3(X_4)$, we have

$$c_3(X_3 - e) \leq c_3(X_3 + X_4)/2 \leq c_3(X_2).$$

Hence, (X_1, X_2, X_3, X_4) is an EFX allocation.

$\square$

4.2   Two Special Agents

Next, we consider the case when there are exact two special agents. Before proving, we show all valuation scenarios for chores of agents firstly (see table 1).

Table 1. Valuation functions for chores.

agents \ chores	e_1	e_2	e_3	e_4	e_5 $\cdots$
1	1	ϵ	ϵ	1	1
2	ϵ	ϵ	ϵ	ϵ	1
3	1	1	ϵ	ϵ	1
4	1	1	ϵ or 1	ϵ or 1	1

Lemma 5. *When there exists three arrows that connect those four agents, an EFX allocation can be computed in polynomial time.*

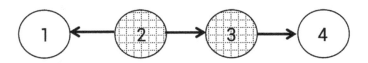

Fig. 4. 2 special agents with 3 arrows that connect 4 agents.

Proof. Those two special agents must be agent 2 and agent 3, due to symmetry, when $e_{2,3}$ and $e_{3,2}$ exist at the same time, the situation is similar to that of only $e_{2,3}$. To simplify the proof, we consider the case where only $e_{2,3}$ exists.(see Fig. 4)

By carefully deciding the ordering of agents, we can compute a Round-Robin allocation (S_1, S_2, S_3, S_4) on items M', satisfying $r_2 < r_1 < r_3 < r_4$, we call this ordering as $\mathcal{L}'$ (see algorithm 4).

- If there exists $e' \in M'$, satisfying $c_1(e') = 1$, $c_2(e') = \epsilon$, $c_3(e') = 1$.
  After executing Round-Robin algorithm, agent 1 and agent 3 is EF1 towards agent 2, agent 4 is EF1 towards agent 3. Similar to the proof above, $(X_1, X_2 + e', X_3 + e_{3,4}, X_4)$ is an EFX allocation.
- If there exists $e' \in M'$, satisfying $c_1(e') = c_2(e') = \epsilon$.
  We only need to prove agent 2 is EFX towards agent 1 and agent 3. After executing Round-Robin algorithm,

$$c_2(X_2) \leq c_2(X_1), \quad c_2(X_2) \leq c_3(X_2).$$

then

$$c_2(X_2 + e_{2,1} + e_{2,3}) - \epsilon = c_2(X_2) + \epsilon \leq c_2(X_1 + e') = c_2(X_1) + \epsilon,$$

So as to agent 3, $(X_1 + e', X_2 + e_{2,1} + e_{2,3}, X_3 + e_{3,4}, X_4)$ is an EFX allocation.

- If neither of these $e' \in M'$ can be found.
  Apparently, agent 2 and agent 3 have identical valuations in M' under our assumption, and

$$for\ every\ e \in M' \cup \{e_{2,1}\},\ c_1(e) = 1,$$

After executing Round-Robin algorithm, agent 1 is EFX towards agent 2. Then, $(X_1, X_2 + e_{2,3}, X_3 + e_3, X_4)$ is an EFX allocation.

□

Lemma 6. *When there dose not exist three arrows that connect those four agents, an EFX allocation can be computed.*

Proof. If we cannot find three arrows that connect those four agents, it means either there is one pair of agents, two pairs of agents, or three agents with the identical valuation functions. The same as at least three special agents, these can be solved using Algorithm 2, or Algorithm 3 to obtain an EFX allocation.

□

Algorithm 4: Algorithm for 2 special agents with 3 arrows that connect 4 agents

---

Input: N, S', **X**, $\mathcal{L}'$, M

1 Initialize: $X_i = \emptyset$ for all i, $M' = M\backslash\{e_{2,1}, e_{2,3}, e_3\}$;

2 **if** *there exists $e' \in M'$, satisfying $c_1(e') = 1$, $c_2(e') = \epsilon$, $c_3(e') = 1$,* **then**

3 $\quad\mid\quad$ $S' = M\backslash\{e', e_3\}$;

4 $\quad\mid\quad$ Computing Round-Robin algorithm with $\mathcal{L}'$;

5 $\quad\mid\quad$ $X_2 = X_2 + e'$, $X_3 = X_3 + e_3$;

6 **if** *there exists $e' \in M'$, satisfying $c_1(e') = c_2(e') = \epsilon$,* **then**

7 $\quad\mid\quad$ $S' = M'\backslash\{e'\}$;

8 $\quad\mid\quad$ Computing Round-Robin algorithm with $\mathcal{L}'$;

9 $\quad\mid\quad$ $X_1 = X_1 + e'$, $X_2 = X_2 + e_{2,1} + e_{2,3}$, $X_3 = X_3 + e_3$;

10 **else**

11 $\quad\mid\quad$ Computing Round-Robin algorithm with $\mathcal{L}'$;

12 $\quad\mid\quad$ $X_2 = X_2 + e_{2,3}$, $X_3 = X_3 + e_3$;

Output: (X_1, X_2, X_3, X_4)

4.3 At Most One Special Agent

Finally, we consider the case when there is at most one special agent. When there is only one special agent, it can be divided into two situations.

One is the special agent located at the endpoint of the line, say agent 1. It means that for any chore, the other three agents have exactly the identical valuation for it. Therefore, we only need to use Algorithm 2 with $\mathcal{L}$:(1,2,3,4). Our results absolutely can meet an EFX allocation.

The other is the special agent is not at the end of the line, say agent 2. Hence, agent 3 and agent 4 have identical valuation functions. We can use Algorithm 3 with $\mathcal{L}$: (1,3,3,2), $p = 3$, $q = 4$ to find an EFX allocation.

When there is no special agent, the situation becomes simpler because all four agents have identical valuation functions.

Building upon the previous analysis, we are now ready to present our main theorem:

Theorem 1. *An algorithm can be devised to compute an EFX allocation of four linear structured agents for bi-valued instances.*

5 Discussion and Conclusion

In our study, we explore the EFX allocation for chores in the context of graph structures. Introducing graphs as a relaxation is a natural approach, as many real-life agents are primarily concerned with interactions among specific agents. This relaxation allows us to achieve results that were not attainable when considering complete graphs. Additionally, demonstrating positive results on natural classes of graphs may contribute to establishing the existence of EFX allocations more broadly.

Our findings reveal the existence of EFX allocation for a bi-valued instance involving four linearly structured agents. An intriguing direction for further investigation would be to explore whether EFX allocation exists for other non-complete graph structures, such as circular graphs. Moreover, graphs offer the opportunity to define and explore several other notions of fairness, such as local proportionality and local max-min share, which could be pursued in future research endeavors.

References

1. Amanatidis, G., Birmpas, G., Filos-Ratsikas, A., Hollender, A., Voudouris, A.A.: Maximum Nash welfare and other stories about EFX. Theoret. Comput. Sci. **863**, 69–85 (2021)
2. Aziz, H., Bouveret, S., Caragiannis, I., Giagkousi, I., Lang, J.: Knowledge, fairness, and social constraints. In: Proceedings of the AAAI Conference on Artificial Intelligence, vol. 32, p. 1 (2018)
3. Beynier, A., et al.: Local envy-freeness in house allocation problems. Auton. Agent. Multi-Agent Syst. **33**, 591–627 (2019)
4. Bhaskar, U., Sricharan, A., Vaish, R.: On approximate envy-freeness for indivisible chores and mixed resources. arXiv preprint arXiv:2012.06788 (2020)
5. Bouveret, S., Cechlárová, K., Elkind, E., Igarashi, A., Peters, D.: Fair division of a graph. arXiv preprint arXiv:1705.10239 (2017)
6. Brams, S.J., Taylor, A.D.: Fair Division: from cake-cutting to dispute resolution. Cambridge University Press (1996)
7. Bredereck, R., Kaczmarczyk, A., Niedermeier, R.: Envy-free allocations respecting social networks. Artif. Intell. **305**, 103664 (2022)
8. Budish, E.: The combinatorial assignment problem: approximate competitive equilibrium from equal incomes. J. Polit. Econ. **119**(6), 1061–1103 (2011)
9. Caragiannis, I., Kurokawa, D., Moulin, H., Procaccia, A.D., Shah, N., Wang, J.: The unreasonable fairness of maximum Nash welfare. ACM Trans. Econ. Comput. (TEAC) **7**(3), 1–32 (2019)
10. Chaudhury, B.R., Garg, J., Mehlhorn, K.: EFX exists for three agents. In: Proceedings of the 21st ACM Conference on Economics and Computation, pp. 1–19 (2020)
11. Dai, S., Gao, G., Liu, S., Lim, B.H., Ning, L., Xu, Y., Zhang, Y.: EFX under budget constraint. In: Frontiers of Algorithmic Wisdom - International Joint Conference, IJTCS-FAW 2022. Lecture Notes in Computer Science, vol. 13461, pp. 3–14. Springer, Cham (2022). https://doi.org/10.1007/978-3-031-20796-9_1
12. Gafni, Y., Huang, X., Lavi, R., Talgam-Cohen, I.: Unified fair allocation of goods and chores via copies. arXiv preprint arXiv:2109.08671 (2021)
13. Igarashi, A., Peters, D.: Pareto-optimal allocation of indivisible goods with connectivity constraints. In: Proceedings of the AAAI Conference on Artificial Intelligence, vol. 33, pp. 2045–2052 (2019)
14. Li, B., Li, Y., Wu, X.: Almost (weighted) proportional allocations for indivisible chores. In: Proceedings of the ACM Web Conference 2022, pp. 122–131 (2022)
15. Lipton, R.J., Markakis, E., Mossel, E., Saberi, A.: On approximately fair allocations of indivisible goods. In: Proceedings of the 5th ACM Conference on Electronic Commerce, pp. 125–131 (2004)

16. Mahara, R.: Extension of additive valuations to general valuations on the existence of EFX. arXiv preprint arXiv:2107.09901 (2021)

17. Payan, J., Sengupta, R., Viswanathan, V.: Relaxations of envy-freeness over graphs (2023)

18. Plaut, B., Roughgarden, T.: Almost envy-freeness with general valuations. SIAM J. Discret. Math. **34**(2), 1039–1068 (2020)

19. Suksompong, W.: Fairly allocating contiguous blocks of indivisible items. Discret. Appl. Math. **260**, 227–236 (2019)

20. Zhou, S., Wu, X.: Approximately EFX allocations for indivisible chores. arXiv preprint arXiv:2109.07313 (2021)

Extreme Graph and Others

Zero-Visibility Cops and Robber Game on Cage Graph

Xiaoli Sun[1,2]($\boxtimes$), Farong Zhong[1,2], and Boting Yang[2]

[1] Department of Computer Science, Zhejiang Normal University, Jinhua, China
hengxiangdaying@163.com
[2] Department of Computer Science, University of Regina, Regina,
Saskatchewan S4S 0A2, Canada

Abstract. We consider zero-visibility cops and robber game that the cops lack of information on the location of the robber at all times, which is a variant of the classical cops and robbers game. First of all, we use the idea of splitting to study properties of cage graphs. Then we apply properties of cage graphs to investigate the lower bounds of cop number and the monotonic zero-visibility cop number of cage graphs. We also propose a searching algorithm to calculate the monotonic zero-visibility cop number of cage graphs.

Keywords: Graph searching · Pursuit-evasion · Cops and robber · Cage graph

1 Introduction

Graph searching, also called pursuit-evasion problem. For a given graph, the core problem is to determine the minimum number of searchers needed to search for robbers [1–4]. In recent years, Graph searching has evolved into a powerful tool that can provide corresponding search models to solve practical problems based on different real-world problems. One of the practical problems, such as mobile robot target hunting and large warehouse retrieval of goods, can be solved by the cops and robbers game. Cops and robber game is one of the graph searching model, which is a perfect information two player game on the graph and introduced independently by Nowakowski and Winkler [5] and Quilliot [6].

Many variants of Cops and robber game have been considered [7]. The Zero-visibility cops and robber game is variant of the Cops and robbers game proposed by Tošić, which has the same setting as Cops and robbers except that the cops have no information about the position of the robber [8]. It means that the robber is invisible to cops. Compared to the perfect information Cops and robbers model, it is harder for cops to catch robber in the zero-visibility cops and robber model. Our goal is to determine the smallest number of cops that can capture the robber definitely.

To the best of our knowledge, despite extensive research on Cops and robbers game, research on zero-visibility cops and robber game remains a huge

© The Author(s), under exclusive license to Springer Nature Switzerland AG 2024
W. Wu and J. Guo (Eds.): COCOA 2023, LNCS 14462, pp. 295–309, 2024.
https://doi.org/10.1007/978-3-031-49614-1_22

challenge. Tošić described graphs which one cop enough to capture the robber and calculated the minimum zero-visibility cop number for the paths, complete graphs, circle and complete bipartite graphs [8]. Additionally, Tang studied the zero-visibility cops and robber game on tree and acyclic graphs, determined the minimum zero-visibility cop number of tree and acyclic graphs and proposed a square time algorithm to compute zero-visibility cop number of tree [9]. There are still some left, *Dereniowski* et al. proved that the zero-visibility cop number of the graph is related to the pathwidth of graph, they also gave the proof that the upper and lower bounds of the zero-visibility cop number are related to the multiple of the path width [10]. *Dereniowski* also gave a linear time algorithm for calculating the zero-visibility cop number on the tree and proved that the problem of determining the zero-visibility cop number of the graph is NP-complete [11].

It can be seen that research on zero-visibility cops and robber game is not abundant. Therefore, we mainly give a study of the zero-visibility cop number on cage graphs and related search algorithm in this paper. It is extremely meaningful to determine the zero-visibility cop number for more different structured graphs.

The structure of the paper is as follows. In Sect. 2 we give some terms and concepts that is related with the paper. In Sect. 3 we make a thorough inquiry about the monotonic zero-visibility cop number of cage graphs. In Sect. 4 zero-visibility searching algorithm on the cage graphs is proposed. And we summarized the paper in Sect. 5.

The corresponding results in the paper are summarized in appendix Table 1.

2   Preliminaries and Basic Definitions

2.1   Graph Theory Notion

All graphs are assumed to be undirected without multiple edges and loops in this paper. We introduce the following terminology regarding this game.

Let $G = (V_G, E_G)$ be a graph with vertex set V_G and edge set E_G. A graph $G' = (V_{G'}, E_{G'})$ is a *subgraph* of G if and only if $V_{G'} \subseteq V_G$ and $E_{G'} \subseteq E_G$. We use uv to represent a edge connecting the vertex u and vertex v. If $uv \in E_G$, we say that u and v are *adjacent*, expressed as $u \sim v$. For any subset X of V_G, the set

$$N_G(X) = \{u \in V_G \mid \exists v \in X \text{ such that } u \sim v\}$$

is the *neighbourhood* of X. If $X = \{u\}$ is a singleton, we use $N_G(u)$ rather than $N_G(\{u\})$ to represent the neighbourhood of u. Clearly, the number of neighbourhood of a vertex u is exactly the *degree or valency* of the vertex u, i.e. $deg_G(u) = |N_G(u)|$. The minimum degree of graph G is written as $\delta(G) = min(|N_G(v)| \mid v \in V_G)$.

A *walk* is an alternating sequence $W = v_0, e_1, v_1, ..., e_k, v_k$ of vertices and edges such that each edge e_i, $1 \leq i \leq k$, has endpoints v_{i-1} and v_i. A *path* is a walk that the vertices except that its first vertex and its last vertex in it are different. We use $p = v_1 v_2 ... v_k$ to denote a path with ends v_1 and v_k. A *cycle* is a path that its first vertex is the same as its last vertex. A *trail* is a walk in which no edge occurs more than once. A *circuit* is a trail whose first vertex is the same as its last. The length of the shortest cycle in the graph is called the *girth*, marked g. The edges that are not in the cycle but its two endpoints are in the cycle are called the *chord*. The induced cycle of G is that does not contain chord. A *matching* in a graph is a set of edges such that no two edges are incident. A *matching* M covers a vertex v if there is an edge in M that has v as an endpoint. The *matching* number of the graph G, $M(G)$, is the size of a maximum matching in the graph G.

A graph is said to be regular of *valency* r if each of its vertices has *valency* r. A regular graph with *valency* r and *girth* g is called a (r, g)-graph. A (r, g)-graph with the least possible number of vertices is called a (r, g)-cage and marked $G_{r,g}$. The number of vertices of a (r, g)-cage is denoted by $|V_{G_{r,g}}|$. Any cage with odd girth g must have at least $1 + r \sum_{i=0}^{(g-3)/2} (r-1)^i$ vertices, and any cage with even girth g must have at least $2 \sum_{i=0}^{(g-2)/2} (r-1)^i$ vertices [12,13].

For $r = 2$, the $(2, g)$-cage is the g-cycle. For $g = 2$, the $(r, 2)$-cage has just two vertices and they are joined by exactly r edges. For $g = 3$, the $(r, 3)$-cage is the $(r + 1)$-clique. Hereafter, we study cage graph with $r = 3, 3 \leq g \leq 12$, $r = 4, 3 \leq g \leq 8$, $g = 4, 3 \leq r \leq 7$ in this paper. For a cage graph with $r = 3$, the neighbor connecting with the chord of v_j, denoted by v_{c_j}. The neighbor connecting with the non-chords of v_j, denoted by v_{n_j}. For a cage graph with $r = 4$, the neighbor of v_j is connected by a chord, which is an edge on the second smallest cycle of v_j, and is denoted as v_{c-long_j}. The neighbor of v_j is connected by a chord, which is an edge on the smallest cycle of v_j, and is denoted as $v_{c-short_j}$.

2.2 Cops and Robber

The zero-visibility cops and robber game is played on an undirected connected graph G and consists of two sides of the confrontation, which a cop player controls the movements of a fixed number of cops and a robber player controls the movement of a single robber. Both cop player and robber player let them move on the graph. The robber has all information about the locations and movements of all cops while the cops have no information about the location and movement of the robber at any time. i.e., the robber is invisible to cops. Cop can only guess the location of the robber based on his previous moves, and the robber has been evading cops tracking at all times. The game is played in a sequence of rounds. Each round consists of a pair of turns, a cops' turn to move, followed by a robber's turn to move.

At the 0*th* round, the cops occupy some of the vertices on the graph, and the robber occupies a vertex without cops. At round $i(i \geq 1)$, the cops move from the current vertex to one of its neighbors, and the robber moves to its neighbor that is not occupied by cop from the current vertex. They can all choose to stay at current vertices. The cops capture the robber if one of them occupies the same vertex as the robber and it happens in a finite number of movements. Otherwise, the robber wins. The zero-visibility cop number of G, denoted by $c_0(G)$, is the minimum number of cops required to capture the robber on G. A $cop - win$ strategy for G is optimal if it uses $c_0(G)$ cops to capture the robber.

The vertex is cleared if we are sure that a vertex is not occupied by the robber. Otherwise, it's contaminated. The set of cleared vertices is denoted as V_C. The set of contaminated vertices is denoted as V_Z. Similarly, the set of edges that robber will be caught immediately or next turn if he passes is denoted as E_C. The set of other edges is denoted as E_Z. If the cleaned vertices in V_C become contaminated again, it is called recontamination. The cleared vertex in the i round is denoted as V_C^i.

For each edge uv and vertices u, v of G, we want to construct a strategy that can clean the graph G. A zero-visibility cops and robber search strategy of a graph is a sequence of movement that clear the whole graph. We call the two vertices u and v guarded from the robber if there is a cop that moves back and forth between u and v. It will be caught either immediately or on the next turn if the robber moves onto either while this is occurring. Considering this activity in the graph cleaning model, although the vertices u and v are possibly being re-contaminated over and over, the contamination will never spread through them. Because they are cleaned before they can possibly re-contaminated any further vertices. We refer to the above activity as vibrating on the edge uv.

In a pursuit-evasion game of this sort, a topic of general interest is that of the monotonicity of strategies. Typically, a strategy is monotonic if recontamination never occurs. In other words, it satisfies $V_Z^i \leq V_Z^{i-1}$. We wish to take advantage of the above strategic element. However, sometimes monotonicity is not guaranteed during the search process. And there may occur some recontamination vertices after each cleanup. Thus, we define a weakly monotonic strategy with length T. In a weakly monotonic strategy, there may occur some cleaned vertices that are re-contaminated after each cleanup. But if on the cops' next round, cops can immediately clean up the re-contaminated vertices, then the contamination is not spread. Thus, if the cop player is following a weakly monotonic strategy and the robber moves onto a vertex that has been previously visited by a cop, he will be caught on the every next turn.

The monotonic zero-visibility cop number of a connected graph G is the minimum order $mc_0 = mc_0(G)$ among successful weakly monotonic strategies on G. It is the smallest cop number required to capture the robber utilising a weakly monotonic strategy. We are exclusively interested in weakly monotonic strategies, so we will simply use the term monotonic with the meaning of weakly monotonic. That is to say, for convenience of description, they are collectively referred to as monotonicity hereinafter. Clearly, we have $c_0(G) \leq mc_0(G)$ for all graphs G.

3   The Monotonic Zero-Visibility Cop Number of Cage Graph

This section includes two subsections: in the first subsection, we will give lower bounds on zero-visibility cop number of cage graph that $r \geq 3$ and $g \geq 3$; in the second subsection, we will apply the lower bounds to give the monotonic zero-visibility cop number of cage graph.

3.1 Lower Bounds of Cop Number

Lemma 1. *For a cage graph G, let $H = G - V_k$. Here $V_k \subseteq V_G$, if it satisfies $|V_H| \leq |E_H|$, then there is at least one cycle in H.*

Proof. If an undirected graph has n vertices and $n-1$ edges, it can be connected. It's obvious that there is no cycle (i.e., spanning tree). But with an additional edge and let there is no vertex with degree less than one , it will form a cycle without considering the heavy edge. □

Property 1. For a cage graph G with $r = 3, g \geq 3$, there are at most $\frac{|V_{G_{3,g}}|}{g}$ cycles with *length* $\geq g$ in the graph G, and it is the smallest number of cycles containing all vertices of G.

Theorem 1. *For a cage graph G with $r = 3, 3 \leq g \leq 12$, $c_0(G_{3,g}) \geq \frac{2|V_{G_{3,g}}|}{g}$.*

Proof. We prove it as follows, where $H_i = G_{3,g} - V_C^{i-1}$, $|N_Z^i[v_n]|$ is the number of contaminated neighbors of v_n .

- (i) When $g = 3$, $|V_{G_{3,3}}| = 4, |E_{G_{3,3}}| = 6, M(G_{3,3}) = 2$, so matching covers $G_{3,3}$. First, we place a cop on each endpoint of edges in matching and let they begin vibrating on the matching, then the cops win in the first round. If not, we use one cop and place the cop on a vertex on the $G_{3,3}$. It's obvious that there is a cycle on $G_{3,3}$, and it's a graph of robber-win. So $c_0(G_{3,3}) = 2 = \frac{|V_{G_{3,3}}|}{2} \geq \frac{2|V_{G_{3,3}}|}{g}$.

- (ii) When $g = 4, |V_{G_{3,4}}| = 6, |E_{G_{3,4}}| = 9, M(G_{3,4}) = 3$. Similar to (i), we can prove that $c_0(G_{3,4}) = 3 = \frac{|V_{G_{3,4}}|}{2} \geq \frac{2|V_{G_{3,4}}|}{g}$.

- (iii) When $g \geq 5$, we assume that $c_0(G_{3,g}) \leq \frac{2|V_{G_{3,g}}|}{g} - 1$, and consider an optimal strategy with at most $\frac{2|V_{G_{3,g}}|}{g} - 1$ cops. We will use induction to show that $|V_C^i| \leq 2\frac{2|V_{G_{3,g}}|}{g} - 2$ for all $i \geq 0$. When $i = 0$, it's easy to see that $|V_C^0| \leq \frac{2|V_{G_{3,g}}|}{g} - 1 < 2\frac{2|V_{G_{3,g}}|}{g} - 2$. We assume that $|V_C^{i-1}| \leq 2\frac{2|V_{G_{3,g}}|}{g} - 2$ holds for round $i - 1$, where $i \geq 1$. There are two cases.

  - Case 1: the remaining vertices and edges in H_i satisfies $|V_Z^i| \leq |E_Z^i|$. Owing to $c_0(G_{3,g}) \leq \frac{2|V_{G_{3,g}}|}{g} - 1 < \frac{|V_{G_{3,g}}|}{2}$ and $|V_C^{i-1}| \leq 2\frac{2|V_{G_{3,g}}|}{g} - 2 \leq 2\frac{2|V_{G_{3,g}}|}{5} - 2$, there is a vertex v_n in H_i that it has at least two adjacent contaminated vertices. In other words, $|N_Z^i[v_n]| \geq 2$. It follows Lemma 1, it exists at least one cycle in H_i. According to **property1**, we can see that it exists at most $\frac{|V_{G_{3,g}}|}{g}$ cycles with *length* $\geq g$ in the graph G. If $\frac{2|V_{G_{3,g}}|}{g} - 1$ cops occupied $\frac{2|V_{G_{3,g}}|}{g} - 1$ vertices on $\frac{|V_{G_{3,g}}|}{g} - 1$ cycles in round

$i - 1$, it exists one cycle without cop. In round i, we move a cop that is adjacent to the cycle to this cycle to clear. It's obvious that one cop can not clear one cycle. If not, in the round $i - 1$, we assume that there are $\frac{2|V_{G_{3,g}}|}{g} - 1$ vertices occupied by cops and there are two or more cops that are adjacent to the cycle. We move these cops to clear H_i in round i. In case of $|V_C^{i-1}| \le 2\frac{2|V_{G_{3,g}}|}{g} - 2$, $|V_Z^i| \ge |V_{G_{3,g}}| - 2\frac{2|V_{G_{3,g}}|}{g} + 2 > \frac{2|V_{G_{3,g}}|}{g} - 1$, and it exists vertices in V_C^{i-1} that get re-contaminated after the robber's turn in round i from above, therefore $|V_C^i| \le |V_C^{i-1}| \le 2\frac{2|V_{G_{3,g}}|}{g} - 2$.

- Case 2: the remaining vertices and edges in H_i satisfies $|V_Z^i| > |E_Z^i|$. Since $c_0(G_{3,g}) \le \frac{2|V_{G_{3,g}}|}{g} - 1 < \frac{|V_{G_{3,g}}|}{2}$, it exists a vertex v_n that is non-adjacent to cops and unoccupied by cops. And in the previous $i - 1$ round, if we move $\frac{2|V_{G_{3,g}}|}{g} - 2$ cops to vibrate on $2\frac{2|V_{G_{3,g}}|}{g} - 4$ vertices with $|V_C^{i-1}| \le 2\frac{2|V_{G_{3,g}}|}{g} - 2$, then $|V_Z^i| = |V_{G_{3,g}}| - 2\frac{2|V_{G_{3,g}}|}{g} + 4 \ge 6$. If it exists one path that contains v_n with length ≥ 2 in H_i, then the remaining one cop can not clear remaining H_i. Similar to case 1, if we move other cop to clear H_i, then $|V_C^i| \le |V_C^{i-1}| \le 2\frac{2|V_{G_{3,g}}|}{g} - 2$.

From the above, we have $|V_C^i| \le 2\frac{2|V_{G_{3,g}}|}{g} - 2$ for all $i \ge 0$ while we use at most $\frac{2|V_{G_{3,g}}|}{g} - 1$ cops, which is a contradiction. Hence, $c_0(G_{3,g}) \ge \frac{2|V_{G_{3,g}}|}{g}$, where $g \ge 3$. This ends the proof.   □

Theorem 2. For a cage graph with $r = 4, 3 \le g \le 8, c_0(G_{4,g}) \ge \frac{2|V_{G_{4,g}}|}{7}$.

Proof. We prove it as follows, where $H_i = G_{3,g} - V_C^{i-1}$.

- when $g = 3$. $|V_{G_{4,3}}| = 5, |E_{G_{4,3}}| = 10, M(G_{4,3}) = 2$, if we place two cops on the one endpoint of edges in matching, and let them begin vibrating on the matching, then there is a vertex not covered by matching. If not, in case of there exists intersect cycles of G, two cops can not capture the robber. So, $c_0(G_{4,3}) > 2$. If we place two cops on the one endpoint of edges in matching, and let they begin vibrating on the matching. Then place the third cop on the vertex that is not covered. We can capture the robber now. So $c_0(G_{4,3}) = 3 > \frac{2|V_{G_{4,3}}|}{7}$.
- when $g = 4$. $|V_{G_{4,4}}| = 8, |E_{G_{4,4}}| = 16, M(G_{4,4}) = 4$, if we place four cops on the one endpoint of edges in matching, and let they begin vibrating on the matching, we can capture the robber in the first round.
- when $g \ge 5$. Assume that $c_0(G_{4,g}) \le \frac{2|V_{G_{4,g}}|}{7} - 1$ in the whole process of search. We use an optimal strategy that makes cop win on $G_{4,g}$ with at most $\frac{2|V_{G_{4,g}}|}{7} - 1$ cops, and we show that $|V_C^i| \le \frac{4|V_{G_{4,g}}|}{7} - 2$ for all $i \ge 0$. When $i = 0$, it's easy to see that $|V_C^0| \le \frac{2|V_{G_{4,g}}|}{7} - 1 \le \frac{4|V_{G_{4,g}}|}{7} - 2$. We assume that $|V_C^{i-1}| \le \frac{4|V_{G_{4,g}}|}{7} - 2$ holds for round $i - 1$, when $i \ge 1$. There are two cases.

- Case 1: the remaining vertices and edges in H_i satisfy $|V_Z^i| \leq |E_Z^i|$. Owing to $c_0(G_{4,g}) \leq \frac{2|V_{G_{4,g}}|}{7} - 1 < \frac{|V_{G_{4,g}}|}{2}$ and $|V_C^{i-1}| \leq \frac{4|V_{G_{4,g}}|}{7} - 2$, it exists a vertex v_n in H_i that satisfies $|N_Z^i[v_n]| \geq 3$. It follows from Lemma 1, it exists at least a cycle in H_i. In round i, if there is a cop that is adjacent to this cycle, then we move the cop to clear H_i. It's obvious that one cop can not clear a cycle. If not, it exists at least two cops are adjacent to this cycle. We move these cops to clear H_i in round i. In this way, we may clear this cycle. However, since $|V_C^{i-1}| \leq \frac{4|V_{G_{4,g}}|}{7} - 2$, $|V_Z^i| \geq |V_{G_{4,g}}| - \frac{4|V_{G_{4,g}}|}{7} + 2 > \frac{2|V_{G_{4,g}}|}{7} - 1$, it exists vertices in V_C^{i-1} that get re-contaminated after the robber's turn in round i from above. Therefore, $|V_C^i| \leq |V_C^{i-1}| \leq \frac{4|V_{G_{4,g}}|}{7} - 2$.
- Case 2: the remaining vertices and edges in H_i satisfy $|V_Z^i| > |E_Z^i|$. Since $c_0(G_{4,g}) \leq \frac{2|V_{G_{4,g}}|}{7} - 1 < \frac{|V_{G_{4,g}}|}{2}$, it exists a vertex v_n that is non-adjacent to cops and is not occupied by cops. And since $|V_C^{i-1}| \leq \frac{4|V_{G_{4,g}}|}{7} - 2$, if we move $\frac{2|V_{G_{4,g}}|}{7} - 2$ cops to vibrate on $\frac{4|V_{G_{4,g}}|}{7} - 4$ vertices, then $|V_Z^i| = |V_{G_{4,g}}| - \frac{4|V_{G_{4,g}}|}{7} + 4 \geq 6$. If there is a path contains v_n with length ≥ 2, then the remaining one cop can not clear remaining H_i. Similar to case 1, if we move other cop to clear H_i, it exists vertices in V_C^{i-1} that get re-contaminated, then $|V_C^i| \leq |V_C^{i-1}| \leq \frac{4|V_{G_{4,g}}|}{7} - 2$.

From the above, we have $|V_C^i| \leq \frac{4|V_{G_{4,g}}|}{7} - 2$ for all $i \geq 0$ while we use at most $\frac{2|V_{G_{4,g}}|}{7} - 1$ cops, which is a contradiction. Hence, $c_0(G_{4,g}) \geq \frac{2|V_{G_{4,g}}|}{7}$, where $g \geq 3$. $\qquad\square$

3.2 Cop Number of Cage Graph

As below, we will give some lemmas about $mc_0(G_{3,g}), mc_0(G_{4,g})$ and $mc_0(G_{r,4})$.

Lemma 2. *For a cage graph with* $r = 3, g = 3, mc_0(G_{3,3}) = 2$.

Lemma 3. *For a cage graph with* $r = 3, g = 4, mc_0(G_{3,4}) = 3$.

The proof of Lemma 2 and Lemma 3 are easy and are omitted.

Lemma 4. *For a cage graph with* $r = 3, g = 5, mc_0(G_{3,5}) = 4$.

Proof. It follows Theorem 1, $mc_0(G_{3,5}) \geq \lfloor \frac{2|V_{G_{3,5}}|}{5} \rfloor \geq 4$.
Here is the monotonic strategy of four cops to clear $G_{3,5}$(see Fig. 1). We place cops c_1, c_2, c_3, c_4 on vertex v_1, then move c_1 from v_1 along $v_1 v_6$ to v_6 and let c_1 begin vibrating on $v_1 v_6$ in whole search process. Since the robber sees the location of the cops, the robber does not choose to move to v_1, v_6. If robber does that, he will be caught immediately or next turn. Therefore, the robber will only choose to move along the path in H_i. We move cop c_2 along $v_1 v_2 v_7$ to v_7, and

let c_2 begin vibrating between v_7v_{10} in whole search process. Similarly, since the robber can see the location of the cop, he will not choose to move to vertex v_7, v_{10}. It's obvious that there is a tree with only six vertices in H_i. Then we move the cop c_3, c_4 along $v_1v_5v_4$ to v_4. Let c_3 begin vibrating on v_4v_3 in whole search process, and let c_4 search along the path of H_i. We can easily clear $G_{3,5}$ from above strategy for four cops. Thus, the lemma holds.   □

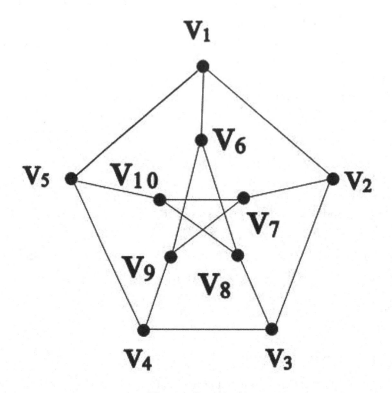

Fig. 1. $Cage(3,5)$.

Lemma 5. *For a cage graph with $r = 3, g = 6$, $mc_0(G_{3,6}) = 5$.*

Proof. It follows Theorem 1, $mc_0(G_{3,6}) \geq \lfloor \frac{2|V_{G_{3,6}}|}{6} \rfloor \geq 4$.
Here is the monotonic strategy to clear $G_{3,6}$(see Fig. 2). We place cops c_1, c_2, c_3, c_4 on vertex v_1, let c_1 move from v_1 along v_1v_{10} to v_{10} and begin vibrating on v_1v_{10} in whole search process. Since the robber sees the location of the cops, the robber does not choose to move to v_1,v_{10}. If robber does that, he will be caught next turn. Therefore the robber will only choose to move along the path in H. We move cop c_2 to v_3. Let c_2 begin vibrating between v_3v_{12} in whole search process. Similarly, since the robber can see the location of the cop, he will not choose to move to vertex v_3,v_{12}. We move cop c_3 to v_6, and let c_3 begin vibrating between v_6v_7 in whole search process. Then it exists a cycle in

H. Now we can only move c_4 to clear H. Obviously, it can not guarantee to capture robber in this way. If not, the vertex in V_C will be re-contaminated. That is to say, $c_0(G_{3,6}) > 4$. If there are five cops, from above strategy, we move cops c_4, c_5 to v_{14}, and then let c_5 search along the cycle in H. We can easily prove that five cops can clear $G_{3,6}$ from above strategy. Thus, Lemma 5 holds. □

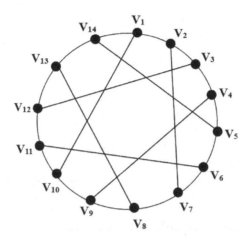

Fig. 2. $Cage(3,6)$.

Due to the limited space, when $7 \leq g \leq 12$, we will not list proof of them one by one. We give the Theorem 3 below to summarize the zero-visibility cop number when $r = 3, 3 \leq g \leq 12$.

Theorem 3. *For a cage graph with* $r = 3, 3 \leq g \leq 12$, *there is* $\frac{2|V_{G_{3,g}}|}{g} \leq mc_0(G_{3,g}) \leq \frac{2|V_{G_{3,g}}|}{g} + 2$.

Lemma 6. *For a cage graph with* $r = 4, g = 3$, $mc_0(G_{4,3}) = 3$.

Proof. It follows Theorem 2, we have $mc_0(G_{4,3}) \geq \lceil \frac{2|V_{G_{4,3}}|}{7} \rceil \geq 2$.
Here is the monotonic strategy of two cops to clear $G_{4,3}$.
We place cops c_1, c_2 on vertex v_1, and let c_1 begin vibrating on v_1v_4 in whole search process. Since the robber sees the location of the cops, the robber does not choose to move to v_1, v_4. If robber do that, he will be caught next turn. Therefore, the robber will only choose to move the path in H. There is a cycle with three vertices in H. Obviously, we can only move c_2 to clear H, and it can not guarantee to capture robber in this way. If not, the vertex in V_C will be re-contaminated. That is to say, $mc_0(G_{4,3}) > 2$. If there are three cops, from above strategy, we move cops c_2, c_3 to v_5, and let c_3 search along the path in H. We can easily prove that three cops can clear $G_{4,3}$ from above strategy. Thus, Lemma 6 holds (Fig. 3).

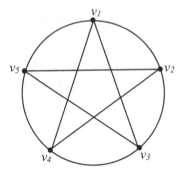

Fig. 3. $Cage(4,3)$.

Lemma 7. *For a cage graph with $r = 4, g = 4$, $mc_0(G_{4,4}) = 4$.*

Proof. It follows Theorem 2, $mc_0(G_{4,4}) \geq \lceil \frac{2|V_{G_{4,4}}|}{7} \rceil \geq 3$.
There is a monotonic strategy of four cops to clear $G_{4,4}$. We place cops c_1, c_2, c_3 on vertex v_1, and let c_1 begin vibrating on v_1v_4 in whole search process. Since the robber sees the location of the cops, the robber does not choose to move to v_1, v_4. If robber do that, he will be caught next turn. Therefore the robber will only choose to move the path in H. We move c_2 to v_8, and let c_2 begin vibrating on v_8v_5 in whole search process. There is a cycle with four vertices in H. Obviously, we can only move c_3 to clear H and it can not guarantee to capture robber in this way. If not, the vertex in V_C will be re-contaminated. That is to say, $mc_0(G_{4,4}) > 3$. If there are four cops, from above strategy, we move cops c_3, c_4 to v_2, and let c_4 search along the path in H. We can easily prove that four cops can clear $G_{4,4}$ from above strategy. Thus, Lemma 7 holds (Fig. 4).

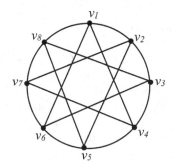

Fig. 4. $Cage(4,4)$.

Due to the limited space, when $5 \leq g \leq 8$, we will not list proof of them one by one. We give the Theorem 4 below to summarize the zero-visibility cop number when $r = 4, 3 \leq g \leq 8$.

Theorem 4. *For a cage graph with $r = 4, 3 \leq g \leq 8$, $\frac{2|V_{G_{4,g}}|}{7} \leq mc_0(G_{4,g}) \leq \frac{2|V_{G_{4,g}}|}{7} + 2$.*

Theorem 5. *For a cage graph with $3 \leq r \leq 7, g = 4$, $mc_0(G_{r,4}) = r$.*

Proof. We place r cops, $c_1, ...c_r$ on v_1, where $1 \leq j \leq |V_{G_{r,4}}|$.
For i from 1 to $r - 2$, j from 1 to $|V_{G_{r,4}}|$,
move c_i from v_j to $v_j v_{j+3}$ and let c_i begin vibrating on $v_j v_{j+3}$.
$i \leftarrow i + 1, j \leftarrow j + 2$.
We move the remaining two cops to clear the remaining cycles in H.   □

4 Algorithm for the Monotonic Zero-Visibility Strategy of Cage

In this section, we propose a zero-visibility search algorithm(ZCRSC-Cleaning a cage graph $G_{r,g}$ in a monotonic manner) for searching cage graphs. In this algorithm, Input: A cage graph $G(r, g)$, Output: the cops' sequence. The idea of our algorithm as follows, firstly, we place n cops $c_1, c_2, ..., c_n$ on v_1, secondly, Starting from the first round, we compare the sizes of $|E_Z|, |V_Z|$ in each round. If $|E_Z| \geq |V_Z|$ and there is a vertex connected to three or more contaminated edges. This means that the remaining graph has multiple cycles that intersect, and we need to call a function(BreakCycle-Cleaning a cage graph $G_{r,g}$ in a monotonic manner)to break these cycles. If $|E_Z| < |V_Z|$ and there is a vertex connected to three or more contaminated edges. This means that the remaining graph has multiple paths that intersect and we need to call a function(IZ-Cleaning a cage graph $G_{r,g}$ in a monotonic manner) to search paths. If $|E_Z| < |V_Z|$ and there is no a vertex connected to three or more contaminated edges. This means that the remaining graph has multiple disjoint paths, we move cops to search every path. If $|E_Z| \geq |V_Z|$ and there is no a vertex connected to three or more contaminated edges. This means that the remaining graph has multiple disjoint cycles, we move cops to search every cycle. Since we have a finite number of vertices and edges, according to the above, we can easy to get that is a linear time complexity algorithm to clean cage graph. We can see appendix for detailed algorithms.

5 Conclusion and Further Discussion

The paper studies zero-visibility cops and robber game on cage graphs. First of all, we solve the lower bounds of zero-visibility cop number on cage graphs. Secondly, we prove that the monotonic zero-visibility cop number of cage graphs. Thirdly, a linear time complexity algorithm has been proposed to clean cage graph in a monotonic manner. In future studies, we will design effective algorithm

about other graph searching problems. For example, zero-visibility cops and
robber game on crown graphs. At the same time, we are also interested to some
problems such as fast searching on cage graphs.

Appendices

Table 1. zero-visibility cop number of Cage graph.

Cage graph $G_{3,g}$				
Name	Vertices	Edges	lower bounds	mc_0
$G_{3,3}$	4	6	$c_0(G_{3,3}) \geq \frac{2|V_{G_{3,3}}|}{3}$	2
$G_{3,4}$	6	9	$c_0(G_{3,4}) \geq \frac{2|V_{G_{3,4}}|}{4}$	3
$G_{3,5}$	10	15	$c_0(G_{3,5}) \geq \frac{2|V_{G_{3,5}}|}{5}$	4
$G_{3,6}$	14	21	$c_0(G_{3,6}) \geq \frac{2|V_{G_{3,6}}|}{6}$	5
$G_{3,7}$	24	36	$c_0(G_{3,7}) \geq \frac{2|V_{G_{3,7}}|}{7}$	6
$G_{3,8}$	30	45	$c_0(G_{3,8}) \geq \frac{2|V_{G_{3,8}}|}{8}$	8
$G_{3,9}$	58	87	$c_0(G_{3,9}) \geq \frac{2|V_{G_{3,9}}|}{9}$	13
$G_{3,10}$	70	105	$c_0(G_{3,10}) \geq \frac{2|V_{G_{3,10}}|}{10}$	16
$G_{3,11}$	112	168	$c_0(G_{3,11}) \geq \frac{2|V_{G_{3,11}}|}{11}$	22
$G_{3,12}$	126	189	$c_0(G_{3,12}) \geq \frac{2|V_{G_{3,12}}|}{12}$	23
Cage graph $G_{4,g}$				
$G_{4,3}$	5	10	$c_0(G_{4,3}) \geq \frac{2|V_{G_{4,3}}|}{7}$	3
$G_{4,4}$	8	16	$c_0(G_{4,4}) \geq \frac{2|V_{G_{4,4}}|}{7}$	4
$G_{4,5}$	19	38	$c_0(G_{4,5}) \geq \frac{2|V_{G_{4,5}}|}{7}$	6
$G_{4,6}$	26	52	$c_0(G_{4,6}) \geq \frac{2|V_{G_{4,6}}|}{7}$	8
$G_{4,7}$	67	134	$c_0(G_{4,7}) \geq \frac{2|V_{G_{4,7}}|}{7}$	19
$G_{4,8}$	80	160	$c_0(G_{4,8}) \geq \frac{2|V_{G_{4,8}}|}{7}$	22
Cage graph $G_{r,4}$				
$G_{3,4}$	6	9	r	3
$G_{4,4}$	8	16	r	4
$G_{5,4}$	10	25	r	5
$G_{6,4}$	12	36	r	6
$G_{7,4}$	14	49	r	7

Algorithm 1. ZCRSC-Cleaning a cage graph $G_{r,g}$ in a monotonic manner

---

Require: A cage graph $G_{r,g}(r = 3, 3 \leq g \leq 12$ and $r = 4, 3 \leq g \leq 8)$
Ensure: The sequence of the searchers'movements S
1: Place n cops on a vertex.
2: **while** $E_Z \neq \emptyset$ **do**
3:   **if** $|E_Z| \geq |V_Z|$ and there is v_j with $|E_Z(v_j)| \geq 3$ **then**
4:     call BreakCycle()
5:   **end if**
6:   **if** $|E_Z| < |V_Z|$ and there is v_j with $|E_Z(v_j)| \geq 3$ **then**
7:     call IZ()
8:   **end if**
9:   **if** $|E_Z| < |V_Z|$ and there is no v_j with $|E_Z(v_j)| \geq 3$ **then**
10:    move cops along each path to search robber.
11:   **end if**
12:   **if** $|E_Z| \geq |V_Z|$ and there is no v_j with $|E_Z(v_j)| \geq 3$ **then**
13:    move cops along each cycle to search robber.
14:   **end if**
15: **end while**
16: **if** $V_Z \neq \emptyset$ **then**
17:   move cops to contaminated vertices to search robber.
18: **end if**
19: return S.

Algorithm 2. BreakCycle-Cleaning a cage graph $G_{r,g}$ in a monotonic manner

---

Require: A cage graph $G_{r,g}(r = 3, 3 \leq g \leq 12$ and $r = 4, 3 \leq g \leq 8)$
Ensure: The sequence of the searchers'movements S
1: **function** BREAKCYCLE( )
2:   **if** $r = 3, g = 5$ **then**
3:     find a vertex v_i from G with $|E_Z(v_i)| = max$ and move cops from v_1 to v_i.
4:   **else**
5:     find a vertex v_i with $|E_Z(v_i)| = max$ and $|E_Z(v_{n_i})| = max$ and move cops from v_1 to v_i
6:   **end if**
7:   **if** there is a vertex v_{ip} in $N_H(v_i)$ with $|E_Z(v_{ip})| = max$ **then**
8:     move cop from v_i along $v_i v_{ip}$ to v_{ip} and let the cop vibrate on $v_i v_{ip}$ until cops win.
9:   **end if**
10: **end function**

Algorithm 3. IZ-Cleaning a cage graph $G_{r,g}$ in a monotonic manner

---

Require: A cage graph $G_{r,g}(r = 3, 3 \leq g \leq 12$ and $r = 4, 3 \leq g \leq 8)$
Ensure: The sequence of the searchers'movements $\mathcal{S}$
1: **function** IZ( )
2:     **if** there are vertices with $|E_Z(v_p)| = 0$ and $|E_Z(v_u)| \geq 2$, and there is no $|E_Z| < |V_Z|$ without the vertices with $|E_Z(v_p)| = 0$ **then**
3:         move cop to the vertex v with $E_Z(v) = max$ and let this cop vibrate on vv_u until cops win.
4:         move another cop along paths to search robber.
5:     **else if** $|V_Z| \leq 6$ **then**
6:         move cops along paths to search robber.
7:     **else**
8:         move cop to the vertex v with $|E_Z(v)| = max$ and move another cop along paths to search robber.
9:     **end if**
10:     return
11: **end function**

References

1. Bonato, A., Yang, B.: Graph searching and related problems. In: Pardalos, P.M., Du, D.-Z., Graham, R.L. (eds.) Handbook of Combinatorial Optimization, pp. 1511–1558. Springer, Heidelberg (2013). https://doi.org/10.1007/978-1-4419-7997-1_76

2. Parsons, T.D.: Pursuit-evasion in a Graph. In: Alavi, Y., Lick, D.R. (eds.) the International Conference on Theory and Applications of Graphs 1976, Lecture Notes in Mathematics, vol. 642, pp. 426–441. Springer, Heidelberg (1978). https://doi.org/10.10007/BFb0070400

3. Megiddo, N., HakimiS, L., Garey, M.R., Johnson, D.S., Papadimitriou, C.H.: The complexity of searching a graph. J. Assoc. Comput. Mach. **35**(1), 18–44 (1988)

4. Kirousis, L.M., Papadimitriou, C.H.: Searching and pebbling. Theor. Comput. Sci. **47**, 205–218 (1986)

5. Nowakowski, R., Winkler, P.: Vertex-to-vertex pursuit in a graph. Discret. Math. **43**(2–3), 235–239 (1983)

6. Quilliot, A.: Problemes de jeux, de point Fixe, de connectivite et de representation sur des graphes, des ensembles ordonnes et des hypergraphes. PhD thesis, Universite de Paris VI (1978)

7. Bonato, A., Nowakowski, R.J.: The Game of Cops and Robbers on Graphs: Student Mathematical Library, American Mathematical Society, Providence, Rhode Island, vol. 61, pp. 191–220 (2011)

8. Tošić.: Vertex to Vertex Search in a Graph. In: Proceedings of the Sixth Yugoslav Seminar on Graph Theory. University of Novi Sad, pp. 43–56 (1985)

9. Tang, A.: Cops and Robber with Bounded Visibility. Masters thesis, Dalhousie University (2004)

10. Dereniowski, D., Dyer, D., Tifenbach, R.M., Yang, B.: Zero-visibility cops and robber and the pathwidth of a graph. J. Comb. Optim. **29**, 541–564 (2015)

11. Dereniowski, D., Dyer, D., Tifenbach, R.M., Yang, B.: The complexity of zero-visibility cops and robber. In: Chen, J., Hopcroft, J.E., Wang, J. (eds.) FAW 2014. LNCS, vol. 8497, pp. 60–70. Springer, Cham (2014). https://doi.org/10.1007/978-3-319-08016-1_6

12. Survey, C.-A.: Pak-Ken Wong. J. Graph Theor. **6**, 1–22 (1982)

13. WIKIPEDIA   Homepage.   http://en.wikipedia.org/wiki/Cage_(graph_theory). Accessed 4 Oct 2018

14. Clarke, N.E., Macgillivray, G.: Characterizations of K-copwin Graphs. J. Discrete Math. **312**(8), 1421–1425 (2012)

Online Facility Assignment for General Layout of Servers on a Line

Tsubasa Harada$^{(\boxtimes)}$ and Toshiya Itoh

Department of Mathematical and Computing Science, Tokyo Institute of Technology,
2 -12 -1 Ookayama, Meguro-ku, Tokyo 152 -8550, Japan
harada.t.ak@m.titech.ac.jp, titoh@c.titech.ac.jp

Abstract. In the online facility assignment on a line OFAL(S, c) with a set S of k servers and a capacity $c : S \to \mathbb{N}$, each server $s \in S$ with a capacity $c(s)$ is placed on a line, and a request arrives on a line one-by-one. The task of an online algorithm is to irrevocably match a current request with one of the servers with vacancies before the next request arrives. An algorithm can match up to $c(s)$ requests to a server $s \in S$. In this paper, we propose a new online algorithm PTCP (Policy Transition at Critical Point) for OFAL(S, c) and show that PTCP is $(2\alpha(S) + 1)$-competitive, where $\alpha(S)$ is informally the ratio of the diameter of S to the maximum distance between two adjacent servers in S. Depending on the layout of servers, $\alpha(S)$ ranges from $O(1)$ to $O(k)$. Among all of known algorithms for OFAL(S, c), this upper bound on the competitive ratio is the best when $\alpha(S)$ is small. We also show that the competitive ratio of any MPFS (Most Preferred Free Servers) algorithm [6] is at least $2\alpha(S) + 1$, where MPFS is a class of algorithms whose competitive ratio does not depend on a capacity c. Recall that the class MPFS includes the natural greedy algorithm and PTCP, etc. Thus, this implies that PTCP is the best for OFAL(S, c) in the class MPFS.

Keywords: Online algorithm · Competitive analysis · Online metric matching · Online matching on a line · Online facility assignment · Greedy algorithm

1 Introduction

The *online facility assignment* (OFA) or *online transportation* problem was introduced by Kalyanasundaram and Pruhs [10]. In this problem, an online algorithm is given a set S of k servers and a capacity $c : S \to \mathbb{N}$, and receives n requests one-by-one in an online fashion. The task of an online algorithm is to match each request immediately with one of the k servers. Note that the number of requests is at most the sum of each server's capacity, i.e., $n \leq \sum_{s \in S} c(s)$. The maximum number of requests that can be matched with a server $s \in S$ is $c(s)$, and the assignment cannot be changed later once it has been decided. The cost of matching a request with a server is determined by the distance between

© The Author(s), under exclusive license to Springer Nature Switzerland AG 2024
W. Wu and J. Guo (Eds.): COCOA 2023, LNCS 14462, pp. 310–322, 2024.
https://doi.org/10.1007/978-3-031-49614-1_23

them. The goal of the problem is to minimize the sum of the costs of matching n requests. When the underlying metric space is restricted to be a line, we refer to such a variant of OFA as OFA on a line (denoted by OFAL).

This problem OFA has many applications. Consider a car-sharing service where there are k car stations and each station s has $c(s)$ available cars. This service must assign users arriving one after another to car stations immediately. It is desirable that all users can use a nearby car station as many as possible. OFAL also can be viewed as the following real-world problem: Consider a rental shop that must rent skis with appropriate length to skiers. What kind of algorithm can be used to reduce the gap between the length of the appropriate skis and the actual rented skis? In this case, each server s is one type of skis, and its capacity $c(s)$ is the number of available skis for rent.

Ahmed et al. [1] dealt with classical competitive analysis for OFAL under the assumption that the servers are evenly spaced. We refer to the setting as OFAL$_{eq}$. Ahmed et al. [1] showed (with rough proofs) that the natural greedy algorithm matching a request with its closest available server is $4k$-competitive and the *Optimal-fill* algorithm is k-competitive for any $k > 2$. On the other hand, Itoh et al. [7] analyzed the competitive ratio for OFAL$_{eq}$ with small $k \geq 2$. They showed that (i) for $k = 2$, the greedy algorithm is 3-competitive and best possible, and (ii) for $k = 3$, 4, and 5, the competitive ratio of any algorithm is at least $1 + \sqrt{6} > 3.449$, $\frac{4+\sqrt{73}}{3} > 4.181$, and $\frac{13}{3} > 4.333$, respectively.

For OFA, Harada et al. [6] introduced a class of algorithms called MPFS (Most Preferred Free Servers) as a natural generalization of the greedy algorithm and showed that the competitive ratio of any MPFS algorithm does not depend on a capacity c. This is referred to as the capacity-insensitive property. In addition, they determine the exact competitive ratio of the greedy algorithm for OFAL$_{eq}$ to be $4k - 5$ ($k \geq 2$) by using the properties of MPFS algorithms. Moreover, they present an MPFS algorithm IDAS (Interior Division for Adjacent Servers) for OFAL and showed that the competitive ratio of IDAS for OFAL$_{eq}$ is $2k - 1$ and best possible among all MPFS algorithms for OFAL$_{eq}$.

1.1   Our Contributions

In this paper, we present a new MPFS algorithm PTCP (Policy Transition at Critical Point) for OFAL with a set S of k servers and a capacity $c : S \rightarrow \mathbb{N}$, and show that the competitive ratio of PTCP is exactly $2\alpha(S) + 1$ and best possible among all MPFS algorithms for OFAL. As we have mentioned, IDAS [6] is best possible only for OFAL$_{eq}$. Informally, $\alpha(S)$ is the ratio of the diameter of S to the maximum distance between two adjacent servers in S (see (1) for details). We emphasize that PTCP has the capacity-insensitive property. Note that $\alpha(S)$ is a constant when the distances between adjacent servers increase exponentially, and becomes up to $k - 1$ when k servers are evenly placed.

We already have three upper bounds on the competitive ratio for OFAL with a set S of k servers and a capacity c. These upper bounds can be optimal according to the specific layout of servers.

(1) $2^k - 1$ (achieved by the greedy algorithm [6,9]),
(2) $2U(S) + 1$ (achieved by IDAS [6]) and
(3) $O(\log c(S))$ (achieved by Robust-Matching [13,14]),

where $U(S)$ is the aspect ratio of S, i.e., the ratio of the diameter of S to the minimum distance between two adjacent servers in S and $c(S) := \sum_{s \in S} c(s)$.

Let us compare the upper bound of PTCP to the above three upper bounds. For the upper bound (1), it follows that $2\alpha(S) + 1 \leq 2k - 1 < 2^k - 1$ if $k \geq 3$ and $2\alpha(S) + 1 = 2^2 - 1 = 3$ if $k = 2$. Then, our new algorithm PTCP performs better than the greedy algorithm when $k \geq 3$ and performs as well as the greedy algorithm when $k = 2$. For the upper bound (2), we always have $\alpha(S) \leq k - 1 \leq U(S)$ and the equalities hold if and only if k servers are evenly placed. Hence, PTCP is better than IDAS except for OFAL$_{eq}$ and performs as well as IDAS for OFAL$_{eq}$. For the upper bound (3), there are cases when PTCP performs worse than Robust-Matching, for example, the case where the servers are evenly placed and the capacity of each server is 1. In this case, we have $O(\log c(S)) = O(\log k) \leq k - 1 = \alpha(S)$. However, we also have another two cases where the performance of PTCP is better than that of Robust-Matching. The first case is when the capacity of each server is very large, especially when $c(S) = \Omega(2^{\alpha(S)})$, and the second case is when $\alpha(S)$ is small, especially when $\alpha(S) = o(\log k)$.

Furthermore, we observe that PTCP is advantageous against the existing algorithms. In fact, we show that there exists a layout of servers such that PTCP performs very well but the greedy algorithm performs very poorly (in Theorem 3), and there exists another layout of servers such that PTCP performs well but the permutation algorithm [1,8] performs poorly (in Theorem 4).

1.2 Related Work

Kalyanasundaram and Pruhs [10] studied OFA under the weakened adversary model where the adversary has only half as many capacities of each server as the online algorithm and the length of a request sequence is at most $c(S)/2$ where $c(S) = \sum_{s \in S} c(s)$. They showed that the greedy algorithm is $\Theta(\min(k, \log c(S)))$-competitive and present an $O(1)$-competitive algorithm under this assumption. Chung et al. [4] also studied OFA under another weakened adversary where the adversary has one less capacity of each server against the online algorithm. Under this model, they presented an $O(\log k)$-competitive deterministic algorithm on an α-HST [4] metric where $\alpha = \Omega(\log k)$ and an $O(\log^3 k)$-competitive randomized algorithm on a general metric.

The special cases of OFA and OFAL, where the capacity of each server is 1, have been known as the *online metric matching* problem (OMM) and the *online matching problem on a line* (OML) respectively. For OMM, Kalyanasundaram and Pruhs [8] and Khuller et al. [11] presented a deterministic online algorithm which is called *Permutation* [8], and showed that it is $(2k - 1)$-competitive and best possible. In addition, Kalyanasundaram and Pruhs [8] also determined the exact competitive ratio of the greedy algorithm to be $2^k - 1$. The best randomized algorithm for OMM so far [3] is $O(\log^2 k)$-competitive and the best lower bound on the competitive ratio [3] is $\Omega(\log k)$. For OML, there have been many active studies [2,5,13,14] and the best upper bound on the competitive ratio [14] is $O(\log k)$, which is achieved by the deterministic algorithm called Robust-Matching [13]. The best lower bound on the competitive ratio [12] is $\Omega(\sqrt{\log k})$.

2   Preliminaries

2.1   Online Facility Assignment Problem

Let (X, d) be a metric space, where X is a (possibly infinite) set of points and $d : X \times X \to \mathbb{R}$ is a distance function. We use $S = \{s_1, \ldots, s_k\}$ to denote the set of k servers and use $\sigma = r_1 \cdots r_n$ to denote a request sequence. For each $1 \leq j \leq k$, a server s_j is characterized by the position $p(s_j) \in X$ and s_j has capacity $c(s_j) \in \mathbb{N}$, i.e., s_j can be matched with at most $c(s_j)$ requests. We assume that $n \leq c(s_1) + \cdots + c(s_k)$. For each $1 \leq i \leq n$, a request r_i is also characterized by the position $p(r_i) \in X$.

The set S is given to an online algorithm in advance, while requests are given one-by-one from r_1 to r_n. At any time of the execution of an algorithm, a server is called *free* if the number of requests matched with it is less than its capacity, and *full* otherwise. When a request r_i is revealed, an online algorithm must match r_i with one of free servers. If r_i is matched with the server s_j, the pair (r_i, s_j) is added to the current matching and the cost $d(p(r_i), p(s_j))$ is incurred for this pair. The cost of the matching is the sum of the costs of all the pairs contained in it. The goal of online algorithms is to minimize the cost of the final matching. We refer to such a problem as the *online facility assignment* problem with servers S and a capacity $c : S \to \mathbb{N}$, which is denoted by OFA(S, c). For the case that $c(s_1) = \cdots = c(s_k) = \ell \geq 1$, it is immediate that $n \leq k\ell$ and we simply use OFA(S, ℓ) to denote the online facility assignment problem with servers S (of uniform capacity ℓ).

2.2   Online Facility Assignment Problem on a Line

By setting $X = \mathbb{R}$, we can regard the online facility assignment problem with servers S as the online facility assignment problem *on a line* with servers S, and we denote such a problem by OFAL(S, c) for a general capacity $c : S \to \mathbb{N}$ and OFAL(S, ℓ) for a uniform capacity $\ell \geq 1$. Without loss of generality, we assume that $p(s_1) < \cdots < p(s_k)$.

To precisely describe the upper bound of the competitive ratio, we introduce the following notation: for any $T = \{t_1, \ldots, t_m\} \subseteq S$ where $p(t_1) < \cdots < p(t_m)$, let

$$L(T) := \frac{p(t_m) - p(t_1)}{\max_u (p(t_{u+1}) - p(t_u))} \text{ and } \alpha(S) := \max_{T \subseteq S} L(T). \tag{1}$$

For convenience, let $L(S) = 0$ and $\alpha(S) = 0$ if $|S| \leq 1$.

In the rest of the paper, we will abuse the notations $r_i \in \mathbb{R}$ and $s_j \in \mathbb{R}$ for OFAL(S, c) instead of $p(r_i) \in \mathbb{R}$ and $p(s_j) \in \mathbb{R}$, respectively, when those are clear from the context.

2.3   Notations and Terminologies

For a request sequence σ, let $|\sigma|$ be the number of requests in σ. For an (online/offline) algorithm $\mathcal{A}$ for OFA(S, c) and a request sequence $\sigma = r_1 \cdots r_n$, we use $s_{\mathcal{A}}(r_i; \sigma)$ to denote the server with which $\mathcal{A}$ matches r_i for each $1 \leq i \leq n$ when $\mathcal{A}$ processes σ. Let $\mathcal{A}(\sigma|S)$ be the total cost incurred when $\mathcal{A}$ processes σ. We use Opt to denote the optimal *offline* algorithm, i.e., Opt knows the entire sequence $\sigma = r_1 \cdots r_n$ in advance and *minimizes* the total cost incurred by Opt to match each request r_i with the server $s_{\mathrm{Opt}}(r_i; \sigma)$. Let $F_i(\mathcal{A})$ be the set of all free servers just after $\mathcal{A}$ matches r_i.

To evaluate the performance of an online algorithm $\mathcal{A}$, we use the (strict) competitive ratio. We say that $\mathcal{A}$ is α-competitive if $\mathcal{A}(\sigma|S) \leq \alpha \cdot \mathrm{Opt}(\sigma|S)$ for any request sequence σ. The competitive ratio $\mathcal{R}(\mathcal{A})$ of $\mathcal{A}$ is defined to be the infimum of $\alpha \geq 1$ such that $\mathcal{A}$ is α-competitive, i.e., $\mathcal{R}(\mathcal{A}) = \inf\{\alpha \geq 1 : \mathcal{A}$ is α-competitive$\}$.

2.4   Technical Lemmas

In this subsection, we introduce some important notions about OFAL(S, c): MPFS (Most Preferred Free Servers) algorithm [6], surrounding-oriented algorithm [2,7], and faithful algorithm [6]. In this paper, we mainly deal with surrounding-oriented and faithful MPFS algorithms. To begin with, we state the definition of an MPFS algorithm and its significant property.

Definition 1. *Let $\mathcal{A}$ be an online algorithm for* OFA(S, c). *We say that $\mathcal{A}$ is an* MPFS (most preferred free servers) *algorithm if it is specified by the following conditions: Let $\sigma = r_1 \ldots r_n$ be a request sequence.*

1. *For each i, the priority of all servers for r_i is determined by only $p(r_i)$,*
2. *$\mathcal{A}$ matches r_i with a server with the highest priority among free servers.*

Let $\mathcal{MPFS}$ be the class of MPFS algorithms. For each MPFS algorithm $\mathcal{A}$, the following strong theorem [6] is known.

Theorem 1. *Let $\mathcal{A} \in \mathcal{MPFS}$ and suppose that $\mathcal{A}$ is α-competitive for* OFA$(S, 1)$. *Then, for any capacity $c : S \to \mathbb{N}$, $\mathcal{A}$ is also α-competitive for* OFA(S, c).

By this theorem, it turns out that there is no need to specify the capacity of each server in evaluating the competitive ratio of an MPFS algorithm.

Definition 2. *Given a request r for* OFAL(S, c), *the* surrounding servers *for r are s^L and s^R, where s^L is the closest free server to the left of r (if any) and s^R is the closest free server to the right of r (if any). If $r = s$ for some $s \in S$ and s is free, then the surrounding server of r is only the server s.*

Next, we present the notion of surrounding-oriented algorithms [2,7] for OFAL(S, c).

Definition 3. *Let $\mathcal{A}$ be an online algorithm for* OFAL(S, c). *We say that $\mathcal{A}$ is* surrounding-oriented *if for every request sequence σ, it matches every request r of σ with one of the surrounding servers of r.*

For surrounding-oriented algorithms, the following useful lemma [2,7] is known.

Lemma 1. *Let $\mathcal{A}$ be an online algorithm for* OFAL(S, c). *Then there exists a surrounding-oriented algorithm $\mathcal{A}'$ for* OFAL(S, c) *such that $\mathcal{A}'(\sigma|S) \leq \mathcal{A}(\sigma|S)$ for any σ.*

By Lemma 1, we assume that any algorithm for OFAL(S, c) is surrounding-oriented in the rest of this paper if otherwise stated.

Finally, we introduce the notion of a faithful algorithm [6] and its useful property. Let $\mathcal{A}$ be an online/offline algorithm for OFAL(S, c) and $\sigma = r_1 \cdots r_n$ and $\tau = q_1 \cdots q_n$ be request sequences. We say that $\tau \neq \sigma$ is *closer* than σ w.r.t. $\mathcal{A}$ if for each i, $r_i \geq q_i \geq s_{\mathcal{A}}(r_i; \sigma)$ or $r_i \leq q_i \leq s_{\mathcal{A}}(r_i; \sigma)$.

Definition 4. *Let $\mathcal{A}$ be an online/offline algorithm for* OFAL(S, c). *For any request sequence $\sigma = r_1 \cdots r_n$ and any request sequence $\tau = q_1 \cdots q_n$ that is closer than σ w.r.t. $\mathcal{A}$, we say that $\mathcal{A}$ is* faithful *if $s_{\mathcal{A}}(r_i; \sigma) = s_{\mathcal{A}}(q_i; \tau)$ for each $1 \leq i \leq n$.*

We say that a request sequence σ is *opposite* w.r.t. $\mathcal{A}$ if every request r in σ is located between $s_{\mathcal{A}}(r_i; \sigma)$ and $s_{\mathrm{Opt}}(r_i; \sigma)$. The following lemma [6] holds for an opposite request sequence w.r.t. $\mathcal{A}$ for OFAL(S, c).

Lemma 2. *Let $\mathcal{A}$ be a faithful online algorithm for* OFAL(S, c). *Then, for any request sequence σ with* Opt$(\sigma|S) > 0$, *there exists an opposite τ w.r.t. $\mathcal{A}$ such that $\mathcal{A}(\sigma|S)/\mathrm{Opt}(\sigma|S) \leq \mathcal{A}(\tau|S)/\mathrm{Opt}(\tau|S)$.*

By the above lemma, it suffices to analyze only opposite request sequences in order to derive the upper bound for the competitive ratio of faithful algorithms.

3   "Hybrid" Algorithm

In this section, we mention the properties of the "hybrid" algorithm. This idea was first used in [5]. Let $\mathcal{A} \in \mathcal{MPFS}$. For an integer $i \geq 1$ and a server $s \in F_{i-1}(\mathcal{A})$, the algorithm $\mathcal{H}_{i,s}^{\mathcal{A}}$ matches the requests $r_1, \ldots, r_{i-1}$ with the same servers as $\mathcal{A}$, r_i with s, and $r_{i+1}, \ldots, r_k$ with some servers according to $\mathcal{A}$. We call $\mathcal{H}_{i,s}^{\mathcal{A}}$ a hybrid algorithm of $\mathcal{A}$. If $s = s_{\mathcal{A}}(r_i; \sigma)$, then $\mathcal{A}$ and $\mathcal{H}_{i,s}^{\mathcal{A}}$ are completely the same. Then, in the rest of the paper, we consider the case $s \neq s_{\mathcal{A}}(r_i; \sigma)$. We abbreviate $\mathcal{H}_{i,s}^{\mathcal{A}}$ as $\mathcal{H}_{i,s}$ when $\mathcal{A}$ is clear from the context.

Lemma 3. *For* OFA$(S, 1)$, *let $\mathcal{A} \in \mathcal{MPFS}$ and $\sigma = r_1 \ldots r_k$ be an request sequence. Suppose $s_{\mathcal{A}}(r_i; \sigma) \neq s$. Then, there exists some $t^* \geq i$, $\{a_t\}_{t=i}^{t^*}$, and $\{h_t\}_{t=i}^{t^*}$ such that*

(1) $F_t(\mathcal{A}) \setminus F_t(\mathcal{H}_{i,s}) = \{a_t\}$ and $F_t(\mathcal{H}_{i,s}) \setminus F_t(\mathcal{A}) = \{h_t\}$ for each $i \leq t \leq t^*$

(2) $F_t(\mathcal{A}) = F_t(\mathcal{H}_{i,s})$ for each $t \geq t^* + 1$.

By the proof of Lemma 3, it is easy to see that the following proposition on $\{a_t\}_{t=i}^{t^*}$ and $\{h_t\}_{t=i}^{t^*}$ holds.

Proposition 1. *For* OFA$(S,1)$ *let* $\mathcal{A} \in \mathcal{MPFS}$ *and* σ *be an request sequence. Suppose* $s_{\mathcal{A}}(r_i; \sigma) \neq s$. *Then, the following conditions hold:*

(P1) $a_t = a_{t+1}$ *or* $h_t = h_{t+1}$ *for each* $i \leq t \leq t^* - 1$,
(P2) If $a_t \neq a_{t+1}$(*resp.* $h_t \neq h_{t+1}$), *then* r_{t+1} *is matched with* a_t(*resp.* h_{t+1}) *by* $\mathcal{A}$ *and with* a_{t+1}(*resp.* h_t) *by* $\mathcal{H}_{i,s}$ *for each* $i \leq t \leq t^* - 1$, *and*
(P3) r_{t^*+1} *is matched with* a_{t^*} *by* $\mathcal{A}$ *and with* h_{t^*} *by* $\mathcal{H}_{i,s}$.

The discussion so far holds for general metrics and any MPFS algorithm $\mathcal{A}$. Next, we state an important lemma that holds for a surrounding-oriented MPFS algorithm on a line metric.

Lemma 4. *For* OFAL$(S,1)$, *let* $\mathcal{A}$ *be a surrounding-oriented MPFS algorithm and* σ *be a request sequence. If* $s_{\mathcal{A}}(r_i; \sigma) \neq s$ *and there is no free server between* $s_{\mathcal{A}}(r_i; \sigma)$ *and* s, *then there is no free server between* a_t *and* h_t *for each* $i \leq t \leq t^*$, *and either* $a_{t^*} \leq \cdots \leq a_i < h_i \leq \cdots \leq h_{t^*}$ *or* $h_{t^*} \leq \cdots \leq h_i < a_i \leq \cdots \leq a_{t^*}$.

4   An Optimal MPFS Algorithm for OFAL

In this section, we present a new MPFS algorithm PTCP (Policy Transition at Critical Point) and show that PTCP is $(2\alpha(S) + 1)$-competitive, where $\alpha(S)$ is given in (1). We consider the following properties of MPFS algorithms for OFAL(S,c).

Definition 5. *Let* $\mathcal{A} \in \mathcal{MPFS}$ *for* OFAL(T,c). *We say that* $C(\mathcal{A},T)$ *holds if* $\mathcal{A}$ *satisfies the following conditions: (C1)* $\mathcal{A}$ *is faithful, (C2)* $\mathcal{A}$ *is surrounding-oriented, and (C3) For* OFAL$(T,1)$, *let* $\sigma = r_1 \ldots r_{|T|}$ *be a request sequence. For* $|T| \geq 2$, *consider the hybrid algorithm* $\mathcal{H}_{i,s}^{\mathcal{A}}$ *where* $s \neq s_{\mathcal{A}}(r_i; \sigma)$ *is a surrounding server[1] of* r_i. *Then,* $|h_{t^*} - r_i| \leq \alpha(S)|r_i - a_i|$.

4.1   A New Algorithm: Policy Transition at Critical Point

To present the algorithm PTCP $\mathcal{A}^*$, we provide several notations. For a set $S = \{s_1, \ldots, s_k\}$ of servers where $s_1 < \cdots < s_k$ and $\max_u(s_{u+1} - s_u) = s_{a+1} - s_a$, let

$$S_1 := \{s_1, \ldots, s_a\}, S_2 := \{s_{a+1}, \ldots s_k\}, \text{ and } x := \frac{(\Delta_2 + D) \cdot D}{(\Delta_1 + D) + (\Delta_2 + D)}, \quad (2)$$

where $\Delta_1 := s_a - s_1$, $\Delta_2 := s_k - s_{a+1}$, and $D := s_{a+1} - s_a$. Note that the value of x is determined by the idea similar to the algorithm IDAS [6].

---

[1] If the number of surrounding servers of r_i is one, then s is one of the free servers which is just to the left/right of $s_{\mathcal{A}}(r_i; \sigma)$.

Let $\mathcal{A}^*[S]$ be $\mathcal{A}^*$ for $OFAL(S, c)$ and $\mathcal{A}^*[S](r, F)$ be a server with which $\mathcal{A}^*[S]$ matches r for a set $F \subseteq S$ of free servers. By using $\mathcal{A}^*[S_1]$ and $\mathcal{A}^*[S_2]$, we inductively define $\mathcal{A}^*[S](r, F)$ as follows:

$$\mathcal{A}^*[S](r, F) = \begin{cases} s & \text{if } F = \{s\}, \\ \mathcal{A}^*[S_1](r, F_1) & \text{if } (r \leq s_a + x \text{ and } F_1 \neq \emptyset) \text{ or } (F_2 = \emptyset), \\ \mathcal{A}^*[S_2](r, F_2) & \text{if } (s_a + x < r \text{ and } F_2 \neq \emptyset) \text{ or } (F_1 = \emptyset), \end{cases} \quad (3)$$

where $F_1 := F \cap S_1$ and $F_2 := F \cap S_2$. $\mathcal{A}^*$ also has the following description.

1. If $r \leq s_a + x$, then match a new request r with a server in S according to $\mathcal{A}^*[S_1]$. When all servers in S_1 are full just before r is revealed, match r with a server in S_2 according to $\mathcal{A}^*[S_2]$.
2. If $r > s_a + x$, then match a new request r with a server in S_2 according to $\mathcal{A}^*[S_2]$. When all servers in S_2 are full just before r is revealed, match r with a server in S_1 according to $\mathcal{A}^*[S_1]$.

The following important lemma holds for $\mathcal{A}^*$ defined in (3).

Lemma 5. *Let S, S_1, and S_2 be sets of servers defined in (2). If $C(\mathcal{A}^*, S_1)$ and $C(\mathcal{A}^*, S_2)$ hold, then $C(\mathcal{A}^*, S)$ also holds.*

Proof. First, we prove that $\mathcal{A}^*[S]$ satisfies (C1). Consider the situation where $\mathcal{A}^*[S]$ matches a request r with a server s and observe what happens when a request q (located between r and s) occurs instead of r. Suppose $s \in S_1$. There are two possible cases: (1) $r \leq s_a + x$, or (2) $r > s_a + x$ and all servers in S_2 are full. For the case (1), $\mathcal{A}^*[S]$ matches r with s according to $\mathcal{A}^*[S_1]$. Since $\mathcal{A}^*[S_1]$ is faithful, it turns out that $\mathcal{A}^*[S]$ matches q with s. For the case (2), s is the rightmost free server and $s < r$ holds. Therefore, $\mathcal{A}^*[S]$ matches q with s. The same discussion can be applied to the case $s \in S_2$. Hence, $\mathcal{A}^*[S]$ is faithful, i.e., $\mathcal{A}^*[S]$ satisfies (C1).

Next, we prove that $\mathcal{A}^*[S]$ satisfies (C2). By contradiction, assume that there exist a request sequence $r_1 \ldots r_n$ and a request r_t such that $\mathcal{A}^*[S]$ matches r_t with a server s that is not a surrounding server of r_t. Let s' be a free server between r_t and s. If $s, s' \in S_1$ (resp. $s, s' \in S_2$), then $\mathcal{A}^*[S]$ matches r_t with s according to $\mathcal{A}^*[S_1]$ (resp. $\mathcal{A}^*[S_2]$). However, this contradicts the fact that $\mathcal{A}^*[S_1]$ (resp. $\mathcal{A}^*[S_2]$) is surrounding-oriented. If $s \in S_1$ and $s' \in S_2$, then we have $s < s_a + x < s' \leq r_t$. Since $s_a + x < r_t$ and there exists a free server $s' \in S_2$, $\mathcal{A}^*[S]$ must match r_t with a free server in S_2 according to $\mathcal{A}^*[S_2]$ and this contradicts the assumption that $\mathcal{A}^*[S]$ matches r_t with $s \in S_1$. The same discussion can be applied to the case where $s \in S_2$ and $s' \in S_1$. Therefore, $\mathcal{A}^*[S]$ is surrounding-oriented, i.e., $\mathcal{A}^*[S]$ satisfies (C2).

Finally, we prove that $\mathcal{A}^*[S]$ satisfies (C3). Fix any request sequence $\sigma = r_1 \ldots r_k$ for $OFAL(S, 1)$. Let $s \neq s_{\mathcal{A}^*[S]}(r_i; \sigma)$ be a surrounding server of r_i and consider the hybrid algorithm $\mathcal{H}_{i,s}$ of $\mathcal{A}^*[S]$. By definition, it follows that $a_i = s$ and $h_i = s_{\mathcal{A}^*[S]}(r_i; \sigma)$. We use the following claim whose proof is omitted.

Claim. If $a_i, h_i \in S_1$ (resp. $a_i, h_i \in S_2$), then $a_t, h_t \in S_1$ (resp. $a_t, h_t \in S_2$) for each $i \leq t \leq t^*$ where t^* is defined in Lemma 3.

If $a_i, h_i \in S_1$, then let σ_1 be a subsequence of σ consisting of all requests r that satisfies $s_{\mathcal{A}^*[S]}(r; \sigma) \in S_1$. By the definition of $\mathcal{A}^*[S]$, each request in σ_1 is matched with a server in S_1 according to $\mathcal{A}^*[S_1]$. By the above claim and Proposition 1, we have $a_t, h_t \in S_1$ for each $i \le t \le t^*$ and all requests that may affect the changes in $\{a_t\}_{t=i}^{t^*}$ and $\{h_t\}_{t=i}^{t^*}$ are included in σ_1. In addition, the way of changes in $\{a_t\}_{t=i}^{t^*}$ and $\{h_t\}_{t=i}^{t^*}$ depends only on the behavior of $\mathcal{A}^*[S_1]$. Then, by the assumption that $C(\mathcal{A}^*, S_1)$ holds, we can see that $\mathcal{A}^*[S]$ satisfies (C3). Analogously, it can be shown that $\mathcal{A}^*[S]$ satisfies (C3) for $a_i, h_i \in S_2$.

Hence, the remaining possible cases are (1) $a_i \in S_1$ and $h_i \in S_2$, and (2) $h_i \in S_1$ and $a_i \in S_2$. For the case (1), by the fact $a_i \le s_a < s_a + x \le r_i \le h_i$, it follows that

$$\frac{|h_{t^*} - r_i|}{|r_i - a_i|} \le \frac{\Delta_2 + D - x}{x} = \frac{\Delta_1 + \Delta_2 + D}{D} = L(S) \le \alpha(S),$$

where the second equality and the last inequality are due to the definition of $L(S)$ and $\alpha(S)$ in (1). For the case (2), by using the fact $h_i \le r_i \le s_a + x < s_{a+1} \le a_i$, we have

$$\frac{|h_{t^*} - r_i|}{|r_i - a_i|} \le \frac{\Delta_1 + x}{D - x} = \frac{\Delta_1 + \Delta_2 + D}{D} = L(S) \le \alpha(S).$$

Therefore, $\mathcal{A}^*[S]$ satisfies (C3).   □

4.2   An Upper Bound on the Competitive Ratio of PTCP

The goal of this subsection is to prove the following theorem, which claims that PTCP is $(2\alpha(S) + 1)$-competitive.

Theorem 2. *For* OFAL(S, c), $\mathcal{A}^*$ *defined in (3) is $(2\alpha(S) + 1)$-competitive, where $c : S \to \mathbb{N}$ is an arbitrary capacity.*

To prove Theorem 2, we introduce a simpler algorithm similar to $\mathcal{A}^*$ and show the important lemma about the algorithm. Let $S = \{s_1, \ldots, s_k\}$ be a set of servers where $s_1 < \cdots < s_k$, and $\mathcal{A} \in \mathcal{MPFS}$ be a $(2\alpha(S) + 1)$-competitive algorithm for which $C(\mathcal{A}, S)$ holds. Let d and x be parameters such that $s_{k+1} = s_k + d$ and $0 < x < d$, and define a new MPFS algorithm $\mathcal{A}_{d,x}$ for OFAL$(S \cup \{s_{k+1}\}, c)$ as follows:

1. If $r \le s_k + x$, then match a new request r with a server in S according to $\mathcal{A}$. When all servers in S are full just before r is revealed, match r with s_{k+1}.
2. If $r > s_k + x$, then match a new request r with s_{k+1}. When s_{k+1} is full just before r is revealed, match r to a server in S according to $\mathcal{A}$.

For the competitive ratio of $\mathcal{A}_{d,x}$ defined above, we have the following lemma.

Lemma 6. *Let $S = \{s_1, \ldots, s_k\}$ and $\tilde{S} = \{s_1, \ldots, s_{k+1}\}$ be sets of servers, where $s_1 < \cdots < s_k < s_{k+1}$, and $\mathcal{A} \in \mathcal{MPFS}$ be a $(2\alpha(S) + 1)$-competitive algorithm for which $C(\mathcal{A}, S)$ holds. Then, for any request sequence σ*

of OFAL($\tilde{S}, c$), $\mathcal{A}_{d,x}(\sigma|\tilde{S}) \leq C_{d,x}(S) \cdot \text{Opt}(\sigma|\tilde{S})$, where $C_{d,x}(S) := \max\{2\alpha(S) + 1, \frac{2d-x}{x}, \frac{2\Delta+d+x}{d-x}\}$, $\Delta := s_k - s_1$, $d = s_{k+1} - s_k$, $0 < x < d$, and $c : \tilde{S} \to \mathbb{N}$ is an arbitrary capacity.

Proof of Theorem 2: The proof is by induction on the number k of servers. For the base case $k = 1$, $\mathcal{A}^*[\{s\}]$ is a trivial algorithm that matches every request to the unique server s. Then, $C(\mathcal{A}^*, \{s\})$ holds and $\mathcal{A}^*[\{s\}]$ is $2\alpha(\{s\}) + 1 = 1$-competitive since $\alpha(S) = 0$ for $|S| \leq 1$.

For the inductive step, assume that $\mathcal{A}^*[T]$ is $(2\alpha(T) + 1)$-competitive and $C(\mathcal{A}^*, T)$ holds for any set T of servers such that $|T| \leq k-1$. Let $S = \{s_1, \ldots, s_k\}$ be a set of $k \geq 2$ servers where $s_1 < \cdots < s_k$ and a be any integer such that $\max_u(s_{u+1} - s_u) = s_{a+1} - s_a$. Note that $S_1, S_2, \Delta_1, \Delta_2$ and x are given by (2). By the induction hypothesis, $\mathcal{A}^*[S_1]$ (resp. $\mathcal{A}^*[S_2]$) is $(2\alpha(S_1) + 1)$-competitive (resp. $(2\alpha(S_2) + 1)$-competitive) and $C(\mathcal{A}^*, S_1)$ (resp. $C(\mathcal{A}^*, S_2)$) holds. Then, by Lemma 5, $C(\mathcal{A}^*, S)$ holds.

Fix a request sequence σ for OFAL($S, 1$) arbitrarily. Note that since $\mathcal{A}^*$ is in $\mathcal{MPFS}$, it suffices to consider request sequences for OFAL($S, 1$). Let m be the number of requests in σ that occur in $(-\infty, s_a + x]$. There are three cases: $m = a$, $m > a$ and $m < a$.

For the case $m = a$, define σ_1 (resp. σ_2) to be a subsequence of σ consisting of all requests r such that $r \leq s_a + x$ (resp. $s_a + x < r$). Since $\mathcal{A}^*$ and Opt[2] match each request in σ_1 (resp. σ_2) with a server in S_1 (resp. S_2), we have that

$$\begin{aligned} \mathcal{A}^*(\sigma|S) &= \mathcal{A}^*(\sigma_1|S_1) + \mathcal{A}^*(\sigma_2|S_2) \\ &\leq (2\alpha(S_1) + 1)\text{Opt}(\sigma_1|S_1) + (2\alpha(S_2) + 1)\text{Opt}(\sigma_2|S_2) \\ &\leq (2\alpha(S) + 1)\text{Opt}(\sigma|S). \end{aligned}$$

For the case $m > a$, define subsequences σ_1 and σ_2 of σ as follows: σ_1 consists of all requests r such that $r \leq s_a + x$ and σ_2 consists of all requests r that is matched with a server in S_2 by $\mathcal{A}^*$. Let R be a set of requests consisting of all requests that belong to both σ_1 and σ_2. Note that R consists of the last $|\sigma_1| - a$ requests in σ_1. We use σ_2' to denote a request sequence obtained by moving the position of each request of σ_2 in R to s_{a+1}. Consider the following operations for σ_1 and σ_2': Operation (1) $\mathcal{A}^*[S_1]_{D,x}$ processes σ_1 with servers $\tilde{S}_1 = S_1 \cup \{s_{a+1}\}$ and a capacity c_1 where $c_1(s_j) = 1$ for each $1 \leq j \leq a$ and $c_1(s_{a+1}) = |R|$, and Operation (2) $\mathcal{A}^*$ processes σ_2' with servers S_2 and a capacity $c_2(s) = 1$ for each $s \in S_2$. By the definition of $\mathcal{A}^*$ and $\mathcal{A}^*[S_1]_{D,x}$, we have

$$\mathcal{A}^*(\sigma|S) = \mathcal{A}^*[S_1]_{D,x}(\sigma_1|\tilde{S}_1) + \mathcal{A}^*(\sigma_2'|S_2) \text{ and}$$
$$\text{Opt}(\sigma|S) = \text{Opt}(\sigma_1|\tilde{S}_1) + \text{Opt}(\sigma_2'|S_2),$$

where $\mathcal{A}^*[S_1]_{D,x}(\sigma_1|\tilde{S}_1)$ denotes the cost of an algorithm $\mathcal{A}^*[S_1]_{D,x}$ for the operation (1) and $\mathcal{A}^*(\sigma_2'|S_2)$ denotes the cost of an algorithm $\mathcal{A}^*$ for the operation (2). We define $\text{Opt}(\sigma_1|\tilde{S}_1)$ and $\text{Opt}(\sigma_2'|S_2)$ analogously. Thus, by Lemma 6,

---

[2]   One of the optimal matchings is obtained by matching i-th request from the left with i-th server from the left for $i = 1, \ldots, k$.

$$\mathcal{A}^*(\sigma|S) = \mathcal{A}[S_1]_{D,x}(\sigma_1|\tilde{S}_1) + \mathcal{A}^*(\sigma_2'|S_2)$$
$$\leq C_{D,x}(S_1)\mathrm{Opt}(\sigma_1|\tilde{S}_1) + (2\alpha(S_2)+1)\mathrm{Opt}(\sigma_2'|S_2)$$
$$\leq C_{D,x}(S_1,S_2)\mathrm{Opt}(\sigma|S), \tag{4}$$

where $C_{D,x}(S_1,S_2) := \max\{2\alpha(S_1)+1, 2\alpha(S_2)+1, \frac{2D-x}{x}, \frac{2\Delta_1+D+x}{D-x}\}$. By substituting $x = D(\Delta_2+D)/(\Delta_1+\Delta_2+2D)$, it follows that $C_{D,x}(S_1,S_2) \leq 2\alpha(S)+1$. Thus, $\mathcal{A}^*(\sigma|S) \leq (2\alpha(S)+1)\mathrm{Opt}(\sigma|S)$ for the case $m > a$.

For the last case $m < a$, we can use the proof for the case $m > a$ by symmetry. By replacing S_1, S_2, Δ_1, and x in (4) with S_2, S_1, Δ_2, and $D-x$ respectively, we have $\mathcal{A}^*(\sigma|S) \leq C_{D,D-x}(S_2,S_1)\mathrm{Opt}(\sigma|S)$ and by substituting $x = D(\Delta_2+D)/(\Delta_1+\Delta_2+2D)$, it follows that $C_{D,D-x}(S_2,S_1) \leq 2\alpha(S)+1$. Therefore, we finally get $\mathcal{A}^*(\sigma|S) \leq (2\alpha(S)+1)\mathrm{Opt}(\sigma|S)$ for the case $m < a$. □

4.3   Comparisons with Other Algorithms

In this subsection, we compare the performance of the PTCP algorithm with other well-known algorithms for $\mathrm{OFAL}(S,c)$, e.g. the greedy algorithm and the permutation algorithm [1,8].

Comparison with the Greedy Algorithm. The greedy algorithm (denoted by $\mathcal{G}$) for $\mathrm{OFAL}(S,c)$ is an algorithm that matches a new request to the nearest free server. For the competitive ratio of $\mathcal{A}^*$ and $\mathcal{G}$, we have the following theorem, which implies that there exists a server layout where $\mathcal{A}^*$ performs very well while $\mathcal{G}$ performs very poorly.

Theorem 3. *Define $S = \{s_1,\ldots,s_k\}$ as follows: $s_1 = 0$ and $s_i = 2^{i-1}$ for $i = 2,\ldots,k$. For the server layout S, $\mathcal{A}^*$ is 5-competitive and the competitive ratio of $\mathcal{G}$ is at least $2^k - 1$.*

Comparison with the Permutation Algorithm. For $\mathrm{OFA}(S,1)$, the permutation algorithm (denoted by $\mathcal{P}$) is known as the best possible algorithm. For $\mathrm{OFAL}(S,c)$, let $\sigma = r_1\ldots r_n$ be a request sequence and M_i be an optimal matching of $r_1,\ldots,r_i$ for $i = 1,\ldots,n$. Define S_i $(i = 1,\ldots,n)$ to be the set of servers included in M_i and S_0 to be $\emptyset$ for convenience. It is known that there is a sequence of optimal matchings $\{M_i\}_{i=1}^n$ such that $S_i \setminus S_{i-1}$ is singleton for $i = 1,\ldots,n$ [8].

When an i-th request r_i is revealed, $\mathcal{P}$ determines the server s such that $S_i \setminus S_{i-1} = \{s\}$ and matches r_i with s. For the competitive ratio of $\mathcal{A}^*$ and $\mathcal{P}$, we also have the following theorem, which implies that there is a server layout where $\mathcal{A}^*$ performs well and $\mathcal{P}$ performs poorly.

Theorem 4. *For any $\epsilon > 0$, define $S = \{s_1,\ldots,s_{2k}\}$ as follows: for $i = 1,\ldots,k$, $s_{k+i} = (1-\delta^i)/(1-\delta)$, $s_{k-i+1} = -s_{k+i}$, where $\delta > 0$ is taken to satisfy $\delta^k + \delta(4k-1) < \epsilon$ and $(1-\delta)^{-1} < 1+\epsilon/2$. For the server layout S, $\mathcal{A}^*$ is $(3+\epsilon)$-competitive and the competitive ratio of $\mathcal{P}$ is at least $4k-1-\epsilon$.*

5 A Lower Bound on the Competitive Ratio of MPFS

In this section, we derive a tight lower bound on the competitive ratio of algorithms in $\mathcal{MPFS}$. In other words, we will show that the following theorem.

Theorem 5. *Let* $\mathcal{A} \in \mathcal{MPFS}$ *for* OFAL(S, c). *Then,* $\mathcal{R}(\mathcal{A}) \geq 2\alpha(S) + 1$.

To prove Theorem 5, the following lemma [6] is useful.

Lemma 7. *Let* $\mathcal{A} \in \mathcal{MPFS}$ *for* OFAL(S, c). *Then, there exists a request sequence* σ *such that* $\mathcal{A}(\sigma|S) \geq (2L(S) + 1)\mathrm{Opt}(\sigma|S)$.

By the definition of $L(S)$ and $\alpha(S)$ in (1), we have $L(S) \leq \max_{T \subseteq S} L(T) = \alpha(S)$. Therefore, Theorem 5 improves Lemma 7.

Proof of Theorem 5: Fix an algorithm $\mathcal{A} \in \mathcal{MPFS}$ for OFAL(S, c) arbitrarily. By Lemma 1, it suffices to consider the case where $\mathcal{A}$ is surrounding-oriented. Let $S' \subseteq S$ be a set of servers such that $L(S') = \max_{T \subseteq S} L(T) = \alpha(S)$.

Define a request sequence σ as follows: for each $s \in S \setminus S'$, we first give $c(s)$ requests on s. Since $\mathcal{A}$ is surrounding-oriented, both $\mathcal{A}$ and Opt match a request on $s \in S \setminus S'$ with s and are incurred no cost at this stage. Next, we give a request sequence σ' for OFAL(S', c) which satisfies the condition of Lemma 7, i.e. $\mathcal{A}(\sigma'|S') \geq (2L(S') + 1)\mathrm{Opt}(\sigma'|S')$. Then, we have

$$\mathcal{A}(\sigma|S) = \mathcal{A}(\sigma'|S') \geq (2L(S') + 1)\mathrm{Opt}(\sigma'|S') = (2\alpha(S) + 1)\mathrm{Opt}(\sigma|S).$$

Since $L(S') = \alpha(S)$ and $\mathrm{Opt}(\sigma'|S') = \mathrm{Opt}(\sigma|S)$, this completes the proof. □

By Theorem 5, the PTCP algorithm $\mathcal{A}^*$ turns out to be best possible among all MPFS algorithms, and thus we have $\mathcal{R}(\mathcal{A}^*) = 2\alpha(S) + 1$.

6 Concluding Remarks and Open Questions

In this paper, we dealt with the online facility assignment problem on a line OFAL(S, c) where S is a set of servers and $c : S \to \mathbb{N}$ is a capacity of each server. We proposed a new MPFS algorithm PTCP (Policy Transition at Critical Point) and showed that for OFAL(S, c), PTCP is $(2\alpha(S) + 1)$-competitive (in Theorem 2), where $\alpha(S)$ is given in (1). We also showed that the competitive ratio of any MPFS algorithm is at least $2\alpha(S) + 1$ (in Theorem 5), i.e., for OFAL(S, c), PTCP is best possible among the MPFS algorithms.

However, it is not known if there is an algorithm $\mathcal{A} \notin \mathcal{MPFS}$ whose competitive ratio is less than $2\alpha(S) + 1$. Moreover, we do not even know whether there exists an algorithm $\mathcal{A} \notin \mathcal{MPFS}$ with the capacity-insensitive property for OFA(S, c) or not. Specifically, it would be interesting to study whether the competitive ratio of the permutation algorithm [1,8] or the Robust-Matching algorithm [13] for OFA(S, c) depends on a capacity c or not.

References

1. Ahmed, A.R., Rahman, M.S., Kobourov, S.: Online facility assignment. Theor. Comput. Sci. **806**, 455–467 (2020)
2. Antoniadis, A., Fischer, C., Tönnis, A.: A collection of lower bounds for online matching on the line. In: Bender, M.A., Farach-Colton, M., Mosteiro, M.A. (eds.) LATIN 2018. LNCS, vol. 10807, pp. 52–65. Springer, Cham (2018). https://doi.org/10.1007/978-3-319-77404-6_5
3. Bansal, N., Buchbinder, N., Gupta, A., Naor, J.S.: An $O(\log^2 k)$-Competitive Algorithm for Metric Bipartite Matching. In: Arge, L., Hoffmann, M., Welzl, E. (eds.) ESA 2007. LNCS, vol. 4698, pp. 522–533. Springer, Heidelberg (2007). https://doi.org/10.1007/978-3-540-75520-3_47
4. Chung, C., Pruhs, K., Uthaisombut, P.: The online transportation problem: on the exponential boost of one extra server. In: Laber, E.S., Bornstein, C., Nogueira, L.T., Faria, L. (eds.) LATIN 2008. LNCS, vol. 4957, pp. 228–239. Springer, Heidelberg (2008). https://doi.org/10.1007/978-3-540-78773-0_20
5. Gupta, A., Lewi, K.: The online metric matching problem for doubling metrics. In: Czumaj, A., Mehlhorn, K., Pitts, A., Wattenhofer, R. (eds.) ICALP 2012. LNCS, vol. 7391, pp. 424–435. Springer, Heidelberg (2012). https://doi.org/10.1007/978-3-642-31594-7_36
6. Harada, T., Itoh, T., Miyazaki, S.: Capacity-insensitive algorithms for online facility assignment problems on a line. online ready in discrete mathematics, algorithms and applications (2023)
7. Itoh, T., Miyazaki, S., Satake, M.: Competitive analysis for two variants of online metric matching problem. Discrete Math. Algorithms Appl. **13**(06), 2150156 (2021)
8. Kalyanasundaram, B., Pruhs, K.: Online weighted matching. J. Algorithms **14**(3), 478–488 (1993)
9. Kalyanasundaram, B., Pruhs, K.: On-line network optimization problems. In: Fiat, A., Woeginger, G.J. (eds.) Online Algorithms. LNCS, vol. 1442, pp. 268–280. Springer, Heidelberg (1998). https://doi.org/10.1007/BFb0029573
10. Kalyanasundaram, B., Pruhs, K.R.: The online transportation problem. SIAM J. Discret. Math. **13**(3), 370–383 (2000)
11. Khuller, S., Mitchell, S.G., Vazirani, V.V.: On-line algorithms for weighted bipartite matching and stable marriages. Theoret. Comput. Sci. **127**(2), 255–267 (1994)
12. Peserico, E., Scquizzato, M.: Matching on the line admits no $o(\sqrt{\log n})$-competitive algorithm. ACM Trans. Algorithms **19**(3), 1–4 (2023)
13. Raghvendra, S.: A robust and optimal online algorithm for minimum metric bipartite matching. In: Approximation, Randomization, and Combinatorial Optimization. Algorithms and Techniques (APPROX/RANDOM 2016). Schloss Dagstuhl-Leibniz-Zentrum für Informatik (2016)
14. Raghvendra, S.: Optimal analysis of an online algorithm for the bipartite matching problem on a line. In: 34th International Symposium on Computational Geometry (SoCG 2018). Schloss Dagstuhl-Leibniz-Zentrum für Informatik (2018)

Guarding Precise and Imprecise Polyhedral Terrains with Segments

Bradley McCoy[1] , Binhai Zhu[1(✉)] , and Aakash Dutt[2]

[1] Gianforte School of Computing, Montana State University,
Bozeman, MT 59717, USA
{bradleymccoy,bhz}@montana.edu
[2] Department of Computer Science, University of Maryland,
College Park, MD 20742, USA
adutt12@umd.edu

Abstract. Guarding polyhedral terrains is a fundamental problem with realistic applications. In this paper we study three such problems when the terrains are precise and imprecise respectively. The first problem we consider is to place a segment pq with a fixed length over a precise input terrain T with n vertices in 2D (resp. 3D), such that pq can see every point on T and the maximum y-coordinate (resp. z-coordinate) of p and q is minimized. For terrains in 2D and 3D we present algorithms running in $O(n)$ and $O(n^3 \log n)$ time respectively. The second problem is to place two horizontal segments p_1q_1 and p_2q_2 of a fixed length and with the minimum y-coordinate to cover a 2D terrain (x-monotone chain), which we solve in $O(n^2 \log n)$ time.

Given a polyhedral terrain T of n vertices in 3D, a shortest watchtower is a vertical segment erected on T such that every point on T is visible from the top of the segment and the length of the segment is minimized. The problem was solved in $O(n \log n)$ time more than 30 years ago. In this paper, we investigate the problem under the imprecise model where each vertex of T is on a given vertical interval. We show that when the location of a watchtower is fixed, the problem in 2D and 3D can be solved with linear programming, which leads to an additive ε-approximation for the general problem. We implement this algorithm using CPLEX which demonstrate the efficiency and accuracy of the algorithm when $n \leq 100$.

Keywords: Polyhedral terrains · visibility problems · shortest watchtower · linear programming

1 Introduction

Polyhedral terrains, which model the surface of mountains, are important objects in GIS, aerial surveillance and path planning, etc. Guarding polyhedral terrains hence has become a fundamental problem in computational geometry since 1988, when Sharir first gave an $O(n \log^2 n)$ time algorithm to compute the shortest watchtower of a polyhedral terrain T of n vertices in 3D (also called a 2.5D

© The Author(s), under exclusive license to Springer Nature Switzerland AG 2024
W. Wu and J. Guo (Eds.): COCOA 2023, LNCS 14462, pp. 323–336, 2024.
https://doi.org/10.1007/978-3-031-49614-1_24

terrain) [16]. Subsequently, Cole and Sharir studied the more general problem of guarding T with a minimum number of vertex guards, for which they proved its NP-hardness [5].

Following Cole and Sharir's research, since the terrain guarding problem with multiple guards is NP-hard, Bose et al. considered using the number of (vertex/edge) guards which are always sufficiently and sometimes necessary to guard a terrain with n vertices (aka. similar to the famous Art Gallery Theorem) [4]. Recently, Duraisamy et al. studied the problem of guarding a terrain collectively with a set of guards of limited vision (e.g., can only see with a viewing angle of π) [7].

Regarding the shortest watchtower problem, Zhu improved Sharir's bound to $O(n \log n)$ [20, 21]. More recently, researchers have considered guarding a terrain with two or more watchtowers [1, 2, 17]. Besides watchtowers (which are vertical segments), Katoh et al. studied guarding a terrain in 2D (also called a 1.5D terrain) with two shortest horizontal segments at a fixed height, for which an $O(n \log^2 n)$ time algorithm was obtained [11]. (*Here, a horizontal segment could be considered as the trajectory of a drone or a helicopter.*) Most recently, Seth et al. studied a generalization of the shortest watchtower problem in 3D which they called the *acrophobic guard watchtower* problem, for which they solved it in $O(n \log n)$ time [15]. At the top of such an acrophobic guard tower there is a horizontal unit square and in 2D that becomes a horizontal unit segment.

In this paper, the first problem investigated is to place a segment pq of fixed length over a terrain such that the maximum z-coordinate (y-coordinate in 2D) of p and q is minimized. (Note that pq could certainly be rotated and translated.) We present $O(n)$ and $O(n^3 \log n)$ time algorithms to solve this problem in 2D and 3D respectively. The second problem we consider is to place two segments $p_1 q_1$ and $p_2 q_2$ of a fixed length on a horizontal line ℓ^* to guard a 1.5D terrain such that the y-coordinate of ℓ^* is minimized. We solve this problem in $O(n^2 \log n)$ time.

In practice, sometimes the data on terrains cannot be precise. Consequently, in another path of research on polyhedral terrains, imprecise (or uncertain) terrains are seriously considered. In 2004, Gray and Evans proposed the imprecise terrain model as one where the vertices have precise x and y-coordinates, but the z-coordinates are uncertain [8]. In fact, it was proposed that in such a model, the z-coordinate of a vertex v is within some interval $[a_v, b_v]$. Many interesting problems have been considered using the imprecise model; for instance, finding shortest paths on such a model, flow computation and smoothing out the terrain [6, 8–10, 13].

In this paper, the third problem we consider is to compute the shortest watchtower on such an uncertain terrain. For a 1.5D imprecise terrain I with n vertical intervals, we first show that the discrete version, i.e., when the x-coordinate of the watchtower is given, can be formulated as a linear programming (LP) problem. Then this solution can be used as a subroutine to obtain an additive ε-approximation for the general problem. We also implement this part using CPLEX to obtain some empirical results.

2   Preliminaries

Given a terrain T in 2D, which is an x-monotone polygonal chain $T = \langle v_1, v_2, ..., v_n \rangle$ (with $x(v_1) < x(v_2) < \cdots < x(v_n)$, hence T is also called a 1.5D terrain), if we compute the intersection of the halfplanes containing the point $(0, +\infty)$, each bounded by the extension line ℓ_i of an edge $e_i = v_i v_{i+1}, i = 1, ..., n-1$, then the intersection of these halfplanes gives us an unbounded convex polygon P (which can be computed in linear time [12]). The shortest watchtower $\overline{uv}$ of T contains two cases:

1. v is a vertex of T, and
2. u is a vertex of P.

Figure 1 shows both cases: the unlabelled vertical segment shows case (1), while the labelled vertical segment shows case (2). We follow the standard *visibility* definition, i.e., two points p and q are *visible* to each other if the segment pq (or (p, q)) does not intersect any edge of T (except it is possibly tangent to some vertex of T). Then, clearly, for case (2), since u is a vertex of P it is visible from any point on T. When all the points on T are visible from a guarding object O (which would be a set of points or a segment, etc.), then we say that T is *guarded* or *covered* by O.

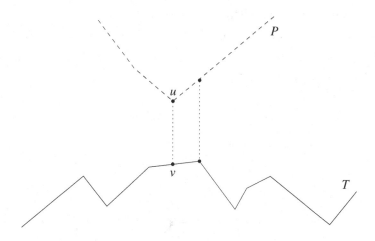

Fig. 1. The shortest watchtower problem for a 1.5D terrain T.

By the monotonicity of T, any halfplane H_i bounded by ℓ_i contains the point $(0, +\infty)$. Hence, we say that all H_i's are *upward* and we assume that all the halfplanes in consideration are upward henceforth. We classify these halfplanes by the slopes of their bounding lines ℓ_i's. If the slope of ℓ_i is in $[0, \pi/2)$, the corresponding halfplane H_i is called a *left* halfplane; otherwise, if the slope of ℓ_i is in $(\pi/2, \pi)$, then H_i is called a *right* halfplane. A pair of halfplanes H_i and H_j form *opposing* halfplanes if one of them is left and the other is right.

When T is a 2.5D terrain with m faces, the above definitions can usually easily generalize, with the twist that H_i, extended from each face f_i, is now a halfspace and P is a convex polyhedron (which can be computed in $O(n \log n)$ time [14]). Certainly, we cannot say H_i being left or right anymore as the direction of a plane is determined by a normal vector in 3D.

Note that, the segment pq can be placed over T when pq is translated and rotated to $\tau(pq) = (\tau(p), \tau(q))$. In Sect. 3, we design polynomial-time algorithms for guarding a 1.5D (resp. 2.5D) terrain T with n vertices in $O(n)$ (resp. $O(n^3 \log n)$) time, by placing pq properly above T such that T is guarded by $\tau(pq)$. In Sect. 4, we solve the problem of guarding a 1.5D terrain with two horizontal segments with a fixed length such that their common y-coordinate is minimized. In Sect. 5, we cover the problem of guarding an imprecise 1.5D terrain with a shortest watchtower.

3 Polynomial-Time Algorithms on Guarding Polyhedral Terrains with a Segment

In this section we solve the problem of placing a segment pq with a fixed length to guard a terrain such that the maximum y-coordinate of the placed segment $\tau(pq)$ is minimized.

3.1 A Linear Time Algorithm for 1.5D Terrains

As a warm-up, we first present a simple linear-time algorithm for 1.5D terrains.

Lemma 1. *Given a 1.5D terrain T and a given segment pq, there is an optimal placement of pq, $\tau(pq)$, such that $\tau(pq)$ is horizontal.*

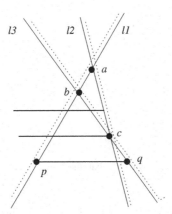

Fig. 2. The optimally placed guarding segment is bounded by a pair of opposing half-planes.

Lemma 2. *Given a 1.5D terrain T and a given segment pq, there is an optimal placement of pq, $\tau(pq)$, such that $\tau(pq)$ is horizontal and is bounded by a pair of opposing halfplanes.*

Proof. We refer to Fig. 2. Following Lemma 1, assume that $\tau(pq)$ is horizontal and it intersects all the upward halfplanes defined by the edges of T. Clearly $\tau(pq)$ does not have to intersect the common intersection of all the halfplanes (e.g., the region bounded by ℓ_i and ℓ_j and above a in Fig. 2), as we could easily move $\tau(pq)$ downward to reduce the maximum y-coordinate of $\tau(p)$ and $\tau(q)$.

Now suppose that $\tau(pq)$ intersects all the halfplanes but it is not bounded by (i.e., touching) any extended line of T. We first move $\tau(pq)$ horizontally, say to the left, such that moving any ε distance further to the left $\tau(q)$ would be out of a right halfplane H_j determined by a line ℓ_j (extended from an edge of T). Up to this point $\tau(pq)$ still intersects all the halfplanes H_l's.

Then, we slide $\tau(pq)$ downward with the constraint that $\tau(q)$ always stays on ℓ_j (which we say $\tau(q)$ is *anchored* at ℓ_j). If the process stops when $\tau(p)$ also hits a line ℓ_i and all H_l's intersect $\tau(pq)$ then we are done, as $\tau(pq)$ is exactly bounded by two opposing halfplanes (defined by ℓ_i and ℓ_j in Fig. 2). If in the process $\tau(q)$ hits a vertex c, which is the intersection of two halfplanes H_i and H_k (determined by ℓ_k), then we need to change the anchor of $\tau(q)$ from ℓ_j to ℓ_k. It is easily seen that since we have at most $O(n)$ lines in $L(T)$, this process must exit after switching $O(n)$ anchor lines and when $\tau(p)$ and $\tau(q)$ are bounded by two opposing halfplanes. □

The above lemma implies a linear time algorithm for solving this problem. Let ℓ be the horizontal copy of pq.

1. For each halfplane H_i, compute the Minkowski sum $H_i \oplus \ell$.
2. Compute the intersection of all $H_i \oplus \ell, i = 1..n$, let the common intersection be $P \oplus \ell$.
3. Compute the lowest position in $P \oplus \ell$ such that a copy of ℓ can be placed.

Each of the three steps takes $O(n)$ time, with the second step implemented using Lee and Preparate's linear-time algorithm [12]. We hence have the following theorem.

Theorem 1. *Given a 1.5D terrain T with n vertices and a given segment pq, an optimal placement of pq, $\tau(pq)$, can be computed in $O(n)$ time.*

We comment that the above algorithm is very similar to the one given by Seth et al. [15]. The difference is that in their problem the segment is constrained to be horizontal while in our case we first need to show that there is an optimal placement of pq which is horizontal.

3.2   An Almost Cubic Time Algorithm for 2.5D Terrains

The above algorithm for 1.5D terrains does not seem to generalize to 2.5D terrains at the first glance. The main reason is that while we could still prove that there is an optimal solution in which $\tau(pq)$, with a fixed length ℓ, is horizontal, its direction could be arbitrary.

Given a 2.5D terrain T, let S_i be the halfspace extended from a face f_i of T. We define *opposing halfspaces* to be a triple of three halfspaces S_i, S_j and

S_k with a common intersection containing a point at $+\infty$, say $(0, +\infty)$. It is straightforward to place $\tau(pq)$ horizontally, below $S_i \cap S_j \cap S_k$ and with the minimum z-coordinate such that $\tau(pq)$ intersects the three halfspaces S_i, S_j and S_k. In fact, the intersection of a horizontal plane G below $S_i \cap S_j \cap S_k$ with S_i, S_j and S_k is a triangle (possibly unbounded), and the problem is to place $\tau(pq)$ to intersect the three edges of this triangle. If the triangle is bounded, the optimal plane G (i.e., with the minimum z-coordinate) must have the property that ℓ is the minimum height of this triangle; otherwise, it is the length of the only bounded edge. See Fig. 3 for an illustration where one can take G_1 as the optimal G.

In Fig. 3, where in (I) the triangle is bounded and in (II) the triangle is unbounded, we have in fact more information included. Note that if G_1, G_2 are horizontal planes and G_2 is above G_1, then if a_2b_2 is a feasible solution for $\tau(pq)$ then we can compute the optimal solution, say a_1b_1 on G_1. The reason is that they form a linear relation, parameterized by the vertical distance between G_1 and G_2.

We can then use a brute-force algorithm to solve the problem: First we extend each face f_i of T to a halfspace S_i. Then we enumerate $O(n^3)$ triples of opposing halfspaces and for each triple compute a horizontal placement of pq, $\tau(pq)$, with the minimum z-coordinate. Finally we check if all halfspaces are intersected by $\tau(pq)$; if so, record the corresponding z-coordinate of $\tau(pq)$. At the end, we return the minimum recorded z-coordinate. The running time is obviously $O(n^4)$.

To improve the trivial brute-force method, we make use of the following result as a subroutine: given n convex polygons in the plane, compute the shortest segment that intersects all the convex polygons in $O(n \log n)$ time [3]. Let us call this algorithm SS. Then we present our algorithm as follows.

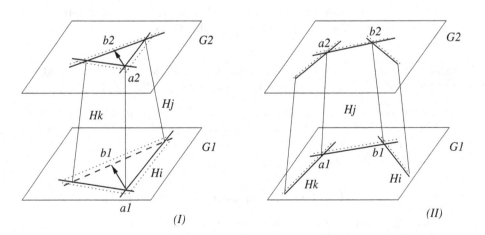

Fig. 3. When restricted to intersect S_i, S_j and S_k, the optimally placed guarding segment can be computed in constant time: (I) when the triangle is bounded; (II) when the triangle is unbounded.

1. Compute the arrangement $\mathcal{A}$ of all halfspaces S_i's, each extended from a face f_i of T. Let the vertices of $\mathcal{A}$ be L, ordered according to their ascending z-coordinates.
2. Use binary search on L to locate two vertices v_j and v_{j+1} such that at $z(v_j)$ the SS-algorithm would report that there is not a segment of length at most $|pq|$ that intersects all the halfplanes $S_i \cap (Z = z(v_j))$, while at $z(v_{j+1})$ the SS algorithm would report that there is a segment of length at most $|pq|$ that intersects all the halfplanes $S_i \cap (Z = z(v_{j+1}))$.
3. Compute the optimal placement of pq, $\tau(pq)$, at a minimum height between $Z = z(v_j)$ and $Z = z(v_{j+1})$ (e.g., G_1 and G_2 as illustrated in Fig. 3).

The arrangement $\mathcal{A}$ could have $O(n^3)$ vertices, hence L can be computed and sorted in $O(n^3 \log n)$ time. As the SS-algorithm takes $O(n \log n)$ time, the whole binary process takes $O(\log n^3) \times O(n \log n) = O(n \log^2 n)$ time. Once v_j and v_{j+1} are found, the remaining step takes linear time. Hence the total running time is $O(n^3 \log n) + O(n \log^2 n) = O(n^3 \log n)$. Therefore, we have the following theorem.

Theorem 2. *Given a 2.5D terrain T with n vertices and a given segment pq, an optimal placement of pq, $\tau(pq)$, can be computed in $O(n^3 \log n)$ time.*

In practice, we could avoid computing L and search in the z-interval $(0, h)$, where h is the lowest z-coordinate of any vertex in $P = \cap_{i=1..m} S_i$.

Corollary 1. *Given a 2.5D terrain T with n vertices and $m = O(n)$ faces, a given segment pq, an optimal placement of pq, $\tau(pq)$, can be computed in $O(n \log n \log h)$ time, where h is the lowest z-coordinate of any vertex in $P = \cap_{i=1..m} S_i$.*

4   A Polynomial-Time Algorithm on Guarding 1.5D Polyhedral Terrains with Two Horizontal Segments

The problem we study in this section is complementary to what has been studied by Katoh et al., where given a 1.5D terrain T of n vertices and a horizontal line H, compute two shortest segments with the same length on H to cover T completely, the problem can be solved in $O(n \log^2 n)$ time using parametric search [11]. Here we consider the complementary problem: given two segments $p_1 q_1$ and $p_2 q_2$ of a fixed length, compute the lowest line H^* such that $p_1 q_1$ and $p_2 q_2$ can be placed on H^* to cover T completely.

We solve this variation in $O(n^2 \log n)$ time, using the algorithm by Katoh et al. (we call it the KWXZ-algorithm) as a subroutine — certainly with additional details to be sorted out. The following is a sketch of the algorithm.

1. Compute the arrangement $\mathcal{A}$ of all halfplanes H_i's, where H_i is extended from the edge e_i of T. Let the vertices of $\mathcal{A}$ be L, ordered according to their ascending y-coordinates.

2. Use binary search on L to locate two vertices v_j and v_{j+1} such that at $y = y(v_j)$ the $KWXZ$-algorithm would report that there are not two segments of length at most $|p_1q_1| = |p_2q_2|$ that can collectively guard T, while at $y = y(v_{j+1})$ the KWXZ-algorithm would report that it is feasible to place two segments p_1q_1 and p_2q_2 on $y = y(v_{j+1})$ to guard T.

3. Compute the optimal horizontal line H^* with the minimum y-coordinate between $y = y(v_j)$ and $y = y(v_{j+1})$ such that p_1q_1 and p_2q_2 can be placed on H^* to guard T.

Similar to the analysis in the previous section, L is of size $O(n^2)$ and can be computed and sorted in $O(n^2 \log n)$ time. The binary search step takes $O(\log n^2) \cdot O(n \log^2 n) = O(n \log^3 n)$ time. Hence, as long as Step 3 can be solved in $O(n^2 \log n)$ time the whole problem can be solved in $O(n^2 \log n)$ time.

We show next that Step 3 can be done in $O(n^2)$ time.

We refer to Fig. 4, based on the binary procedure, we know an optimal solution must exist between $y = y(v_j)$ and $y = y(v_{j+1})$. Hence, the central problem is to find a horizontal line H^* with its y-coordinate being in $(y(v_j), y(v_{j+1}))$.

First of all, it is easily seen that if p_1q_1 is placed to the left of p_2q_2 then p_1q_1 must be bounded by the leftmost left halfspace H_p, say defined by line $y = k_1x + b_1$. Similarly, p_2q_2 is bounded by the rightmost right halfspace H_q, defined by line $y = k_2x + b_2$. Now assume that p_1 is on H_p and $p_1 = (x_1, y_1)$, similarly, p_2 is on H_q and $p_2 = (x_2, y_2)$. If we enforce that $y_1 = y_2 = h$ (h being a variable in $(y(v_j), y(v_{j+1}))$) and let $\ell = |p_1q_1| = |p_2q_2|$, then $p_1 = ((h-b_1)/k_1, h)$, $q_1 = ((h - b_1)/k_1 + \ell, h)$, $p_2 = ((h - b_2)/k_2, h)$ and $q_2 = ((h - b_2)/k_2 - \ell, h)$.

Now, h is determined when q_1 and q_2 collectively guard some edges between $x(q_1')$ and $x(q_2')$, which come from the placement of p_1q_1 and p_2q_2 at the height of $y = y(v_j)$. In Fig. 4 we illustrate such an edge $e = (p_e, q_e)$, which is on the line $y = k_3x + b_3$. Then, we have the following lemma.

Lemma 3. *Let $e_1, e_2, ..., e_t$ be the list of edges between $x(q_1')$ and $x(q_2')$ which are collectively covered by q_1 and q_2 in an optimal solution, then there must be e_k, $1 \le k \le t$, such that e_k can be partitioned into two parts $e_k = e_{k,1} \cdot e_{k,2}$ such that $e_{k,2}$ is only covered by q_1 and $e_{k,1}$ is covered by q_2.*

Proof. If some part of e_k is not covered by q_1 and q_2, then the solution is infeasible. If e_k is composed of three parts $e_{k,1} \cdot e_{k,3} \cdot e_{k,2}$ such that $e_{k,2}$ is only covered by q_1, $e_{k,1}$ is covered by q_2 and $e_{k,3}$ is covered by both q_1 and q_2, then we can lower p_1q_1 and p_2q_2 until $e_{k,3}$ becomes a single point.   □

With the above lemma, we proceed as follows. We use e to denote any e_k, $1 \le k \le t$. Let $\langle u_1, u_2, ..., u_l (= p_e) \rangle$ be the part of the shortest path to the right of

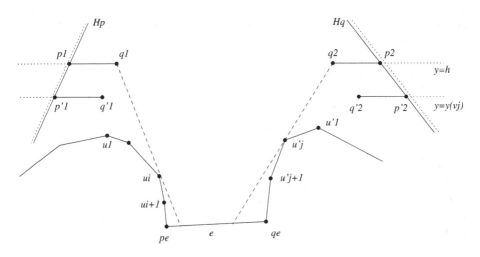

Fig. 4. Computing the minimum height h: when p_1q_1 and p_2q_2 have to guard an edge e collectively.

q_1' from v_1 to p_e. When q_1 cannot see e completely, it must be blocked by some $u_i = (a_i, b_i)$. The line of q_1u_i has an equation

$$\frac{y - b_i}{x - a_i} = \frac{k_1h - k_1b_i}{h - b_1 + k_1\ell - k_1a_i},$$

which, when intersecting e (with an equation $y = k_3x + b_3$) would give the x-coordinate of the intersection on e as

$$x = \frac{f_{i,e}(h)}{g_{i,e}(h)},$$

where $f_{i,e}, g_{i,e}$ are linear functions on h. (For instance, $g_{i,e} = k_1h - k_1b_i - k_3$.)

Consequently, when we extend the lines $u_iu_{i+1}, i = 1..l$, they naturally divide e into $l + 1$ red intervals and each interval would give a function in the form of $x = \frac{f_{i,e}(h)}{g_{i,e}(h)}$.

Symmetrically, let $\langle u_1', u_2', ..., u_{l'}' \rangle$ be the part of the shortest path to the left of q_2' from v_n to q_e. If q_2 cannot see e completely, it must be blocked by some $u_j' = (a_j', b_j')$. The line q_2u_j' intersects e and results in an equation like

$$x = \frac{f_{j,e}'(h)}{g_{j,e}'(h)},$$

with $f_{j,e}'$ and $g_{j,e}'$ being linear functions on h.

Now to decide if q_1 and q_2 can guard e completely is reduced to deciding if

$$\frac{f_{i,e}(h)}{g_{i,e}(h)} \leq \frac{f_{j,e}'(h)}{g_{j,e}'(h)},$$

which is very much equivalent to solving a quadratic equation on h. We show below that this can be done in $O(n)$ time for each e.

First we extend the lines $u'_j u'_{j+1}, j = 1..l'$, to intersect and divide e into $l'+1$ blue intervals. We need to find some red and blue intervals which overlap and an h can be found such that

$$\frac{f_{i,e}(h)}{g_{i,e}(h)} = \frac{f'_{j,e}(h)}{g'_{j,e}(h)}.$$

The testing whether $\frac{f_{i,e}(h)}{g_{i,e}(h)} \le \frac{f'_{j,e}(h)}{g'_{j,e}(h)}$ is easily done by checking this equality.

The difficulty here is that even when e is fixed, we cannot enumerate all pairs of i and j, which has a size of $O(n^2)$. We here use prune-and-search to obtain a linear time algorithm. The idea is to first merge the red and blue intervals into a list of sorted base intervals of size $O(l + l') = O(n)$, then pick the $\lfloor (l + l')/2 \rfloor$-th base interval and check if there is some h satisfying $\frac{f_{i,e}(h)}{g_{i,e}(h)} = \frac{f'_{j,e}(h)}{g'_{j,e}(h)}$, recall that i determines (a_i, b_i) and j determines (a'_j, b'_j). If so, then we are done. If $\frac{f_{i,e}(h)}{g_{i,e}(h)} < \frac{f'_{j,e}(h)}{g'_{j,e}(h)}$ then we can eliminate the leftmost $(l + l')/2$ base intervals and recurse; otherwise, we eliminate the rightmost $(l + l')/2$ base intervals and recurse. This recursion continues until the base interval, where the right h satisfies $\frac{f_{i,e}(h)}{g_{i,e}(h)} = \frac{f'_{j,e}(h)}{g'_{j,e}(h)}$ is found or reports that such an h does not exist. The running time of this process is

$$T(n) = T(n/2) + O(1) = O(n).$$

Since we could have $O(n)$ number of e's, the total cost after the $O(n^2 \log n)$ sorting is $O(n^2)$, then the following theorem follows.

Theorem 3. *Given a 1.5D terrain T with n vertices and two given segments $p_1 q_1$ and $p_2 q_2$ of the same fixed length, an optimal (lowest) horizontal line can be computed in $O(n^2 \log n)$ time, such that $p_1 q_1$ and $p_2 q_2$ can be placed on the line to collectively guard T.*

We note that this idea seems to be hard to generalize to 2.5D terrains, i.e., placing $p_1 q_1$ and $p_2 q_2$ on the lowest plane to cover T. The reason is that in 3D when one computes the visibility region from some line above T, the line of sight could be a quadratic surface [19], so unless we specify the starting points, say p_1 and p_2, and maybe even the directions of the placed segments, it seems hard to generalize the ideas for 1.5D terrain to 2.5D terrains.

In the next section, we briefly discuss how to compute an approximate shortest watchtower to guard a 1.5D imprecise terrain I.

5   Guarding a 1.5D Imprecise Terrain

Given a set of vertical intervals $I = \{\langle (x_i, c_i), (x_i, d_i) \rangle | c_i \le d_i, x_i\text{'s are sorted}$ from left to right, $i = 1..n\}$, the *shortest watchtower problem on an imprecise terrain* is to compute a terrain T composed of vertices $\{(x_i, y_i)|c_i \leq y_i \leq d_i, i = 1..n\}$ such that a shortest watchtower uv, where v is put on T, u can see every point of T and $|uv|$ is minimized over all possible T's. When the x-coordinate of u and v are fixed, we call the corresponding problem *discrete shortest watchtower problem on an imprecise terrain*. See Fig. 5 for an example.

Let $u = (x, y)$. The constraint that u is above the line through the j-th edge $e_j = \langle (x_j, y_j), (x_{j+1}, y_{j+1}) \rangle$ is

$$y - y_j \geq \frac{y_{j+1} - y_j}{x_{j+1} - x_j}(x - x_j),$$

which is not a linear inequality as x, y, y_j, y_{j+1} are all variables. To make it linear, we need to fix x — to have a discrete version. Now, suppose that x is a fixed constant between x_i and x_{i+1}, we have the following LP formulation. (Note that the constraint (1) is linear now.) For convenience, denote $[n] = \{1, ..., n\}$.

$$\min \quad y - \frac{y_{i+1} - y_i}{x_{i+1} - x_i}(x - x_i) - y_i \qquad \text{(LP1)}$$

$$\text{s.t.} \quad y - y_j \geq \frac{y_{j+1} - y_j}{x_{j+1} - x_j}(x - x_j), \qquad \forall j \in [n] \qquad (1)$$

$$c_j \leq y_j \leq d_j, \qquad \forall j \in [n]$$

$$x \text{ is a fixed constant in } (x_i, x_{i+1})$$

Following Vaidya's algorithm [18], LP can be solved in $O((m+n)^{1.5}nL)$ time, where m is the number of constraints, n is the number of variables and L is the precision measure (in bits). Hence we have the following lemma.

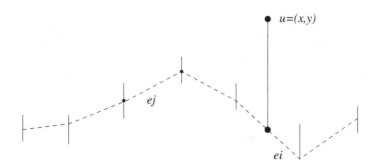

Fig. 5. A 1.5D imprecise terrain T and the visibility computation regarding u.

Lemma 4. *The discrete shortest watchtower problem on an imprecise terrain specified by n vertical intervals can be solved in $O(n^{2.5}L)$ time, where L is the precision measure in bits.*

If x is not fixed, we could obtain a close (additive) approximation for the shortest watchtower problem by running through all the ϵ-intervals from x_1 to x_n. Note that we need to change the objective functions $n-1$ times, which correspond to the scenarios when the tower is above $e_i = \langle (x_i, y_i), (x_{i+1}, y_{i+1}) \rangle, i = 1..n-1$.

Let $k_{\max}$ be the maximum absolute value of the slope of e_i's. We set $\epsilon = \varepsilon/(2k_{\max})$, then we cut the range $[x_1, x_n]$ into $\lceil \frac{x_n - x_1}{\varepsilon/(2k_{\max})} \rceil = \lceil \frac{2(x_n - x_1)k_{\max}}{\varepsilon} \rceil$ intervals of length $\varepsilon/k_{\max}$ (except possibly the last one). Then we just run LP1 when x is set to be among these interval points, one by one. (Recall that the objective function could be adjusted when x is moved from $[x_{i-1}, x_i)$ to $[x_i, x_{i+1})$). It is easily seen that if the optimal solution is OPT, then the additive error generated by the algorithm is at most $2k_{\max} \cdot \epsilon = \varepsilon$. We then have the following theorem.

Theorem 4. *The shortest watchtower problem on an imprecise terrain specified by n vertical intervals can be approximated with an additive error of ε, the algorithm runs in $O(n^{2.5}L \cdot \lceil \frac{(x_n - x_1)k_{\max}}{\varepsilon} \rceil)$ time, where L is the precision measure in bits, $k_{\max}$ is the maximum absolute value of the slope of e_i's in the imprecise terrain, and x_1 and x_n are the leftmost and rightmost x-coordinates of the terrain vertices respectively.*

We have several comments regarding this result.

1. The algorithm seems to be hard to generalize to the regular multiplicative approximation algorithm. In fact, if OPT=0, the current algorithm would incur an infinite (multiplicative) factor — unless the optimal solution is exactly located at some interval point.
2. The algorithm can be generalized to 2.5D terrains, when an underlying triangulation is given for those given vertical intervals. The plane through a triangle of three points (x_1, y_1, z_1), (x_2, y_2, z_2) and (x_3, y_3, z_3), where x_i' and y_i's are constants and z_i's are variables, would define an equation like $n_i X + n_j Y + n_k Z = K$. When we fix X and Y, then a point (X, Y, Z) is above this plane is realized as $n_i X + n_j Y + n_k Z \geq K$, which is linear as n_k is not involving the variables z_i's.
3. We implement this algorithm in 2D using Python and CPLEX, the program runs reasonably well and returns a close approximate solution for $n \leq 100$ and decently small ε.

6   Concluding Remarks

We study three problems on guarding a terrain with (horizontal) segments. Several questions are still open. For instance, for the first problem in 3D, can we obtain a sub-cubic algorithm? For the second problem, can something non-trivial be done to obtain some result in 3D? For the third problem on guarding imprecise terrains in 2D, is the problem NP-hard? And what can be done if we are in 3D and the underlying triangulation is not given?

Acknowledgments. This research is supported by NSF under grant CNS-2243010.

References

1. Agarwal, P.K., et al.: Guarding a terrain by two watchtowers. Algorithmica **58**(2), 352–390 (2010)
2. Bespamyatnikh, S., Chen, Z., Wang, K., Zhu, B.: On the planar two-watchtower problem. In: Wang, J. (ed.) COCOON 2001. LNCS, vol. 2108, pp. 121–130. Springer, Heidelberg (2001). https://doi.org/10.1007/3-540-44679-6_14
3. Bhattacharya, B.K., Czyzowicz, J., Egyed, P., Toussaint, G.T., Stojmenovic, I., Urrutia, J.: Computing shortest transversals of sets. Int. J. Comput. Geom. Appl. **2**(4), 417–442 (1992)
4. Bose, P., Shermer, T.C., Toussaint, G.T., Zhu, B.: Guarding polyhedral terrains. Comput. Geom. Theo. Appl. **7**, 173–185 (1997)
5. Cole, R., Sharir, M.: Visibility problems for polyhedral terrains. J. Symb. Comput. **7**(1), 11–30 (1989)
6. Driemel, A., Haverkort, H.J., Löffler, M., Silveira, R.I.: Flow computations on imprecise terrains. J. Comput. Geom. **4**(1), 38–78 (2013)
7. Duraisamy, N., et al.: Half-guarding weakly-visible polygons and terrains. In: Dawar, A., Guruswami, V., editors, 42nd IARCS Annual Conference on Foundations of Software Technology and Theoretical Computer Science (FSTTCS'2022), December 18–20, 2022, IIT Madras, Chennai, India, volume 250 of LIPIcs, pp. 18:1–18:17. Schloss Dagstuhl - Leibniz-Zentrum für Informatik (2022)
8. Gray, C., Evans, W.S.: Optimistic shortest paths on uncertain terrains. In: Proceedings of the 16th Canadian Conference on Computational Geometry (CCCG'2004), Concordia University, Montréal, Québec, Canada, August 9–11, 2004, pp. 68–71 (2004)
9. Gray, C., Kammer, F., Löffler, M., Silveira, R.I.: Removing local extrema from imprecise terrains. Comput. Geom. Theo. Appl. **45**(7), 334–349 (2012)
10. Gray, C., Löffler, M., Silveira, R.I.: Smoothing imprecise 1.5d terrains. Int. J. Comput. Geom. Appl. **20**(4), 381–414 (2010)
11. Katoh, N., Wang, W., Yinfeng, X., Zhu, B.: Parametric search: three new applications. Front. Math. China **5**(2), 65–73 (2010)
12. Lee, D.T., Preparata, F.P.: An optimal algorithm for finding the kernel of a polygon. J. ACM **26**(3), 415–421 (1979)
13. Lubiw, A., Stroud, G.: Computing realistic terrains from imprecise elevations. In: Y. Bahoo., Georgiou, K., editors, Proceedings of the 34th Canadian Conference on Computational Geometry, (CCCG'2022), Toronto Metropolitan University, Toronto, Ontario, Canada, August 25–27, 2022, pp. 227–234 (2022)
14. Preparata, F.P., Muller, D.E.: Finding the intersection of n half-spaces in time O(n log n). Theor. Comput. Sci. **8**, 45–55 (1979)
15. Seth, R., Maheshwari, A., Nandy, S.C.: Acrophobic guard watchtower problem. Comput. Geom. Theo. Appl. **109**, 101918 (2023)
16. Sharir, M.: The shortest watchtower and related problems for polyhedral terrains. Inf. Process. Lett. **29**(5), 265–270 (1988)
17. Tripathi, N., Pal, M., De, M., Das, G., Nandy, S.C.: Guarding polyhedral terrain by k-watchtowers. In: Chen, J., Lu, P. (eds.) FAW 2018. LNCS, vol. 10823, pp. 112–125. Springer, Cham (2018). https://doi.org/10.1007/978-3-319-78455-7_9
18. Vaidya, P.M.: Speeding-up linear programming using fast matrix multiplication (extended abstract). In: 30th Annual Symposium on Foundations of Computer Science, Research Triangle Park, North Carolina, USA, 30 October - 1 November 1989, pp. 332–337. IEEE Computer Society (1989)

19. Wang, C., Zhu, B.: Three dimensional weak visibility: complexity and algorithms. Theor. Comput. Sci. **234**(1–2), 219–232 (2000)
20. Zhu, B.: Improved algorithms for computing the shortest watchtower of polyhedral terrains. In: Proceedings of the 3rd Canadian Conference on Computational Geometry (CCCG'1992), St. John's, Newfoundland, Canada, August 1992, pp. 286–291. Memorial University of Newfoundland (1992)
21. Zhu, B.: Computing the shortest watchtower of a polyhedral terrain in O(n log n) time. Comput. Geom. Theo. Appl. **8**, 181–193 (1997)

The Bag-Based Search: A Meta-Algorithm to Construct Tractable Logical Circuits for Graphs Based on Tree Decomposition

Masakazu Ishihata[✉]

NTT Communication Science Laboratories, Kyoto, Japan
masakazu.ishihata@ntt.com

Abstract. Tractable logical circuits (TLCs) have attracted more attention in the AI field as bases of knowledge representation and tractable probabilistic modeling. We propose the bag-based search (BBS), a new meta-algorithm for constructing a TLC that accepts all subgraphs of a given input graph that satisfies a target graph property. We implemented BBS examples for various graph properties, including independent set, k-edgeset, dominating set, k-matchings, and spanning trees, and applied them to artificial and real-world graphs. The experimental results showed that BBS generated significantly smaller circuits than ZDDs obtained by the frontier-based search (FBS).

1 Introduction

Tractable logical circuits (TLCs) have played a central role in the AI field for a long time as a language of knowledge representation [9], and recently, have received more attention as a fundamental tool of tractable probabilistic modeling (TPM) [6]. A probabilistic circuit (PC) is a logical circuit that represents a probability distribution, and a logical circuit is *tractable* for a target query if it satisfies some structural properties required for answering the target query in polynomial time in its size. In the context of TPM, *smoothness*, *determinism*, and *structured-decomposability* have been recognized as essential structural properties because PCs satisfying those properties provide various basic queries, including marginal and conditional inference, moments of distributions, maximum a-posteriori inference, expectations [6], and various advanced queries, such as expected predictions [15], SHAP scores [3], and some information-theoretic queries [31]. By making full use of the various queries described above, PCs have been applied to not only probabilistic modeling but also a wide variety of applications, including lossless compression [18], fair prediction [27], and more.

One of the most promising applications of PCs is modeling *structured probabilistic space* [4], distributions over discrete structures, e.g., combinations, permutations, and graphs. For instance, PCs have been used for modeling distributions over subgraphs to ranking routes on maps [5], maximizing the influence spread [19], constructing reliable communication networks [23,30], evaluating the scan statistics [12], and more. Hence, constructing smooth, deterministic, and structured-decomposable PCs for graphs is increasingly essential to use the above wide range of queries when modeling structured probabilistic space. Whereas several empirically efficient and accurate methods have been proposed for learning the structure of PCs for non-structured probabilistic space [17], there has been little discussion on how to construct PCs for graph structures.

Theoretically, constructing a deterministic and structured-decomposable circuit for the graph property described in monadic second-order logic is fixed-parameter tractable (FPT) of treewidth [1], and smoothing such circuits is almost linear in their size [28]. However, there is still a large gap between theory and practice: the theoretical FPT algorithm is too complicated to implement and practically inefficient because of an unrealistic huge constant factor of treewidth. Several practically efficient construction methods have been proposed to fill this gap. The frontier-based search (FBS) [14, 16] is the most widely used method to efficiently construct circuits that accept subgraphs satisfying various graph properties, including trees, paths, matchings, and connected components. FBS constructs a zero-suppressed decision diagram (ZDD) [21], a deterministic and structured-decomposed circuit but non-smooth where its decomposition structure is restricted in a linear form (the detail will be explained later). Recently, the FBS-like constructions for sentential decision diagrams (SDDs) [8] and zero-suppressed SDDs (ZSDDs) [24] have also been proposed [22, 25, 26, 29]. A (Z)SDD is non-smooth in common with ZDDs, but its restriction on decomposition is relaxed in a tree form.

This paper aims to go one step further; we propose a new FBS-like method directly constructing smooth, deterministic, and structured-decomposable circuits of various graph properties that is easy to implement and practically efficient. The contributions of this paper are three-fold. First, we propose the bag-based search (BBS), a new meta-algorithm for constructing a TLC that accepts all subgraphs of an input graph satisfying a target graph property. BBS is dynamic programming (DP) on a tree decomposition of the input graph and constructs TLCs of various properties by slightly modifying its DP update; i.e., BBS is easy-to-implement. Second, we show specific BBS examples, including independent set, k-edgeset, dominating set, k-matching, and spanning tree. Third, we empirically show that circuits constructed by BBS are much smaller than ZDDs obtained by FBS in many real-world instances; i.e., BBS is practically efficient.

2 Preliminary

This paper aims to construct a TLC of a Boolean function, which accepts all subgraphs of an input graph that satisfy a target property, in a dynamic programming manner on a tree decomposition (TD) of the input graph. We handle four types of graphs: G is the input graph, C is the TLC to be constructed, S is the vtree (explain later in detail) respected by C, and T is a TD of G. To avoid confusion, we define these graphs as follows: $G \triangleq \langle V, E \rangle$ where V is a *vertex* set and E is an *edge* set, $C \triangleq \langle G, W \rangle$ where G is a *gate* set and W is a *wire* set, $S \triangleq \langle P, L \rangle$ where P is a *point* set and L is a *line* set, and $T \triangleq \langle N, A \rangle$ where N is a *node* set and A is an *arc* set.

2.1 Subgraphs and Their Boolean Representation

We first introduce the general notation of graphs and use the same notation for G, C, S, and T. G is *undirected* if $E \subseteq \{e \subseteq V \mid |e| = 2\}$ is an undirected edge set and is *directed* if $E \subseteq \{(u,v) \mid \{u,v\} \subseteq V\}$ is a directed edge set. For any directed G and $u \in V$, $\mathrm{cld}(u) \triangleq \{c \in V \mid (u,c) \in E)\}$ is the *children* of u and $\mathrm{par}(u) \triangleq \{p \in V \mid (p,u) \in E\}$ is the *parents* of u. u is a *leaf*, *root*, and *internal* if $\mathrm{cld}(u) = \emptyset$, $\mathrm{par}(u) = \emptyset$, and $\mathrm{cld}(u) \neq \emptyset$, respectively. G is *rooted* if G has exactly one root, and is *acyclic* if G has no directed cycle. For any $U \subseteq V$ and $F \subseteq E$, let $E[U] \triangleq \{e \in E \mid e \subseteq U\}$ and $V[F] \triangleq \{v \in V \mid \exists e \in F, v \in e\}$. Given any G, U and F, $\mathrm{G}[U] \triangleq \langle U, E[U] \rangle$ is a *vertex-induced subgraph*, $\mathrm{G}[F] \triangleq \langle V[F], F \rangle$ is an *edge-induced subgraph*, and $\mathrm{G}[U,F] \triangleq \langle U, F \rangle$ is a *subgraph* if $V[F] \subseteq U$.

We next introduce a Boolean representation of subgraphs. For any $u \in V$ and $e \in E$, let X_u and X_e be Boolean variables and x_u and $x_e \in \{0,1\}$ be their realizations. Let $\boldsymbol{X}_V \triangleq \{X_u \mid u \in V\}$, $\boldsymbol{X}_E \triangleq \{X_e \mid e \in E\}$, and $\boldsymbol{X}_G \triangleq \boldsymbol{X}_V \cup \boldsymbol{X}_E$ be variable sets. Similarly, let $\boldsymbol{x}_V \triangleq \{x_u \mid u \in V\}$, $\boldsymbol{x}_E \triangleq \{x_e \mid e \in E\}$, and $\boldsymbol{x}_G \triangleq \boldsymbol{x}_V \cup \boldsymbol{x}_E$ be their realizations. Then, $\boldsymbol{X}_V$ and $\boldsymbol{X}_E$ represent a subset of V and E denoted by $V(\boldsymbol{X}_V) \triangleq \{u \in V \mid X_u = 1\}$ and $E(\boldsymbol{X}_E) \triangleq \{e \in E \mid X_e = 1\}$, respectively. In addition, we say that $\boldsymbol{X}_V$, $\boldsymbol{X}_E$, and $\boldsymbol{X}_G$ also represent $\mathrm{G}[U]$, $\mathrm{G}[F]$, and $\mathrm{G}[U,F]$, respectively, where $U = V(\boldsymbol{X}_V)$ and $F = E(\boldsymbol{X}_E)$. In the rest of this paper, we explain the case of $\boldsymbol{X}_G$ as the general case of $\boldsymbol{X}_V$ and $\boldsymbol{X}_E$.

Let $f_{\mathcal{P}}(\boldsymbol{X}_G)$ be a Boolean function that represents a particular graph property $\mathcal{P}$; in other words, $f_{\mathcal{P}}(\boldsymbol{X}_G) = 1 \Leftrightarrow \mathrm{G}[\boldsymbol{X}_G]$ satisfies the target property $\mathcal{P}$. For example, let G be undirected and consider the following two properties: (1) $U \subseteq V$ is an *independent set* (IS) of G if $E[U] = \emptyset$, and (2) $F \subseteq E$ is a *spanning tree* (ST) of G if $\mathrm{G}[V,F]$ is connected and has no cycle. Then, f_{IS} represents IS if $f_{\mathrm{IS}}(\boldsymbol{X}_V) = 1 \Leftrightarrow V(\boldsymbol{X}_V)$ is an IS, and f_{ST} represents ST if $f_{\mathrm{ST}}(\boldsymbol{X}_E) = 1 \Leftrightarrow E(\boldsymbol{X}_E)$ is an ST.

2.2 Tractable Logical Circuits (TLCs)

A logical circuit is a rooted directed acyclic graph (DAG) representing a Boolean function. Let $f(\boldsymbol{X}_G)$ be a Boolean function and $\mathrm{C} \triangleq \langle G, A \rangle$ be a rooted DAG representing f whose each gate $g \in G$ has a *type* denoted by $\mathrm{type}(g)$. When g is a leaf, $\mathrm{type}(g)$ is a *literal*: $\mathrm{type}(g) = X$ or $\neg X$ ($X \in \boldsymbol{X}_G$). When g is an internal, $\mathrm{type}(g)$ is a logical operator $\otimes$ (AND) or $\oplus$ (OR). g is referred to as a $\otimes$-*gate*, $\oplus$-*gate*, and *input* if $\mathrm{type}(g)$ is $\otimes$, $\oplus$, and a literal, respectively. Each g represents a Boolean function denoted by $f_g(\boldsymbol{X}_g)$ where $\boldsymbol{X}_g \subseteq \boldsymbol{X}_G$ is the variable set of g: each input represents its literal, each $\otimes/\oplus$-gate performs conjunction/disjunction of its children, and the root r satisfies $f_r(\boldsymbol{X}_r) = f(\boldsymbol{X}_G)$. Consequently, f_g and $\boldsymbol{X}_g$ are recursively defined as follows:

$$\boldsymbol{X}_g \triangleq \begin{cases} \{X\} & \mathrm{type}(g) \in \{X, \neg X\} \\ \bigcup \{\boldsymbol{X}_{g'} \mid g' \in \mathrm{cld}(g)\} & \mathrm{type}(g) \in \{\otimes, \oplus\} \end{cases}$$

$$f_g(\boldsymbol{X}_g) \triangleq \begin{cases} X & \mathrm{type}(g) = X \\ \neg X & \mathrm{type}(g) = \neg X \\ \bigwedge \{f_{g'}(\boldsymbol{X}_{g'}) \mid g' \in \mathrm{cld}(g)\} & \mathrm{type}(g) = \otimes \\ \bigvee \{f_{g'}(\boldsymbol{X}_{g'}) \mid g' \in \mathrm{cld}(g)\} & \mathrm{type}(g) = \oplus \end{cases}$$

For the sake of simplicity, we assume that there is at most one leaf g such that $\mathrm{type}(g) = X$ (resp. $\neg X$) for each $X \in \boldsymbol{X}_G$, and use $g(X,1)$ (resp. $g(X,0)$) to denote such g. For any $t \in \{\otimes, \oplus\}$ and $g', g'' \in G$, we introduce an operation $\mathrm{apply}(t, g', g'')$ that returns $g \in G$ such that $\mathrm{type}(g) = t$ and $\mathrm{cld}(g) = \{g', g''\}$ if it exists; otherwise it creates such g before returning. One can construct a target logical circuit by recursively applying $\mathrm{apply}(t, g', g'')$ for the **initial circuit** $C_0 \triangleq \langle \{g(X, x) \mid X \in \boldsymbol{X}_{\mathrm{G}}, x \in \{0,1\}\}, \emptyset \rangle$.

We next define the *tractableness* of circuits. Let $\mathcal{X}_g \triangleq \{\boldsymbol{X}_{g'} \mid g' \in \mathrm{cld}(g)\}$ be the family of variable sets of g's children and $G_t \triangleq \{g \in G \mid \mathrm{type}(g) = t\}$. C is **smooth** if $\forall g \in G_\oplus, |\mathcal{X}_g| = 1$: the variable sets are the same for all children of every $\oplus$-gate. C is **deterministic** if the Boolean functions are mutually exclusive between the two children of every $\oplus$-gate; i.e., $\forall g \in G_\oplus, \forall \{g', g''\} \subseteq \mathrm{cld}(g), \neg(f_{g'}(\boldsymbol{X}_{g'}) \wedge f_{g''}(\boldsymbol{X}_{g''}))$. C is **decomposable** if $\mathcal{X}_g$ forms a partition of $\boldsymbol{X}_g$ for every $\otimes$-gate g; i.e., the variable sets are disjoint between the two children of every $\otimes$-gate. A *vtree* of $\boldsymbol{X}_{\mathrm{G}}$, denoted by $\mathrm{S} \triangleq \langle P, L \rangle$, is a directed full binary tree whose leaves have a one-to-one correspondence with the variables in $\boldsymbol{X}_{\mathrm{G}}$. For any leaf point $p \in P$, let $\boldsymbol{X}_p \triangleq \{X\}$ where $X \in \boldsymbol{X}_{\mathrm{G}}$ is the variable corresponding to p, and for any internal $p \in P$, let $\boldsymbol{X}_p \triangleq \bigcup\{\boldsymbol{X}_{p'} \mid p' \in \mathrm{cld}(p)\}$ be the variable set under p; i.e., $\mathcal{X}_p \triangleq \{\boldsymbol{X}_{p'} \mid p' \in \mathrm{cld}(p)\}$ forms a partition of $\boldsymbol{X}_p$. For any $g \in G_\otimes$ and $p \in P$, g *respects* p if $\mathcal{X}_g = \mathcal{X}_p$; i.e., g and p represent the same partition. C *respects* S if all $\otimes$-gate represents some point in S, and C is **structured-decomposable** if there exists a vtree S that C respects. C is **tractable** if C is smooth, deterministic, and structured-decomposable. We aim to construct tractable C of $f_\mathcal{P}(\boldsymbol{X}_{\mathrm{G}})$ for various properties $\mathcal{P}$ using $\mathrm{apply}(t, g', g'')$ in a dynamic programming manner on a tree decomposition of G explained in the following.

2.3 Tree Decompositions (TDs)

A TD of G is a tree whose each node has a corresponding subset of V called a *bag*. Let $\mathrm{T} \triangleq \langle N, A \rangle$ be a tree, $\mathcal{U} \triangleq \{U_n \subseteq V \mid n \in N\}$ be the family of bags, and $N_u \triangleq \{n \in N \mid u \in U_n\}$ be the set of nodes that includes $u \in V$. Then, T is a TD of G if it satisfies the following conditions: (1) $\forall u \in V, \exists n \in N, u \in U_n$, (2) $\forall e \in E, \exists n \in N, e \subseteq U_n$, and (3) $\forall u \in V, T[N_u]$ is connected. The width of T is defined as $w_\mathrm{T} \triangleq \max\{|U_n| \mid n \in N\} - 1$. T is said to be a *path decomposition* (PD) if T forms a path. Let $\mathcal{T}_\mathrm{G}$ and $\mathcal{P}_\mathrm{G}$ be all possible TDs and PDs of G, and $\mathrm{tw}_\mathrm{G} \triangleq \min\{w_T \mid T \in \mathcal{T}_\mathrm{G}\}$ and $\mathrm{pw}_\mathrm{G} \triangleq \min\{w_T \mid T \in \mathcal{P}_\mathrm{G}\}$ be the *treewidth* and *pathwidth* of G. By definition, $\mathrm{tw}_\mathrm{G} = 1$ if G is a tree, $\mathrm{pw}_\mathrm{G} = 1$ if G is a path, and $\mathrm{tw}_\mathrm{G} \leq \mathrm{pw}_\mathrm{G}$ holds for any G.

A *very nice tree decomposition* (VNTD) [2] is a variant of TDs whose each node n additionally has a corresponding subset $F_n \subseteq E$. We refer to U_n and F_n as the *vbag* and *ebag*, respectively, and denote the family of ebags by $\mathcal{F} \triangleq \{F_n \subseteq E \mid n \in N\}$. Then, a rooted directed tree T is a VNTD of G if it satisfies the followings: (1) T and $\mathcal{U}$ form a TD of G, (2) $\forall n \in N, V[F_n] \subseteq U_n$, (3) $U_n = F_n = \emptyset$ if n is the root or a leaf, and (4) n is one of the following five types if n is an internal:

1. **Introduce vertex (IV):** n has exactly one child m such that $F_n = F_m$ and $U_n = U_m \cup \{u\}$ for some $u \in V \backslash U_m$,
2. **Introduce edge (IE):** n has exactly one child m such that $U_n = U_m$ and $F_n = F_m \cup \{e\}$ for some $e \in E \backslash F_m$,
3. **Forget vertex (FV):** n has exactly one child m such that $F_n = F_m$ and $U_n = U_m \setminus \{u\}$ for some $u \in U_m$,

4. **Forget edge (FE)**: n has exactly one child m such that $U_n = U_m$ and $F_n = F_m \setminus \{e\}$ for some $e \in F_m$,
5. **Join (J)**: n has exactly two children m and m' such that $U_n = U_m = U_{m'}$ and $F_n = F_m = F_{m'}$.

Consequently, for any $n \in N$, U_n and F_n forms an induced subgraph of G denoted by $B_n \triangleq G[U_n, F_n]$. For any internal $n \in N$, Let $\text{type}(n) \in \{\text{IV}, \text{IE}, \text{FV}, \text{FE}, \text{J}\}$ be the type of n, and let u_n and e_n be the vertex and edge introduced/forgotten at n if it exists. On a VNTD T, each vertex and edge is forgotten just once [2]. For any $n \in N$, let X_n be the variable set that $X_v \in X_n$ (resp. $X_e \in X_n$) if-and-only-if v (resp. e) is forgotten at/under n. Hence, $X_r = X_G$ holds for the root r, and $X_m \cap X_{m'} = \emptyset$ holds for any join n where $\text{cld}(n) = \{m, m'\}$. Given a TD T with width w_T, one can make it very nice without changing its width in $O(w_T^2 \max\{|V|, |N|\})$ time, and the size of the VNTD is at most $O(w_T|N|)$ [7].

3 The Bag-Based Search (BBS)

This paper aims to construct a TLC C of $f_{\mathcal{P}}(X_G)$ of an input graph G where $\mathcal{P}$ is a target graph property, and proposes the *bag-based search* (BBS) that is a meta-algorithm to construct C for various property $\mathcal{P}$ in a dynamic programming (DP) manner on a VNTD T of G. BBS employs two types of DP states, *M-state* (meta-state) and *S-state* (specific-update), and two types of DP updates, *M-update* and *S-update*; the M-state and M-update are common to all properties, while the S-state and S-update are different for each property. Thus, BBS is easy-to-implement because one can implement various BBS examples by modifying only S-states and S-updates. Some concrete examples of S-states and S-updates for various graph properties are shown in Sect. 4.

3.1 S-States and S-Updates

Let $X_{B_n} \triangleq X_{V_n} \cup X_{F_n}$ be a Boolean representation of a subgraph of B_n, x_{B_n} be its realization, and s_n be an S-state at n. Note that x_{B_n} and s_n may have multiple instantiations for the same n. BBS generates a new instantiation of s_n by the following S-updates depending on $\text{type}(n)$;

$$
\begin{array}{ll}
\texttt{Supdate}(x_{B_n}, s_m \mid \text{type}(n)) & \text{type}(n) \in \{\text{IV}, \text{IE}\}, \\
\texttt{Supdate}(x_{B_m}, s_m \mid \text{type}(n)) & \text{type}(n) \in \{\text{FV}, \text{FE}\}, \\
\texttt{Supdate}(x_{B_n}, s_m, s_{m'} \mid \text{type}(n)) & \text{type}(n) = \text{J},
\end{array}
$$

where $m, m' \in \text{cld}(n)$. Each update returns a new instantiation of s_n or a special symbol $\perp$ that means pruning. Let the default S-state be $\emptyset$, and let the default S-update always return the input; no pruning occurs. The BBS with the default S-states and S-updates constructs a TLC of the tautological Boolean function $f(X_G) \triangleq 1$. One can implement BBS examples by modifying the default S-state and S-updates to the specialized ones for the target property $\mathcal{P}$.

Algorithm 1 Mupdate$(n, \text{cld}(n) \mid \text{type}(n))$

1: **function** MUPDATE$(n, \{m\} \mid \text{IV})$ ▷ *Introduce Vertex*
2:    **for all** $\langle \boldsymbol{x}_{\mathrm{B}_m}, \boldsymbol{s}_m, g_m \rangle \in \mathbf{T}_m$ and $x_{u_n} \in \{0, 1\}$ **do**
3:       $\boldsymbol{x}_{\mathrm{B}_n} \leftarrow \boldsymbol{x}_{\mathrm{B}_m} \cup \{x_{u_n}\}$
4:       $\boldsymbol{s}_n \leftarrow \text{Supdate}(\boldsymbol{x}_{\mathrm{B}_n}, \boldsymbol{s}_m \mid \text{IV})$
5:       $\mathbf{T}_n(\boldsymbol{x}_{\mathrm{B}_n}, \boldsymbol{s}_n) \leftarrow g_m$ if $\boldsymbol{s}_n \neq \bot$
6: **function** MUPDATE$(n, \{m\} \mid \text{IE})$ ▷ *Introduce Edge*
7:    **for all** $\langle \boldsymbol{x}_{\mathrm{B}_m}, \boldsymbol{s}_m, g_m \rangle \in \mathbf{T}_m$ and $x_{e_n} \in \{0, 1\}$ **do**
8:       $\boldsymbol{x}_{\mathrm{B}_n} \leftarrow \boldsymbol{x}_{\mathrm{B}_m} \cup \{x_{e_n}\}$
9:       $\boldsymbol{s}_n \leftarrow \text{Supdate}(\boldsymbol{x}_{\mathrm{B}_n}, \boldsymbol{s}_m \mid \text{IE})$
10:      $\mathbf{T}_n(\boldsymbol{x}_{\mathrm{B}_n}, \boldsymbol{s}_n) \leftarrow g_m$ if $\boldsymbol{s}_n \neq \bot$
11: **function** MUPDATE$(n, \{m\} \mid \text{FV})$ ▷ *Forget Vertex*
12:    **for all** $\langle \boldsymbol{x}_{\mathrm{B}_m}, \boldsymbol{s}_m, g_m \rangle \in \mathbf{T}_m$ **do**
13:       $\boldsymbol{x}_{\mathrm{B}_n} \leftarrow \boldsymbol{x}_{\mathrm{B}_m} \setminus \{x_{u_n}\}$
14:       $\boldsymbol{s}_n \leftarrow \text{Supdate}(\boldsymbol{x}_{\mathrm{B}_m}, \boldsymbol{s}_m \mid \text{FV})$
15:       **if** $\boldsymbol{s}_n \neq \bot$ **then**
16:          $g \leftarrow \text{apply}(\otimes, g_m, g(X_{u_n}, x_{u_n}))$
17:          $g \leftarrow \text{apply}(\oplus, g, g')$ if $\exists \langle \boldsymbol{x}_{\mathrm{B}_n}, \boldsymbol{s}_n, g' \rangle \in \mathbf{T}_n$
18:          $\mathbf{T}_n(\boldsymbol{x}_{\mathrm{B}_n}, \boldsymbol{s}_n) \leftarrow g$
19: **function** MUPDATE$(n, \{m\} \mid \text{FE})$ ▷ *Forget Edge*
20:    **for all** $\langle \boldsymbol{x}_{\mathrm{B}_m}, \boldsymbol{s}_m, g_m \rangle \in \mathbf{T}_m$ **do**
21:       $\boldsymbol{x}_{\mathrm{B}_n} \leftarrow \boldsymbol{x}_{\mathrm{B}_m} \setminus \{x_{e_n}\}$
22:       $\boldsymbol{s}_n \leftarrow \text{Supdate}(\boldsymbol{x}_{\mathrm{B}_m}, \boldsymbol{s}_m \mid \text{FE})$
23:       **if** $\boldsymbol{s}_n \neq \bot$ **then**
24:          $g \leftarrow \text{apply}(\otimes, g_m, g(X_{e_n}, x_{e_n}))$
25:          $g \leftarrow \text{apply}(\oplus, g, g')$ if $\exists \langle \boldsymbol{x}_{\mathrm{B}_n}, \boldsymbol{s}_n, g' \rangle \in \mathbf{T}_n$
26:          $\mathbf{T}_n(\boldsymbol{x}_{\mathrm{B}_n}, \boldsymbol{s}_n) \leftarrow g$
27: **function** MUPDATE$(n, \{m, m'\} \mid \text{J})$ ▷ *Join*
28:    **for all** $\langle \boldsymbol{x}_{\mathrm{B}_m}, \boldsymbol{s}_m, g_m \rangle \in \mathbf{T}_m$ such that $\exists \langle \boldsymbol{x}_{\mathrm{B}_m}, \boldsymbol{s}_{m'}, g_{m'} \rangle \in \mathbf{T}_{m'}$ **do**
29:       $\boldsymbol{x}_{\mathrm{B}_n} \leftarrow \boldsymbol{x}_{\mathrm{B}_m}$
30:       $\boldsymbol{s}_n \leftarrow \text{Supdate}(\boldsymbol{x}_{\mathrm{B}_n}, \boldsymbol{s}_m, \boldsymbol{s}_{m'} \mid \text{J})$
31:       $\mathbf{T}_n(\boldsymbol{x}_{\mathrm{B}_n}, \boldsymbol{s}_n) \leftarrow \text{apply}(\otimes, g_m, g_{m'})$ if $\boldsymbol{s}_n \neq \bot$

3.2 M-States and M-Updates

Let $\boldsymbol{m}_n \triangleq \langle \boldsymbol{x}_{\mathrm{B}_n}, \boldsymbol{s}_n, g_n \rangle$ be an M-state where $\boldsymbol{x}_{\mathrm{B}_n}$ is a realization of $\boldsymbol{X}_{\mathrm{B}_n}$, $\boldsymbol{s}_n$ is an instantiation of the S-state, and $g_n \in G$ is a gate of C under construction. BBS manages instantiations of $\boldsymbol{m}_n$ by DP table $\mathbf{T}_n : (\boldsymbol{x}_{\mathrm{B}_n}, \boldsymbol{s}_n) \mapsto g_n$, and constructs $\mathbf{T}_n$ from $\mathbf{T}_m$ ($m \in \text{cld}(n)$) by M-update Mupdate$(n, \text{cld}(n) \mid \text{type}(n))$ shown in Algorithm 1, where $\langle \boldsymbol{x}_{\mathrm{B}_n}, \boldsymbol{s}_n, g_n \rangle \in \mathbf{T}_n$ denotes $\mathbf{T}_n(\boldsymbol{x}_{\mathrm{B}_n}, \boldsymbol{s}_n) = g_n$. The M-update obtains a new $\boldsymbol{x}_{\mathrm{B}_n}$ by adding (resp. deleting) x_{u_n} or x_{e_n} to (resp. from) $\boldsymbol{x}_{\mathrm{B}_m}$ if type$(n) \in \{\text{IV}, \text{IE}\}$ (resp. $\{\text{FV}, \text{FE}\}$), and computes new $\boldsymbol{s}_n$ by the corresponding S-update. When type$(n) \in \{\text{FV}, \text{FE}\}$, it constructs new gate $g = \text{apply}(\otimes, g_m, g(X, x))$, where X is the variable of the forgotten vertex/edge at n and x is its realization. Then, it gets a disjunction of new gates with the same $\boldsymbol{x}_{\mathrm{B}_n}$ and $\boldsymbol{s}_n$. When type$(n) = \text{J}$, it constructs new gate $g = \text{apply}(\otimes, g_m, g_{m'})$ where g_m and $g_{m'}$ correspond to the same $\boldsymbol{x}_{\mathrm{B}_n}$.

Algorithm 2 Construct(G)

---

function CONSTRUCT(G)   ▷ *The main part of BBS*
    Construct a VNTD T of G and the initial circuit C = C_0.
    ▷ *Initialize leaves*   ◁
    $\mathbf{T}_n(\emptyset, \emptyset) \leftarrow$ true **for all** leaves $n \in N$
    ▷ *Update internals*   ◁
    for all internals $n \in N$ in a reverse topological order **do**
        Mupdate(n, cld(n) | type(n))
    ▷ *Construct the root of the TLC*   ◁
    $r \leftarrow$ the root of T
    $g \leftarrow$ true
    $g \leftarrow$ apply($\oplus, g, g'$) **for** $g' : \langle \boldsymbol{x}_{B_r}, \boldsymbol{s}_r, g' \rangle \in \mathbf{T}_r$
    return C

3.3 The Bag-Based Search (BBS)

The main DP part of BBS is shown in Algorithm 2. First, it constructs a VNTD T of the input graph G and the initial circuit C. Then, it initializes the DP table $\mathbf{T}_n$ of each leaf $n \in N$ with one element $\boldsymbol{m}_n = \langle \emptyset, \emptyset, \text{true} \rangle$, and constructs $\mathbf{T}_n$ of each internal $n \in N$ by the M-update in a DP manner. Finally, it constructs the root gate g that is a disjunction of all gates in $\mathbf{T}_r$ where r is the root of T, and returns C as a TLC of $f_{\mathcal{P}}(\boldsymbol{X}_G)$ where $\mathcal{P}$ is the target property.

The complexity of BBS depends on the concrete definition of the S-state and S-update. Let $f(n)$ be the domain size (the number of possible instantiations) of S-state $\boldsymbol{s}_n$ and $g(n)$ be the time complexity of the S-update for n. Then, the maximum number of elements in the DP table $\mathbf{T}_n$ is $O(f(n)2^{|U_n|+|F_n|})$ and the time complexity of constructing $\mathbf{T}_n$ is $O(g(n)f(m)2^{|U_m|+|F_m|})$ if type(n) $\neq$ J; otherwise $O(g(n)f(m)f(m')2^{|U_n|+|F_n|})$ where $m, m' \in$ cld(n). Let $f \triangleq \max\{f(n) \mid n \in N\}$, $g \triangleq \max\{g(n) \mid n \in N\}$, and $w \triangleq \max\{|U_n| + |F_n| \mid n \in N\}$. Then, BBS's space and time complexity is $O(|N|f2^w)$ and $O(|N|gf^2 2^w)$, respectively. Basically, $f(n)$ and $g(n)$ are functions of $|U_n|$ and $|F_n|$, not $|V|$ and $|E|$ (see examples of Sect. 4). In such cases, BBS is FPT of treewidth w_G; i.e., f, g, and w are regarded as constants if w_G is a constant.

3.4 Tractableness of Generated Circuits by BBS

Suppose C is the circuit generated by BBS and represents $f_{\mathcal{P}}(G)$ that is NOT a contradiction. We here prove that C is tractable.

Corollary 1. *For any $n \in N$, every instantiation of g_n in the DP table $\mathbf{T}_n$ is a Boolean function of $\boldsymbol{X}_n$ and contains at least one literal of each $X \in \boldsymbol{X}_n$.*

Proof. In BBS, each literal $g(X, x)$ is only introduced in Mupdate(n, | type(n)) cld(n) such that type(n) $\in$ {FV, FE}; namely, a literal of X_{v_n} (resp. X_{e_n}) is added to g_n at n if-and-only-if n forgets v_n (resp. e_n). Hence, by the definition of $\boldsymbol{X}_n$, the variable set of g_n must correspond to $\boldsymbol{X}_n$.

Corollary 2. *For any $n \in N$, instantiations of g_n in the DP table $\mathbf{T}_n$ are mutually exclusive.*

Proof. When n is a leaf, $\mathbf{T}_n$ has exactly one instantiation of g_n: $g_n = $ true. Hence, Corollary 2 holds for n of leaves. When n is an internal, we assume Corollary 2 holds for each $m \in \mathrm{cld}(n)$ and prove that it holds for n. When $\mathrm{type}(n) \in \{\mathrm{IV}, \mathrm{IE}\}$, instantiations of g_n are obviously exclusive because they are copies of g_m. When $\mathrm{type}(n) \in \{\mathrm{FV}, \mathrm{FE}\}$, for each g_m, BBS generates new gates $g = \mathrm{apply}(\otimes, g_m, g(X, x))$ where X corresponds to the forgotten vertex or edge at n and x is its realization. The new gates are mutually exclusive because all g_m are exclusive, and X is a new variable. BBS constructs an OR gate of two new gates if they correspond to the same x_{B_n} and s_n; i.e., every new gate g never becomes inputs of different OR gates. Hence, the OR gates must also be mutually exclusive. When $\mathrm{type}(n) = \mathrm{J}$, BBS generates new gates $g = \mathrm{apply}(\otimes, g_m, g'_m)$, where g_m and g'_m share no variable because $\mathbf{X}_m \cap \mathbf{X}_{m'} = \emptyset$; Hence, the new gates must be exclusive because all g_m (or g'_m) in them are mutually exclusive. Consequently, for each $n \in N$, instantiations of g_n are mutually exclusive by mathematical induction.

Corollary 3. *For any $g \in G_\oplus$, every $g' \in \mathrm{cld}(g)$ shares the same variable set.*

Proof. BBS operates $\mathrm{apply}(\oplus, g, g')$ in $\mathtt{Construct}(\mathrm{G})$ or $\mathtt{Mupdate}(n, \mid \mathrm{type}(n))$ $\mathrm{cld}(n)$ ($\mathrm{type}(n) \in \{\mathrm{FV}, \mathrm{FE}\}$) where g and g' are chosen from the same DP table. Corollary 1 guarantees that every gate in the same DP table $\mathbf{T}_n$ has the same variable set $\mathbf{X}_n$. Hence, the children of every $\oplus$-gate g share the same variable set $\mathbf{X}_g$.

Lemma 1. C *is smooth.*

Proof. For any $g \in G_\oplus$ and $g' \in \mathrm{cld}(g)$, $\mathbf{X}_g = \mathbf{X}_{g'}$ (i.e., $\mathcal{X}_g = \{\mathbf{X}_g\}$) holds by Corollary 3; hence, C is smooth.

Lemma 2. C *is deterministic.*

Proof. For any $g \in G_\oplus$, g' and g'' are mutually exclusive where $\mathrm{cld}(g) = \{g', g''\}$ by Corollary 2; hence, C is deterministic.

Lemma 3. C *is decomposable.*

Proof. In BBS, $\mathtt{Mupdate}(n, \mathrm{cld}(n) \mid \mathrm{type}(n))$ operates $\mathrm{apply}(\otimes, g, g')$ only when $\mathrm{type}(n) \in \{\mathrm{FV}, \mathrm{FE}, \mathrm{J}\}$. When $\mathrm{type}(n) \in \{\mathrm{FV}, \mathrm{FE}\}$, it operates $\mathrm{apply}(\otimes, g_m, g(X, x))$ where X is the variable corresponding to the forgotten vertex or edge at n. Because each vertex/edge is forgotten exactly once in a VNTD, g_m never contains X; i.e., the variable sets of g_m and $g(X, x)$ must be disjoint. When $\mathrm{type}(n) = \mathrm{J}$, it operates $\mathrm{apply}(\otimes, g_m, g_{m'})$ where $\mathrm{cld}(n) = \{m, m'\}$ and $\mathbf{X}_m$ and $\mathbf{X}_{m'}$ are disjoint; i.e., g_m and $g_{m'}$ must be disjoint. Hence, for any $g \in G_\otimes$, $\mathcal{X}_g$ is a partition of X_g; i.e., C is decomposable.

Lemma 4. C *is structured-decomposable.*

Proof. Let's consider the vtree S constructed by below:

Table 1. The list of properties, their variable sets, and their domains of S-states

Property $\mathcal{P}$	Variable Set	Domain of s_n
Independent Set (IS)	$\boldsymbol{X}_V$	$\{\emptyset\}$
k-edgeset (k-ES)	$\boldsymbol{X}_V$	$\{0, \ldots, k\}$
Dominating Set (DS)	$\boldsymbol{X}_V$	$\{0, 1\}^{U_n}$
k-matching (k-MT)	$\boldsymbol{X}_E$	$\{0, \ldots, k\}^{U_n}$
Spanning Tree (ST)	$\boldsymbol{X}_E$	$\{1, \ldots, |U_n|\}^{U_n}$

1. Let S be a copy of T and for any point $p \in P$, $n_p \in N$ be the original node in T corresponding to p.
2. Contract all $p \in P$ that $\text{type}(n_p) \in \{\text{IV}, \text{IE}\}$; i.e., $\forall p \in P, \text{type}(n_p) \in \{\text{FV}, \text{FE}, \text{J}\}$ holds on the resulting S.
3. For any $p \in P$ such that $\text{type}(n_p) \in \{\text{FV}, \text{FE}\}$, add a new child q to $\text{cld}(p)$ and correspond X to q where X is the variable of the forgotten vertex or edge at n_p.

For any $g \in G_\otimes$, let n_g be the node that g is generated. Then, each g respects some p such that $n_p = n_g$; namely, C respects the above vtree S. C is decomposable by Lemma 3. Hence, C is structured-decomposable.

Theorem 1. C *is tractable.*

Proof. It follows from Lemmas 1, 2, and 4.

4   BBS Examples

We show some examples of S-states and S-updates of some graph properties, including independent set (IS), spanning tree (ST), k-edgeset (k-ES; $U \subseteq V$ is a k-ES of G if $E[U] \leq k$), dominating set (DS; $U \subseteq V$ is a DS of G if $\forall v \in V \setminus U, \exists u \in U, \{u, v\} \in E$), and k-matching (k-MT; $F \subseteq E$ is a k-MT of G if $\forall u \in V[F]$, the degree of u on $G[F]$ is less than or equal to k).

Table 1 shows the variable set and the domain of the S-state for each property, and the following examples are the S-state and S-update of each property. In the following examples, s_n is initialized as a copy of s_m. Due to space limitations, pseudo-codes of S-updates for some examples are shown in the Appendix. We also omit the proof of the correctness of each specific algorithm because one can obtain it by slightly modifying the proof of the corresponding TD-based algorithm shown in [7].

Independent Set (IS): The pseudo-code of the modified $\mathtt{Supdate}(\boldsymbol{x}_{U_n}, \emptyset \mid \text{IE})$ for IS is shown in Appendix, and the other S-updates are the same as default. The S-state of

IS is $\emptyset$; i.e., the S-updates of IS require no additional information other than $\boldsymbol{x}_{U_n}$. When $e_n = \{u, v\}$, $\texttt{Supdate}(\boldsymbol{x}_{U_n}, \emptyset \mid \text{IE})$ returns $\bot$ if $\boldsymbol{x}_u = 1 \wedge \boldsymbol{x}_v = 1$: two vertices in U_n are connected.

k-**Edgeset** (k-**ES**): The pseudo-codes of the modified $\texttt{Supdate}(\boldsymbol{x}_{U_n}, \boldsymbol{s}_m \mid \text{IE})$ and $\texttt{Supdate}(\boldsymbol{x}_{U_n}, \boldsymbol{s}_m, \boldsymbol{s}_{m'} \mid \text{J})$ for k-ES are shown in Appendix, and the other S-updates are the same as default. The S-state of k-ES is an Integer $\boldsymbol{s}_n \in \{0, \ldots, k\}$ that indicates the current number of induced edges. Consider introducing $e_n = \{u, v\}$. When $\boldsymbol{x}_u = 0$ or $\boldsymbol{x}_v = 0$, e_n is not induced. When $\boldsymbol{x}_u = 1 \wedge \boldsymbol{x}_v = 1$, e_n is a new induced edge and $\texttt{Supdate}(\boldsymbol{x}_{U_n}, \boldsymbol{s}_m \mid \text{IE})$ increments $\boldsymbol{s}_n$ by one and returns $\boldsymbol{s}_n$ if $\boldsymbol{s}_n \leq k$; otherwise it returns $\bot$: pruning takes place. $\texttt{Supdate}(\boldsymbol{x}_{U_n}, \boldsymbol{s}_m, \boldsymbol{s}_{m'} \mid \text{J})$ computes $\boldsymbol{s}_n = \boldsymbol{s}_m + \boldsymbol{s}_{m'} - c$ where c is the number of induced edges in the current $\mathrm{B}_n[\boldsymbol{x}_{\mathrm{B}_n}]$ that corresponds to the number of doubly counted induced edges (i.e., Inclusion-Exclusion Principle). Then, it returns $\boldsymbol{s}_n$ if $\boldsymbol{s}_n \leq k$; otherwise it returns $\bot$.

Dominating Set (DS): The pseudo-code of the modified S-updates for DS are shown in Appendix. The S-state of DS is a Boolean array $\boldsymbol{s}_n \in \{0, 1\}^{U_n}$ where $\boldsymbol{s}_n[u] = 1$ ($u \in U_n$) indicates that u is *dominated*. $\texttt{Supdate}(\boldsymbol{x}_{U_n}, \boldsymbol{s}_m \mid \text{IV})$ initializes $\boldsymbol{s}_n[u_n] = 0$: u_n is not dominated yet. When $e_n = \{u, v\}$, $\texttt{Supdate}(\boldsymbol{x}_{U_n}, \boldsymbol{s}_m \mid \text{IE})$ sets $\boldsymbol{s}_n[u] = 1$ if $\boldsymbol{x}_v = 1$: u is dominated by v. $\texttt{Supdate}(\boldsymbol{x}_{U_n}, \boldsymbol{s}_m, \boldsymbol{s}_{m'} \mid \text{J})$ computes an element-wise OR of $\boldsymbol{s}_m$ and $\boldsymbol{s}_{m'}$ as a new $\boldsymbol{s}_n$. $\texttt{Supdate}(\boldsymbol{x}_{U_m}, \boldsymbol{s}_m \mid \text{FV})$ returns $\bot$ if $d_{u_n} = 0$: u_n is forgotten without dominated; otherwise it returns $\boldsymbol{s}_n$.

k-**Matching** (k-**MT**): The pseudo-codes of the modified S-updates for k-MT are shown in Appendix[1]. The S-tate of k-MT is an Integer array $\boldsymbol{s}_n \in \{0, \ldots, k\}^{U_n}$ where $\boldsymbol{s}_n[u]$ ($u \in U_n$) indicates the current degree of u. $\texttt{Supdate}(\boldsymbol{x}_{F_n}, \boldsymbol{s}_m \mid \text{IV})$ initializes $\boldsymbol{s}_n[u_n] = 0$: u_n is an isolated vertex. When $e_n = \{u, v\}$, $\texttt{Supdate}(\boldsymbol{x}_{F_n}, \boldsymbol{s}_m \mid \text{IE})$ increments $\boldsymbol{s}_n[u]$ and $\boldsymbol{s}_n[v]$ by one and returns $\boldsymbol{s}_n$ if $(\boldsymbol{s}_n[u] \leq k) \wedge (\boldsymbol{s}_n[v] \leq k)$; otherwise it returns $\bot$. $\texttt{Supdate}(\boldsymbol{x}_{F_n}, \boldsymbol{s}_m, \boldsymbol{s}_{m'} \mid \text{J})$ computes $\boldsymbol{s}_n[u] = \boldsymbol{s}_m[u] + \boldsymbol{s}_m[u] - d_u$ where d_u is the degree of u in $\mathrm{B}[E(\boldsymbol{x}_{F_n})]$ that corresponds to the number of doubly counted degree of u. $\texttt{Supdate}(\boldsymbol{x}_{F_m}, \boldsymbol{s}_m \mid \text{FV})$ returns $\boldsymbol{s}_n$ if $\forall u \in U_n, \boldsymbol{s}_n[u] \leq k$; otherwise it returns $\bot$.

Spanning Tree (ST): The pseudo-code of the modified S-updates for ST are shown in Appendix. The S-state of ST is an Integer array $\boldsymbol{s}_n \in \{0, \ldots, |U_n|\}^{U_n}$ that represents hidden connectivity on U_n (a partition of U_n) whose blocks correspond to connected components. $\texttt{Supdate}(\boldsymbol{x}_{F_n}, \boldsymbol{s}_m \mid \text{IV})$ adds u_n to $\boldsymbol{s}_n$ an isolated vertex. $\texttt{Supdate}(\boldsymbol{x}_{F_n}, \boldsymbol{s}_m \mid \text{IE})$ returns $\bot$ if $\mathrm{B}_n[E(\boldsymbol{x}_{F_n})]$ considering the hidden connectivity $\boldsymbol{s}_n$ has a cycle. When $e_n = \{u, v\}$ and $\boldsymbol{x}_{e_n} = 1$, $\texttt{Supdate}(\boldsymbol{x}_{\mathrm{B}_n}, \boldsymbol{s}_m \mid \text{FE})$ unites u and v on $\boldsymbol{s}_n$: the forgotten connection e_n is stored in $\boldsymbol{s}_n$. $\texttt{Supdate}(\boldsymbol{x}_{\mathrm{B}_n}, \boldsymbol{s}_m, \boldsymbol{s}_{m'} \mid \text{J})$ returns $\bot$ if $\mathrm{B}_n[E(\boldsymbol{x}_{U_n})]$ considering $\boldsymbol{s}_m$ and $\boldsymbol{s}_{m'}$ has a cycle; otherwise it returns the merge of $\boldsymbol{s}_m$ and $\boldsymbol{s}_{m'}$ as $\boldsymbol{s}_n$.

5   Experiments

We implemented our BBS and the conventional FBS and compared the size of TLCs and ZDDs. FBS is the most widely used method for constructing logical circuits of

---

[1] Appendix is provided on the author's website.

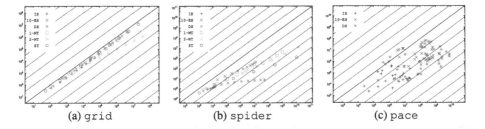

Fig. 1. The comparison results of the FBS and BBS. Each point corresponds to a pair of a graph G and a property $\mathcal{P}$ where its x-axis (resp. y-axis) indicates the size of the circuit of $f_{\mathcal{P}}(\boldsymbol{X}_{\mathrm{G}})$ generated by the FBS (resp. BBS); i.e., each point under the solid black line, indicating $y = x$, means that the BBS constructs a smaller circuit than the FBS. Note that both the x- and y-axis are logarithmic axes and each dotted black line indicates $y = 10^{\pm d}x$ where $d \in \{1, 2, \dots\}$.

$f_{\mathcal{P}}(\boldsymbol{X}_{\mathrm{G}})$ for various properties $\mathcal{P}$. FBS generally constructs a reduced ZDD that is a structured-decomposed and deterministic logical circuit but is non-smooth; however, we got smooth ZDDs in the experiment by omitting its reduction step. A ZDD is structured-decomposed, but its structure has a restriction that a vtree S respected by a ZDD must be *linear*: Each internal point $p \in P$ has two children, and at least one child is a leaf; hence, linear S has no join. The restriction is derived from FBS being based on PD, not TD. This experiment aims to see how this difference affects the sizes of the constructed TLCs and ZDDs.

5.1   Experimental Setting

We implemented the proposed BBS and the conventional FBS in C++ and built them with g++ 13.0.0 with the O3 option. We conducted experiments on a computer with an Intel Core i7 CPU (3.2 GHz), 64 GB main memory, and macOS Big Sur. We implemented specific FBSs for IS, k-ES, DS, k-MT, and ST using TdZdd [13].

We used a greedy heuristic search to find a TD/PD with a small width. We implemented three criteria for TD: degree, fillin, and degree+fillin used in [20], and five criteria for PD: DFS, BFS, NDS, LUD, and RFS used in [11]. We applied greedy heuristics with each criterion for each input graph, and employed the TD and PD with the smallest width in the experiment.

We constructed three datasets `grid`, `spider`, and `pace`. The `grid` consists of (w, h)-grids with $w = 1, \dots, 10$ and $h = 10$ where the (w, h)-grid is a grid graph with width w and height h; i.e., $w \times h$ vertices. The treewidth and pathwidth of the (w, h)-grid ($w \le h$) are both w, and the greedy heuristics obtained an optimal TD and PD. The `spider` consists of (k, ℓ)-spiders with $k = 1, \dots, 10$ and $\ell = 10$ where the (k, ℓ)-spider is a spider-like k-tree with $k + \ell^2$ vertices formed by the following steps: (1) let $C = \{1, \dots, k\}$, (2) let $U = C$, (3) add a new vertex u to U and make U a clique, (4) remove the oldest vertex from U, (5) repeat (3)-(4) ℓ times, and (6) repeat (2)-(5) ℓ times. The treewidth and pathwidth of the (k, ℓ)-spider ($k \le \ell$) are k and $2k$, respectively, and the greedy heuristics obtained an optimal TD and PD. The `pace` consists of some instances for the *exact treewidth challenge* in PACE2017 [10] where all instances are based on real-world data, and their true treewidth is known. We applied the greedy heuristics for each instance and selected ones whose smallest width of PD is less than or equal to 20 as pace. As a result, the pace consists of 30 real-world graphs, and the details of those graphs are given in Appendix.

5.2 Experimental Results

Figure 1 shows the comparison results of BBS and FBS. Each point corresponds to a pair of a graph and a property, and its x- and y-axis indicate the size of the constructed ZDD and TLC, respectively. We here define the size of a TLC as the number of generated wires by BBS and that of a ZDD as six times the number of generated vertices by FBS because one ZDD vertex is equivalent to three logical gates with two inputs (namely, six wires) [9]. Further details of the experiment are given in the Appendix.

We applied BBS and FBS to each (w, h)-grid and (k, ℓ)-spider with each properties IS, 10-ES, DS, 1-MT, 2-MT, and ST. In case (w, h)-grid, both BBS and FBS successfully constructed the TLC and ZDD for each w and property, and their size is shown in Fig. 1(a). Because the treewidth and pathwidth of the (w, h)-grid is both w, there was no significant difference between TLC and ZDD sizes; however, in the most significant case, ZDD was almost 30 times larger than TLC. In case (k, ℓ)-spider, FBS failed in constructing ZDDs for 2-MT with $k \geq 8$ and for ST with $k \geq 7$ due to lack of memory. Figure 1(b) shows the TLC and ZDD sizes except for the above failure cases. Because the pathwidth of the (k, ℓ)-spider is $2k$ while its treewidth is k, the ZDD size tends to be much larger than the TLC size, and in the most significant case, ZDD was almost 8,000 times larger than TLC. This result indicates that BBS is significantly more efficient than FBS when the treewidth of the input graphs is smaller than its pathwidth.

We applied BBS and FBS to every 30 graphs in pace with three properties IS, 10-ES, and DS, where we omitted k-MT and ST due to lack of memory, and the results are shown in Fig. 1(c). TLC was larger than ZDD only in eight cases of 90, and the most significant difference was less than three times. In contrast, TLC was smaller in the rest 82 cases, and ZDD was more than 6,000 times larger than TLC in the most significant case. This result suggested that BBS is significantly more efficient than FBS for real-world graphs.

6   Conclusion

We proposed BBS, an easy-to-implement and practically efficient meta-algorithm for constructing TLCs of various graph properties. BBS is based on DP on a VNTD of the input graph and constructs a TLC using two types of updates, M- and S-update. M-update is common to all properties, whereas S-update differs for each. We showed concrete BBS examples for various properties; independent set, k-edgeset, dominating set, k-matching, and spanning tree, and applied them to various graphs. The experimental results suggested that BBS constructs smaller circuits than FBS in many cases.

TLCs have been widely applied for modeling structured probabilistic space, and the computational costs of logical and probabilistic queries on TLCs are polynomial to their size. Hence, constructing smaller TLCs accelerates all applications that use such queries of TLCs; our BBS is expected to reduce the execution time of such applications.

Designing FBS for the target graph constraint is a thorny task because one has to design and implement its appropriate update that generally contains complex conditional branches and exception handling. BBS is also expected to alleviate the suffering because one only has to implement four types of S-updates that can be intuitively associated with four types of vertex of VNTD; i.e., our BBS is easy to design and implement.

References

1. Amarilli, A., Bourhis, P., Jachiet, L., Mengel, S.: A circuit-based approach to efficient enumeration. In: ICALP (2017)
2. Bannach, M., Berndt, S.: Practical access to dynamic programming on tree decompositions. In: ESA (2018)
3. Van den Broeck, G., Lykov, A., Schleich, M., Suciu, D.: On the tractability of SHAP explanations. In: AAAI (2021)
4. Choi, A., den Broeck, G.V., Darwiche, A.: Tractable learning for structured probability spaces: a case study in learning preference distributions. In: IJCAI (2015)
5. Choi, A., Shen, Y., Darwiche, A.: Tractability in structured probability spaces. In: NIPS (2017)
6. Choi, Y., Vergari, A., Van den Broeck, G.: Probabilistic circuits: A unifying framework for tractable probabilistic models. In: Technical report, UCLA (2020)
7. Cygan, M., et al.: Parameterized Algorithms. Springer (2015)
8. Darwiche, A.: SDD: a new canonical representation of propositional knowledge bases. In: IJCAI (2011)
9. Darwiche, A., Marquis, P.: A knowledge compilation map. J. Artif. Intell. Res. **17**, 229–264 (2002)
10. Dell, H., Husfeldt, T., Jansen, B.M.P., Kaski, P., Komusiewicz, C., Rosamond, F.A.: The first parameterized algorithms and computational experiments challenge. In: IPEC (2017)
11. Inoue, Y., Minato, S.: Acceleration of ZDD construction for subgraph enumeration via pathwidth optimization. In: TCS-TR-A16-80, Hokkaido University (2016)
12. Ishihata, M., Maehara, T.: Exact bernoulli scan statistics using binary decision diagrams. In: IJCAI (2019)
13. Iwashita, H.: Tdzdd: a top-down/breadth-first decision diagram manipulation framework. https://github.com/kunisura/TdZdd
14. Kawahara, J., Inoue, T., Iwashita, H., Minato, S.: Frontier-based search for enumerating all constrained subgraphs with compressed representation. IEICE Trans. **100-A**(9), 1773–1784 (2017)
15. Khosravi, P., Choi, Y., Liang, Y., Vergari, A., Van den Broeck, G.: On tractable computation of expected predictions. In: NeurIPS (2019)
16. Knuth, D.E.: The art of computer programming, volume 4A: combinatorial algorithms, part 1. Pearson Education India (2011)
17. Liu, A., Van den Broeck, G.: Tractable regularization of probabilistic circuits. In: NeurIPS (2021)
18. Liu, A., Mandt, S., Van den Broeck, G.: Lossless compression with probabilistic circuits. In: ICLR (2022)
19. Maehara, T., Suzuki, H., Ishihata, M.: Exact computation of influence spread by binary decision diagrams. In: WWW (2017)

20. Maniu, S., Senellart, P., Jog, S.: An experimental study of the treewidth of real-world graph data. In: ICDT (2019)
21. Minato, S.: Zero-suppressed BDDs for set manipulation in combinatorial problems. In: DAC (1993)
22. Nakahata, Y., Nishino, M., Kawahara, J., Ichi Minato, S.: Enumerating all subgraphs under given constraints using zero-suppressed sentential decision diagrams. In: SEA (2020)
23. Nishino, M., Inoue, T., Yasuda, N., Minato, S., Nagata, M.: Optimizing network reliability via best-first search over decision diagrams. In: INFOCOM (2018)
24. Nishino, M., Yasuda, N., Minato, S., Nagata, M.: Zero-suppressed sentential decision diagrams. In: AAAI (2016)
25. Nishino, M., Yasuda, N., Minato, S., Nagata, M.: Compiling graph substructures into sentential decision diagrams. In: AAAI (2017)
26. Oztok, U., Darwiche, A.: A top-down compiler for sentential decision diagrams. In: IJCAI (2015)
27. Selvam, N.R., Van den Broeck, G., Choi, Y.: Certifying fairness of probabilistic circuits. In: AAAI (2023)
28. Shih, A., Van den Broeck, G., Beame, P., Amarilli, A.: Smoothing structured decomposable circuits. In: NeurIPS (2019)
29. Sugaya, T., Nishino, M., Yasuda, N., Minato, S.: Tree decomposition-based approach for compiling independent sets. J. Inf. Process. **28**, 354–368 (2020)
30. Suzuki, H., Ishihata, M., Minato, S.: Designing survivable networks with zero-suppressed binary decision diagrams. In: WALCOM (2020)
31. Vergari, A., Choi, Y., Liu, A., Teso, S., Van den Broeck, G.: A compositional atlas of tractable circuit operations for probabilistic inference. In: NeurIPS (2021)

On Problems Related to Absent Subsequences

Zdenek Tronicek$^{(\boxtimes)}$ (iD)

State University of New York, Oneonta, NY, USA
zdenek.tronicek@oneonta.edu

Abstract. The paper introduces the absent subsequence automaton as a compact representation of shortest absent subsequences and minimal absent subsequences and describes its application to various related problems. It also reveals interesting combinatorial properties of minimal absent subsequences and derives an algorithm for computing the number of minimal absent subsequences.

Keywords: Absent subsequences · Subsequence automaton · Directed Acyclic Subsequence Graph · Distinguishing words

1 Introduction

A *subsequence* of a string T is any string that can be obtained by removing zero or more symbols from T. In other words, a string $S = s_1 s_2 \ldots s_m$ is a subsequence of a string $T = t_1 t_2 \ldots t_n$ if there exist indices $i_1, i_2, \ldots, i_m$ such that $1 \leq i_1 < i_2 < \ldots < i_m \leq n$ and $s_1 = t_{i_1}, s_2 = t_{i_2}, \ldots, s_m = t_{i_m}$. Problems related to subsequences have been researched for decades and there are dozens of papers on them (see a survey [14] by Kosche *et al.* for details).

One fundamental task related to subsequences is the subsequence matching problem: given two strings, a pattern S of length m and a text T of length n, we are to decide whether S is a subsequence of T. If S is a subsequence of T, we may also want to find indices $i_1, i_2, \ldots, i_m$ such that $s_1 = t_{i_1}, s_2 = t_{i_2}, \ldots, s_m = t_{i_m}$. One approach is to preprocess the pattern to a deterministic finite automaton that searches for a subsequence S. Another approach is to preprocess the text to a deterministic finite automaton that accepts all subsequences of T. This finite automaton was first mentioned by Hebrard and Crochemore [8] who also described a right-to-left algorithm for building. Baeza-Yates [1] called the finite automaton "Directed Acyclic Subsequence Graph" and described how it can be extended to multiple texts. Crochemore *et al.* [3] described a left-to-right algorithm for building the Directed Acyclic Subsequence Graph for one text as well as for multiple texts. The complete finite automaton for a text of length n has $n + 2$ states and can be built in $O(n\sigma)$ time where σ is the size of the alphabet. Hoshino *et al.* [9] described an online algorithm for building the subsequence automaton for multiple texts. Bille *et al.* [2] considered a compact representation of the subsequence automaton based on default transitions, which do not consume any input

W. Wu and J. Guo (Eds.): COCOA 2023, LNCS 14462, pp. 351–363, 2024.
https://doi.org/10.1007/978-3-031-49614-1_26

symbol. This clever representation provides a trade-off between the size of the finite automaton and the delay when processing the input symbols. If we accept the delay $O(\log \sigma)$, the size of the finite automaton can be reduced to $O(n \log \sigma)$. Concerning the size of the subsequence automaton for multiple texts, it has been proved that in the worst case, the number of states is exponential in the number of texts (see [4] and [17]).

An *absent subsequence* of a string T is any string that is not a subsequence of T. A *shortest absent subsequence* (SAS) of T is an absent subsequence of T of the minimum length. Let SAS(T) denote the set of shortest absent subsequences of T. A *minimal absent subsequence* (MAS) of T is an absent subsequence of T whose every subsequence is not absent. Let MAS(T) denote the set of minimal absent subsequences of T. A trivial fact is that SAS(T) $\subseteq$ MAS(T). As an example, let's consider $T = ababb$. Then SAS(T) = $\{aaa, baa, bba\}$ and MAS(T) = SAS(T) $\cup$ $\{aabbb, bbbb\}$. This shows that not every MAS is a SAS.

Absent subsequences were introduced by Kosche *et al.* [13] as an analogy to absent substrings. They defined minimal absent subsequences and shortest absent subsequences and showed several combinatorial and algorithmic results related to them. Their motivation to deal with absent subsequences stemmed from the problems related to Simon's congruence $\sim_k$, which is defined as follows: given two strings u and v, u is equivalent to v (we write $u \sim_k v$) if and only if u and v have the same set of subsequences of length at most k.

Fleischer and Kufleitner [5] researched the problem of testing whether two strings are $\sim_k$-equivalent. The problem is related to the notion of a distinguishing string of u and v, which is the shortest string that is a subsequence of one string and an absent subsequence of the other string. Gawrychowski *et al.* [7] investigated a similar problem: given two strings u and v, we are to find the largest k for which $u \sim_k v$. Kim *et al.* [10] showed how to build, for a given string u, a finite automaton that accepts any string congruent to u under Simon's congruence and designed an algorithm that, given a string u, an integer k, and a language L, decides whether there exists a string $v \in L$ such that $u \sim_k v$. Kim *et al.* [11] described how to find all substrings of a text that are congruent to a pattern under Simon's congruence.

Kosche *et al.* [12] considered subsequences in bounded ranges: given two strings u and v and an integer p, we are to decide whether there exists a substring of v of length p that contains u as a subsequence. If such a substring exists, we say that u is a *p-subsequence* of v. They extended the definition of absent subsequences to absent p-subsequences and researched problems related to them.

Garel [6] dealt with minimal separators of two strings. Given two strings u and v, a separator of u and v is a string that is a subsequence of exactly one of u and v. She described an algorithm for building a finite automaton that recognizes the shortest separators of two strings.

The scientific contribution of this paper is the introduction of the absent subsequence automaton as a compact representation of shortest absent subsequences and minimal absent subsequences, the description of combinatorial properties of minimal absent subsequences, and the derivation of a dynamic programming algorithm for computing the number of minimal absent subsequences.

The rest of the paper is organized as follows. Section 2 defines the absent subsequence automaton and its transition graph, Sect. 3 relates the transition graph to shortest absent subsequences, Sect. 4 relates the transition graph to minimal absent subsequences, reveals interesting combinatorial properties of minimal absent subsequences, and derives an algorithm for computing the number of minimal absent subsequences, Sect. 5 describes an application of absent subsequence automata to the problem of finding separators of two strings, and Sect. 6 concludes.

2 Absent Subsequence Automaton

A finite automaton is a 5-tuple $(Q, \Sigma, \delta, q_0, F)$, where Q is a finite set of states, Σ is an input alphabet, $\delta : Q \times \Sigma \rightarrow Q$ is a transition function, q_0 is the initial state, and $F \subseteq Q$ is the set of final states. The notation $[i, j]$ means the closed interval of integers i through j, σ is the size of Σ, and ε is the empty word.

Given a string $T = t_1 t_2 \ldots t_n$ over an alphabet Σ, the subsequence automaton for T is the deterministic finite automaton $(Q, \Sigma, \delta, q_0, F)$ where

- $Q = \{q_0, q_1, \ldots, q_{n+1}\}$,
- $F = \{q_0, q_1, \ldots, q_n\}$, and
- $\delta(q_i, a) = q_j$ where $j = \min(\{j \mid j > i \text{ and } a = t_j\} \cup \{n+1\})$ for each $a \in \Sigma$ and each $i \in [0, n+1]$.

The complete automaton has $n + 1$ final states and one nonfinal (sink) state. It can be built in $O(n\sigma)$ time and represented in $O(n\sigma)$ space.

Definition 1. *Given a string T over an alphabet Σ, the absent subsequence automaton for T is the deterministic finite automaton $(Q, \Sigma, \delta, q_0, F_a)$ where Q, Σ, and δ are defined the same way as in the subsequence automaton and $F_a = \{q_{n+1}\}$.*

Informally, we can get the absent subsequence automaton from the subsequence automaton by changing final states to nonfinal and vice-versa (see Fig. 1).

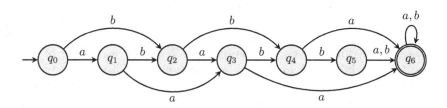

Fig. 1. The absent subsequence automaton for *ababb*.

Lemma 1. *The absent subsequence automaton for T accepts a string S if and only if S is an absent subsequence of T.*

Proof. This directly follows from the definition of the absent subsequence automaton. □

Definition 2. *Given a finite automaton $A = (Q, \Sigma, \delta, q_0, F)$, the transition graph of A is $G = (V, E)$ where $V = Q$ is a set of vertices and $E = \{(p, a, q) : \delta(p, a) = q, \forall p, q \in Q, a \in \Sigma\}$ is a set of edges.*

3   Shortest Absent Subsequences

Since the absent subsequence automaton for T accepts all absent subsequences of T, it must accept also the shortest absent subsequences of T and the minimal absent subsequences of T.

Theorem 1. *A string S is a SAS of T if and only if it is the labeling of a shortest path from q_0 to q_{n+1} in the transition graph of the absent subsequence automaton for T.*

Proof. ⇒: S is an absent subsequence and thus S must be the labeling of a path from q_0 to q_{n+1} in the transition graph. Let's assume there exists a path P from q_0 to q_{n+1} that is shorter than S. Then the labeling of this path is an absent subsequence that is shorter than S, which contradicts the assumption that S is a SAS of T.

⇐: S is the labeling of a path from q_0 to q_{n+1} and thus S is an absent subsequence. Let's assume there exists an absent subsequence P that is shorter than S. Then P must be accepted by the absent subsequence automaton and thus there must be a path from q_0 to q_{n+1} labeled with P in the transition graph. Since P is shorter than S, the path labeled with P must be shorter than the path labeled with S, which contradicts the assumption that S is the labeling of a shortest path from q_0 to q_{n+1}. □

Example 1. In the absent subsequence automaton in Fig. 1, there are three different shortest paths from q_0 to q_6. The labelings of these paths are *aaa*, *baa*, and *bba*, which are the shortest absent subsequences of *ababb*.

The transition graph of the absent subsequence automaton can be used to solve various problems related to absent subsequences. If we store at each vertex q_i the length of the shortest path from q_0 to q_i (it can be computed by breadth-first search), we can answer the question "Is $S = s_1 s_2 \ldots s_m$ a SAS of T?" as follows: we start in q_0 and go through a transition for s_1 if it goes to a vertex with a greater distance. Then we go through a transition for s_2 if it goes to a vertex with a greater distance, and so on until s_m. If we end up in q_{n+1}, we can conclude that S is a SAS. Otherwise, the last visited vertex determines the longest prefix of S that is also a prefix of a SAS.

Since the transition graph contains all shortest absent subsequences, we can use a graph traversal to count them and enumerate them. Counting the shortest absent subsequences of T requires $O(n\sigma)$ time and enumerating them requires $O(n\sigma + M)$ time, where M is the total length of shortest absent subsequences.

4   Minimal Absent Subsequences

Definition 3. *Let $T = t_1 t_2 \ldots t_n$ be a string over an alphabet Σ. The alphabet of T, denoted by $\Sigma(T)$, is a set of symbols used in T, i.e., $\Sigma(T) = \{a \in \Sigma \mid \exists i \in [1, n] : a = t_i\}$.*

The following lemma formalizes that we can find the alphabet of $t_i \ldots t_j$ by inspecting edges of q_{i-1} in the transition graph.

Lemma 2. *Let $T = t_1 t_2 \ldots t_n$ be a string over an alphabet Σ and G_T be the transition graph of the absent subsequence automaton for T. Then $\Sigma(t_i \ldots t_j) = \{a \in \Sigma \mid \exists k \in [i, j] : \delta(q_{i-1}, a) = q_k\}$ for $1 \leq i \leq j \leq n$.*

Proof. This directly follows from the definition of the transition function δ.   □

The ALPHABET function computes the alphabet of a string that is specified by two vertices p and r of the transition graph. If $p = q_i$ and $r = q_j$ for some $0 \leq i < j \leq n$, the function returns the alphabet of $t_i \ldots t_j$.

Algorithm 1 *

ALPHABET(p, r)

1: $alph = \emptyset$
2: **for** $a \in \Sigma$ **do**
3:     $q = \delta(p, a)$
4:     **if** $q.num \leq r.num$ **then**
5:         $alph = alph \cup \{a\}$
6:     **end if**
7: **end for**
8: **return** $alph$

Given a vertex v of the transition graph, let $level(v)$ denote the length of the longest simple path from q_0 to v. (A simple path contains each vertex at most once.)

Theorem 2. *Let $T = t_1 t_2 \ldots t_n$ be a string over an alphabet Σ and G_T be the transition graph of the absent subsequence automaton for T. Let $v_0, v_1, \ldots, v_m$ be the vertices of a simple path from q_0 to q_{n+1} in G_T and $S = s_1 s_2 \ldots s_m$ be the labeling of this path. Then S is a MAS of T if and only if $s_i \in \Sigma(t_{j+1} \ldots t_k)$ where $j = level(v_{i-2})$ and $k = level(v_{i-1})$ for each $i \in [2, m]$.*

Proof. $\Rightarrow$: Assume there exists $r \in [2, m]$ such that $s_r \notin \Sigma(t_{j+1} \ldots t_k)$ where $j = level(v_{r-2})$ and $k = level(v_{r-1})$. Let S' be a string derived from S by removing s_{r-1}, i.e., $S' = s_1 \ldots s_{r-2} s_r \ldots s_m$. The path $s_1 \ldots s_{r-2}$ goes to vertex v_{r-2}. Since $s_r \notin \Sigma(t_{j+1} \ldots t_k)$, the transition from v_{r-2} for s_r goes to v_r and the path continues to q_{n+1} the same way as for S. This means that S' is an absent subsequence of T, which contradicts the assumption that S is a MAS of T.

$\Leftarrow$: Since S is the labeling of a path from q_0 to q_{n+1}, it is an absent subsequence. To prove that it is minimal, we need to prove that any proper subsequence of S is a subsequence of T. Let S_m be a string derived from S by removing s_m, i.e., $S_m = s_1 s_2 \ldots s_{m-1}$. Since $v_0 v_1 \ldots v_m$ is a simple path and $v_m = q_{n+1}$, v_{m-1} is not q_{n+1}. S_m goes to v_{m-1} and thus S_m is a subsequence of T. Let S_{m-1} be a string derived from S by removing s_{m-1}, i.e., $S_{m-1} = s_1 s_2 \ldots s_{m-2} s_m$. The path $s_1 s_2 \ldots s_{m-2}$ goes to v_{m-2}. Since $s_m \in \Sigma(t_{j+1} \ldots t_k)$ where $j = level(v_{m-2})$ and $k = level(v_{m-1})$, the transition for s_m from v_{m-2} goes to q_ℓ where $\ell \in [j+1, k]$. Thus, S_{m-1} is a subsequence of T. Similarly, we can show that S_{m-2} is a subsequence of T, and so on. $\qquad\square$

Example 2. In the absent subsequence automaton in Fig. 1, there are five paths from q_0 to q_6 that satisfy conditions specified in theorem 2. The labelings of these paths are aaa, baa, bba, $bbbb$, and $aabbb$, which are the minimal absent subsequences of $ababb$.

Theorem 2 provides direction on how to find minimal absent subsequences in the transition graph. We can check whether a string is a MAS, find a longest MAS, and enumerate all minimal absent subsequences. The FIND-A-LONGEST-MAS algorithm is based on the well-known algorithm for the longest path in an acyclic graph, which begins with the topological sort. Since we know the topological ordering of vertices (it is $q_0, q_1, \ldots, q_{n+1}$), we can immediately process them in that order. There can be multiple longest minimal absent subsequences, but only one is found. The algorithm requires $O(n\sigma)$ time.

The following two lemmas show how minimal absent subsequences change when we remove the first symbol of T and when we append a new symbol at the end of T. This knowledge can be used to efficiently compute minimal absent subsequences for a sliding window (see also [16] for changes required in the subsequence automaton).

Lemma 3. *Let $T = t_1 t_2 \ldots t_n$ be a string over an alphabet Σ and $T' = t_2 t_3 \ldots t_n$ be T with the first symbol removed. Let $S = s_1 s_2 \ldots s_m$ be a MAS of T. If $s_1 = t_1$, $S' = s_2 s_3 \ldots s_m$ (i.e., S' is S with the first symbol removed) is a MAS of T'. If $s_1 \neq t_1$, S is a MAS of T' if either (i) $\delta(q_0, s_1) = q_{n+1}$ or (ii) $\delta(q_0, s_1) = q_k$ and $s_2 \in \Sigma(t_2 \ldots t_k)$ for some $k \in [2, n]$.*

Proof. S is a MAS of T and so there is a path labeled with S from q_0 to q_{n+1} in the transition graph G_T. If $s_1 = t_1$, there is a path labeled with $s_2 s_3 \ldots s_m$ from q_1 to q_{n+1}, which implies that $s_2 s_3 \ldots s_m$ is a MAS of T'. If $s_1 \neq t_1$, since q_0 and q_1 have transitions to the same vertex for each $a \in \Sigma \setminus \{t_1\}$, there is a path labeled with S from q_1 to q_{n+1}. Then, either (i) $m = 1$ or (ii) $m > 1$. If $m = 1$, there is a transition from q_1 to q_{n+1} labeled with s_1, which implies that S is a MAS of T'. If $m > 1$, according to Theorem 2, S is a MAS of T' if $s_2 \in \Sigma(t_2 \ldots t_k)$. $\qquad\square$

Example 3. Let $T = abab$ and $T' = bab$ be strings over alphabet $\Sigma = \{a, b\}$. The minimal absent subsequences of T are aaa, $aabb$, baa, bba, and bbb and the minimal absent subsequences of T' are aa, abb, bba, and bbb.

Algorithm 2 *

FIND-A-LONGEST-MAS(G)

```
 1: for i = 0 to n + 1 do
 2:     q_i.dist = 0
 3: end for
 4: q_0.pred = null
 5: for a ∈ Σ do
 6:     p = δ(q_0, a)
 7:     p.dist = 1
 8:     p.pred = q_0
 9:     p.symbol = a
10: end for
11: for i = 1 to n do
12:     Σ_i = ALPHABET(q_i.pred, q_i)
13:     for each a ∈ Σ_i do
14:         p = δ(q_i, a)
15:         if p.dist < q_i.dist + 1 then
16:             p.dist = q_i.dist + 1
17:             p.pred = q_i
18:             p.symbol = a
19:         end if
20:     end for
21: end for
22: lmas = null
23: p = q_{n+1}
24: while p.pred ≠ null do
25:     lmas = concat(p.symbol, lmas)
26:     p = p.pred
27: end while
28: return lmas
```

Lemma 4. *Let $T = t_1t_2 \ldots t_n$ be a string over an alphabet Σ and $T' = Tt_{n+1}$ be T with a symbol t_{n+1} appended. Let $S = s_1s_2 \ldots s_m$ be a MAS of T and $t_1t_2 \ldots t_{k-1}$ be the shortest prefix of T that contains $s_1s_2 \ldots s_{m-1}$ as a subsequence. If $s_m = t_{n+1}$, $S' = Sb$ (i.e., S' is S with a new symbol appended) is a MAS of T' for each $b \in \Sigma(t_k \ldots t_n)$. If $s_m \neq t_{n+1}$, S is a MAS of T'.*

Proof. S is a MAS of T and so there is a path labeled with S from q_0 to q_{n+1} in the transition graph G_T. If $s_m = t_{n+1}$, the path goes to q_{n+1} in the transition graph $G_{T'}$ and needs to be extended to q_{n+2}. According to Theorem 2, if $b \in \Sigma(t_k \ldots t_n)$, Sb is a minimal absent subsequence of T'. If $s_m \neq t_{n+1}$, the path goes to q_{n+2} in the transition graph $G_{T'}$ and so S is a MAS of T'.   □

Example 4. Let $T = abab$ and $T' = ababa$ be strings over alphabet $\Sigma = \{a, b\}$. The minimal absent subsequences of T are aaa, $aabb$, baa, bba, and bbb and the minimal absent subsequences of T' are $aaaa$, $aaab$, $aabb$, $baaa$, $baab$, $bbaa$, and bbb.

Let $\Sigma = \{a, b\}$ be an alphabet and A_n be a string over Σ defined as follows:

$$A_n = \begin{cases} (ab)^k, & \text{if } n = 2k, \\ (ab)^k a, & \text{if } n = 2k+1, \text{ for some integer } k \geq 0. \end{cases}$$

We are going to investigate the number of minimal absent subsequences of A_n. To simplify the notation, let's define a string B_n as A_{n+1} without the first a and let's denote $mas(A_n)$ the number of minimal absent subsequences of A_n and $mas_c(A_n)$ the number of minimal absent subsequences of A_n that begin with c, for some $c \in \Sigma$. Let's assume $n \geq 3$ and focus on the first three vertices of the transition graph of the absent subsequence automaton for A_n (see Fig. 2). If a MAS begins with a, we use the transition from q_0 to q_1 and then, since $\Sigma(a) = \{a\}$, we can continue from q_1 only with a. If a MAS begins with b, we use the transition from q_0 to q_2 and then, since $\Sigma(ab) = \{a, b\}$, we can continue from q_2 with a or b. Thus, $mas(A_n) = mas_a(B_{n-1}) + mas(A_{n-2})$. From q_1, we continue with a transition for a to q_3, and from q_3 we can continue with a or b. Therefore, $mas_a(B_{n-1}) = mas(B_{n-3})$. We get $mas(A_n) = mas(B_{n-3}) + mas(A_{n-2})$. From symmetry, $mas(A_n) = mas(B_n)$ and thus

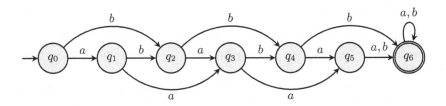

Fig. 2. The absent subsequence automaton for A_5.

$$mas(A_n) = mas(A_{n-3}) + mas(A_{n-2}).$$

This recurrence equation defines a sequence $2, 2, 3, 4, 5, 7, 9, 12, 16, \ldots$, which is the Padovan sequence (sequence A000931 in [15]) with offset 8. The recurrence equation can be solved by the characteristic root technique, which provides a solution in the form $mas(A_n) = \alpha r_1^n + \beta r_2^n + \gamma r_3^n$ where α, β, and γ are constants and r_1, r_2, and r_3 are roots of the characteristic equation $x^3 - x - 1 = 0$ (one of them is real and two are complex). The real root is known as the plastic number

$$\rho = \sqrt[3]{\frac{9 + \sqrt{69}}{18}} + \sqrt[3]{\frac{9 - \sqrt{69}}{18}} = 1.32471\ldots$$

and the complex roots are conjugates with an absolute value of less than 1. In other words, the number of minimal absent subsequences of A_n grows exponentially with n.

Similarly, we can derive equations for the number of minimal absent subsequences of a string over a general alphabet. Let $T = t_1 t_2 \ldots t_n$ be a string over an alphabet Σ and let $mas_{-a}(T)$ denote the number of minimal absent subsequences of T that end with a. Let's discuss how the minimal absent subsequences change when we append a symbol a at the end of T. A MAS of T that ends with a becomes a subsequence of Ta and needs to be extended to be a MAS of Ta. It can always be extended with a and sometimes also with other symbols. This implies that the number of minimal absent subsequences that end with a does not change:

$$mas_{-a}(Ta) = mas_{-a}(T).$$

A MAS of T that ends with $b \neq a$ is also a MAS of Ta. However, new minimal absent subsequences that end with b may appear. Let's denote $last(a)$ the last occurrence of symbol a in T:

$$last(a) = \max(\{i \mid 1 \leq i \leq n : a = t_i\} \cup \{0\}).$$

When we append a symbol a to T and there is an occurrence of b between the last occurrence of a and the end of T, a MAS of $t_1 \ldots t_{last(b)-1}$ that ends with a can be extended with b:

$$mas_{-b}(Ta) = \begin{cases} mas_{-b}(T) + mas_{-a}(t_1 \ldots t_{last(b)-1}), & \text{if } last(b) > last(a), \\ mas_{-b}(T), & \text{otherwise.} \end{cases}$$

These equations are used in the COUNT-MAS function, which computes the number of minimal absent subsequences by dynamic programming. The parameters of the function are T and its length.

Let $\Sigma = \{a_1, a_2, \ldots, a_\sigma\}$ be an alphabet and C_n be a string over Σ defined recurrently as follows:

$$C_n = \begin{cases} \varepsilon, & \text{if } n = 0, \\ a_1 a_2 \ldots a_n, & \text{if } 1 \leq n \leq \sigma, \\ a_1 a_2 \ldots a_\sigma a_\sigma a_{\sigma-1} \ldots a_{2\sigma-n+1}, & \text{if } \sigma < n \leq 2\sigma, \\ a_1 a_2 \ldots a_\sigma a_\sigma a_{\sigma-1} \ldots a_1 C_{n-2\sigma}, & \text{if } 2\sigma < n. \end{cases}$$

Example 5. Let's enumerate $C_0, C_1, \ldots, C_9$ over $\Sigma = \{a, b, c\}$: ε, a, ab, abc, $abcc$, $abccb$, $abccba$, $abccbaa$, $abccbaab$, $abccbaabc$.

We are going to investigate the set of minimal absent subsequences of C_n and show that if n is a multiple of σ (i.e., $n = k\sigma$ for some $k \geq 0$), all minimal absent subsequences are shortest absent subsequences. Let's sketch a proof by induction. Since $\text{MAS}(\varepsilon) = \{a_1, a_2, \ldots, a_\sigma\}$, it holds for $k = 0$. For the inductive step, let's investigate how minimal absent subsequences change when we append a symbol to C_n. Let's assume that C_n ends with a_1 (the case when C_n ends with a_σ is analogous). When we append a_1, the minimal absent subsequences that end with a_1 become subsequences of C_{n+1}, and all other minimal absent subsequences remain absent. Let's denote S a MAS of C_n that ends with a_1. In the transition graph for C_{n+1}, the path labeled with S goes to q_{n+1} with the

---

Algorithm 3 *

COUNT-MAS(T, n)

```
 1: for each a ∈ Σ do
 2:     last[a] = 0
 3: end for
 4: for each a ∈ Σ do
 5:     mas[a][0] = 1
 6: end for
 7: for i = 1 to n do
 8:     for each a ∈ Σ do
 9:         if a ≠ tᵢ and last[a] > last[tᵢ] then
10:             d = last[a] − 1
11:             mas[a][i] = mas[a][i − 1] + mas[tᵢ][d]
12:         else
13:             mas[a][i] = mas[a][i − 1]
14:         end if
15:     end for
16:     last[tᵢ] = i
17: end for
18: m = 0
19: for each a ∈ Σ do
20:     m = m + mas[a][n]
21: end for
22: return m
```

---

last transition from q_n to q_{n+1}. Since $\Sigma(a_1) = \{a_1\}$, we can continue from q_{n+1} only with a_1. Thus, by appending a_1 to C_n, the minimal absent subsequences that end with a_1 need to be extended with a_1. Let's continue and append a_2 to C_{n+1} and denote S' a MAS of C_{n+1} that ends with a_2. In the transition graph for C_{n+2}, the path labeled with S' goes to q_{n+2} with the last transition from q_{n-1}, q_n or q_{n+1}. Since $\Sigma(a_1 a_1 a_2) = \{a_1, a_2\}$, we can continue from q_{n+2} only with a_1 or a_2. Thus, by appending a_2 to C_{n+1}, the minimal absent subsequences that end with a_2 need to be extended with a_1 or a_2. We can append a_3 to C_{n+2} and show that the minimal absent subsequences that end with a_3 need to be extended with a_1, a_2 or a_3, and so on. Thus, when we append a new symbol, only the minimal absent subsequences that end with that symbol need to be extended and they are extended only with symbols that were already appended. In other words, when appending $a_i, 1 \leq i \leq \sigma$, we never extend a MAS that was extended when we appended $a_j, 1 \leq j < i$. So, when we append a_σ, all minimal absent subsequences of $C_{n+\sigma}$ were created by appending one symbol to a MAS of C_n and thus all minimal absent subsequences of $C_{n+\sigma}$ have the same length and are the shortest absent subsequences of $C_{n+\sigma}$.

5   Distinguishing Words

Let S and T be strings over an alphabet Σ. A string w distinguishes S and T if w is a subsequence of one and only one of S and T. A string w that distinguishes S and T is called a separator of S and T (see also [6]). We can use the absent subsequence automaton for S and the absent subsequence automaton for T to check whether a string w is a separator of S and T because each separator is accepted by exactly one of these two finite automata. We can also use the finite automata to build a finite automaton that accepts all separators of S and T.

Theorem 3. *Let S and T be strings over an alphabet Σ. Let $(Q_S, \Sigma, \delta_S, q_{S0}, F_S)$ be the absent subsequence automaton for S and $(Q_T, \Sigma, \delta_T, q_{T0}, F_T)$ be the absent subsequence automaton for T. We define Q, δ, q_0, and F as follows:*
$Q = \{(p_S, p_T) \mid p_S \in Q_S \text{ and } p_T \in Q_T\}$,
$\delta((p_S, p_T), a) = (\delta_S(p_S, a), \delta_T(p_T, a))$ for all $a \in \Sigma$,
$q_0 = (q_{S0}, q_{T0})$,
$F = \{(p_S, p_T) \mid p_S \in F_S \text{ and } p_T \notin F_T\} \cup \{(p_S, p_T) \mid p_S \notin F_S \text{ and } p_T \in F_T\}$.
Then a finite automaton $A_{sep} = (Q, \Sigma, \delta, q_0, F)$ accepts a string w if and only if w distinguishes S and T.

Proof. The theorem directly follows from the definition of a separator.   □

Since the number of states of the finite automaton accepting distinguishing strings of S and T is $\Omega(mn)$ in the worst case (see [4] and [17] for details), we can view the two absent subsequence automata as a compact representation of the finite automaton accepting all separators.

A string w that distinguishes S and T is a minimal separator of S and T if no proper subsequence of w is a separator of S and T. Informally, we can say that w is minimal if no string created from w by removing one or more symbols is a separator of S and T. A string w that distinguishes S and T is a shortest separator of S and T if no shorter separator of S and T exists. Since the finite automaton A_{sep} accepts all separators of S and T, it must also accept the shortest separators of S and T. Thus, we can find the shortest separators by a breadth-first search of the A_{sep} transition graph. The following lemma provides direction on how to find the shortest separators of S and T if we already know the minimal absent subsequences of S and T.

Lemma 5. *Let S and T be strings over an alphabet Σ. If w is a shortest separator of S and T then $w \in MAS(S) \cup MAS(T)$.*

Proof. Assume w is a subsequence of S and an absent subsequence of T. If $w \notin MAS(T)$, we can remove some symbols from w and get a shorter absent subsequence of T, which is still a subsequence of S. This contradicts the assumption that w is a shortest separator.   □

6 Conclusion

The paper has introduced the absent subsequence automaton as a compact representation of shortest absent subsequences and minimal absent subsequences and described how the finite automaton can facilitate solving various problems related to absent subsequences. It has also revealed an interesting connection between the number of minimal absent subsequences and the Padovan sequence and derived a dynamic programming algorithm for computing the number of minimal absent subsequences.

References

1. Baeza-Yates, R.A.: Searching subsequences. Theor. Comput. Sci. **78**(2), 363–376 (1991)
2. Bille, P., Gørtz, I.L., Skjoldjensen, F.R.: Subsequence automata with default transitions. J. Disc. Algor. **44**, 48–55 (2017)
3. Crochemore, M., Melichar, B., Troníček, Z.: Directed acyclic subsequence graph–overview. J. Disc. Algor. **1**(3–4), 255–280 (2003)
4. Crochemore, M., Troníček, Z.: On the size of DASG for multiple texts. In: Laender, A.H.F., Oliveira, A.L. (eds.) SPIRE 2002. LNCS, vol. 2476, pp. 58–64. Springer, Heidelberg (2002). https://doi.org/10.1007/3-540-45735-6_6
5. Fleischer, L., Kufleitner, M.: Testing Simon's congruence. In: International Symposium on Mathematical Foundations of Computer Science (MFCS), pp. 62:1–62:13 (2018)
6. Garel, E.: Minimal separators of two words. In: Apostolico, A., Crochemore, M., Galil, Z., Manber, U. (eds.) CPM 1993. LNCS, vol. 684, pp. 35–53. Springer, Heidelberg (1993). https://doi.org/10.1007/BFb0029795
7. Gawrychowski, P., Kosche, M., Koß, T., Manea, F., Siemer, S.: Efficiently testing Simon's congruence. In: International Symposium on Theoretical Aspects of Computer Science (STACS), pp. 34:1–34:18 (2021)
8. Hebrard, J.J., Crochemore, M.: Calcul de la distance par les sous-mots. RAIRO-Theor. Inf. Appl. **20**(4), 441–456 (1986)
9. Hoshino, H., Shinohara, A., Takeda, M., Arikawa, S.: Online construction of subsequence automata for multiple texts. In: Symposium on String Processing and Information Retrieval (SPIRE), pp. 146–152. IEEE (2000)
10. Kim, S., Han, Y.S., Ko, S.K., Salomaa, K.: On Simon's congruence closure of a string. In: Han, Y.S., Vaszil, G. (eds.) DCFS 2022. LNCS, vol. 13439, pp. 127–141. Springer, Heidelberg (2022). https://doi.org/10.1007/978-3-031-13257-5_10
11. Kim, S., Ko, S.K., Han, Y.S.: Simon's congruence pattern matching. In: International Symposium on Algorithms and Computation (ISAAC). Schloss Dagstuhl-Leibniz-Zentrum für Informatik (2022)
12. Kosche, M., Koß, T., Manea, F., Pak, V.: Subsequences in bounded ranges: matching and analysis problems. In: Lin, A.W., Zetzsche, G., Potapov, I. (eds.) RP 2022. LNCS, vol. 13608, pp. 140–159. Springer, Heidelberg (2022). https://doi.org/10.1007/978-3-031-19135-0_10
13. Kosche, M., Koß, T., Manea, F., Siemer, S.: Absent subsequences in words. In: Bell, P.C., Totzke, P., Potapov, I. (eds.) RP 2021. LNCS, vol. 13035, pp. 115–131. Springer, Cham (2021). https://doi.org/10.1007/978-3-030-89716-1_8

14. Kosche, M., Koß, T., Manea, F., Siemer, S.: Combinatorial algorithms for subsequence matching: a survey. In: International Workshop on Non-Classical Models of Automata and Applications (NCMA), vol. 367, pp. 11–27 (2022)

15. Sloane, N.J., et al.: The on-line encyclopedia of integer sequences (2018). https://oeis.org/

16. Troníček, Z.: Operations on DASG. In: Champarnaud, J.-M., Ziadi, D., Maurel, D. (eds.) WIA 1998. LNCS, vol. 1660, pp. 82–91. Springer, Heidelberg (1999). https://doi.org/10.1007/3-540-48057-9_7

17. Troníček, Z., Shinohara, A.: The size of subsequence automaton. Theor. Comput. Sci. **341**(1–3), 379–384 (2005)

Some Combinatorial Algorithms on the Dominating Number of Anti-rank k Hypergraphs

Zhuo Diao[1(✉)] and Zhongzheng Tang[2]

[1] School of Statistics and Mathematics, Central University of Finance and Economics, Beijing 100081, China
diaozhuo@amss.ac.cn
[2] School of Science, Beijing University of Posts and Telecommunications, Beijing 100876, China
tangzhongzheng@amss.ac.cn

Abstract. Given a hypergraph $H(V, E)$, a set of vertices $S \subseteq V$ is a dominating set if every vertex $v \in V \setminus S$ is adjacent to at least one vertex in S. The dominating number $\gamma(H)$ is the minimum cardinality of a dominating set in H. H is anti-rank k if each edge contains at least k vertices. In this paper, we prove some upper bounds of dominating number for anti-rank k hypergraphs without isolated vertices. (i) For anti-rank 3 hypergraphs, $\gamma(H) \leq \frac{n}{3}$. (ii) For anti-rank 4 hypergraphs, $\gamma(H) \leq \frac{n}{4}$. (iii) For anti-rank k hypergraphs, $\gamma(H) \leq \frac{n}{k}$ for many special hypergraphs. (iv) For the classical random hypergraph model $\mathcal{H}(n, p)$ and $H \in \mathcal{H}(n, p)$, for any positive number $\varepsilon > 0$, $\gamma(H) \leq (1 + \varepsilon)\frac{n}{k}$ holds with high probability. These results are a generalization of Ore's Theorem on simple graphs, which states $\gamma(G) \leq \frac{n}{2}$.

Keywords: anti-rank k hypergraphs · dominating number · upper bound

1 Introduction

A hypergraph is a generalization of a graph in which an edge can join any number of vertices. A simple hypergraph is a hypergraph without multiple edges. Let $H = (V, E)$ be a simple hypergraph with vertex set V and edge set E. As for a graph, the order of H, denoted by n, is the number of vertices. The number of edges is denoted by m.

For each vertex $v \in V$, the degree $d(v)$ is the number of edges in E that contains v. We say v is an isolated vertex of H if $d(v) = 0$. Hypergraph H is k-regular if each vertex's degree is k ($d(v) = k, \forall v \in V$). The maximum degree of H is $\Delta(H) = \max_{v \in V} d(v)$. The minimum degree of H is $\delta(H) = \min_{v \in V} d(v)$.

Supported by National Natural Science Foundation of China under Grant No.11901605, No.12101069, the disciplinary funding of Central University of Finance and Economics.

Hypergraph H is **anti-rank** k if each edge contains at least k vertices ($|e| \geq k$, $\forall e \in E$). Hypergraph H is k-uniform if each edge contains exactly k vertices ($|e| = k$, $\forall e \in E$).

Let $k \geq 2$ be an integer. A cycle of length k, denoted as k-cycle, is a vertex-edge sequence $C = v_1 e_1 v_2 e_2 \cdots v_k e_k v_1$ with: (1)$\{e_1, e_2, \ldots, e_k\}$ are distinct edges of H. (2)$\{v_1, v_2, \ldots, v_k\}$ are distinct vertices of H. (3)$\{v_i, v_{i+1}\} \subseteq e_i$ for each $i \in [k]$, here $v_{k+1} = v_1$. We consider the cycle C as a subhypergraph of H with vertex set $\{v_i, i \in [k]\}$ and edge set $\{e_j, j \in [k]\}$. Similarily, a path of length k, denoted as k-path, is a vertex-edge sequence $P = v_1 e_1 v_2 e_2 \cdots v_k e_k v_{k+1}$ with: (1)$\{e_1, e_2, \ldots, e_k\}$ are distinct edges of H. (2)$\{v_1, v_2, \ldots, v_{k+1}\}$ are distinct vertices of H. (3)$\{v_i, v_{i+1}\} \subseteq e_i$ for each $i \in [k]$. We consider the path P as a sub-hypergraph of H with vertex set $\{v_i, i \in [k+1]\}$ and edge set $\{e_j, j \in [k]\}$. A hypergraph $H = (V, E)$ is called *connected* if any two of its vertices are linked by a path in H. A hypergraph $H = (V, E)$ is called a *hyertree* if H is connected and acyclic, not containing any cycles, denoted by $T(V, E)$.

For any vertex set $S \subseteq V$, we write $H \backslash S$ for the subhypergraph of H obtained from H by deleting all vertices in S and all edges incident with some vertices in S. For any edge set $A \subseteq E$, we write $H \backslash A$ for the subhypergraph of H obtained from H by deleting all edges in A and keeping vertices. If S is a singleton set $\{s\}$, we write $H \backslash s$ instead of $H \backslash \{s\}$.

Given a hypergraph $H(V, E)$, some classical parameters are listed as follows:

- A set of vertices $S \subseteq V$ is a dominating set if every vertex $v \in V \backslash S$ is adjacent to at least one vertex in S. The dominating number $\gamma(H)$ is the minimum cardinality of a dominating set in H.
- A set of vertices $S \subseteq V$ is an independent set if every two vertices in S, there is no edge connecting the two. The independent number $\alpha(H)$ is the maximum cardinality of an independent set in H.
- A set of vertices $S \subseteq V$ is a transversal if every edge is incident with at least a vertex in S. The transversal number is the minimum cardinality of a transversal, denoted by $\tau(H)$.
- A set of edges $A \subseteq E$ is a matching if every two distinct edges have no common vertex. A perfect matching is a matching that covers every vertex of the hypergraph. The matching number is the maximum cardinality of a matching, denoted by $\nu(H)$.

In this paper, we focus on the dominating number in anti-rank k hypergraphs.

1.1   Related Works

Dominations in hypergraphs are well studied in the literatures [1–15].

Ore [14] established the following upper bound on the domination number of a simple graph in terms of its order.

Theorem 1. *(Ore's Theorem) If G is an isolated-free graph of order n, then $\gamma(G) \leq \frac{n}{2}$.*

There are many different versions of proofs for Ore's Theorem. An Ore's lemma states that the complement $V \setminus D$ of any minimal dominating set D in an isolate-free graph $G(V, E)$ is also a dominating set. The Ore's Theorem is an immediate consequence of this lemma. The Ore's lemma also applies to hypergraphs and the Ore's Theorem on hypergraphs is stated as follows:

Theorem 2. *If $H(V, E)$ is an isolated-free hypergraph of order n, then $\gamma(H) \leq \frac{n}{2}$.*

It is interesting to ask the question: $H(V, E)$ is a k-uniform hypergraph without isolated vertices. Does $\gamma(H) \leq \frac{n}{k}$ hold? For $k = 2$, the result is the Ore's Theorem. C. Bujtás et al. [13] proved the following theorem and its corollaries, which verifies the question for $k \in \{3, 4\}$. For $k = 5$, C. Bujtás et al. constructed a 5-uniform hypergraph H of order n without isolated vertices satisfying $\gamma(H) = \frac{2n}{9} > \frac{n}{5}$. There are some counterexamples for $k \geq 5$, but it is reasonable to believe $\gamma(H) \leq \frac{n}{k}$ for almost all k-uniform isolated-free hypergraphs.

Theorem 3. *If H is a k-uniform hypergraph of order n and size m without isolated vertices and $k \geq 3$. Then $\gamma(H) \leq \frac{n + \lfloor \frac{k-3}{2} \rfloor m}{\lfloor \frac{3(k-1)}{2} \rfloor}$ and this bound is sharp.*

Corollary 1. *For $k \in \{3, 4\}$. Let H is a k-uniform hypergraph of order n without isolated vertices. Then $\gamma(H) \leq \frac{n}{k}$.*

1.2   Our Results

In this paper, for anti-rank k isolated-free hypergraph $H(V, E)$, some combinatorial algorithms on the dominating number are designed:

- In Sect. 2, for isolated-vertex-free anti-rank 3 hypergraph $H(V, E)$, a combinatorial algorithm is designed to get a dominating set of no more than $n/3$.
- In Sect. 3, for isolated-vertex-free anti-rank 4 hypergraph $H(V, E)$, a combinatorial algorithm is designed to get a dominating set of no more than $n/4$.
- In Sect. 4, for isolated-vertex-free anti-rank k hypergraphs, we prove that the dominating number is no more than n/k in many special hypergraphs.
- In Sect. 5, for the classical random hypergraph model $\mathcal{H}(n, p)$ and $H \in \mathcal{H}(n, p)$, given any positive number $0 < \varepsilon < 1$, we prove $\gamma(H) \leq (1 + \varepsilon)\frac{n}{k}$ holds with high probability.

2   The Anti-rank 3 Hypergraphs

In this section, we focus on isolated-vertex-free anti-rank 3 hypergraphs with n vertices. We introduce an algorithm designed to identify a dominating set of size no greater than $n/3$. Algorithm 1 is directly implied by Theorem 4.

Theorem 4. *$H(V, E)$ is an isolated-vertex-free anti-rank 3 hypergraph. Then the dominating number $\gamma(H) \leq \frac{n}{3}$ where n is the number of vertices.*

Proof. We prove the theorem by contradiction. Take out a counterexample with the minimum number of edges. $H(V, E)$ is an isolated-vertex-free anti-rank 3 hypergraph with order n, $\gamma(H) > \frac{n}{3}$. Let us break the proof into a series of claims.

Claim 1. *Every edge contains at least a vertex with degree one.*

If there is an edge $e \in E$ and every vertex $v \in e$ has degree at least two. Then $H \setminus e$ is also an isolated-vertex-free anti-rank 3 hypergraph with order n. Because H is a counterexample with the minimum number of edges, we have $\gamma(H \setminus e) \leq \frac{n}{3}$. $H \setminus e$ is a spanning subhypergraph of H. Thus $\gamma(H) \leq \gamma(H \setminus e) \leq \frac{n}{3}$. This is a contradiction with $\gamma(H) > \frac{n}{3}$.

Claim 2. *For every vertex $v \in V$, the degree of v is at most one.*

If there is a vertex $u \in V$ with $d(u) \geq 2$. Take out arbitrarily two edges e_1, e_2 adjacent to u. According to Claim 1, every edge e_i contains at least a vertex v_i with degree one, as shown in Fig. 1. v_1, v_2 are two 1-degree vertices in H, thus $H \setminus u$ has at least two isolated vertices v_1, v_2. Deleting all the isolated vertices in $H \setminus u$ and denote the subhypergraph as H'. Because u, v_1, v_2 are deleted, the order of H' is at most $n - 3$. Because H is a counterexample with the minimum number of edges, we have $\gamma(H') \leq \frac{n-3}{3}$. Take out a minimum dominating set S' of H', then $S' \cup \{u\}$ is a dominating set of H. Thus we have $\gamma(H) \leq \gamma(H') + 1 \leq \frac{n-3}{3} + 1 \leq \frac{n}{3}$. This is a contradiction with $\gamma(H) > \frac{n}{3}$.

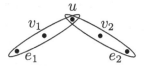

Fig. 1. Schematic diagram of the proof of Claim 2

According to Claim 2, each component of H is a single edge and every edge contains at least three vertices. Obviously $\gamma(H) \leq \frac{n}{3}$. This is a contradiction with $\gamma(H) > \frac{n}{3}$. □

For 3-uniform hypergraphs, according to Theorem 4, the next corollary is instant:

Corollary 2. $H(V, E)$ *is an isolated-vertex-free 3-uniform hypergraph with order n. Then $\gamma(H) \leq \frac{n}{3}$.*

Remark 1. Theorem 4 implies a recursive combinatorial algorithm, as demonstrated in Algorithm 1. The core idea of this recursive algorithm is that if we can find a dominating set of size no more than $n'/3$ in a small hypergraph with order n', then a dominating set of size no more than $n/3$ can be obtained in a larger hypergraph with order n. This algorithm involves two basic operations: edge deletion and vertex deletion.

Algorithm 1. Dominating set of anti-rank 3 hypergraphs

Input: An isolated-vertex-free anti-rank 3 hypergraph H with n vertices.

Output: A dominating set with size no more than $\frac{n}{3}$.

1: **while** there is an edge e with $d(v) \geq 2$ for each $v \in e$ **do**

2:    $H \leftarrow H \backslash e$

3: **if** H contains a vertex u with $d(u) \geq 2$ **then**

4:    Denote the isolated vertex set of $H \backslash u$ as I.

5:    **return** ALG1$(H \backslash u \backslash I) \cup \{u\}$

6: **else**

7:    Let D be the vertex set formed by selecting a vertex from each edge.

8:    **return** D

---

3   The Anti-rank 4 Hypergraphs

In this section, we examine isolated-vertex-free anti-rank 4 hypergraphs containing n vertices. We present an algorithm to identify a dominating set whose size does not exceed $n/4$. Algorithm 2 directly follows from Theorem 5.

Theorem 5. *$H(V, E)$ is an isolated-vertex-free anti-rank 4 hypergraph. Then the dominating number $\gamma(H) \leq \frac{n}{4}$ where n is the number of vertices.*

Proof. We prove the theorem by contradiction. Take out a counterexample with the minimum number of edges. $H(V, E)$ is an isolated-vertex-free anti-rank 4 hypergraph with order n, $\gamma(H) > \frac{n}{4}$. Let us break the proof into a series of claims.

Claim 3. *Every edge contains at least a vertex with degree one.*

If there is an edge $e \in E$ and every vertex $v \in e$ has degree at least two. Then $H \backslash e$ is also an isolated-vertex-free anti-rank 4 hypergraph with order n. Because H is a counterexample with the minimum number of edges, we have $\gamma(H \backslash e) \leq \frac{n}{4}$. $H \backslash e$ is a spanning subhypergraph of H. Thus $\gamma(H) \leq \gamma(H \backslash e) \leq \frac{n}{4}$. This is a contradiction with $\gamma(H) > \frac{n}{4}$.

Claim 4. *For every vertex $v \in V$, the degree of v is at most two.*

If there is a vertex $u \in V$ with $d(u) \geq 3$. Take out arbitrarily two edges e_1, e_2, e_3 adjacent to u. According to Claim 3, every edge e_i contains at least a vertex v_i with degree one, as shown in Fig. 2. v_1, v_2, v_3 are three 1-degree vertices in H, thus $H \backslash u$ has at least three isolated vertices v_1, v_2, v_3. Deleting all the isolated vertices in $H \backslash u$ and denote the subhypergraph as H'. Because u, v_1, v_2, v_3 are deleted, the order of H' is at most $n - 4$. Because H is a counterexample with the minimum number of edges, we have $\gamma(H') \leq \frac{n-4}{4}$. Take out a minimum

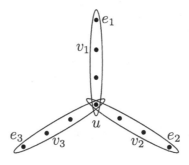

Fig. 2. Schematic diagram of the proof of Claim 4

dominating set S' of H', then $S' \cup \{u\}$ is a dominating set of H. Thus we have $\gamma(H) \leq \gamma(H') + 1 \leq \frac{n-4}{4} + 1 \leq \frac{n}{4}$. This is a contradiction with $\gamma(H) > \frac{n}{4}$.

Claim 5. *H has some cycles.*

Assume H is acyclic, by König Property, the transversal number is equal to the matching number, thus $\tau(H) = \nu(H)$. $H(V, E)$ is isolated-vertex-free, any transversal is also a dominating set, thus $\gamma(H) \leq \tau(H)$. Every edge contains at least four vertices, thus $\nu(H) \leq \frac{n}{4}$. Above all, we have $\gamma(H) \leq \tau(H) = \nu(H) \leq \frac{n}{4}$, a contradiction with $\gamma(H) > \frac{n}{4}$.

Take out a minimum cycle $C = v_1 e_1 v_2 ... v_t e_t v_1$ in H, here $\{v_i, 1 \leq i \leq t\}$ are distinct vertices, $\{e_i, 1 \leq i \leq t\}$ are distinct edges. The vertices v_i, v_{i+1} are adjacent to edge e_i for each $i \in [t]$. C is a minimum cycle in H, thus for each nonadjacent edges pair $\{e_i, e_j\}$ in C, e_i and e_j have no common vertices.

Claim 6. *Each edge e_i in the cycle C contains exactly one vertex with degree 1.*

Assume e_i has more than one vertex with degree 1. By Claim 3, u, v are two 1-degree vertices in e_i and w is a 1-degree vertex in e_{i+1}, as shown in Fig. 3. u, v, w are three 1-degree vertices in H, thus $H \setminus v_{i+1}$ has at least three isolated vertices u, v, w. Deleting all the isolated vertices in $H \setminus v_{i+1}$ and denote the subhypergraph as H'. Because u, v, w, v_{i+1} are deleted, the order of H' is at most $n - 4$. Because H is a counterexample with the minimum number of edges, we have $\gamma(H') \leq \frac{n-4}{4}$. Take out a minimum dominating set S' of H', then $S' \cup \{v_{i+1}\}$ is a dominating set of H. Thus we have $\gamma(H) \leq \gamma(H') + 1 \leq \frac{n-4}{4} + 1 \leq \frac{n}{4}$. This is a contradiction with $\gamma(H) > \frac{n}{4}$.

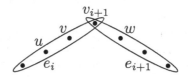

Fig. 3. Schematic diagram of the proof of Claim 6

Claim 7. *The length of C is not even.*

Assume C is an even cycle with length $t = 2k$. The vertices $\{v_i, 1 \leq i \leq 2k\}$ can be partitioned into two sets $V_o = \{v_i \mid 1 \leq i \leq 2k, \ i \equiv 1 \pmod 2\}$ and $V_e = \{v_i \mid 1 \leq i \leq 2k, \ i \equiv 0 \pmod 2\}$, here $|V_o| = |V_e| = k$. By Claim 3, for $1 \leq i \leq 2k$, u_i is a 1-degree vertex in e_i, as shown in Fig. 4. Consider the subhypergraph $H \setminus V_o$:

- In $H \setminus V_o$, all the edges $\{e_i \mid 1 \leq i \leq 2k\}$ of C are deleted, thus $\{u_i \mid 1 \leq i \leq 2k\}$ are isolated vertices.
- In $H \setminus V_{odd}$, by Claim 4, all the vertices in V_e are isolated vertices.

In $H \setminus V_o$, V_o are deleted, V_e and $\{u_i \mid 1 \leq i \leq 2k\}$ are isolated vertices. Deleting all the isolated vertices in $H \setminus V_o$ and denote the subhypergraph as H'. Thus the order of H' is at most $n - 4k$. Because H is a counterexample with the minimum number of edges, we have $\gamma(H') \leq \frac{n-4k}{4}$. Take out a minimum dominating set S' of H', then $S' \cup V_o$ is a dominating set of H. Thus we have $\gamma(H) \leq \gamma(H') + k \leq \frac{n-4k}{4} + k \leq \frac{n}{4}$. This is a contradiction with $\gamma(H) > \frac{n}{4}$.

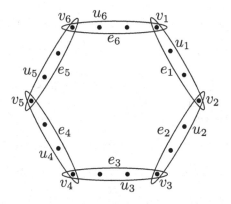

Fig. 4. Schematic diagram of the proof of Claim 7

Claim 8. *The length of C is not odd.*

Assume C is an odd cycle with length $t = 2k + 1$. The vertices $\{v_i \mid 1 \leq i \leq 2k+1\}$ can be partitioned into two sets $V_o = \{v_i \mid 1 \leq i \leq 2k+1, \ i \equiv 1 \pmod 2\}$ and $V_e = \{v_i \mid 1 \leq i \leq 2k+1, \ i \equiv 0 \pmod 2\}$, here $|V_o| = k + 1, |V_e| = k$. By Claim 3, for $1 \leq i \leq 2k + 1$, u_i is a 1-degree vertex in e_i. By Claim 6, in the edge e_{2k+1}, the vertices other than u_{2k+1} are two degree. e_{2k+1} has at least four vertices, we can take out a 2-degree vertex u in e_{2k+1} with $u \notin \{v_{2k}, v_{2k+1}\}$. e is the edge adjacent to u other than e_{2k+1} and v is a 1-degree vertex in e, as shown in Fig. 5. Consider the subhypergraph $H \setminus (V_e \cup \{u\})$:

- In $H \setminus (V_e \cup \{u\})$, all the edges $\{e_i \mid 1 \leq i \leq 2k+1\}$ of C are deleted, thus $\{u_i \mid 1 \leq i \leq 2k+1\}$ are isolated vertices.
- In $H \setminus (V_e \cup \{u\})$, by Claim 4, all the vertices in V_o are isolated vertices.
- In $H \setminus (V_e \cup \{u\})$, e is deleted and v is an isolated vertex.

In $H \setminus (V_e \cup \{u\})$, V_e and u are deleted, V_o, $\{u_i \mid 1 \leq i \leq 2k+1\}$ and v are isolated vertices. Deleting all the isolated vertices in $H \setminus (V_e \cup \{u\})$ and denote the subhypergraph as H'. Thus the order of H' is at most $n - 4k - 4$. Because H is a counterexample with the minimum number of edges, we have $\gamma(H') \leq \frac{n-4k-4}{4}$. Take out a minimum dominating set S' of H', then $S' \cup V_e \cup \{u\}$ is a dominating set of H. Thus we have $\gamma(H) \leq \gamma(H') + k + 1 \leq \frac{n-4k-4}{4} + k + 1 \leq \frac{n}{4}$. This is a contradiction with $\gamma(H) > \frac{n}{4}$.

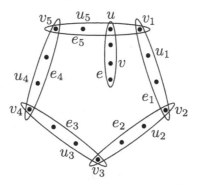

Fig. 5. Schematic diagram of the proof of Claim 8

According to Claim 5, Claim 7 and Claim 8, there is a contradiction. Thus our assumption doesn't hold and the theorem is proved.   □

For 4-uniform hypergraphs, according to Theorem 5, the next corollary is instant:

Corollary 3. $H(V, E)$ *is an isolated-vertex-free 4-uniform hypergraph with order n. Then $\gamma(H) \leq \frac{n}{4}$.*

Remark 2. Theorem 5 implies a recursive combinatorial algorithm, as demonstrated in Algorithm 2. The core idea of this recursive algorithm is that if we can find a dominating set of size no more than $n'/4$ in a small hypergraph with order n', then a dominating set of size no more than $n/4$ can be obtained in a larger hypergraph with order n. This algorithm involves three basic operations: edge deletion, vertex deletion, and cycle deletion.

Algorithm 2. Dominating set of anti-rank 4 hypergraphs

Input: An isolated-vertex-free anti-rank 4 hypergraph H with n vertices.

Output: A dominating set with size no more than $\frac{n}{4}$.

1: **while** there is an edge e with $d(v) \geq 2$ for each $v \in e$ **do**

2:     $H \leftarrow H \backslash e$

3: **if** H contains a vertex u with $d(u) \geq 3$ **then**

4:     Denote the isolated vertex set of $H \backslash u$ as I_1.

5:     **return** ALG2($H \backslash u \backslash I_1$)$\cup\{u\}$

6: **if** H is acyclic **then**

7:     Find a minimum transversal set S by Algorithm 3 in [16].

8:     **return** S

9: Find a minimum cycle $C = v_1 e_1 v_2 \cdots v_t e_t v_1$ in H. Set $V_o = \{v_i \mid 1 \leq i \leq t, \ i \equiv 1 \ (\mathrm{mod}\ 2)\}$ and $V_e = \{v_i \mid 1 \leq i \leq t, \ i \equiv 0 \ (\mathrm{mod}\ 2)\}$.

10: **if** there exists $e_i \in C$ containing at least 2 vertices with degree 1 **then**

11:     Denote the isolated vertex set of $H \backslash v_{i+1}$ as I_2.

12:     **return** ALG2($H \backslash v_{i+1} \backslash I_2$)$\cup\{v_{i+1}\}$

13: **if** C is a even cycle **then**

14:     Denote the isolated vertex set of $H \backslash V_o$ as I_3.

15:     **return** ALG2($H \backslash V_o \backslash I_3$)$\cup V_o$

16: **else**

17:     Take $u \notin \{v_{t-1}, v_t\}$ with degree 2 from e_t.

18:     Denote the isolated vertex set of $H \backslash (V_e \cup \{u\})$ as I_4.

19:     **return** ALG2($H \backslash (V_e \cup \{u\}) \backslash I_4$)$\cup(V_e \cup \{u\})$

---

4 The Anti-rank k Hypergraphs

In this section, we consider isolated-vertex-free anti-rank k hypergraphs with n vertices. In many hypergraphs, n/k is the upper bound of the dominating number. The content is structured as follows:

- In Theorem 6, for isolated-vertex-free anti-rank k hypergraph with the maximum degree Δ and the minimum degree δ, a combinatorial algorithm is designed to get a dominating set of no more than $\frac{\Delta n}{\delta k}$. Specially, for regular hypergraphs, the bound is n/k.
- Theorem 7 asserts that for every isolated-vertex-free anti-rank k hypergraph possessing the König Property, the dominating number does not exceed n/k.

- Theorem 8 states that for every isolated-vertex-free anti-rank k hypergraph with perfect matching, the dominating number is no more than n/k.
- Theorem 9 indicates that in any isolated-vertex-free anti-rank k hypergraph with the strong coloring property, the dominating number is at most n/k.

Theorem 6. $H(V, E)$ *is an isolated-vertex-free anti-rank k hypergraph with order n. Then the dominating number $\gamma(H) \leq \frac{\Delta n}{\delta k}$, where Δ is the maximum degree and δ is the minimum degree.*

Proof. Let S be an arbitrary maximal independent set of H. It's important to note that every maximal independent set is a dominating set. Therefore, $\gamma(H) \leq |S|$. Given that S is an independent set, the sets $\{e \mid v \in e\}$, $v \in S$ are disjoint. This implies $|S|\delta \leq \sum_{v \in S} d(v) \leq m$. Furthermore, when summing over $\sum_{v \in V} d(v)$, each edge is counted at least k times. This results in $\Delta n \geq \sum_{v \in V} d(v) \geq km$. Taking into account the above inequalities, we conclude that:

$$\gamma(H) \leq |S| \leq \frac{m}{\delta} \leq \frac{\Delta n}{\delta k}.$$

$\square$

Remark 3. Theorem 6 implies a combinatorial algorithm for determining a dominating set with size no more than $\frac{\Delta}{\delta} \frac{n}{k}$.

For k-uniform hypergraphs, the following two corollaries emerge directly from Theorem 6:

Corollary 4. $H(V, E)$ *is an isolated-vertex-free k-uniform hypergraph with order n. Then the dominating number $\gamma(H) \leq \frac{\Delta}{\delta} \frac{n}{k}$, here Δ is the maximum degree and δ is the minimum degree.*

Corollary 5. $H(V, E)$ *is an isolated-vertex-free k-uniform regular hypergraph with order n. Then the dominating number $\gamma(H) \leq \frac{n}{k}$.*

Theorem 7. $H(V, E)$ *is an isolated-vertex-free anti-rank k hypergraph with order n. If the König Property holds in H: the transversal number equals to the matching number, $\tau(H) = \nu(H)$, then the dominating number $\gamma(H) \leq \frac{n}{k}$.*

Proof. $H(V, E)$ is an isolated-vertex-free hypergraph, any transversal is also a dominating set. Thus $\gamma(H) \leq \tau(H)$. Every edge in H contains at least k vertices. Thus $\nu(H) \leq \frac{n}{k}$. Combining the König Property, we have $\gamma(H) \leq \tau(H) = \nu(H) \leq \frac{n}{k}$.
$\square$

Theorem 8. $H(V, E)$ *is an isolated-vertex-free anti-rank k hypergraph with order n. If H has a perfect matching, then the dominating number $\gamma(H) \leq \frac{n}{k}$.*

Proof. Let M represent a perfect matching of H, and let H' be the spanning subhypergraph of H induced by M. It follows that $\gamma(H) \leq \gamma(H')$. Consequently, the dominating number $\gamma(H) \leq \gamma(H') = |M| \leq \frac{n}{k}$.
$\square$

A strong coloring is a function that maps vertex set to a set of colors such that all vertices within any single hyperedge are mapped to distinct colors. The strong chromatic number $\chi_s(H)$ is the smallest number of colors needed to strongly color H.

Theorem 9. $H(V, E)$ *is an isolated-vertex-free anti-rank k hypergraph with order n. If there is an isolated-vertex-free spanning subhypergraph H' of H and the strong coloring number $\chi_s(H') = k$, then the dominating number $\gamma(H) \leq \frac{n}{k}$.*

Proof. H' is a spanning subhypergraph of H, thus $\gamma(H) \leq \gamma(H')$. The strong coloring number of H' is k, so H' is k-uniform. Consider a strong k-vertex-color S, the vertices are divided into k independent sets $\{S_i, 1 \leq i \leq k\}$. Because H' is an isolated-vertex-free k-uniform hypergraph, every edge contains exactly one vertex in $\{S_i, 1 \leq i \leq k\}$. Thus any independent set $\{S_i, 1 \leq i \leq k\}$ is a transversal set of H', also a dominating set of H'. Above all we have $\gamma(H) \leq \gamma(H') \leq \min\{|S_i|, 1 \leq i \leq k\} \leq \frac{n}{k}$.   □

5   The Random Hypergraph Model

In this section, we demonstrate that in the classical random hypergraph model $\mathcal{H}(n, p)$ and $H \in \mathcal{H}(n, p)$, for any positive number $\varepsilon > 0$, $\gamma(H) \leq (1 + \varepsilon)n/k$ holds with high probability.

In extending the random graph model $\mathcal{G}(n, p)$, we introduce a random k-uniform hypergraph model, denoted $\mathcal{H}(n, p)$, where $k \geq 5$. Given $0 \leq p \leq 1$, in $\mathcal{H}(n, p)$ model, the probability $\mathbf{Pr}[\{v_1, v_2, \ldots, v_k\} \in H] = p$ holds for all distinct vertices $v_1, v_2, \ldots, v_k$ in H, and these probabilities are mutually independent. We employ probabilistic methods to determine the probabilistic properties of the dominating number's upper bound in random k-uniform hypergraphs. Firstly, we introduce some symbols in asymptotic analysis as follows.

- $f(n) = \Omega(g(n))$: $\exists \, c > 0, n_0 \in \mathbb{N}_+, \forall n \geq n_0, 0 \leq cg(n) \leq f(n)$.
- $f(n) = o(g(n))$: $\forall \, c > 0, \exists \, n_0 \in \mathbb{N}_+, \forall n \geq n_0, 0 \leq f(n) < cg(n)$.

Next, we establish a concentration property for vertex degree.

Lemma 1. *If $H \in \mathcal{H}(n, p)$ and $p \in \Omega(1/n^{k-2})$, then for any $0 < \varepsilon < 1$, it holds that*

$$\mathbf{Pr}[\Delta(H)/\delta(H) \leq 1 + \varepsilon] = 1 - o(1).$$

Proof. Consider an arbitrary vertex v. For each $k - 1$ set $\{v_1, v_2, \ldots, v_{k-1}\} \subseteq V \backslash v$, let $X_{v_1 v_2 \cdots v_{k-1}}$ be the random variable defined by $X_{v_1 v_2 \cdots v_{k-1}} = 1$ if $\{v, v_1, \ldots, v_{k-1}\} \in E(H)$ and 0 otherwise. $X_{v_1 v_2 \cdots v_{k-1}}, \{v_1, v_2, \ldots, v_{k-1}\} \subseteq V \backslash v$ are independent 0-1 variables satisfying $\mathbf{E}[X_{v_1 v_2 \cdots v_{k-1}}] = p$, and the degree of v, written as $d(v)$ satisfies $d(v) = \sum_{\{v_1, v_2, \ldots, v_{k-1}\} \subseteq V \backslash v} X_{v_1 v_2 \cdots v_{k-1}}$ and $\mathbf{E}[d(v)] =$

$\binom{n-1}{k-1}p = \mu$. We know that $\mu \to \infty$ when $n \to \infty$. For given ε, using Chernoff's Inequality and Union Bound Inequality [17], we obtain

$$\mathbf{Pr}[\Delta(H) \geq (1+\varepsilon/3)\mu] = \mathbf{Pr}[\exists v, \ d(v) \geq (1+\varepsilon/3)\mu]$$

$$\leq \sum_{v \in V} \mathbf{Pr}[d(v) \geq (1+\varepsilon/3)\mu] \leq n \cdot \exp(-\frac{\varepsilon^2 \mu}{27}) = o(1).$$

$$\mathbf{Pr}[\delta(H) \leq (1-\varepsilon/3)\mu] = \mathbf{Pr}[\exists v, \ d(v) \leq (1-\varepsilon/3)\mu]$$

$$\leq \sum_{v \in V} \mathbf{Pr}[d(v) \leq (1-\varepsilon/3)\mu] \leq n \cdot \exp(-\frac{\varepsilon^2 \mu}{18}) = o(1).$$

Then, we have $\mathbf{Pr}[\Delta(H) \leq (1+\varepsilon/3)\mu, \ \delta(H) \geq (1-\varepsilon/3)\mu] = 1 - o(1)$. Given that $\Delta(H) \leq (1+\varepsilon/3)\mu$ and $\delta(H) \geq (1-\varepsilon/3)\mu$, it follows that $\Delta(H)/\delta(H) \leq (1+\varepsilon/3)/(1-\varepsilon/3) \leq 1+\varepsilon$. Consequently, $\mathbf{Pr}[\Delta(H)/\delta(H) \leq 1+\varepsilon] = 1 - o(1)$. □

Theorem 10. *If $H \in \mathcal{H}(n,p)$ and $p \in \Omega(1/n^{k-2})$, then for any $0 < \varepsilon < 1$, it holds that*

$$\mathbf{Pr}[\gamma(H) \leq (1+\varepsilon)\frac{n}{k}] = 1 - o(1).$$

Proof. According to Lemma 1, consider

$$\mathbf{Pr}[\delta(H) = 0] \leq \mathbf{Pr}[\delta(H) \leq (1-\varepsilon/3)\mu] = o(1),$$

which means that H is isolated-vertex-free with high probability. According to Theorem 6 and Lemma 1, we have

$$\mathbf{Pr}[\gamma(H) \leq (1+\varepsilon)\frac{n}{k}] \geq \mathbf{Pr}[\frac{\Delta}{\delta}\frac{n}{k} \leq (1+\varepsilon)\frac{n}{k}]$$

$$= \mathbf{Pr}[\frac{\Delta}{\delta} \leq 1+\varepsilon] = 1 - o(1).$$

□

Acknowledgement. The authors are very indebted to three anonymous referees for their invaluable suggestions and comments.

References

1. Henning, M.A., Yeo, A.: Hypergraphs with large transversal number and with edge sizes at least 3. J. Graph Theory **59**(4), 326–348 (2008)
2. Henning, M.A., Löwenstein, C.: Hypergraphs with large domination number and with edge sizes at least three. Disc. Appl. Math. **160**(12), 1757–1765 (2012)
3. Jose, B.K., Tuza, Z.: Hypergraph domination and strong independence. Appl. Anal. Disc. Math. **3**(2), 347–358 (2009)

4. Cockayne, E.J., Hedetniemi, S.T., Slater, P.J.: Matchings and transversals in hypergraphs, domination and independence-in trees. J. Comb. Theory Series B **26**(1), 78–80 (1979)

5. Behr, A., Camarinopoulos, L.: On the domination of hypergraphs by their edges. Disc. Math. **187**(1–3), 31–38 (1998)

6. Acharya, B.D.: Domination in hypergraphs ii. new directions. In Proceedings of International Conference-ICDM, pp. 1–16 (2008)

7. Acharya, B.D.: Domination in hypergraphs. AKCE Int. J. Graphs Comb. **4**(2), 117–126 (2007)

8. Bujtás, C., Henning, M.A., Tuza, Z.: Transversal game on hypergraphs and the 34-conjecture on the total domination game. SIAM J. Disc. Math. **30**(3), 1830–1847 (2016)

9. Bujtás, C., Patkós, B., Tuza, Z., Vizer, M.: Domination game on uniform hypergraphs. Disc. Appl. Math. **258**, 65–75 (2019)

10. Dong, Y., Shan, E., Kang, L., Li, S.: Domination in intersecting hypergraphs. Disc. Appl. Math. **251**, 155–159 (2018)

11. Kang, L., Li, S., Dong, Y., Shan, E.: Matching and domination numbers in r-uniform hypergraphs. J. Comb. Optim. **34**, 656–659 (2017)

12. Shan, E., Dong, Y., Kang, L., Li, S.: Extremal hypergraphs for matching number and domination number. Disc. Appl. Math. **236**, 415–421 (2018)

13. Bujtás, C., Henning, M.A., Tuza, Z.: Transversals and domination in uniform hypergraphs. Eur. J. Comb. **33**(1), 62–71 (2012)

14. Ore, O.: Theory of graphs. In: Colloquium Publications. American Mathematical Society (1962)

15. Cyman, J.: The outer-connected domination number of a graph. Aust. J. Comb. **38**, 35–46 (2007)

16. Chen, Z., Chen, B., Tang, Z., Diao, Z.: A sharp upper bound for the transversal number of k-uniform connected hypergraphs with given size. J. Comb. Optim. **45**(1), 37 (2023)

17. Mitzenmacher, M., Upfal, E.: Probability and Computing: Randomization and Probabilistic Techniques in Algorithms and Data Analysis. Cambridge University Press, Cambridge (2017)

Parameterized and Exact-Exponential Algorithms for the Read-Once Integer Refutation Problem in UTVPI Constraints

K. Subramani$^{(\boxtimes)}$ and Piotr Wojciechowski

LDCSEE, West Virginia University, Morgantown, WV, USA
{k.subramani,pwojciec}@mail.wvu.edu

Abstract. In this paper, we discuss parameterized and exact-exponential algorithms for the read-once integer refutability problem in Unit Two Variable Per Inequality (UTVPI) constraint systems. UTVPI constraint systems (UCSs) arise in a number of domains including operations research, program verification and abstract interpretation. The integer feasibility problem in UCSs is polynomial time solvable and there exist several algorithms for the same. This paper is concerned with *refutations* of integer feasibility in UCSs. Inasmuch as the integer feasibility problem is in **P**, there exist polynomial time algorithms to establish unrestricted refutations of integer feasibility. The focus of this paper is on a specific class of refutations called read-once refutations. Previous research has established that the problem of determining the existence of read-once refutations of integer feasibility in UCSs is **NP-hard**. This paper extends that research by examining the read-once refutability from the parameterized perspective. Using the number of refutation steps in the shortest read-once refutation, we establish fixed-parameter tractability. We also show that no polynomial size kernel exists for this problem. From the exact perspective, we design a non-trivial exponential algorithm.

1 Introduction

This paper is concerned with the read-once integer refutation problem in Unit Two Variable Per Inequality (UTVPI) constraints. The problem of checking the integer feasibility of a UTVPI constraint system (UCS), henceforth the UTVPI Integer Feasibility (UTIF) problem, occurs in a number of disparate domains such as abstract interpretation [1,11], array bounds checking [10], packing and scheduling [7] and so on. As per the literature, there exist several algorithms for the UTIF problem [10,19]. Inasmuch as those decision procedures, they provide unrestricted refutations for infeasible instances. Recall that a refutation is a "negative" certificate in that it *certifies* the infeasibility of an infeasible instance. Unrestricted refutations are difficult to manage and "visualize". In contrast,

This research was supported in part by the Defense Advanced Research Projects Agency through grant HR001123S0001-FP-004.

read-once refutations (RORs) are easy to understand and visualize [8]. In this paper, we focus on the problem of checking whether a UCS has an ROR in the cutting plane proof (refutation) system [6] (UTVROR). Previous research has established that the UTVROR problem is **NP-hard** [16]. As with any **NP-hard** problem, it is worthwhile to study the problem from a fine-grained perspective [3], in order to derive sufficient conditions for the existence of efficient algorithms. We use the number of inference steps in an optimal length refutation as our parameter of interest. Using this parameter, we derive a number of interesting results for the UTVROR problem. In particular, we show that the problem is fixed-parameter tractable (**FPT**). We also establish that no polynomial kernelizations can exist, unless **coNP** $\subseteq$ **NP/poly**. We also design exact exponential algorithms for the UTVROR problem.

2 Statement of Problem

In this section, we formally describe the constraint system, the refutation system and the refutation type of interest.

2.1 Constraint System

Definition 1. *A linear constraint of the form* $a_i \cdot x_i + a_j \cdot x_j \leq b_{ij}$ *is said to be a Unit Two Variable Per Inequality (UTVPI) constraint, if* $a_i, a_j \in \{0, 1, -1\}$ *and* $b_{ij} \in \mathbb{Z}$.

A conjunction of UTVPI constraints is called a UTVPI constraint system (UCS) and can be written in matrix form as: $\mathbf{A} \cdot \mathbf{x} \leq \mathbf{b}$. We assume without loss of generality that $a_i, a_j \neq 0$ in each constraint of our UCS [18]. The following problem is called the UTVPI Integer Feasibility (UVIF) problem: Given a UCS $\mathbf{A} \cdot \mathbf{x} \leq \mathbf{b}$, does it enclose a lattice point, i.e., a vector $\mathbf{a}$, all of whose components are integers?

2.2 Refutation System

We utilize a refutation system consisting of two inference rules to establish integer infeasibility.

1. Transitive rule -

$$\frac{a_i \cdot x_i + a_j \cdot x_j \leq b_{ij} \qquad\qquad -a_j \cdot x_j + a_k \cdot x_k \leq b_{jk}}{a_i \cdot x_i + a_k \cdot x_k \leq b_{ij} + b_{jk}}$$

   Observe that the transitive rule preserves linear (all) solutions. Note that this rule is a restricted version of the addition rule for linear inequalities.
2. Tightening rule -

$$\frac{a_i \cdot x_i + a_j \cdot x_j \leq b_{ij} \qquad\qquad a_i \cdot x_i - a_j \cdot x_j \leq b'_{ij}}{a_i \cdot x_i \leq \left\lfloor \frac{b_{ij} + b'_{ij}}{2} \right\rfloor}$$

Together, the transitive and tightening rules form a variation of the cutting plane refutation system with the tightening rule being used to add new cuts to the system. Recall that a cut is a constraint that is added to the system; a cut removes linear solutions, while preserving integer solutions.

We will refer to the above refutation system as the UCUT refutation system (since we apply it exclusively to UCSs). The UCUT refutation system is sound, complete and efficient [17,19].

Definition 2. *An* integer refutation *of a UCS* **U***, is a sequence of applications of the transitive and tightening inference rules that results in a constraint of the form* $0 \leq b$*, where* $b < 0$*.*

2.3 Read-Once Refutations

Definition 3. *A* read-once *integer refutation (ROR) is a refutation in which each constraint is used only once either as part of the transitive rule or as part of the tightening rule.*

However, a constraint can be re-derived as long it can be derived from an unused set of input constraints.

The read-once integer refutation problem in UCSs (UTVROR) is defined as follows: Given a UCS **U** : **A** · **x** ≤ **b**, does it have a read-once refutation of integer feasibility, i.e., a read-once integer refutation?

The principal contributions of this paper are as follows:

1. A $O(2^k \cdot k^{O(\log k)} \cdot n \cdot m \cdot \log m)$ **FPT** algorithm for the UTVROR problem parameterized by k, the length of the refutation.
2. A likely lower bound on the size of kernelizations for the UTVROR problem, when it is parameterized by the length of the refutation.
3. An $O^*(2^{\frac{m}{3}})$ exact exponential algorithm for the UTVROR problem.
4. A $2^{o(n)}$ likely lower bound on the running times of exact exponential algorithms for the UTVROR problem.

3 Motivation and Related Work

UTVPI constraints occur in a number of practical domains such as abstract interpretation [2], program verification [11], array bounds checking [10], and packing and covering [7]. Both the linear and integer feasibility problems in UCSs have been investigated by a number of researchers. Algorithms for the linear feasibility problem in UCSs have been described in [11,18] and [10]. For the integer feasibility problem, an optimal, certifying algorithm is detailed in [19].

The read-once refutation model was introduced in [8]. In that paper, the constraint system was Boolean formulas in CNF and the refutation system was resolution [13]. The authors made a convincing case for the study of read-once refutations. In essence, they argued that weakening the refutation system could potentially result in "short" refutations. Unfortunately, they discovered that asking if a 3CNF formula has a read-once resolution refutation is **NP-hard**. Their result was strengthened in [9], where it was shown that the read-once refutation problem is **NP-hard** even for UCSs.

On the continuous side, i.e., linear programs, the refutation system of choice is the ADD refutations system (see [16]). Under this refutation system, it has been shown that difference constraint systems always have a read-once refutation and that such a refutation can be found in polynomial time [14]. In [15], it is shown that UCSs do not always have a read-once refutation of linear feasibility. However, if one exists, it can be detected in polynomial time.

In the case of integer feasibility, variants of the cutting plane refutation system have been studied. It was shown in [17] that the problem of detecting a read-once refutation in UCSs in the cutting plane refutation system is **NP-hard**.

4   A Fixed-Parameter Algorithm

In this section, we describe an **FPT** algorithm for finding a read-once refutation of a UCS **U**. This algorithm is parameterized by the number of input constraints that can be used in a read-once refutation.

First, we provide a randomized algorithm (Algorithm 4.1) for solving the UTVROR problem for a UCS **U**, with m constraints over n variables. This algorithm will be derandomized later.

Input: UCS **U**

Output: true, if **U** has a read-once refutation and **false** otherwise.

1: **procedure** UTVROR-RAND(**U**)
2:     **for** (each variable x_i in **U**) **do**
3:         Create the empty sets $\mathbf{U}_i^+$ and $\mathbf{U}_i^-$.
4:         **for** (each constraint $\mathbf{U}_j \in \mathbf{U}$) **do**
5:             Uniformly and at random choose whether $\mathbf{U}_j$ is added to the set $\mathbf{U}_i^+$ or the set $\mathbf{U}_i^-$.
6:         **if** ($x_i \leq b$ can be derived from $\mathbf{U}_i^+$, $-x_i \leq b'$ can be derived from $\mathbf{U}_i^-$, and $b + b' < 0$) **then**
7:             **return true**.
8:     **return false**.

Algorithm 4.1: Randomized algorithm for the UTVROR problem

Theorem 1. *If Algorithm 4.1 returns **true**, then **U** has a read-once refutation.*

Proof. If Algorithm 4.1 returns **true**, then for some x_i, there exist sets $\mathbf{U}_i^+$ and $\mathbf{U}_i^-$ such that a constraint of the form $x_i \leq b$ can be derived from $\mathbf{U}_i^+$ and the constraint $-x_i \leq b'$ can be derived from $\mathbf{U}_i^-$. This can be determined by using an integer closure algorithm for UTVPI constraints [20]. If the constraint $x_i \leq b$

can be derived from $\mathbf{U}_i^+$, then it can be derived using each constraint at most once. The same holds for deriving $-x_i \leq b'$ from $\mathbf{U}_i^-$.

Since $\mathbf{U}_i^+$ and $\mathbf{U}_i^-$ are disjoint, the derivations of $x_i \leq b$ and $-x_i \leq b'$. can be combined with an inference step to derive $0 \leq b + b' < 0$. This is a read-once refutation of $\mathbf{U}$.   □

Theorem 2. *If* $\mathbf{U}$ *has a read-once refutation that uses at most* k *input constraints, then Algorithm 4.1 will return* **true** *with probability at least* $\frac{1}{2^k}$.

Proof. Let R be a read-once refutation of $\mathbf{U}$ of that uses at most k input constraints. Since R is a read-once refutation of $\mathbf{U}$, $x_i \leq b$ and $-x_i \leq b'$ must be derived from disjoint subsets of $\mathbf{U}$. Let $R^+ \subseteq \mathbf{U}$ be the set of input constraints used to derive $x_i \leq b$ and let $R^- \subseteq \mathbf{U}$ be the set of input constraints used to derive $-x_i \leq b'$. Since R uses at most k input constraints, $|R^+| + |R^-| \leq k$.

Algorithm 4.1 will find R if every constraint in R^+ is added to the set $\mathbf{U}_i^+$ and every constraint in R^- is added to the set $\mathbf{U}_i^-$. The probability of this happening is $\frac{1}{2^{|R^+|+|R^-|}} \geq \frac{1}{2^k}$.   □

To obtain an **FPT** algorithm for solving the UTVROR problem we will derandomize Algorithm 4.1. This derandomization utilizes (m, k)-universal sets [12]. These sets are defined as follows:

Definition 4. *Let* S *be a set of size* m. *An* (m, k)-*universal set is a family* $\mathbf{F}$ *of subsets of* S, *such that for any set* $R \subseteq S$ *of size* k, *the family* $\{A \cap R : A \in \mathbf{F}\}$ *contains every subset of* R.

Note that we can construct an (m, k)-universal set for $\mathbf{U}$ of $O(2^k \cdot k^{O(\log k)} \cdot \log m)$ size in $O(2^k \cdot k^{O(\log k)} \cdot m \cdot \log m)$ time [12].

Let R be a read-once refutation of $\mathbf{U}$ that uses at most k input constraints. Let $\mathbf{U}_R$ be the set of input constraints used by R. Since R is a read-once refutation of $\mathbf{U}$, $x_i \leq b$ and $-x_i \leq b'$ must be derived from disjoint subsets of $\mathbf{U}_R$. Let $R^+ \subseteq \mathbf{U}_R$ be the set of input constraints used to derive $x_i \leq b$ and let $R^- \subseteq \mathbf{U}_R$ be the set of input constraints used to derive $-x_i \leq b'$. Note that $R^+ \cup R^- = \mathbf{U}_R$.

Let $\mathbf{F}$ be an (m, k)-universal set for $\mathbf{U}$. Let $r = |\mathbf{U}_R|$. Since $r \leq k$, $\mathbf{F}$ is also an (m, r)-universal set for $\mathbf{U}$. Thus, for some $A \in \mathbf{F}$, $A \cap \mathbf{U}_R = R^+$. This means that $(\mathbf{U} \setminus A) \cap \mathbf{U}_R = R^-$. Thus, $R^+ \subseteq A$ and $R^- \subseteq \mathbf{U} \setminus A$. Consequently, the constraint $x_i \leq b$ can be derived from A and the constraint $-x_i \leq b'$ can be derived from $\mathbf{U} \setminus A$. Note that both these derivations are read-once.

It follows that Algorithm 4.2 checks if $\mathbf{U}$ has a read-once refutation using at most k input constraints, with k and $\mathbf{U}$ as inputs.

Constructing $\mathbf{F}$ takes $O(2^k \cdot k^{O(\log k)} \cdot m \cdot \log m)$ time. Additionally, processing each of the $(2^k \cdot k^{O(\log k)} \cdot \log m)$ elements of $\mathbf{F}$ takes $O(m \cdot n)$ time. Thus, Algorithm 4.2 runs in time $O(2^k \cdot k^{O(\log k)} \cdot n \cdot m \cdot \log m)$. In other words, Algorithm 4.2 is an **FPT** algorithm for the UTVROR problem.

Instead of storing the entire (m, k)-universal set $\mathbf{F}$, we can utilize a generator that creates each set in $\mathbf{F}$.

Input: UCS **U** and integer k

Output: true, if **U** has a read-once refutation using at most k input constraints, **false** otherwise.

1: **procedure** UTVROR-PARAM(**U**, k)
2:    Construct an (m, k)-universal set **F** for **U**.
3:    **for** (each $A \in$ **F** and each variable x_i) **do**
4:        **if** ($x_i \leq b$ can be derived from A, $-x_i \leq b'$ can be derived from **U** $\setminus A$, and $b + b' < 0$) **then**
5:            **return true.**
6:    **return false.**

Algorithm 4.2: Parameterized algorithm for the UTVROR problem

5 Kernelization Lower Bounds

In this section, we establish that a polynomial size kernelization cannot exist for the UTVROR problem, unless **coNP** $\subseteq$ **NP/poly**. This result applies when the problem is parameterized by the length of the shortest refutation. This is done through the use of an OR-distillation [4].

Definition 5. *Let P and Q be a pair of problems and let $t : \mathbb{N} \to \mathbb{N} \setminus \{0\}$ be a polynomially bounded function. Then a t-**bounded OR-distillation** from P into Q is an algorithm that for every s, given as input $t(s)$ strings $x_1, \ldots, x_{t(s)}$ with $|x_j| = s$ for all j:*

1. *Runs in polynomial time, and*
2. *Outputs a string y of length at most $t(s) \cdot \log s$ such that y is a **yes** instance of Q if and only if x_j is a **yes** instance of P for some $j \in \{1, \ldots, t(s)\}$.*

If any **NP-hard** problem has a t-bounded OR-distillation, then **coNP** $\subseteq$ **NP/poly** [4]. If **coNP** $\subseteq$ **NP/poly**, then $\Sigma_3^P = \Pi_3^P$ [21]. Thus, the polynomial hierarchy would collapse to the third level.

Theorem 3. *The UTVROR problem does not have a polynomial sized kernel unless* **coNP** $\subseteq$ **NP/poly**.

Proof. We will prove this by showing that if the UTVROR problem has a polynomial sized kernel, then there exists a t-bounded OR-distillation from the UTVROR problem into itself.

For each l, let $\mathbf{U}_l$ be a UCS with n variables and m constraints such that, for each constraint $a_i \cdot x_i + a_j \cdot x_j \leq b$, $b \leq B_{max}$ for a fixed integer B_{max}. We have that $s = |\mathbf{U}_l| = m \cdot (\log n + \log B_{max})$.

Assume that for some constant c, the UTVROR problem has a kernel of size k^c. Let $t(s) = s^c$. Note that $t(s)$ is a polynomial.

For each $j = 1 \ldots t(s)$, let $\mathbf{U}_l$ be a UCS with n variables and m constraints such that $|\mathbf{U}_l| = s$. From these UCSs, we can create a new UCS network $\mathbf{U}$ with $t(s) \cdot n$ variables and $t(s) \cdot m$ constraints such that $\mathbf{U}$ is a disjoint union of $\mathbf{U}_1, ..., \mathbf{U}_{t(S)}$.

Note that no constraint in $\mathbf{U}$ corresponding to a constraint in $\mathbf{U}_l$ shares variables with a constraint in $\mathbf{U}$ corresponding to a constraint in $\mathbf{U}_{l'}$, $l' \neq l$. Thus, any refutation of $\mathbf{U}$ corresponds to a refutation of $\mathbf{U}_l$ for some $l \in \{1, \ldots, t(s)\}$. Consequently, $\mathbf{U}$ has a read-once refutation of length k, if and only if $\mathbf{U}_j$ has a read-once refutation of length k for some $j \in \{1, \ldots, t(s)\}$.

Let $\mathbf{U}'$ be a kernel of $\mathbf{U}$ such that $|\mathbf{U}'| \leq k^c$. Since $k \leq m \leq s$, we have that $|\mathbf{U}'| \leq k^c \leq s^c = t(s)$. Additionally, $\mathbf{U}'$ has a read-once refutation of length k if and only if $\mathbf{U}_j$ has a read-once refutation of length k for some $j \in \{1, \ldots, t(s)\}$. Thus, we have a t-bounded OR-distillation from the UTVROR problem to itself. This cannot happen unless $\mathbf{coNP} \subseteq \mathbf{NP/poly}$. $\quad\square$

6 Lower Bounds on Exponential Algorithms

In this section, we show that the UTVROR problem cannot be solved in time $2^{o(n)}$ unless the Exponential Time Hypothesis (ETH) fails. This is accomplished by a reduction from 3SAT. Let Φ be a 3CNF formula with m' clauses over n' variables. From Φ, we construct a UCS $\mathbf{U}$ as follows:

1. Create the variables x_0 and y_0.
2. Create the constraint $-x_0 - y_0 \leq -1$.
3. For each variable x_i in Φ: (a) Create the variable x_i. (b) Create the variables $y_{i,0}^+$ through $y_{i,k_i^+}^+$, where k_i^+ is the number of clauses that use the literal x_i. (c) Create the constraints $y_{i,j-1}^+ - y_{i,j}^+ \leq 0$ for $j = 1, \ldots, k_i^+$. Additionally, create the constraints $x_{i-1} - x_{i,0}^+ \leq 0$ and $x_{i,k_i^+}^+ - x_i \leq 0$. (d) Create the variables $y_{i,0}^-$ through $y_{i,k_i^-}^-$, where k_i^- is the number of clauses that use the literal $\neg x_i$. (e) Create the constraints $y_{i,j-1}^- - y_{i,j}^- \leq 0$ for $j = 1, \ldots, k_i^-$. Additionally, create the constraints $x_{i-1} - x_{i,0}^- \leq 0$ and $x_{i,k_i^-}^- - x_i \leq 0$.
4. Create the constraint $x_0 + x_{n'} \leq 1$.
5. For each clause ϕ_r in Φ, create the variable y_r.
6. Create the constraint $y_{m'} - x_0 \leq 0$.
7. For each clause ϕ_j in Φ, and each variable x_i in ϕ_j: (a) If ϕ_r is the jth clause to use the literal x_i, create the constraints $y_{r-1} - x_{i,j-1}^+ \leq 0$ and $x_{i,j}^+ - y_r \leq 0$. (b) If ϕ_r is the jth clause to use the literal $\neg x_i$, create the constraints $y_{r-1} - x_{i,j-1}^- \leq 0$ and $x_{i,j}^- - y_r \leq 0$.

Note that $\mathbf{U}$ has $n = (3 + 2 \cdot n' + 2 \cdot m')$ variables.

Lemma 1. *Let Φ be a 3CNF formula with m' clauses over n' variables, and let $\mathbf{U}$ be the corresponding UCS. Φ is satisfiable, if and only if $\mathbf{U}$ has a read-once integer refutation.*

A proof of Lemma 1 can be found in the journal version of the paper. Using Lemma 1, we can establish a lower bound on the running time of any exact exponential algorithm for the UTVROR problem.

Theorem 4. *Unless the ETH fails, the UTVROR problem cannot be solved in time $2^{o(n)}$.*

Proof. Let Φ be a 3CNF formula with m' clauses over n' variables. Using the construction above, we are able to construct a UCS $\mathbf{U}$ from Φ with $n = (3 + 2 \cdot n' + 2 \cdot m')$ variables. From Lemma 1, $\mathbf{U}$ has a read-once refutation, if and only if, Φ is satisfiable. Thus, if the WPVCT problem can be solved in $2^{o(n)}$ time, then 3SAT can be solved in $2^{o(m'+n')}$ time. This, through the Sparcification Lemma, violates the ETH.   □

7   An Exact Exponential Algorithm

In this section, we describe an exact exponential algorithm for the UTVROR problem. We first make the following assumptions about the structure of $\mathbf{U}$: For each variable x_i in $\mathbf{U}$, at least one constraint in $\mathbf{U}$ uses the term x_i and at least one constraint in $\mathbf{U}$ uses the term $-x_i$: If x_i appears as only a positive term or only a negative term in $\mathbf{U}$, then no refutation of $\mathbf{U}$ contains any input constraint that uses x_i. Thus, all such constraints can be removed from $\mathbf{U}$ without affecting the existence of read-once refutations.

From these observations, we get the following reduction rules for the UTVROR problem. 1. If x_i appears as only a positive term, then remove all constraints containing x_i from $\mathbf{U}$. 2. If x_i appears as only a negative term, then remove all constraints containing x_i from $\mathbf{U}$.

We now introduce the concept of a decomposable read-once refutation.

Definition 6. *A read-once refutation R of a UCS $\mathbf{U}$ is **decomposable**, if for some terms l_i and l_j, R can be divided into: 1. A derivation R_i of the constraint $l_i \leq b_1$, such that no two input constraints in R_i share a non-l_i term. 2. A derivation R_j of the constraint $l_j \leq b_2$, such that no two input constraints in R_j share a non-l_j term. 3. A derivation R_{ij} of $-l_i - l_j \leq b_3$, such that no two input constraints in R_{ij} share any terms. Note that we can have $R_{ij} = \emptyset$. This can happen when $l_i = -l_j$.*

We have the following result on decomposable read-once refutations in UCSs.

Theorem 5. *Let $\mathbf{U}$ be a UCS. If $\mathbf{U}$ has a read-once refutation, then it has a decomposable read-once refutation.*

Proof. Let R be a read-once refutation of $\mathbf{U}$, and let deriving $0 \leq b + b'$ from $x_1 \leq b$ and $-x_1 \leq b'$ be the last step of R. Observe that R can be divided into a read-once derivation of $x_k \leq b$ (say, R_k^+), and a read-once derivation of $-x_k \leq b'$ (say, R_k^-).

We first focus on R_k^+, the read-once derivation of $x_k \leq b$.

We can decompose R_k^+ as follows: 1. A derivation, R_1, of the constraint $a_i \cdot x_i \leq b_i$. 2. A derivation, R_2, of the constraint $x_k - a_i \cdot x_i \leq b_{ik}$.

We know that no two variables are used by both R_1 and R_2 (except for x_i). Let R_i be the set of constraints used by R_1.

From our choice of x_k, $\mathbf{U} \setminus R_i$ contains a derivation of $-x_k \leq b'$. Additionally, $\mathbf{U} \setminus R_i$ contains the derivation R_2 of $x_k - a_i \cdot x_i \leq b_{ik}$. Thus, $\mathbf{U} \setminus R_i$ contains a derivation of $-a_i \cdot x_i \leq b' + b_{ik}$.

Using the same argument as before (replacing x_i with x_j and x_k with $-a_i \cdot x_i$), we can divide the derivation of $-a_i \cdot x_i \leq b' + b_{ik}$ in $\mathbf{U} \setminus R_i$ into a derivation of $a_j \cdot x_j \leq b_j$ for some variable x_j, such that no two input constraints used in this derivation share a non-x_j term and a derivation of $-a_i \cdot x_i - a_j \cdot x_j \leq b_{ij}$ such that no two input constraints used in this derivation share a term. Let R_j and R_{ij} denote the sets of input constraints used in these derivations respectively. The sets R_i, R_j, and R_{ij} form the decomposable read-once refutation R'. $\square$

Thus, we only need to look for decomposable read-once refutations of $\mathbf{U}$.

We make the following additional observation about the structure of read-once refutations of UCSs.

Theorem 6. *Let $\mathbf{U}$ be a UCS and let R be a shortest decomposable read-once refutation of $\mathbf{U}$. Additionally, let R_i, R_j, and R_{ij} be the components of R. Furthermore, let $a_i \cdot x_i$ and $a_j \cdot x_j$ denote the terms referred to in Definition 6. No variable is used by both a constraint in R_j and a constraint in R_{ij}.*

Proof. Assume that there exists a variable x_k that is used in both R_{ij} and R_j. Thus, R_j can be divided into a derivation of $a_j \cdot x_j + x_k \leq b$ and a derivation of $a_j \cdot x_j - x_k \leq b'$. Similarly R_{ij} can be divided into either: 1. A derivation of $-a_i \cdot x_i + x_k \leq b''$ and a derivation of $-a_j \cdot x_j - x_k \leq b'''$. In this case, we can get the following decomposable read-once refutation of $\mathbf{U}$: (a) Derive $a_i \cdot x_i \leq b_i$ from R_i. (b) Derive $-x_k \leq \lfloor \frac{b' + b'''}{2} \rfloor$ from $a_j \cdot x_j - x_k \leq b'$ and $-a_j \cdot x_j - x_k \leq b'''$. (c) Derive a contradiction from $a_i \cdot x_i \leq b_i$, $x_k \leq \lfloor \frac{b' + b'''}{2} \rfloor$ and, $-a_i \cdot x_i + x_k \leq b''$. Observe that this new derivation is shorter than the original one, because no constraint from the derivation of $a_j \cdot x_j + x_k \leq b$ has been used in the new derivation. 2. A derivation of $-a_i \cdot x_i - x_k \leq b''$ and a derivation of $-a_j \cdot x_j + x_k \leq b'''$. In this case, we can get the following decomposable read-once refutation of $\mathbf{U}$: (a) Derive $a_i \cdot x_i \leq b_i$ from R_i. (b) Derive $x_k \leq \lfloor \frac{b + b''}{2} \rfloor$ from $a_j \cdot x_j + x_k \leq b$ and $-a_j \cdot x_j + x_k \leq b'''$. (c) Derive a contradiction from $a_i \cdot x_i \leq b_i$, $x_k \leq \lfloor \frac{b + b''}{2} \rfloor$ and, $-a_i \cdot x_i - x_k \leq b''$. Observe that this new derivation is shorter than the original one, because no constraint from the derivation of $a_j \cdot x_j - x_k \leq b'$ has been used in the new derivation. In either case, we get a contradiction to the choice of R as the shortest decomposable read-once refutation of $\mathbf{U}$. $\square$

Algorithm 7.1 finds a decomposable read-once refutation of $\mathbf{U}$. Note that on Line 10, Algorithm 7.1 chooses to add the constraint $a_g \cdot x_g + a_h \cdot x_h \leq b$ to the set $\mathbf{U}_i^+$ or to the set $\mathbf{U}_i^-$. Let C represent the set of all possible sequences of choices made by Algorithm 7.1 on Line 10.

Input: UCS **U**

Output: true if **U** has a read-once refutation, **false** otherwise.

1: **procedure** UTVROR-EFF(**U**)
2:    **for** (each pair of terms $a_i \cdot x_i \neq a_j \cdot x_j$) **do**
3:       Let $\mathbf{U}_i^+$ be the constraints used for the derivation of $a_i \cdot x_i \leq b_1$.
4:       Let $\mathbf{U}_i^-$ be the constraints used for the derivation of $-a_i \cdot x_i \leq b_2$.
5:       Add all constraints using the term $a_i \cdot x_i$ to $\mathbf{U}_i^+$.
6:       Add all constraints using the term $a_j \cdot x_j$ to $\mathbf{U}_i^-$.
7:       **if** $(a_i \cdot x_i \neq -a_j \cdot x_j)$ **then**
8:          Add all constraints using the term $-a_i \cdot x_i$ or $-a_j \cdot x_j$ to $\mathbf{U}_i^-$.
9:       **for** (each constraint $a_g \cdot x_g + a_h \cdot x_h \leq b \in \mathbf{U} \setminus (\mathbf{U}_i^+ \cup \mathbf{U}_i^-)$) **do**
10:          Add $a_g \cdot x_g + a_h \cdot x_h \leq b$ to $\mathbf{U}_i^+$ or $\mathbf{U}_i^-$.
11:          **if** $(a_g \cdot x_g + a_h \cdot x_h \leq b$ was added to $\mathbf{U}_i^+)$ **then**
12:             **if** (only one constraint uses the term $-a_g \cdot x_g$ $(-a_h \cdot x_h)$) **then**
13:                Add that constraint to $\mathbf{U}_i^+$.
14:                Add all other constraints using $a_g \cdot x_g$ $(a_h \cdot x_h)$ to $\mathbf{U}_i^+$.
15:             **else if** $(a_g \cdot x_g + a_h \cdot x_h \leq b$ is the only constraint to use the term $a_g \cdot x_g$ $(a_h \cdot x_h))$ **then**
16:                Add all the constraints using $-a_g \cdot x_g$ (or $-a_h \cdot x_h$) to $\mathbf{U}_i^+$.
17:             **else**
18:                Add all other constraints using $a_g \cdot x_g$ or $a_h \cdot x_h$ to $\mathbf{U}_i^-$.
19:          **else**
20:             **if** (only one constraint uses the term $-a_g \cdot x_g$ $(-a_h \cdot x_h)$) **then**
21:                Add that constraint to $\mathbf{U}_i^-$.
22:                Add all other constraints using $a_g \cdot x_g$ $(a_h \cdot x_h)$ to $\mathbf{U}_i^-$.
23:             **else if** $(a_g \cdot x_g + a_h \cdot x_h \leq b$ is the only constraint to use the term $a_g \cdot x_g$ $(a_h \cdot x_h))$ **then**
24:                Add all constraints using $-a_g \cdot x_g$ $(-a_h \cdot x_h)$ to $\mathbf{U}_i^-$.
25:             **else**
26:                Add all other constraints using $a_g \cdot x_g$ or $a_h \cdot x_h$ to $\mathbf{U}_i^+$.
27:          **if** ($x_i \leq b_1$ can be derived from $\mathbf{U}_i^+$ and $-x_1 \leq b_2$ can be derived from $\mathbf{U}_i^-$ such that $b_1 + b_2 < 0$) **then**
28:             **return true.**
29:    **return false.**

Algorithm 7.1: A more efficient exact exponential algorithm for the UTVROR problem

Lemma 2. *Let* **U** *be a UCS. If, for some sequence of choices* $c \in C$, *Algorithm 7.1 returns* **true**, *then* **U** *has a read-once refutation.*

Proof. If, for some sequence of choices $c \in C$, Algorithm 7.1 returns **true**, then for some variable x_i there exist disjoint sets $\mathbf{U}_i^+$ and $\mathbf{U}_i^-$ such that the constraint $x_i \leq b_1$ can be derived from $\mathbf{U}_i^+$ and the constraint $-x_i \leq b_2$ can be derived from $\mathbf{U}_i^-$. This means that **U** has a read-once refutation [9]. □

Lemma 3. *Let* **U** *be a UCS with a read-once refutation. For some sequence of choices* $c \in C$, *Algorithm 7.1 will return* **true**.

The proof of Lemma 3 can be found in the journal version of the paper.

Theorem 7. *Let* **U** *be a UCS. Running Algorithm 7.1 for every sequence of choices in* C *will return* **true**, *if and only if* **U** *has a read-once refutation.*

Proof. From Lemma 2, if Algorithm 7.1 returns **true** for some sequence of choices $c \in C$, then **U** has a read-once refutation. From Lemma 3, if **U** has a read-once refutation, then some sequence of choices $c \in C$, will result in Algorithm 7.1 returning **true**. □

For each iteration of the **for** loop on Line 9, at least 3 constraints are assigned to either $\mathbf{U}_i^+$ or $\mathbf{U}_i^-$. These constraints are $a_g \cdot x_g + a_h \cdot x_h \leq b$, one constraint using the term $a_g \cdot x_g$ or $-a_g \cdot x_g$, and one constraint using the term $a_h \cdot x_h$ or $-a_h \cdot x_h$. Finding and processing these constraints can be done in $O(m)$ time.

Thus, running Algorithm 7.1 over all possible choices is governed by the recurrence relation $T(m) \leq T(m-3) + T(m-3) + O(m)$. Using the techniques in [5], it can be shown that Algorithm 7.1 runs in time $O^*(2^{\frac{m}{3}})$. This can be approximated as $O^*(1.26^m)$.

Once a sequence of choices c is made we do not need that sequence again. Thus, we never need to store the entirety of C. Thus, in each run of Algorithm 7.1, we only need space to store the $\frac{m}{3}$ choices in c, the sets $\mathbf{U}_i^+$ and $\mathbf{U}_i^-$, and the UCS **U**. This can be done in $O(m+n)$ space.

8 Conclusion

This paper examined the UTVROR problem from the perspective of parameterized and exact-exponential algorithms. Previous research had established the **NP-hardness** of this problem and hence this study is justified. We designed an **FPT** algorithm. We also argued that the problem does not admit a polynomial sized kernel, unless a well-accepted complexity conjecture fails. Finally, we designed a non-trivial exact exponential algorithm.

References

1. Bagnara, R., Hill, P.M., Zaffanella, E.: Weakly-relational shapes for numeric abstractions: improved algorithms and proofs of correctness. Formal Methods Syst. Des. **35**(3), 279–323 (2009)
2. Cousot, P., Cousot, R.: Abstract interpretation: a unified lattice model for static analysis of programs by construction or approximation of fixpoints. In: POPL, pp. 238–252 (1977)
3. Cygan, M., Fomin, F.V., Kowalik, L, Lokshtanov, D., Marx, D., Pilipczuk, M., Pilipczuk, M., Saurabh, S.: Parameterized Algorithms. Springer, Cham (2015). https://doi.org/10.1007/978-3-319-21275-3
4. Fomin, F.V., Lokshtanov, D., Saurabh, S., Zehavi, M.: Kernelization: Theory of Parameterized Preprocessing. Cambridge University Press, Cambridge (2019)
5. Fomin, F.V., Kratsch, D.: Exact Exponential Algorithms. TTCSAES, 1st edn. Springer, Heidelberg (2010). https://doi.org/10.1007/978-3-642-16533-7
6. Gomory, R.E.: Solving linear programming problems in integers. Combinat. Anal. **10**, 211–215 (1960)
7. Hochbaum, D., Megiddo, N., Naor, J., Tamir, A.: Tight bounds and 2-approximation algorithms for integer programs with two variables per inequality. Math. Program. **62**, 63–92 (1993)
8. Iwama, K., Miyano, E.: Intractability of read-once resolution. In: Proceedings of the 10th Annual Conference on Structure in Complexity Theory (SCTC 1995), Los Alamitos, CA, USA, June 1995, pp. 29–36. IEEE Computer Society Press (1995)
9. Büning, H.K., Wojciechowski, P.J., Subramani, K.: Finding read-once resolution refutations in systems of 2CNF clauses. Theor. Comput. Sci. **729**, 42–56 (2018)
10. Lahiri, S.K., Musuvathi, M.: An efficient decision procedure for UTVPI constraints. In: Gramlich, B. (ed.) FroCoS 2005. LNCS (LNAI), vol. 3717, pp. 168–183. Springer, Heidelberg (2005). https://doi.org/10.1007/11559306_9
11. Miné, A.: The octagon abstract domain. Higher-Order Symb. Comput. **19**(1), 31–100 (2006)
12. Naor, M., Schulman, L.J., Srinivasan, A.: Splitters and near-optimal derandomization. In: Proceedings of IEEE 36th Annual Foundations of Computer Science, pp. 182–191 (1995)
13. John Alan Robinson: A machine-oriented logic based on the resolution principle. J. ACM **12**(1), 23–41 (1965)
14. Subramani, K.: Optimal length resolution refutations of difference constraint systems. J. Autom. Reas. (JAR) **43**(2), 121–137 (2009)
15. Subramani, K., Wojciechowki, P.: A polynomial time algorithm for read-once certification of linear infeasibility in UTVPI constraints. Algorithmica **81**(7), 2765–2794 (2019)
16. Subramani, K., Wojciechowki, P.: Integer feasibility and refutations in UTVPI constraints using bit-scaling. Algorithmica **85**, 610–637 (2022)
17. Subramani, K., Wojciechowski, P.J.: A bit-scaling algorithm for integer feasibility in UTVPI constraints. In: Combinatorial Algorithms - 27th International Workshop, IWOCA 2016, Helsinki, Finland, 17–19 August 2016, Proceedings, vol. 9843, pp. 321–333 (2016)
18. Subramani, K., Wojciechowski, P.J.: A combinatorial certifying algorithm for linear feasibility in UTVPI constraints. Algorithmica **78**(1), 166–208 (2017)
19. Subramani, K., Wojciechowski, P.J.: A certifying algorithm for lattice point feasibility in a system of UTVPI constraints. J. Combinat. Optim. **35**(2), 389–408 (2018)

20. Subramani, K., Wojciechowski, P.J.: On integer closure in a system of unit two variable per inequality constraints. Ann. Math. Artif. Intell. **88**(10), 1101–1118 (2020)
21. Yap, C.K.: Some consequences of non-uniform conditions on uniform classes. Theor. Comput. Sci. **26**(3), 287–300 (1983)

Critical $(P_5, dart)$-Free Graphs

Wen Xia[1,2], Jorik Jooken[3], Jan Goedgebeur[3,4], and Shenwei Huang[1,2(✉)]

[1] College of Computer Science, Nankai University, Tianjin 300071, China
[2] Tianjin Key Laboratory of Network and Data Security Technology,
Nankai University, Tianjin 300071, China
shenweihuang@nankai.edu.cn
[3] Department of Computer Science, KU Leuven Campus Kulak-Kortrijk,
8500 Kortrijk, Belgium
{jorik.jooken,jan.goedgebeur}@kuleuven.be
[4] Department of Applied Mathematics, Computer Science and Statistics,
Ghent University, 9000 Ghent, Belgium

Abstract. Given two graphs H_1 and H_2, a graph is (H_1, H_2)-free if it contains no induced subgraph isomorphic to H_1 nor H_2. A dart is the graph obtained from a diamond by adding a new vertex and making it adjacent to exactly one vertex with degree 3 in the diamond.

In this paper, we show that there are finitely many k-vertex-critical $(P_5, dart)$-free graphs for $k \geq 1$. To prove these results, we use induction on k and perform a careful structural analysis via Strong Perfect Graph Theorem combined with the pigeonhole principle based on the properties of vertex-critical graphs. Moreover, for $k \in \{5, 6, 7\}$ we characterize all k-vertex-critical $(P_5, dart)$-free graphs using a computer generation algorithm. Our results imply the existence of a polynomial-time certifying algorithm to decide the k-colorability of $(P_5, dart)$-free graphs for $k \geq 1$ where the certificate is either a k-coloring or a $(k+1)$-vertex-critical induced subgraph.

Keywords: Graph coloring · k-critical graphs · Strong perfect graph theorem · Polynomial-time algorithms

1 Introduction

All graphs in this paper are finite and simple. A k-*coloring* of a graph G is a function $\phi : V(G) \longrightarrow \{1, \ldots, k\}$ such that $\phi(u) \neq \phi(v)$ whenever $uv \in E(G)$. Equivalently, a k-coloring of G can be viewed as a partition of $V(G)$ into k stable sets. If a k-coloring exists, we say that G is k-*colorable*. The *chromatic number* of G, denoted by $\chi(G)$, is the minimum number k such that G is k-colorable. A

The research of Jan Goedgebeur was supported by Internal Funds of KU Leuven. Jorik Jooken is supported by a Postdoctoral Fellowship of the Research Foundation Flanders (FWO) with contract number 1222524N. Shenwei Huang(the corresponding author) is supported by National Natural Science Foundation of China (12171256).

graph G is k-*chromatic* if $\chi(G) = k$. A graph G is k-*critical* if it is k-chromatic and $\chi(G - e) < \chi(G)$ for any edge $e \in E(G)$. For instance, K_2 is the only 2-critical graph and odd cycles are the only 3-critical graphs. A graph is *critical* if it is k-critical for some integer $k \geq 1$. Vertex-criticality is a weaker notion. A graph G is k-*vertex-critical* if $\chi(G) = k$ and $\chi(G - v) < k$ for any $v \in V(G)$.

For a fixed $k \geq 3$, it has long been known that determining the k-colorability of a general graph is an NP-complete problem [22]. However, the situation changes if one restricts the structure of the graphs under consideration.

Let $\mathcal{H}$ be a set of graphs. A graph G is $\mathcal{H}$-*free* if it does not contain any member in $\mathcal{H}$ as an induced subgraph. When $\mathcal{H}$ consists of a single graph H or two graphs H_1 and H_2, we write H-free and (H_1, H_2)-free instead of $\{H\}$-free and $\{H_1, H_2\}$-free, respectively. We say that G is k-vertex-critical $\mathcal{H}$-free if it is k-vertex-critical and $\mathcal{H}$-free. In this paper, we study k-vertex-critical $\mathcal{H}$ -free graphs. The following problem arouses our interest: Given a set $\mathcal{H}$ of graphs and an integer $k \geq 1$, are there finitely many k-vertex-critical $\mathcal{H}$-free graphs? This question is very important because the finiteness of the set has a fundamental algorithmic implication.

Theorem 1 (Folklore). *If the set of all k-vertex-critical $\mathcal{H}$-free graphs is finite, then there is a polynomial-time algorithm to determine whether an $\mathcal{H}$-free graph is $(k-1)$-colorable.* □

Let K_n be the complete graph on n vertices. Let P_t and C_t denote the path and the cycle on t vertices, respectively. The complement of G is denoted by $\overline{G}$. For two graphs G and H, we use $G + H$ to denote the disjoint union of G and H. For a positive integer, we use rG to denote the disjoint union of r copies of G. For $s, r \geq 1$, let $K_{r,s}$ be the complete bipartite graph with one part of size r and the other part of size s. Our research is mainly motivated by the following two theorems.

Theorem 2 ([16]). *For any fixed integer $k \geq 5$, there are infinitely many k-vertex-critical P_5-free graphs.*

It is natural to consider which subclasses of P_5-free graphs have finitely many k-vertex-critical graphs. In 2021, Cameron, Goedgebeur, Huang and Shi [8] obtained the following dichotomy result.

Theorem 3 ([8]). *Let H be a graph of order 4 and $k \geq 5$ be a fixed integer. Then there are infinitely many k-vertex-critical (P_5, H)-free graphs if and only if H is $2P_2$ or $P_1 + K_3$.*

This theorem completely solves the finiteness problem of k-vertex-critical (P_5, H)-free graphs for $|H| = 4$. In [8], the authors also posed the natural question of which five-vertex graphs H lead to finitely many k-vertex-critical (P_5, H)-free graphs.

It is known that there are exactly 13 5-vertex-critical (P_5, C_5)-free graphs [16]. Recently, Cameron and Hoàng constructed infinite families of k-vertex-critical (P_5, C_5)-free graphs for $k \geq 6$ [7]. It has been proven that there are finitely many k-vertex-critical $(P_5, banner)$-free graphs for $k = 5$ [18] and 6 [5] and there are finitely many k-vertex-critical $(P_5, \overline{P_5})$-free graphs for fixed k [10]. Hell and Huang proved that there are finitely many k-vertex-critical (P_6, C_4)-free graphs [14]. This was later generalized to $(P_t, K_{r,s})$-free graphs in the context of H-coloring [21]. This gives an affirmative answer for $H = K_{2,3}$. In [4], Cai, Goedgebeur and Huang showed that there are finitely many k-vertex-critical (P_5, gem)-free graphs and finitely many k-vertex-critical $(P_5, \overline{P_3 + P_2})$-free graphs. Later, Cameron and Hoàng [6] gave a better bound on the order of k-vertex-critical (P_5, gem)-free graphs and determined all such graphs for $k \leq 7$. Moreover, it has been proven that there are finitely many 5-vertex-critical $(P_5, bull)$-free graphs [17] and finitely many 5-vertex-critical $(P_5, chair)$-free graphs [19].

Our Contributions. A dart (see Fig. 1) is the graph obtained from a diamond by adding a new vertex and making it adjacent to exactly one vertex with degree 3 in the diamond. Our main result is as follows.

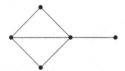

Fig. 1. The dart graph.

Theorem 4. *For every fixed integer $k \geq 1$, there are finitely many k-vertex-critical $(P_5, dart)$-free graphs.*

We prove a Ramsey-type statement (see Lemma 4) which allows us to prove our main result by induction on k. This is another example of a result that is proved by induction on k for the finitess problem besides the one in [10].

We perform a careful structural analysis via Strong Perfect Graph Theorem combined with the pigeonhole principle based on the properties of vertex-critical graphs. Moreover, for $k \in \{5, 6, 7\}$ we computationally determine a list of all k-vertex-critical $(P_5, dart)$-free graphs.

Our results imply the existence of a polynomial-time certifying algorithm to decide the k-colorability of $(P_5, dart)$-free graphs for $k \geq 1$. (An algorithm is *certifying* if, along with the answer given by the algorithm, it also gives a certificate which allows to verify in polynomial time that the output of the algorithm is indeed correct; in case of k-coloring the certificate is either a k-coloring or a $(k + 1)$-vertex-critical induced subgraph.)

Theorem 5. *For every fixed integer $k \geq 1$, there is a polynomial-time certifying algorithm to decide the k-colorability of $(P_5, dart)$-free graphs.*

Proof. Let G be a $(P_5, dart)$-free graph. We first run the polynomial-time algorithm for determining whether a P_5-free graph is k-colorable from [15] for G. If the answer is yes, the algorithm outputs a k-coloring of G. Otherwise G is not k-colorable. In this case, G must contain a $(k+1)$-vertex-critical $(P_5, dart)$-free graph as an induced subgraph. For each $(k+1)$-vertex-critical $(P_5, dart)$-free graph H, it takes polynomial-time to check whether G contains H. Since there are only finitely many such graphs by Theorem 4, we can do this for every such graph H and the total running time is still polynomial.

The remainder of the paper is organized as follows. We present some preliminaries in Sect. 2 and give structural properties around an induced C_5 in a (P_5,dart)-free graph in Sect. 3. We show that there are finitely many k-vertex-critical (P_5,dart)-free graphs for all $k \geq 1$ in Sect. 4 and computationally determine an exhaustive list of such graphs for $k \in \{5, 6, 7\}$ in Sect. 5. Finally, we give a conclusion in Sect. 6.

2   Preliminaries

For general graph theory notation we follow [1]. For $k \geq 4$, an induced cycle of length k is called a k-*hole*. A k-hole is an *odd hole* (respectively *even hole*) if k is odd (respectively even). A k-*antihole* is the complement of a k-hole. Odd and even antiholes are defined analogously.

Let $G = (V, E)$ be a graph. If $uv \in E(G)$, we say that u and v are *neighbors* or *adjacent*, otherwise u and v are *nonneighbors* or *nonadjacent*. The *neighborhood* of a vertex v, denoted by $N_G(v)$, is the set of neighbors of v. For a set $X \subseteq V(G)$, let $N_G(X) = \cup_{v \in X} N_G(v) \setminus X$. We shall omit the subscript whenever the context is clear. For $x \in V(G)$ and $S \subseteq V(G)$, we denote by $N_S(x)$ the set of neighbors of x that are in S, i.e., $N_S(x) = N_G(x) \cap S$. For two sets $X, S \subseteq V(G)$, let $N_S(X) = \cup_{v \in X} N_S(v) \setminus X$.

For $X, Y \subseteq V(G)$, we say that X is *complete* (resp. *anticomplete*) to Y if every vertex in X is adjacent (resp. nonadjacent) to every vertex in Y. If $X = \{x\}$, we write "x is complete (resp. anticomplete) to Y" instead of "$\{x\}$ is complete (resp. anticomplete) to Y". If a vertex v is neither complete nor anticomplete to a set S, we say that v is *mixed* on S. For a vertex $v \in V$ and an edge $xy \in E$, if v is mixed on $\{x, y\}$, we say that v is *mixed* on xy. For a set $H \subseteq V(G)$, if no vertex in $V(G) \setminus H$ is mixed on H, we say that H is a *homogeneous set*, otherwise H is a *nonhomogeneous set*.

A vertex subset $S \subseteq V(G)$ is *stable* if no two vertices in S are adjacent. A *clique* is the complement of a stable set. Two nonadjacent vertices u and v are said to be *comparable* if $N(v) \subseteq N(u)$ or $N(u) \subseteq N(v)$. For an induced subgraph A of G, we write $G - A$ instead of $G - V(A)$. For $S \subseteq V$, the subgraph *induced* by S is denoted by $G[S]$. We say that a vertex w *distinguishes* two vertices u and v if w is adjacent to exactly one of u and v.

We proceed with a few useful results that will be used later. The first folk-lore property of vertex-critical graph is that such graphs contain no comparable vertices. A generalization of this property was presented in [8].

Lemma 1 ([8]). *Let G be a k-vertex-critical graph. Then G has no two nonempty disjoint subsets X and Y of $V(G)$ that satisfy all the following conditions.*

- *X and Y are anticomplete to each other.*
- *$\chi(G[X]) \leq \chi(G[Y])$.*
- *Y is complete to $N(X)$.*

Lemma 2 ([17]). *Let G be a 5-vertex-critical P_5-free graph and S be a homogeneous set of $V(G)$. For each component A of $G[S]$,*

- *(i) if $\chi(A) = 1$, then A is a K_1;*
- *(ii) if $\chi(A) = 2$, then A is a K_2;*
- *(iii) if $\chi(A) = 3$, then A is a K_3 or a C_5.*

We extend Lemma 2 to all critical graphs, which may be of independent interest.

Lemma 3 ($\spadesuit^1$). *Let G be a k-vertex-critical graph and S be a homogeneous set of $V(G)$. For each component A of $G[S]$, if $\chi(A) = m$ with $m < k$, then A is an m-vertex-critical graph.*

Next, we prove an important lemma, which will be used frequently in the proof of our results.

Lemma 4. *Let G be a dart-free graph and c be a fixed integer. Let S, T be two disjoint subsets of $V(G)$ such that $|S| \leq c$ and $\overline{G[T \cup S]}$ is connected. If every vertex in S is adjacent to at least one of w_1 and w_2 where w_1 and w_2 are two arbitrary nonadjacent vertices in T, and there exists a vertex $w \in V(G) \setminus (S \cup T)$ such that w is complete to $S \cup T$, then $|T| \leq c(\chi(T))^1 + \ldots + c(\chi(T))^{2\chi(T)+1}$.*

Proof. Let $N_0 = S$, $N_i = \{v \mid v \in T \setminus \cup_{j=0}^{i-1} N_j,\ v$ has a nonneighbor in $N_{i-1}\}$ where $i \geq 1$. Let $u \in N_{i-1}$ and $v, v' \in N_i$ with $vv' \notin E(G)$. Since $\overline{G[T \cup S]}$ is connected, $N_i \neq \emptyset$.

First, we show that u is adjacent to at least one of v and v'. If $i = 1$, we are done. Now we consider the case of $i \geq 2$. Suppose that u is adjacent to neither v nor v'. Let $u' \in N_{i-2}$ be the nonneighbor of u. Then u' is complete to v and v' by the definition of N_i. Since w is complete to $S \cup T$. Then $\{v, v', u', w, u\}$ induces a dart. So u is adjacent to at least one of v and v'.

---

[1] The proofs of theorems and lemmas marked with a $\spadesuit$ have been omitted in the interest of space. They can be found in an extended version of this paper [24].

Next, we show that $|N_i| \leq c(\chi(T))^i$. We have $\chi(N_i) \leq \chi(T)$. Hence, we can partition N_i into at most $\chi(T)$ stable sets, because u is adjacent to at least one of v and v'. Thus, each vertex in N_{i-1} has at most $\chi(T)$ nonneighbors in N_i. Since $|N_0| = |S| \leq c$, we have $|N_i| \leq c(\chi(T))^i$.

Finally, we show that $i \leq 2\chi(T) + 1$. Suppose not. By the definition of N_i, N_i and N_j are complete when $|i - j| > 1$. Then we take a vertex $u_i \in N_i$ where i is even. Now, $\{u_2, u_4, \dots, u_{2\chi(T)+2}\}$ induces a $K_{\chi(T)+1}$, a contradiction. This shows $i \leq 2\chi(T) + 1$.

Therefore, $|T| \leq c(\chi(T))^1 + \dots + c(\chi(T))^{2\chi(T)+1}$.

The following theorem tells us there are finitely many 4-vertex-critical P_5-free graphs.

Theorem 6 ([2,23]). *If $G = (V, E)$ is a 4-vertex-critical P_5-free graph, then $|V| \leq 13$.*

A property on bipartite graphs is shown as follows.

Lemma 5 ([11]). *Let G be a connected bipartite graph. If G contains a $2K_2$, then G must contain a P_5.*

The clique number of G, denoted by $\omega(G)$, is the size of a largest clique in G. A graph G is *perfect* if $\chi(H) = \omega(H)$ for every induced subgraph H of G. Another result we use is the famous Strong Perfect Graph Theorem.

Theorem 7 (The Strong Perfect Graph Theorem [9]). *A graph is perfect if and only if it contains no odd holes or odd antiholes.*

3 Structure Around 5-Hole

Let $G = (V, E)$ be a graph and H be an induced subgraph of G. We partition $V \setminus V(H)$ into subsets with respect to H as follows: for any $X \subseteq V(H)$, we denote by $S(X)$ the set of vertices in $V \setminus V(H)$ that have X as their neighborhood among $V(H)$, i.e.,

$$S(X) = \{v \in V \setminus V(H) : N_{V(H)}(v) = X\}.$$

For $0 \leq m \leq |V(H)|$, we denote by S_m the set of vertices in $V \setminus V(H)$ that have exactly m neighbors in $V(H)$. Note that $S_m = \bigcup_{X \subseteq V(H):|X|=m} S(X)$.

Let G be a $(P_5, dart)$-free graph and $C = v_1, v_2, v_3, v_4, v_5$ be an induced C_5 in G. We partition $V \setminus C$ with respect to C as follows, where all indices below are modulo five.

$$\begin{aligned}
S_0 &= \{v \in V \setminus V(C) : N_C(v) = \emptyset\}, \\
S_2(i) &= \{v \in V \setminus V(C) : N_C(v) = \{v_{i-1}, v_{i+1}\}\}, \\
S_3^1(i) &= \{v \in V \setminus V(C) : N_C(v) = \{v_{i-1}, v_i, v_{i+1}\}\}, \\
S_3^2(i) &= \{v \in V \setminus V(C) : N_C(v) = \{v_{i-2}, v_i, v_{i+2}\}\}, \\
S_4(i) &= \{v \in V \setminus V(C) : N_C(v) = \{v_{i-2}, v_{i-1}, v_{i+1}, v_{i+2}\}\}, \\
S_5 &= \{v \in V \setminus V(C) : N_C(v) = V(C)\}.
\end{aligned}$$

Let $S_2 = \bigcup_{i=1}^{5} S_2(i)$, $S_3^1 = \bigcup_{i=1}^{5} S_3^1(i)$, $S_3^2 = \bigcup_{i=1}^{5} S_3^2(i)$ and $S_4 = \bigcup_{i=1}^{5} S_4(i)$. Since G is P_5-free, we have that

$$V(G) = S_0 \cup S_2 \cup S_3^1 \cup S_3^2 \cup S_4 \cup S_5.$$

We now list a number of useful properties of these sets. The proofs can be found in an extended version of this paper [24].

(1) S_0 is anticomplete to $S_2 \cup S_3^1$.
(2) For each $1 \leq i \leq 5$, $S_3^2(i)$ is not mixed on any edge of S_0.
(3) S_0 is anticomplete to $S_4 \cup S_5$.
(4) Let A be a component of S_0, then A is homogeneous and P_3-free.
(5) For each $1 \leq i \leq 5$, $S_3^1(i)$ is not mixed on any edge of $S_2(i)$.
(6) For each $1 \leq i \leq 5$, $S_2(i)$ is complete to $S_3^1(i+1) \cup S_3^1(i-1)$.
(7) For each $1 \leq i \leq 5$, $S_2(i)$ is anticomplete to $S_3^1(i+2) \cup S_3^1(i-2)$.
(8) For each $1 \leq i \leq 5$, $S_2(i)$ is complete to $S_3^2(i)$.
(9) For each $1 \leq i \leq 5$, $S_2(i)$ is anticomplete to $S_3^2 \setminus S_3^2(i)$.
(10) For each $1 \leq i \leq 5$, $S_2(i)$ is anticomplete to $S_4(i)$.
(11) For each $1 \leq i \leq 5$, $S_2(i)$ is complete to $S_4 \setminus S_4(i)$.
(12) For each $1 \leq i \leq 5$, $S_2(i)$ is complete to $S_2(i+1) \cup S_2(i-1)$.
(13) For each $1 \leq i \leq 5$, $S_2(i+2) \cup S_2(i-2)$ is not mixed on any edge of $S_2(i)$.
(14) If $S_2 \neq \emptyset$, then $S_5 = \emptyset$. If $S_5 \neq \emptyset$, then $S_2 = \emptyset$.
(15) Let A be a component of $S_2(i)$. Then A is homogeneous and A is P_3-free.
(16) For each $1 \leq i \leq 5$, S_5 is complete to $S_3^2(i)$.
(17) Let $u, u' \in S_5$ with $uu' \notin E(G)$. Then every vertex in $S_3^1 \cup S_4$ is adjacent to at least one of u and u'.
(18) For $1 \leq i \leq 5$, $S_3^1(i)$, $S_3^2(i)$ and $S_4(i)$ is a clique, respectively.

4 The Proof of Theorem 4

Proof. We prove the theorem by induction on k. If $1 \leq k \leq 4$, there are finitely many k-vertex-critical $(P_5, dart)$-free graphs by Theorem 6. In the following, we assume that $k \geq 5$ and there are finitely many i-vertex-critical graphs for $i \leq k - 1$. Now, we consider the case of k.

Let $G = (V, E)$ be a k-vertex-critical $(P_5, dart)$-free graph. We show that $|G|$ is bounded. Let $\mathcal{L} = \{K_k, \overline{C_{2k-1}}\}$. If G has a subgraph isomorphic to a member $L \in \mathcal{L}$, then $|V(G)| = |V(L)|$ by the definition of vertex-critical and so we are done. So, we assume in the following G has no induced subgraph isomorphic to a member in $\mathcal{L}$. Then G is imperfect. Since G is k-vertex-critical and $\chi(\overline{C_{2t+1}}) \geq k+1$ if $t \geq k$, it follows that G does not contain $\overline{C_{2t+1}}$ for $t \geq k$. Moreover, since G is P_5-free, it does not contain C_{2t+1} for $t \geq 3$. It then follows from Theorem 7, G must contain some $\overline{C_{2t+1}}$ for $2 \leq t \leq k - 2$. To finish the proof, we only need to prove the following two lemmas.

Lemma 6. *If G contains an induced C_5, then G has finite order.*

Proof. Let $C = v_1, v_2, v_3, v_4, v_5$ be an induced C_5. We partition $V(G)$ with respect to C.

Since G is K_k-free, combined with **(18)**, we have $S_3^1(i)$, $S_3^2(i)$ and $S_4(i)$ is K_{k-2}-free, respectively. Then $|S_3^1(i)| \leq k-3$, $|S_3^2(i)| \leq k-3$ and $|S_4(i)| \leq k-3$. Therefore, $|S_3^1| \leq 5k-15$, $|S_3^2| \leq 5k-15$ and $|S_4| \leq 5k-15$. Thus, in the following, we only need to bound S_0, S_2 and S_5.

We first bound S_0.

Claim 4.01. *Let A be a component of S_0, then $\chi(A) \leq k-2$.*

Proof. Suppose that $\chi(A) \geq k-1$. Since G is connected, there must exist $v \in N(A)$ and $v \in S_3^2$ by **(1)**–**(3)**. Since A is homogeneous, v is complete to A. Then $\chi(G[V(A) \cup \{v\}]) \geq k$. Since $G[V(A) \cup \{v\}] \subset G$, it contradicts with G is k-vertex-critical.

Let A be a component of G, we call A a K_i-component if $A \cong K_i$ where $i \geq 1$.

Claim 4.02. *Each component of S_0 is a K_m where $1 \leq m \leq k-2$. For each $1 \leq m \leq k-2$, the number of K_m-components is not more than 2^{5k-15}.*

Proof. By **(4)**, S_0 is a disjoint union of cliques. Combined with Claim 4.01, it follows that each component of S_0 is a K_m where $1 \leq m \leq k-2$.

Suppose that the number of K_1-components in S_0 is more than $2^{5k-15} \geq 2^{|S_3^2|}$. The pigeonhole principle shows that there are two K_1-components u, v having the same neighborhood in S_3^2. Since S_0 is anticomplete to $S_2 \cup S_3^1 \cup S_4 \cup S_5$. Then u, v have the same neighborhood in $V(G)$. This contradicts with Lemma 1.

Similarly, we can show that the number of K_2-components $,\ldots,$ K_{k-2}-components is not more than 2^{5k-15}, respectively.

By Claim 4.02, it follows that S_0 is bounded. Next, we bound S_2 and S_5. Note that at least one of S_2 and S_5 is an empty set by **(14)**. In the following, we first assume that $S_5 \neq \emptyset$. Then $S_2 = \emptyset$.

Claim 4.03. *S_5 is bounded.*

Proof. Let $N_0 = S_3^1 \cup S_4$, $N_i = \{v | v \in S_5 \setminus \cup_{j=0}^{i-1} N_j$, v has a nonneighbor in $N_{i-1}\}$ where $i \geq 1$.

If $\overline{G[S_5 \cup S_3^1 \cup S_4]}$ is connected. Since $|S_3^1 \cup S_4| \leq 10k-30$ and there exist $w \in V(C)$ such that w is complete to $S_3^1 \cup S_4 \cup S_5$. By **(17)**, every vertex in $S_3^1 \cup S_4$ is adjacent to at least one of w_1 and w_2 where $w_1, w_2 \in S_5$ and $w_1 w_2 \notin E(G)$. Then $|S_5|$ is a function of $\chi(S_5)$ by Lemma 4. Since $\chi(S_5) \leq k-3$, it follows that S_5 is bounded.

If $\overline{G[S_5 \cup S_3^1 \cup S_4]}$ is not connected. Then there exists an integer $j \geq 0$ such that $N_0, N_1, \ldots, N_j \neq \emptyset$ but $N_{j+1} = \emptyset$. Then $S_5 - \cup_{i=0}^{j} N_i$ is complete to $\cup_{i=0}^{j} N_i$, and so $S_5 - \cup_{i=0}^{j} N_i$ is a homogeneous set. Since $\chi(S_5 - \cup_{i=0}^{j} N_i) \leq k-3$, each component of $S_5 - \cup_{i=0}^{j} N_i$ is an m-vertex-critical graph with $1 \leq m \leq k-3$ by Lemma 3. By the inductive hypothesis, it follows that there are finitely many

m-vertex-critical (P_5, $dart$)-free graphs with $1 \leq m \leq k - 3$. By the pigeonhole principle, the number of each kind of graph is not more than 2. So, $S_5 - \cup_{i=0}^{j} N_i$ is bounded. For each N_i with $1 \leq i \leq j$, N_i is bounded by Lemma 4. Therefore, S_5 is bounded.

Thus, $|G|$ is bounded if $S_5 \neq \emptyset$. Next, we assume that $S_2 \neq \emptyset$. Then $S_5 = \emptyset$ by (14).

Claim 4.04. *Each component of $S_2(i)$ is a K_m where $1 \leq m \leq k - 2$. For each $1 \leq m \leq k - 2$, the number of K_m-components is not more than $2^{k-2} + 1$.*

Proof. By (15), $S_2(i)$ is a disjoint union of cliques. If $S_2(i)$ contains a K_{k-1}, then G contains a K_k, a contradiction. So each component of $S_2(i)$ is a K_m where $1 \leq m \leq k - 2$.

Suppose that the number of K_1-components in $G[S_2(1)]$ is more than $2^{k-2} + 1 \geq 2 \cdot 2^{|S_3^1(1)|} + 1$. By the pigeonhole principle, there are y_1, y_2, y_3 having the same neighborhood in $S_3^1(1)$. For any $y_i \neq y_j$ with $i, j \in \{1, 2, 3\}$, since y_i and y_j are not comparable, there must exist $y_i' \in N(y_i) \setminus N(y_j)$, $y_j' \in N(y_j) \setminus N(y_i)$. By (5)–(13), $y_i', y_j' \in S_2(3) \cup S_2(4)$.

If $y_i', y_j' \in S_2(3)$, then $y_i' y_j' \in E(G)$. For otherwise $\{y_i', y_i, v_5, y_j, y_j'\}$ is an induced P_5. Then $\{y_i', y_j', y_j, v_2, v_1\}$ is a dart, a contradiction. By symmetry, $y_i', y_j' \notin S_2(4)$. In the following, by symmetry, we assume that $y_i' \in S_2(3)$, $y_j' \in S_2(4)$. Then $y_i' y_j' \in E(G)$, otherwise $\{y_j, y_j', v_3, v_4, y_i'\}$ is an induced P_5. Let $y_l \neq y_i, y_j$ with $1 \leq l \leq 3$. Then $y_l y_i' \notin E(G)$, otherwise $\{y_i, y_i', y_l, v_2, v_1\}$ is a dart. By symmetry, $y_l y_j' \notin E(G)$. Since y_i and y_l are not comparable, there must exist $y_l' \in N(y_l) \setminus N(y_i)$. Then $y_l' \in S_2(4)$, $y_j y_l' \notin E(G)$ and $y_l y_j' \notin E(G)$. Then $y_l' y_j' \in E(G)$, otherwise $\{y_l, y_l', v_3, y_j', y_j\}$ is an induced P_5. Then $\{y_j, y_j', y_l', v_5, v_1\}$ is a dart, a contradiction.

Similarly, we can show that the number of K_2-components , . . . , K_{k-2}-components is not more than $2^{k-2} + 1$, respectively.

Therefore, $|G|$ is bounded if $S_2 \neq \emptyset$. This completes the proof of Lemma 6.

Lemma 7 (♠). *If G contains an induced $\overline{C_{2t+1}}$ for $3 \leq t \leq k - 2$, then G has finite order.*

By Lemma 6–Lemma 7, it follows that Theorem 4 holds.

5   Complete Characterization for $k \in \{5, 6, 7\}$

In Sect. 4, we proved that there are finitely many k-vertex-critical (P_5, $dart$)-free graphs by showing the existence of an upper bound for the order of such graphs for every integer $k \geq 1$. These bounds are not necessarily sharp. In the current section, we show sharp upper bounds for $k \in \{5, 6, 7\}$ by computationally determining an exhaustive list of all k-vertex-critical (P_5, $dart$)-free graphs.

We created two independent implementations of the algorithm from Goedgebeur and Schaudt [13], which we extended for the (P_5, $dart$)-free case. The source code of these implementations can be downloaded from [12] and [20]. The algorithm expects three parameters as an input: an integer $k \geq 1$, a set of graphs $\mathcal{H}$ and a graph I. It generates all k-vertex-critical $\mathcal{H}$-free graphs that contain I as an induced subgraph. The pseudocode is given in Algorithm 1. The algorithm is not guaranteed to terminate (e.g. there could be infinitely many such graphs). However, if the algorithm terminates, it is guaranteed that the generated graphs are exhaustive. The algorithm works by recursively extending a graph with one vertex and adding edges between this new vertex and already existing vertices. In each step of the recursion, the algorithm uses powerful pruning rules that restrict the ways in which the edges are added. These pruning rules allow the algorithm to terminate some branches of the recursion and are the reason why the algorithm itself can terminate in some cases. For example, a graph which is not $(k-1)$-colorable cannot appear as a proper induced subgraph of a k-vertex-critical graph, so the algorithm does not need to extend such graphs. In general, the pruning rules that are used by the algorithm are much more sophisticated than this. We refer the interested reader to [13] for further details about the correctness of the algorithm and the different pruning rules.

Algorithm 1. Extend(An integer k, A set of graphs $\mathcal{H}$, A graph I)

1: **if** I is $\mathcal{H}$-free AND not generated before **then**
2:    **if** I is not $(k-1)$-colorable **then**
3:       **if** I is a k-vertex-critical graph **then**
4:          output I
5:       **end if**
6:    **else**
7:       **for** every graph I' obtained by adding a new vertex u to I and edges between u and vertices in $V(I)$ in all possible ways that are permitted by the pruning rules **do**
8:          Extend($k,\mathcal{H},I'$)
9:       **end for**
10:   **end if**
11: **end if**

We now prove the following characterization theorem:

Theorem 8. *There are exactly 184 5-vertex-critical $(P_5, dart)$-free graphs and the largest such graphs have order 13. There are exactly 18,029 6-vertex-critical $(P_5, dart)$-free graphs and the largest such graphs have order 16. There are exactly 6,367,701 7-vertex-critical $(P_5, dart)$-free graphs and the largest such graphs have order 19.*

Proof. We saw in Sect. 4 that every k-vertex-critical $(P_5, dart)$-free graph is either K_k, $\overline{C_{2k-1}}$ or contains C_5 as an induced subgraph or contains $\overline{C_{2t+1}}$ as an induced subgraph for some $2 \leq t \leq k-2$. If Algorithm 1 is called with the parameters $k \in \{5, 6, 7\}$, $\mathcal{H} = \{P_5, dart\}$ and I one of these graphs, the algorithm terminates in less than a second for $k = 5$, less than a minute for $k = 6$

and a few hours for $k = 7$ (for all choices of k and I). The counts of these graphs are reported in Table 1. The results of the two independent implementations of this algorithm (cf. [12] and [20]) are in complete agreement with each other.

Table 1. The number of k-critical and k-vertex-critical $(P_5, dart)$-free graphs (for $k \in \{5, 6, 7\}$).

Vertices	5	6	7	8	9	10	11	Total
5-critical	1		1	1	7	1		
5-vertex-critical	1		1	6	172	1		
6-critical		1		1	1	6	33	
6-vertex-critical		1		1	6	171	17,834	
7-critical			1	1	1	6		
7-vertex-critical			1	1	6	171		

Vertices	12	13	14	15	16	19		Total
5-critical		3						14
5-vertex-critical		3						184
6-critical	2	1			13			58
6-vertex-critical	2	1			13			18,029
7-critical	28	250	6	2	1	35		331
7-vertex-critical	17,834	6,349,644	6	2	1	35		6,367,701

Table 1 gives an overview of the number of k-critical and k-vertex-critical $(P_5, dart)$-free graphs for $k \in \{5, 6, 7\}$. A graph G is k-critical $(P_5, dart)$-free if it is k-chromatic, $(P_5, dart)$-free and every $(P_5, dart)$-free proper subgraph of G is $(k - 1)$-colorable. The graphs from Table 1 can be obtained from the meta-directory of the *House of Graphs* [3] at https://houseofgraphs.org/meta-directory/critical-h-free. Moreover, the k-critical graphs from Table 1 can also be inspected in the searchable database of the *House of Graphs* [3] by searching for the keywords "critical (P5,dart)-free". The 5-critical $(P_5, dart)$-free graphs are shown in Fig. 2.

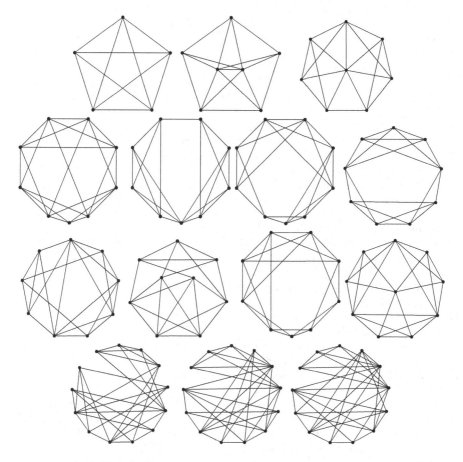

Fig. 2. All 14 5-critical $(P_5, dart)$-free graphs.

6   Conclusion

In this paper, we have proved that there are finitely many k-vertex-critical $(P_5, dart)$-free graphs for $k \geq 1$ and computationally determined an exhaustive list of such graphs for $k \in \{5, 6, 7\}$. Our results gave an affirmative answer to the problem posed in [8] for $H = dart$. In the future, it is natural to investigate the finiteness of the set of k-vertex-critical (P_5, H)-free graphs for other graphs H of order 5, see [7].

References

1. Bondy, J.A., Murty, U.S.R.: Graph Theory. Springer, Heidelberg (2008)
2. Bruce, D., Hoàng, C.T., Sawada, J.: A certifying algorithm for 3-colorability of P_5-free graphs. In: Dong, Y., Du, D.-Z., Ibarra, O. (eds.) ISAAC 2009. LNCS, vol. 5878, pp. 594–604. Springer, Heidelberg (2009). https://doi.org/10.1007/978-3-642-10631-6_61

3. Coolsaet, K., D'hondt, S., Goedgebeur, J.: House of graphs 2.0: a database of interesting graphs and more. Discrete Appl. Math. **325**, 97–107 (2023)
4. Cai, Q., Goedgebeur, J., Huang, S.: Some results on k-critical P_5-free graphs. Discrete Appl. Math. **334**, 91–100 (2023)
5. Cai, Q., Huang, S., Li, T., Shi, Y.: Vertex-critical (P_5, banner)-free graphs. In: Chen, Y., Deng, X., Lu, M. (eds.) FAW 2019. LNCS, vol. 11458, pp. 111–120. Springer, Cham (2019). https://doi.org/10.1007/978-3-030-18126-0_10
6. Cameron, B., Hoàng, C.T.: A refinement on the structure of vertex-critical (P_5, gem)-free graphs. Theoret. Comput. Sci. **961**, 113936 (2023)
7. Cameron, B. and Hoàng, C. T.: Infinite families of k-vertex-critical (P_5, C_5)-free graphs. arXiv arXiv:2306.03376v1 [math.CO] (2023)
8. Cameron, K., Goedgebeur, J., Huang, S., Shi, Y.: k-critical graphs in P_5-free graphs. Theoret. Comput. Sci. **864**, 80–91 (2021)
9. Chudnovsky, M., Robertson, N., Seymour, P., Thomas, R.: The strong perfect graph theorem. Ann. Math. **164**, 51–229 (2006)
10. Dhaliwal, H.S., Hamel, A.M., Hoàng, C.T., Maffray, F., McConnell, T.J.D., Panait, S.A.: On color-critical (P_5, co-P_5)-free graphs. Discrete Appl. Math. **216**, 142–148 (2017)
11. Fouquet, J.L.: A decomposition for a class of (P_5, $\overline{P_5}$)-free graphs. Discrete Math. **121**, 75–83 (1993)
12. Goedgebeur, J.: Homepage of generator for k-critical $\mathcal{H}$-free graphs. https://caagt.ugent.be/criticalpfree/
13. Goedgebeur, J., Schaudt, O.: Exhaustive generation of k-critical $\mathcal{H}$-free graphs. J. Graph Theor. **87**, 188–207 (2018)
14. Hell, P., Huang, S.: Complexity of coloring graphs without paths and cycles. Discrete Appl. Math. **216**, 211–232 (2017)
15. Hoàng, C.T., Kamiński, M., Lozin, V.V., Sawada, J., Shu, X.: Deciding k-colorability of P_5-free graphs in polynomial time. Algorithmica **57**, 74–81 (2010)
16. Hoàng, C. T., Moore, B., Recoskiez, D., Sawada, J., Vatshelle. M.: Constructions of k-critical P_5-free graphs. Discrete Appl. Math. **182**, 91–98, (2015)
17. Huang, S., Li, J., Xia, W.: Critical (P_5, bull)-free graphs. Discrete Appl. Math. **334**, 15–25 (2023)
18. Huang, S., Li, T., Shi, Y.: Critical (P_6, banner)-free graphs. Discrete Appl. Math. **258**, 143–151 (2019)
19. Huang, S., Li, Z.: Vertex-critical (P_5, chair)-free graphs. Discrete Appl. Math. **341**, 9–15 (2023)
20. Jooken, J.: GitHub page containing generator for k-vertex-critical $\mathcal{H}$-free graphs. https://github.com/JorikJooken/kVertexCriticalGraphs
21. Kamiński, M., Pstrucha, A.: Certifying coloring algorithms for graphs without long induced paths. Discrete Appl. Math. **261**, 258–267 (2019)
22. Karp, R.M.: Reducibility among combinatorial problems. In: Miller, R.E., Thatcher, J.W., Bohlinger, J.D. (eds.) Complexity of Computer Computations. The IBM Research Symposia Series. Springer, Boston (1972). https://doi.org/10.1007/978-1-4684-2001-2_9
23. Maffray, F., Morel, G.: On 3-colorable P_5-free graphs. SIAM J. Discrete Math. **26**, 1682–1708 (2012)
24. Xia, W., Jooken, J., Goedgebeur, J., Huang, S.: Critical (P_5, dart)-Free Graphs. arXiv arXiv:2308.03414v2 [math.CO] (2023)

Graph Clustering Through Users' Properties and Social Influence

Jianxiong Guo[1], Zhehao Zhu[2], Yucen Gao[2], and Xiaofeng Gao[2(✉)]

[1] Advanced Institute of Natural Sciences, Beijing Normal University,
Zhuhai 519087, China
`jianxiongguo@bnu.edu.cn`
[2] Department of Computer Science and Engineering, Shanghai Jiao Tong University,
Shanghai, China
`zhehaozhu11@gmail.com`, `guo_ke@sjtu.edu.cn`, `gao-xf@cs.sjtu.edu.cn`

Abstract. Clustering is a basic technology in data mining, and similarity measurement plays a crucial role in it. The existing clustering algorithms, especially those for social networks, pay more attention to users' properties while ignoring the global measurement across social relationships. In this paper, a new clustering algorithm is proposed, which not only considers the distance of users' properties but also considers users' social influence. Social influence can be further divided into mutual influence and self influence. With mutual influence, we can deal with users' interests and measure their similarities by introducing areas and activities, thus better weighing the influence between them in an indirect way. Separately, we formulate a new propagation model, PR-Threshold++, by merging the PageRank algorithm and Linear Threshold model, to model the self influence. Based on that, we design a novel similarity by exploiting users' distance, mutual influence, and self influence. Finally, we adjust K-medoids according to our similarity and use real-world datasets to evaluate their performance in intensive simulations.

Keywords: Graph Clustering · Social Influence · K-medoids · Data Mining

1 Introduction

Clustering is one of the most commonly used methods in data mining. It hopes to divide data points into several groups, and make sure instances in the same group are similar to each other but different from those in other groups [6]. Clustering has been widely used in a variety of fields, such as machine learning [4,8], pattern classification [9,12], image segmentation [3,14], and so on. For

This work was supported in part by the National Key R&D Program of China [2020YFB1707900], the National Natural Science Foundation of China (NSFC) [62202055, 62272302, 62172276], and Shanghai Municipal Science and Technology Major Project [2021SHZDZX0102].

W. Wu and J. Guo (Eds.): COCOA 2023, LNCS 14462, pp. 403–415, 2024.
https://doi.org/10.1007/978-3-031-49614-1_30

example, a general method for clustering is called "centroid-based clustering". Its key point is to find the cluster center, which is related to the number of clusters K. The most important part of clustering is how to define features, and then group instances with similar features together. They realize grouping, try to improve the quality of clustering by measuring similarity and repeat until they converge. Therefore, when it comes to clustering, the similarity measure should be predefined according to the requirements of real-world applications.

However, in the existing similarity measure of users in social networks, they tended to obtain information directly from graph structure or make use of the inherent properties such as users' age and gender. In other words, the similarity measures used for clustering in the past were usually single-dimensional. In fact, this is not enough. For example, global factors such as social influence were usually neglected. Social influence is the influence of a group of users through their words and behaviors. By analyzing and mining social networks, we can understand how people communicate with each other and how the information spreads in social networks. Naturally, when defining similarity, we should take the influence analysis into account and the similarity measure should be multi-dimensional. Therefore, our design is based on the following assumptions. Firstly, we divide users' interests into several areas, and each area has several activities. Then, the users' interests can be taken advantage of to find out the activities they participate in. For one thing, we need to make use of the direct connection between users. For another thing, some messages are contained in the users' participation in different areas and activities. Secondly, the influence of each user should be contained instead of being discarded because the influential people for an area or activity can be considered similar to some extent.

Based on the above Observations, We can Summarize our Main Innovations as Follows. Firstly, the influence in social networks can be divided into mutual influence and self influence by us. Mutual influence depends on users' interests, and it will be quantified by studying the areas and activities that users are engaged in. Inspired by this, we novelly define the *MatchScore* and *MismatchScore* between any two users, so as to weigh the similarity in the mutual influence part and optimize the influence propagation in the self influence part. Secondly, self influence is more about the accumulation of influence in social networks, and the results are generated by the users themselves. In the process of influence propagation, both PageRank (PR) algorithm and linear threshold (LT) model in the influence maximization (IM) problem [2,5] ignore the topic when propagating messages. In order to solve these challenges, we merge them together and propose a new influence propagation model, PR-Threshold++, where the weights of edges are defined by using the above matching scores. Thirdly, by comprehensively considering users' properties, mutual influence, and self influence, we propose a multi-dimensional similarity measure, which is totally different from the existing single-dimensional ones. Finally, under our proposed similarity measure, we design a new clustering algorithm, Clustering with Influence Analysis (CIAA), based on K-medoids clustering [1]. In addition, we also prove that our CIAA algorithm can achieve the clustering that contains various factors compared to the common K-medoids algorithm [7,13]. At the same time, we put forward a new index, sparsity, to evaluate our results, which indicate the correctness and effectiveness of our proposed clustering.

2   Problem Statement

A social network usually relies on graph, which is denoted by $G(V, E)$ where V represents node (user) set and E represents edge set. There are two types of edges: undirected (v_1, v_2) and directed $\langle v_1, v_2 \rangle$. In the directed graphs, the node v's in-degree is the number of edges whose destination is v, and out-degree is the number of edges whose source is v. There are two kinds of graph in our problem: Distance graph and Propagation graph. A Distance graph is defined as $G_d(V, E, D)$, where D represents the set of distances corresponding to set E and is calculated according to the adopted method. A Propagation graph includes User network and Activity network, which is used to deal with self influence and mutual influence respectively. A User Network is defined as $G_u(V, E, P)$, where P is the set of influence probability which goes with each edge in E. In other words, each element $p \in P$ decides the probability whether this edge exists or not. The Activity network is of great significance. It consists of several independent subsets, each of which represents an area (a kind of activities). Each element in an area is a specific activity. For example, an Activity network might include areas like music, while in the area of music, it has activities like playing flute, blues, and so on.

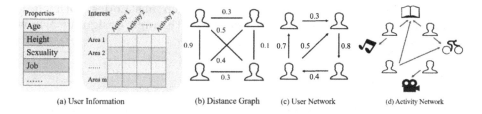

(a) User Information   (b) Distance Graph   (c) User Network   (d) Activity Network

Fig. 1. Structure of Graphs and Networks

We take Fig. 1 to illustrate our problem in detail. As shown in Fig. 1, the (a) shows the format of user information. There are two kinds of knowledge about her for each user. The first part is her properties like age, height, and job, which makes it convenient to calculate the distance between different users. The second part is her interests, where areas range from 1 to m and in each area, activities range from 1 to n. The (b) offers an example of Distance graph, where the value on each undirected edge represents the distance between two users calculated by their properties. The (c) shows us a case of User network, in which users are connected through directed edges. Each user has her own influence ability, thus the possibility p of each edge is used in influence propagation and we temporarily dismiss it here. The (d) is an example of Activity network. To better explain, we combine Activity network with User network, where the red edges from users to activities represent that the user is fond of its corresponding area. For example, there are four areas like music, sport, reading, and movie.

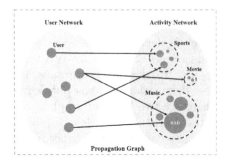 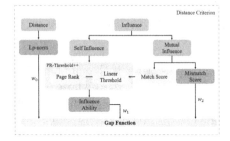

Fig. 2. Propagation Graph and Bipartite Relationship

Fig. 3. Compositions of Similarity Measure

Activity network is denoted by the set $S_a = S_1 \cup S_2 \cup \cdots \cup S_m$, where S_k represents the set of the kth area and contains at most n different activities. Thus, Activity network S_a can be described as a graph with no edge, like the right part of Fig. 2. When we temporarily ignore the directed edge in the User network, a bipartite graph is established from users in the User network to activities in the Active network, which is shown in Fig. 2. **The core problem is to partition users V into K disjoint clusters, and ensure the clustering results in densely connected groups and each has nodes with similar influence ability and activity behaviors through designing a new similarity measure.**

3   Similarity Measure

In clustering, it is important to distinguish whether two instances are similar. In this section, we are going to take three factors into consideration **Distance**, **Mutual Influence**, and **Self Influence**. As Shown in Fig. 3, we will use L_2-norm to calculate the distance between user's properties. Given two instance $X = (x_1, \cdots, x_d)$ and $Y = (y_1, \cdots, y_d)$, we have $Dist(X, Y) = \sqrt{\sum_{i=1}^{d}(x_i - y_i)^2}$. In the influence part, we will use $MatchScore$ and $MismatchScore$ to deal with Mutual influence, then put forward PR-Threshold++ that combines PageRank, LT model, and $MatchScore$ to calculate Self influence. Finally, the components in orange boxes together form the Gap function.

3.1   Mutual Influence

In the mutual influence part, we should determine their matching degree according to their interests. We put forward two indicators to measure the similarity and dissimilarity respectively, $MatchScore$ (MaS) and $MismatchScore$ (MiS). Here, we define the **importance (popularity)** of an area or activity with the rate they make up. For example, if 10 person-times (10 out of 100 people) are interested in music, the importance of music will be graded as 0.1. The"person-time" is statistical counts, not a person. The degree to which two users match each other depends on how many common activities they share within the scope of our statistics.

We set a matrix $A_{m \times n}$ for every user. In such a matrix, m represents the number of areas and n represents the number of activities in an area. Then, $A_u[x][y] = 1$ if and only if the user u is fond of activity y in the area x; otherwise, we have $A_u[x][y] = 0$. The extent to which two nodes match, for example, u and v, will be calucated as Eq. (1):

$$MaS(u,v) = \sum_{i=1}^{M} \sum_{j=1}^{N} (A_u[i][j] \& A_v[i][j]) \times z[i][j] \qquad (1)$$

where $\&$ is logical AND. The $z[i][j]$ represents the importance of area i or activity j in the area i. For example, $z[music][\cdot] = 0.1$ is defined according to the above example, which can be considered as a weighted index. Correspondingly, the extent to which u and v mismatch will be calculated as Eq. (2):

$$MiS(u,v) = \sum_{i=1}^{M} \sum_{j=1}^{N} (A_u[i][j] \oplus A_v[i][j]) \times z[i][j] \qquad (2)$$

where $\oplus$ is logical Exclusive OR. Here, MaS is mainly used in self influence and MiS is mainly used to measure the gap in mutual influence, but it also plays a role in influence propagation.

3.2 Self Influence

In self influence part, we try to measure the ability of each node to spread messages and influence others in the network. PageRank is an algorithm for evaluating the importance and quality of a webpage in a search system. From these points of view, the influence exerted by a node is related to its neighboring nodes, similar to the idea of social network analysis.

To the PageRank algorithm, we first calculate the PR values of those nodes who are not destinations of any other nodes. Then, we can further use them as source, and calculate the PR values of those nodes that are only connected to the source nodes. Afterwards, the process will repeat until every node has its own PR value. For each user u, its PR value is defined as

$$PR(u) = \sum_{v \in I_u} PR(v)/L(v), \qquad (3)$$

where I_u is the set of inbound links of page u and $L(v)$ is the number of outbound links of v. We have to get the topological orders of nodes, thus the time complexity is $O(|V|^2)$.

In fact, the process in PageRank algorithm is closer to the accumulation of influence. It is carried out in a fixed way, and the interactions between the

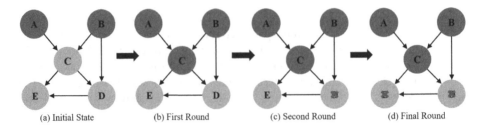

| (a) Initial State | (b) First Round | (c) Second Round | (d) Final Round |

Fig. 4. Example of Linear Threshold Model

nodes can not simulate diffusion in social networks. Thus, we turn to the LT model [2], which can be summarized as follows (a little different from its original definition): 1) Each directed edge (u, v) has a weight $w(u, v) \in [0, 1]$; 2) Each node v has a threshold θ_v randomly chosen in $[0, 1]$; 3) At the beginning, only nodes in the seed set S_0 are active; 4) In round t, the inactive nodes $v \in V \setminus S_{t-1}$ will be activated by its in-neighbors if their influence sum exceeds v's threshold, then adding the newly activated node in this round to S_{t-1}.

Let us take Fig. 4 as an example. At the beginning, node A and node B are active, their influence values are 0.5 and 0.2 respectively. The threshold θ is set as 0.5. In the first round, node C can be activated because $Inf_C = \frac{Inf_A}{out-degree_A} + \frac{Inf_B}{out-degree_B} = 0.5 + \frac{0.2}{2} = 0.6 > 0.5$. In the second round, node D cannot be activated because $Inf_D = \frac{Inf_B}{out-degree_B} + \frac{Inf_C}{out-degree_C} = \frac{0.6}{2} + \frac{0.2}{2} = 0.4 < 0.5$. In the third round, node E cannot be activated as well because $Inf_E = \frac{Inf_C}{out-degree_C} = \frac{0.6}{2} = 0.3 < 0.5$. In fact, we find some shared characteristics of these two models: Firstly, in PR Algorithm, it arranges the weight of each edge according to the out-degree of the source of edges. And in LT model, it randomly decides the weight of each edge theoretically but it takes a similar method as PR in real applications. Secondly, the influence goes along the edges in both two models, which means the propagation ways are the same. The LT model defined above is a little different from its original definition in [2] because we use influence values of nodes. That is to say, we can calculate the influence ability with the help of PR algorithm and then simulate the propagation in LT model.

Because of too many random operations, LT model also has some shortcomings. It is mainly summarized as the following two points: (1) The influence degree of active users on inactive users is uncertain, because the weight of each edge is determined randomly; and (2) The threshold of each node are also randomly assigned. In order to improve the problems in LT model and help the whole model get better performance, we made the following adjustments. As for the first drawback, with the help of mutual influence, we can weigh the similarity between users by MaS, thus deciding how much the inactive users are affected. We will compare the MaS and MiS, as well as the reciprocal of out-degree of source. The rule of comparison is: if $MaS \geq MiS$, then the influence will propagate in the greater one between the similarity (MaS) and the reciprocal of out-degree of source. Otherwise, we compare MiS and the reciprocal of out-degree of source. If MiS is bigger, then the propagation weight will be zero; Otherwise, the propagation will be the reciprocal of out-degree of source. The weight in LT model between u and v is shown as Eq. (4).

$$w(u,v) = \begin{cases} \max\{MaS(u,v), \frac{1}{od(u)}\}, MaS(u,v) \geq MiS(u,v) \\ \text{cmp}\{MiS(u,v), \frac{1}{od(u)}\}, MaS(u,v) < MiS(u,v), \end{cases} \qquad (4)$$

where $od(u)$ means out-degree of user u and function cmp defined as Eq. (5):

$$\text{cmp}\left\{MiS(u,v), \frac{1}{od(u)}\right\} = \begin{cases} 0, MiS(u,v) \geq \frac{1}{od(u)} \\ \frac{1}{od(u)}, MiS(u,v) < \frac{1}{od(u)}. \end{cases} \qquad (5)$$

As for the second drawback, we have already made the threshold of subsequent activation after initialization should not be lower than the PR values of the first 20% users, thus ensuring that influence ability of users in active set will not be too weak.

To sum up, the new propagation model, PR-Threshold++, has been formulated, which combines PR algorithm and LT model to measure the ability to spread influence. It can be used to calculate the self influence of any user. Given any user u, the $SelfInfluence(u)$ can be calculated by following steps:

1. Initialize each node with PR value of $1/|V|$.
2. Compute the PR value of all nodes until convergence based on Eqn. (3).
3. Initialize $S_0 = \{u\}$.
4. Repeat from $t = \{1, 2, \cdots\}$ until no activation happens: For any node v that is newly activated in round $t-1$, it can diffuse its influence to its inactive out-neighbor $v' \in N^+(v) \cup (V\backslash S_{t-1})$. Here, the weigh $w(v,v')$ is defined in Equ. (4). If we have $\sum_{x \in N^-(v') \cap S_{t-1}} w(x,v') \geq \theta_{v'}$, then v' can be activated and $S_t = S_{t-1} \cup \{v'\}$. Here, $N^+(\cdot)$ and $N^-(\cdot)$ are out- and in-neighbor set.
5. Return the final number of active nodes $|S_t|$.

3.3 The Combination of Different Parts

Because we emphasize the influence factor, we can even map self influence and mutual influence to multiple times of the average gap, which leads to greater differences in the calculation of similarity between influences. This kind of combination is much better since both parts can make a difference. However, it is worth mentioning that the number of times cannot be too large. On the contrary, after mapping, ideally, the distance is greater than the difference between users' influence in the same cluster, but less than the difference between users' influence from two different clusters. Along with this thought, we can find out the coefficient of **Distance**, **Self Influence** and **Mutual Influence**. We denote them by w_0, w_1, and w_2 respectively. And we can draw a conclusion about the final similarity measure between user u and user v as

$$Gap(u,v) = w_0 \cdot Dist(u,v) + w_1 \cdot |SeG(u,v)| + w_2 \cdot |MiS(u,v)| \qquad (6)$$

where $SelfGap(u,v)$ is abbreviated as $SeG(u,v)$, which is to represent the value of $|SelfInfluence(u) - SelfInfluence(v)|$ as shown in Fig. 3.

4   Clustering with Influence Analysis Algorithm

In this section, we are going to talk about the clustering algorithm that we proposed: Clustering with Influence Analysis Algorithm (CIAA). We will go from several aspects, including the general algorithm, as well as some details we have changed and adjusted according to our similarity measure. On the whole, we can choose a clustering algorithm from K-means and K-medoids. In this paper, since function Gap contains influence factors other than distance, it is more likely to have noise. Although K-medoids run slowly, we still choose it to avoid the negative results brought by K-means. Finally, we choose K-medoids as the template to design our CIAA as shown in Algorithm 1.

Algorithm 1: CIAA CLUSTERING

Input: Social network (both user's properties and Graph G)

Output: K clusters and its composition

1 Run PR-Threshold++, obtain the influence ability of each node ;

2 Run DENCLUE and get initial cluster centroids, for example
   $C = \{c_1, c_2, \cdots, c_k\}$;

3 **repeat**

4 | For every node v, compute its Gap to the k cluster centers respectively according to Eq. (6) ;

5 | Allocate the instance to its nearest cluster;

6 | For each c_i, update cluster center according to Eq. (7);

7 **until** *Cluster centers no change or the iterations reach upper bound;*

Initialization: The main work in the initialization part is to decide which nodes to be centroids. As to how to choose initial centroids, it is an important problem in clustering algorithms because it will directly affect the results. Here, we adopt DENCLUE [10], which chooses centroids from the densest area. Thus, the first step is to compute the density for each user node in the social graph. The density of a certain node can be computed by $D(u_i) = \sum_{u_j \in V, u_j \neq u_i} 1/Gap(u_i, u_j)$, where the Gap is defined in Eqn. (6). If a user node has a large density, it is close to surrounding users in distance and has adequate connections with them. In the meanwhile, it shares highly similar areas or activities with the surrounding neighbors. Thus, we choose the top K user node with the highest density as initial centroids, which is denoted by $\{c_1, c_2, \cdots, c_k\}$.

Cluster Objectives: The clustering objective is to minimize the sum of distance within every cluster. However, we have to adjust the distance calculation since we not only consider the distance part but take the influence part into account as well, which has been defined in Eqn. (6). Besides, the property of convergence can be easily inherited from K-medoids.

Node Assignment and Updating Process: The goal of our clustering algorithm is to minimize the Gap. When it comes to node assignment, we will allocate a node to the cluster whose centroid has the minimum Gap to the current node.

To explain it specifically, when a number of nodes are assigned to its closest cluster centroid, we have to re-determine and update the centroid of the cluster. To find such a node, we need to compute the *Gap* between a chosen node and other nodes, then the centroid will be updated as the node with minimum *Gap*. For example, for node u_i, the *Gap* between it and other nodes is

$$Sum = \sum_{u_j \in V_c, u_i \neq u_j} Gap(u_i, u_j), \tag{7}$$

where U_c represents the user set of the cluster that u_i and u_j in. In the meantime, we would like to find out the cluster's interest label. To realize it, we only need to evaluate the most popular area or activity in this cluster.

5   Experiment and Performance

In this section, we would like to carry out the experiments and analyze the performance of our CIAA clustering. To declare, all executing time in this section is in microseconds. The experiments can be separated into several parts to demonstrate the superiority of our algorithm from different angles. First, the SSE refers to the sum of squared estimate of errors. In clustering, it can be conveniently calculated as $SSE = \sum_{i=1}^{k} \sum_{p \in C_i} |p - m_i|^2$, where C_i refers to the ith cluster, p is data point in the C_i, and m_i is the mean value of all the sample points in C_i. Therefore, the smaller SSE is, the more accurate the results are and the better the clustering algorithm is.

The Way to Choose Proper K. To decide the proper value of K, a commonly used method is "Elbow Method". As the number of clusters (K) increases, the degree of aggregation of each cluster gradually increases and the clusters are getting finer-grained, thus the SSE will become smaller. When K is under the real number of clusters, the SSE will decrease significantly because the degree of aggregation inside each cluster increases significantly as K increases. However, when K reaches the real number of clusters, the aggregation payback will go down rapidly even though K keeps going up. Therefore, the SSE will turn into a steady trend. For a randomly generated dataset, the K begins from 1. Before K is set as 3, the SSE goes down rapidly. But the speed lowers dawn and goes in a steady trend after that. Thus, we should set the number of clusters as three, which goes with the fact. The variation tendency between SSE and K is shown in Fig. 5.

Sparsity. To analyze the performance of our CIAA algorithm, the SSE is one of the chosen quantitative standards. However, since we have changed the similarity measure in CIAA, it is not proper to compare their performances under this measurement. Thus, we design a new index to characterize the clustering effects, *Sparsity*, which is defined as $Sparsity = IntelGap/InnerGap$, where *IntelGap* refers to the average *Gap* between the final centroids of each cluster and *InnerGap* refers to the average *Gap* between the nodes which are in the

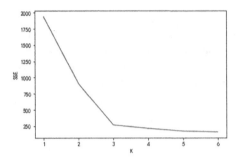

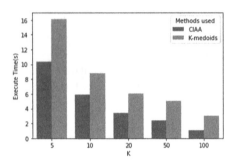

Fig. 5. Relationship between SSE and K **Fig. 6.** Comparison of Execute Time

same cluster. This index ignores the difference brought about by the inconsistency between various measurements. Since we take K-medoids model, the time complexity is a great part that we care about. Therefore, we make comparisons in the execution time. Then, we observe whether the objective function in CIAA will help the clustering process converge faster. Besides, we will look into the inner structure and composition of the clusters to understand the clustering results.

The dataset is based on *Cora* [11], it consists of two separate files, which are *Content* file and *Site* file. In the *Site* file, it describes the relationship between papers in the form of $< referred\_paper\_id >$ and $< paper\_id >$, and the edge goes from the right to the left. The experimental results of their corresponding running time and sparsity are shown in Fig. 6 and Fig. 7. Shown as Fig. 6, in the running time, CIAA can help clusters converge faster. No matter how many clusters there are, CIAA consumes much less time than K-medoids. As we predicted before, since the introduction of social influence by CIAA, the value of SSE cannot be on the same scale, where the SSE of CIAA is five to nine times of that in K-medoids. The SSE can be controlled by adjusting the weights in *Gap* function. Then, we change our mind and turn to Sparsity. Shown as Fig. 7, we compare these two algorithms. It is obvious that the two models are quite close in Sparsity, which reflects the structures and compositions of the clusters are almost the same in density. Although the sparsity of CIAA is slightly smaller than that of K-medoids when K is five, CIAA performs better than K-medoids when K increases to 10, 20, 50, and 100, which indicates that it achieves a good performance in clustering.

Composition of Clusters. We realize that when the number of clusters is large enough, the social graph will be partitioned into many pieces and the clusters can be very fine-grained. In this case, the number of nodes in each cluster will be quite small. Hence, we would like to check whether the influence of the nodes in the same cluster is quite close under this circumstance. The trends of composition and reduction ratio are shown in Fig. 8.

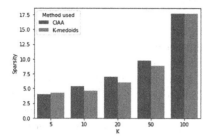

Fig. 7. Comparison of Sparsity

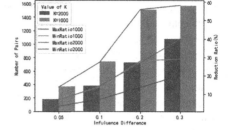

Fig. 8. Composition of Clusters and Reduction Ratio

Firstly, we set the number of clusters as 1000 since there are 2700 nodes in the graph in all. The reason is the average number of nodes in each cluster will be 2.7 by doing so and it is convenient for us to dive into the inside composition. Therefore, in this situation, we can compare the nodes in the same cluster and check the difference in their influence abilities. Furthermore, if two users who are in the same clusters have a similar influence, we can consider merging them. Shown as Fig. 8, the number of pairs between which the difference in influence ability is less than 0.05 is 372. If we enlarge the range, the number of pairs that are less than the range will increase. For these pairs of nodes, their *Gap* are small enough. If we merge them, then we can reduce the size of nodes in the whole social graph. The maximum reduction ratio is achieved if all the nodes are merged into one of them. However, it is quite rare in real applications. The method we take is to merge these nodes pair by pair, which corresponds to the minimum reduction ratio. Even in this case, the reduction ratio is close to 30% if we allow merging nodes whose influence difference is less than 0.3. When we set the number of clusters as 2000, all indexes in Fig. 8 decrease compared to that in $K = 1000$. This is because the average number of nodes in a cluster is 1.35, thus there is only one node in most clusters, which cannot be merged.

To sum up, we should properly set the number of clusters if we want to reduce the size of nodes by merging them together. A good suggestion is to make the average number of nodes in a single cluster to be slightly more than 2. Here, we validate that when the number of clusters is large enough, nodes in the same cluster tend to be very close in *Gap* values. If we want to promote something in the whole social network, we can choose a representative among the nodes with a close *Gap* value, thus saving the cost. Take IM for example, we can design heuristic seed selection strategies based on this clustering, and thus accelerate the process within this reduced graph.

6   Conclusion

To summarize, we propose a new clustering algorithm, CIAA, in this paper, which can be effectively applied to clustering in social networks. Different from the previous work, we not only consider users' properties, but also pay attention to social influence from a global perspective, where the influence exerted by users' direct relationships and influence generated through indirect ways like common interests. The distance part in traditional clustering algorithm still exists in our mind, and we study social influence by mutual influence and self influence. With these taken into consideration, we put forward a new PR-Threshold++ model and make many improvements. The PR-Threshold++ can be divided into two steps: influence accumulation via PageRank and influence diffusion through adjusted LT Model. Last but not least, we design a new similarity measure by combining the distance factor with the influence. We incorporate this similarity into K-medoids and make corresponding modifications. Through experiments and analysis, we conclude that CIAA really helps us achieve multi-dimensional clustering and get good performance.

References

1. Han, J., Kamber, M., Pei, J.: Data Mining: Concepts and Techniques, 3rd edition. Morgan Kaufmann (2011)
2. Kempe, D., Kleinberg, J., Tardos, E.: Maximizing the spread of influence through a social network. International Conference on Knowledge Discovery and Data Mining (ACM SIGKDD), pp. 137–146 (2003)
3. Kim, W., Kanezaki, A., Tanaka, M.: Unsupervised learning of image segmentation based on differentiable feature clustering. IEEE Trans. Image Process. **29**, 8055–8068 (2020)
4. Knattrup, Y., Kubecka, J., Ayoubi, D., Elm, J.: Clusterome: a comprehensive data set of atmospheric molecular clusters for machine learning applications. ACS Omega **8**(28), 25155–25164 (2023)
5. Li, Y., Gao, H., Gao, Y., Guo, J., Wu, W.: A survey on influence maximization: from an ml-based combinatorial optimization. ACM Trans. Knowl. Discov. Data **17**(9), 133:1–133:50 (2023)
6. Mishra, P.K., Verma, S.K.: A survey on clustering in wireless sensor network. In: 2020 11th International Conference on Computing, Communication and Networking Technologies (ICCCNT), pp. 1–5. IEEE (2020)
7. Park, H.S., Jun, C.H.: A simple and fast algorithm for k-medoids clustering. Expert Syst. Appl. **36**(2, Part 2), 3336–3341 (2009)
8. Parker, A.J., Barnard, A.S.: Selecting appropriate clustering methods for materials science applications of machine learning. Adv. Theory Simul. **2**(12), 1900145 (2019)
9. Ran, X., Zhou, X., Lei, M., Tepsan, W., Deng, W.: A novel k-means clustering algorithm with a noise algorithm for capturing urban hotspots. Appl. Sci. **11**(23), 11202 (2021)
10. Rehioui, H., Idrissi, A., Abourezq, M., Zegrari, F.: DENCLUE-IM: a new approach for big data clustering. Int. Conf. Ambient Syst. Netw. Technol. (ANT) **83**, 560–567 (2016)
11. Sen, P., Namata, G., Bilgic, M., Getoor, L., Galligher, B., Eliassi-Rad, T.: Collective classification in network data. AI Mag. **29**(3), 93–93 (2008)
12. Taunk, K., De, S., Verma, S., Swetapadma, A.: A brief review of nearest neighbor algorithm for learning and classification. In: 2019 International Conference on Intelligent Computing and Control Systems (ICCS), pp. 1255–1260. IEEE (2019)

13. Velmurugan, T., Santhanam, T.: Computational complexity between k-means and k-medoids clustering algorithms for normal and uniform distributions of data points. J. Comput. Sci. **6**(3), 363–368 (2010)
14. Zhang, H., Li, H., Chen, N., Chen, S., Liu, J.: Novel fuzzy clustering algorithm with variable multi-pixel fitting spatial information for image segmentation. Pattern Recogn. **121**, 108201 (2022)

Machine Learning, Blockchain
and Others

Incorporating Neural Point Process-Based Temporal Feature for Rumor Detection

Runzhe Li, Zhipeng Jiang, Suixiang Gao, and Wenguo Yang$^{(\boxtimes)}$

School of Mathematical Sciences, University of Chinese Academy of Sciences,
Beijing, China
lirunzhe20@mails.ucas.ac.cn, {jiangzhipeng,sxgao,yangwg}@ucas.ac.cn

Abstract. Social network platforms have facilitated information exchange, but also made the spread of rumors more convenient and rapid. Rumors on the internet leave behind multiple pieces of information with each repost, with temporal information playing a crucial role. Existing studies have focused on extracting various features to discern the veracity of rumors, but the direct analysis of repost timing has been overlooked. In this paper, we present a comprehensive rumor detection method by incorporating temporal features derived from a neural point process model. Our approach investigates the divergences in temporal patterns of reposts between true and false rumors. Moreover, our proposed features can be easily integrated with existing rumor detection methods based on alternative features. We conduct experiments on two publicly available datasets to validate the effectiveness of temporal features. The results demonstrate our proposed model outperforms competing methods.

Keywords: Social networks · Deep learning · Rumor detection

1 Introduction

In the era of high-speed information dissemination through social media and online platforms such as Twitter, rumors have become an epidemic and influential force in shaping public opinion. The rapid increase of rumors poses a significant challenge, as they can spread swiftly, often lacking verifiable evidence or factual basis. Consequently, identifying and combating rumors has become a critical endeavor, requiring effective and efficient methods for rumor detection.

In recent years, deep learning models have emerged as the dominant approach for rumor detection, replacing traditional machine learning methods such as Decision Tree [1]. Concurrently, the range of available information and data types has expanded significantly. For instance, Ma et al. [2] processed the text content of tweets using Recurrent Neural Network (RNN) and Long Short-Term

Supported by the National Key R&D Program of China under grant 2022YFA1003900 and by the National Natural Science Foundation of China under grant numbers 12071459 and 11991022.

Memory (LSTM) to detect rumors, while Bian et al. [3] employed a bi-directional graph neural network to handle rumor propagation structure through user relationship chains. Lu et al. [4] utilized multiple neural networks, including RNN, Convolutional Neural Network (CNN), and attention mechanism, to incorporate various information sources, such as text, user characteristics, and relationships between users, resulting in improved detection performance. However, most articles do not incorporate modules specifically designed to handle temporal information, and the utilization of temporal information often converts to the study of propagation sequences or cascades [1,5-7]. Research has indicated notable differences in the temporal characteristic between the spread of true and false rumors [6]. Moreover, unlike text and personal information, time information in social media is difficult to manipulate or tamper with, making it a robust characteristic that researchers can leverage to determine the veracity of rumors.

Point process is used to describe the underlying temporal mechanism govern a series of events occurring in chronological order [8]. The Hawkes process [9], as a specific type of point process, captures the unique mechanism of mutual excitation between events, making it suitable for capturing the complex temporal dynamics involved in the rumor propagation. Therefore, point process methods have recently been employed in rumor detection tasks. Naumzik et al. [10] present a probabilistic model that classifies true and false rumors based on the underlying spreading process on Twitter. Zeng et al. [11] utilized neural Hawkes process [12] to determine the optimal stopping time for rumor detection models, ensuring both speed and reliable results. While the parameter-defined point process has obvious limitations in the context of social networks, neural networks have shown better performance. However, there has been no direct research utilizing neural point processes to generate temporal features for rumor detection tasks.

Our research focuses on leveraging two fundamental and simple pieces of information: the original text and the timestamps of reposts. We firmly believe that delving deeper into the analysis of temporal information can offer valuable insights, and temporal point processes provide a robust theoretical foundation for our investigation. In this work, our primary contributions are as follows:

- We employ a neural point process model to capture the underlying temporal pattern behind the propagation of true rumors, making the simple timestamps generated during rumor propagation more meaningful.
- We construct several temporal features for rumor detection tasks, which can easily be combined with existing models, augmenting their detection ability.
- Through extensive experiments conducted on publicly available datasets, we demonstrate the effectiveness of our temporal features and show that our proposed model outperforms state-of-the-art models.

2   Related Work

The emergence of deep learning technology has propelled the development of rumor detection and point process methods to a new peak. In this section, we introduce several relevant deep learning methods.

2.1   Rumor Detection Methods with Multiple Features

To extensively explore diverse sources of information and capture high-level features, many rumor detection methods based on deep learning models have recently been proposed. Graph-aware Co-Attention Networks [4] utilized a mixed model consisting of five different modules to generate prediction results. Bi-GCN [3] extracted propagation and dispersion features through Top-Down graph convolutional networks and Bottom-Up graph convolutional networks, respectively. In addition, Song et al. [13] introduced Credible Early Detection (CED) by combining RNN and CNN to efficiently detect rumors at an early stage. Tan et al. [14] conducted a comprehensive review of the research status in rumor detection from multiple perspectives, including feature selection methods. However, above mentioned methods are not efficient in directly capturing features from temporal information. Moreover, due to the limited interpretability of deep learning models, the latent features used to determine the veracity of rumors fail to provide additional information, resulting in a lack of theoretical support. Recently, Zhang et al. [15] proposed Dual Emotion Features for distinguishing true and false news, our temporal feature is inspired by their work.

2.2   Deep Learning Approaches for Point Process

In the context of information diffusion, point processes have been employed to capture the temporal dynamics of rumor spread and information propagation in social networks. Recently, Deep learning approaches for point processes have gained significant attention, since they offer powerful tools for capturing complex patterns and dependencies in the data [16]. For instance, Mei et al. [12] utilized a neurally self-modulating multivariate point process with a continuous-time LSTM, allowing past events to influence future event intensities. Similarly, [17] used a feed-forward neural network to model the integral of the intensity function, enabling exact evaluation of the log-likelihood function without numerical approximations. Zhang et al. [18] designed a model called self-attentive Hawkes process (SAHP) bring more interpretable than RNN-based models. Moreover, [19] showed that existing research on neural point processes primarily focuses on their applications in web-related domains, such as modeling user behavior on social media platforms. This contextual background is why we opt for a neural network model to generate temporal features.

3   Preliminaries

We introduce several fundamental concepts that are essential for understanding the related problem. Firstly, we provide a concise problem description as follows.

3.1   Problem Statement

The goal of the rumor detection task is to predict the labels (true or false) of rumors based on certain known information. Let $C = \{c_1, c_2, \ldots, c_m\}$ be a set of rumor instances, where c_i represents the i-th rumor and m is the total number of instances. Each rumor instance $c_i = \{y_i, \mathbf{o}_i, S_i\}$ is composed of the ground-truth label $y_i \in \{0, 1\}$, the text content of the original post $\mathbf{o}_i$, and a set of relevant repost timestamps in chronological order $S_i = \{t_1, t_2, \ldots, t_{|S_i|}\}$. The label y_i indicates c_i is false if $y_i = 1$ or true otherwise. $|S_i|$ denotes the number of relevant reposts, and t_j represents the timestamp of the j-th repost, which is defined as the relative time of the j-th repost with respect to the original post, such that $t_{j-1} < t_j$ for $j \geq 1$, with the convention $t_0 = 0$. The objective is to learn a rumor detection classifier $f : f(\mathbf{o}_i, S_i) \rightarrow y_i$ that maximizes the prediction accuracy for any given instance c_i.

3.2   Neural Temporal Point Process

Temporal point process (TPP) is a probabilistic model that captures the behavior of sequences of discrete events observed in continuous time, such as information spreading in social media. The intensity function [20] is a convenient and intuitive way of specifying TPP, which is defined as

$$\lambda(t) = \lim_{h \to 0^+} \frac{P(N(t, t+h] > 0)}{h} \tag{1}$$

where $N(t, t+h]$ indicates the count of the number of events in an interval of length h. $\lambda(t)$ intuitively measures the rate at which the events occur.

In most cases, events in a TPP mutually influence each other. Therefore, we need to consider the historical information before time t, denoted as $H_t = \{t_i\}_{t_i < t}$, when define a conditional intensity function

$$\lambda(t \mid H_t) = \lim_{\Delta \mapsto 0} \frac{P(\text{ one event occur in } [t, t+\Delta) \mid H_t)}{\Delta} \tag{2}$$

Hawkes process defines a classical parameterized form for the conditional intensity function as $\lambda(t \mid H_t) = \mu + \alpha \sum_{t_i < t} \exp(-(t - t_i))$, where μ and α are positive parameters.

However, utilizing neural network models to describe TPP is more suitable for complex real-world scenarios than simply using parameterized forms to define conditional intensity functions. RMTPP [21] uses a recurrent neural network to calculate the influence of the history up to the i-th event $\mathbf{h}_i$ by

$$\mathbf{h}_i = \max\left\{\mathbf{W}^y \mathbf{y}_i + \mathbf{W}^t \mathbf{t}_i + \mathbf{W}^h \mathbf{h}_{i-1} + \mathbf{b}^h, 0\right\} \tag{3}$$

where $\mathbf{y}_i$ and $\mathbf{t}_i$ are the mark (additional information accompanying an event) feature and time feature of the i-th event, $\{\mathbf{W}^y, \mathbf{W}^t, \mathbf{W}^h, \mathbf{b}^h\}$ are learnable parameters. Based on $\mathbf{h}_i$, the conditional intensity function can be formulated by a exponential function

$$\lambda(t \mid H_t) = \exp\left(\mathbf{v}^{t^\top} \cdot \mathbf{h}_i + w^t(t - t_i) + b^t\right) \tag{4}$$

where $\mathbf{v}^t$ is a column vector, and w^t, b^t are scalars. According to the conditional intensity function, the conditional density function can be specified by

$$f(t \mid H_t) = \lambda(t \mid H_t) \exp\left(-\int_{t_i}^{t} \lambda(s \mid H_t)\,ds\right) \tag{5}$$

Then, the timing for the next event can be estimated by using the expectation

$$\hat{t}_{i+1} = \int_{t_i}^{\infty} t \cdot f(t \mid H_t)\,dt \tag{6}$$

3.3 Temporal Pattern in Rumor Propagation

On social media platforms, rumors propagate through user actions, such as clicking the share button, which are recorded with timestamps, forming a time sequence $\{t_1, t_2, \ldots, t_n\}$ along with the rumor cascade. The cascade exhibits complex and specific statistical temporal characteristics, such as the self-exciting or rich-get-richer effect, where the accumulation of more shares indicates a higher likelihood of further reposts.

The previously mentioned Hawkes process is a point process used to describe the self-exciting nature and has been employed to capture the temporal pattern of rumor cascades. Prior research on rumor detection assumes the existence of distinct Hawkes processes, M_T and M_F, underlying true and false rumors, respectively. Empirical validation has demonstrated the feasibility and effectiveness of this assumption. Furthermore, authoritative research has indicated significant differences in the propagation characteristics between true and false rumors. Therefore, the temporal structure characteristics are well-suited for detecting rumors and, objectively, they are more challenging to manipulate compared to common features such as textual and user-based features.

However, in our preliminary research, when attempting to utilize a neural point process model to fit the latent point processes corresponding to true and false rumors, we observed that true rumors exhibit a more consistent temporal pattern, while learning a universal temporal pattern for false rumors proved difficult. This challenge may arise from the fact that false rumors contain various fabricated information without a unified false characteristic, leading to a lack of consistent temporal pattern. To address this issue, in our upcoming research, we assume that true rumors exhibit a relatively consistent temporal pattern, while false rumors deviate from this pattern to some extent.

4 Model

In this section, we propose a novel model for rumor detection by incorporating neural point process-based temporal features. The core idea of our approach is to explore the underlying information contained in each simple timestamp throughout the rumor propagation process. Figure 1 illustrates the procedure of extracting temporal features and integrating them with textual features to predict the rumor label.

We first discuss how to generate the temporal features of an instance c_i.

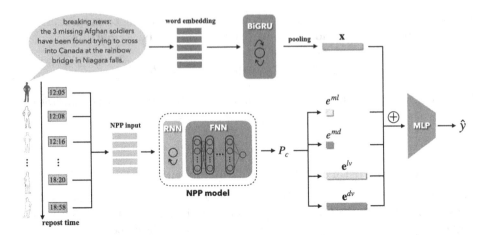

Fig. 1. Model framework. To simplify the explanation, we have omitted the subscript i. The integrated temporal feature **e** is composed of four components and is concatenated with the textual feature **x** for the final prediction.

4.1   Temporal Feature Construction

To capture the underlying statistical distribution of cascades in the propagation of true rumors, we employ a neural point process model [17] to fit this distribution. The objective of the model is to compute the cumulative hazard function, which is defined as follows

$$\Phi\left(t - t_i \mid \boldsymbol{h}_i\right) = \int_0^{t-t_i} \phi\left(s \mid \boldsymbol{h}_i\right) ds \qquad (7)$$

where $\boldsymbol{h}_i$ is the hidden state of the RNN which captures the historical information up to time t_i. ϕ is a non-negative function referred to as the hazard function and the following equation hold numerically

$$\phi\left(t - t_i \mid \boldsymbol{h}_i\right) = \lambda\left(t \mid H_t\right) \qquad (8)$$

By directly estimating (7) instead of (8), we can avoid the numerical loss caused by approximating the integral, leading to more accurate results.

After training the model, we obtain a neural point process R_T that captures the underlying distribution of true rumor reposts. Subsequently, we fix all model parameters and calculate the joint likelihood P_{c_i} for the time sequence $\{t_1, ..., t_n\}$ of each rumor cascade c_i by

$$P_{c_i}\left(t_1, \ldots, t_n\right) = \prod_{i=1}^{n} \lambda\left(t_i \mid H_t\right) \cdot \exp\left\{-\int_0^t \lambda\left(s \mid H_s\right) ds\right\} \qquad (9)$$

Since we can determine the proximity of a sequence to the true distribution R_T by P_{c_i}, naturally, the temporal feature can be defined as it. Consequently, we obtain the likelihood vector $\mathbf{e}_i^{lv} = (l_1, ..., l_z) \in \mathbb{R}^z$ and the mean likelihood feature $e_i^{ml} = \frac{1}{z} \sum_{i=1}^{z} l_i \in \mathbb{R}$. Here, $l_i = P_{c_i}(t_i, \ldots, t_{i+k})$ represents the likelihood in an observed k-events subsequence.

Considering the predictive performance of the NPP model is also a crucial metric, we can further define the temporal feature as a distance vector $\mathbf{e}_i^{dv} = (d_1, ..., d_z) \in \mathbb{R}^z$ which describes the absolute error between the actual time of each repost and the model's predicted repost time. Here, $d_j = \|\hat{t}_j - t_j\|$. Similarly, the mean distance feature $e_i^{md} = \frac{1}{z} \sum_{i=1}^{z} d_i \in \mathbb{R}$.

Finally, our integrated temporal feature is a concatenated vector that includes the likelihood vector, mean likelihood feature, distance vector, and mean distance feature as follows

$$\mathbf{e}_i = [\mathbf{e}_i^{lv}, e_i^{ml}, \mathbf{e}_i^{dv}, e_i^{md}] \in \mathbb{R}^{2z+2} \tag{10}$$

Note that for the variable z, assuming we have observed a total of n reposts and the NPP model has an RNN with a window length set to k, then $z = n + 1 - k$.

4.2   Rumor Detection Module

Recognizing the limitations of relying solely on temporal features for achieving highly accurate predictions, it is necessary to integrate a rumor detection module capable of handling textual information. By adopting a combined approach that leverages both temporal and textual features, we can derive a comprehensive framework for making the final prediction label.

Textual Feature Generation. For the raw text $\mathbf{o}_i$ of the original post, we need to transform words and sentences into textual feature $\mathbf{x}_i$. To accomplish this, we employ GloVe [22] to obtain word embeddings for each word, which are then fed into a BiGRU (Bi-directional Gated Recurrent Unit) model [15]. Subsequently, mean pooling is applied to derive the textual feature $\mathbf{x}_i$.

Model Prediction. Our objective is to predict whether rumor c_i is true or false by utilizing both the temporal feature $\mathbf{e}_i$ and textual feature $\mathbf{x}_i$. These features are concatenated and then input into a multi-layer feedforward neural network to generate the binary prediction vector $\hat{\mathbf{y}}_i$ as follows

$$\hat{\mathbf{y}}_i = \text{softmax}\left(\text{ReLU}\left(\mathbf{W}_f[\mathbf{e}_i, \mathbf{x}_i] + \mathbf{b}_f\right)\right) \tag{11}$$

where $\mathbf{W}_f$ is the matrix of learnable parameters and $\mathbf{b}_f$ is the bias. Note that $\hat{\mathbf{y}}_i = [\hat{p}_0, \hat{p}_1]$, where $\hat{p}_0$ and $\hat{p}_1$ correspond to the predicted probabilities of label y_i being 0 and 1, respectively. Thus, it holds that $\hat{p}_0 + \hat{p}_1 = 1$.

We train the rumor detection module by minimizing the cross-entropy value, which is defined as follows

$$\mathcal{L}(\Theta) = -y_i \log(\hat{p}_0) - (1 - y_i) \log(\hat{p}_1) \tag{12}$$

where Θ denotes all the learnable parameters in the neural network.

5   Experiments

In this section, we first validate the effectiveness of temporal features. Next, we compare our approach with several baseline methods. Finally, we provide illustrative examples to demonstrate that temporal features can enhance and refine text-based rumor detection task.

5.1   Settings and Datasets

Datasets. We evaluate our proposed method on two real-world datasets: Twitter15 and Twitter16 [23]. These datasets comprise original posts and their corresponding sequences of reposts. Our study focuses solely on differentiating between true and false (fake) labels. The labels of each instance are based on the veracity tag of the corresponding article from rumor debunking websites such as snopes.com and Emergent.info [23]. For each dataset, we allocate half of the data for training the point process model. The training and testing set ratio is set at 4:1, with 20 percent of the training set being partitioned as the validation set. The remaining half of each dataset is divided into training, validation, and testing sets for the rumor detection task, with a ratio of 3:1:1.

Competing Methods. We compare our proposed method with several state-of-the-art baselines, including:

- SVM-TS [24]: A linear SVM classifier that constructs a time-series model by using handcrafted features.
- CSI [25]: An RNN-based model that integrates three key features of rumor: article text, user response, and source user behavior.
- mGRU [2]: A modified gated recurrent unit model for detecting rumors on microblogging platforms, which mainly captures contextual information of reposts.
- dEFEND [26]: An explainable rumor detection model that utilizes a sentence-comment co-attention sub-network to analyze both post contents and user comments.
- BiGRU: A bidirectional gated recurrent unit model is used to better capture the semantic information embedded in the original post text. The BiGRU model fused with corresponding temporal features will be represented as + *feature*.

Evaluation Metrics. We use four metrics to evaluate our model: Accuracy, Precision, Recall, and F1 score. Referring to the experiments conducted in [15], we adopt the macro F1 score to evaluate the effectiveness of our temporal features.

5.2   Experimental Results

Effectiveness of Temporal Features. We aim to validate the effectiveness of our temporal features as crucial information for distinguishing between true and false rumors. To achieve this, we utilize different temporal features as inputs to a multi-layer perceptron (MLP) model for rumor detection task. We choose the mean delay time as the temporal feature for comparison. The results on three datasets, namely Twitter15 (T15) and Twitter16 (T16), are presented in Table 1. T16_e presents experiments on the entire Twitter16 dataset, here we trained the NPP model using a portion of the Twitter15 data and then generated temporal features for the Twitter16 data. In Table 1, we can see that the effectiveness of the four different features varies across different datasets, with the overall integrated temporal feature demonstrating the highest efficacy across all datasets. Additionally, compared to T16, the results on T16_e demonstrate stability, indicating that the temporal dynamics behind real rumor propagation learned from T15 can be transferred across different datasets, consistent with the viewpoint in [16].

Table 1. Macro F1 scores on the MLP model.

Temporal Features	Datasets		
	T15	T16	T16_e
Mean Delay Time	0.448	0.658	0.670
Mean Likelihood Feature	0.526	0.714	0.717
Mean Distance Feature	0.474	0.760	0.764
Likelihood Vector	0.512	0.690	0.682
Distance Vector	0.483	0.664	0.687
Integrated Temporal Vector	**0.587**	**0.783**	**0.767**

Table 2. Main results on two datasets.

Method	Twitter15				Twitter16			
	F1	Rec	Pre	Acc	F1	Rec	Pre	Acc
SVM-TS	0.519	0.519	0.520	0.520	0.692	0.691	0.693	0.693
mGRU	0.510	0.515	0.515	0.555	0.556	0.562	0.560	0.661
CSI	0.717	0.687	0.699	0.699	0.630	0.631	0.632	0.661
dEFEND	0.654	0.661	0.658	0.738	0.631	0.638	0.637	0.702
BiGRU	0.765	0.816	0.721	0.750	0.737	0.667	0.824	0.762
$+ e^{ml}$	0.800	0.789	0.811	0.803	0.884	0.905	0.864	0.881
$+ e^{md}$	0.825	0.808	0.786	0.816	0.844	0.905	0.792	0.833
$+ e^{lv}$	0.811	0.789	0.833	0.816	0.850	0.810	0.895	0.857
$+ e^{dv}$	0.872	0.895	0.850	0.868	0.878	0.857	**0.900**	0.881
$+ e$	**0.897**	**0.921**	**0.875**	**0.895**	**0.909**	**0.952**	0.870	**0.905**

Main Results. The performance of our proposed method and the compared methods are summarized in Table 2. We can clearly find that the BiGRU model combined with integrated temporal feature **e** outperforms the competing methods across most metrics on both datasets. Even two global temporal features (mean likelihood feature e^{ml} and mean distance feature e^{md}) with the dimension of 1 significantly improve the detection accuracy of BiGRU by 7.07% and 8.80% on Twitter15 and 15.6% and 9.32% on Twitter16. However, due to the limitation of information scale, they did not reach the optimal level. The BiGRU model with the addition of integrated temporal feature **e** showed improvements of 19.3% and 18.8% on Twitter15 and Twitter16 respectively. These results prove the effectiveness of our proposed temporal features for rumor detection. Finally, we provide two examples in Fig. 2 to demonstrate the contribution of temporal features in rumor detection. These two false rumors were incorrectly classified as true by the BiGRU model that only utilized textual features **x** as input. However, after incorporating temporal feature **e**, they were correctly identified with high confidence. This demonstrates that temporal features can help for the limitations of textual features in discerning specific narratives from another perspective.

Y'all, I just read that ABC paid Darren Wilson $500k for the interview. Destroying Black life remains a lucrative American career. #Ferguson.				A 15-year-old who "swatted" a gamer has been convicted of domestic terrorism and sentenced to 25 years to life in federal prison. [URL]		
Feature	True	False		Feature	True	False
$\mathbf{x}_i$	**0.511**	0.489		$\mathbf{x}_i$	**0.894**	0.106
$\mathbf{x}_i + \mathbf{e}_i$	0.040	**0.960**		$\mathbf{x}_i + \mathbf{e}_i$	0.126	**0.874**

Fig. 2. Two false rumors misclassified by the BiGRU model using only textual feature $\mathbf{x}_i$ were correctly identified after incorporating temporal feature $\mathbf{e}_i$.

6   Conclusions

In this study, we explored the feasibility of directly utilizing temporal information for rumor detection. We employed a neural point process model to capture the underlying temporal patterns behind real rumor propagation. Then our proposed temporal features are generated via it. Evaluation results on two public datasets show the effectiveness and the reasonable explainability of our temporal features. In future work, we will explore incorporating additional information into temporal features, such as user information and structural information. We will also investigate the effectiveness of temporal features in early rumor detection tasks.

References

1. Castillo, C., Mendoza, M., Poblete, B.: Information credibility on twitter. In: Proceedings of the 20th International Conference on World Wide Web, pp. 675–684. Association for Computing Machinery, Hyderabad (2011). https://doi.org/10.1145/1963405.1963500

2. Ma, J., et al.: Detecting rumors from microblogs with recurrent neural networks. In: Proceedings of the Twenty-Fifth International Joint Conference on Artificial Intelligence, pp. 3818–3824. AAAI Press, New York (2016). https://doi.org/10.5555/3061053.3061153

3. Bian, T., et al.: Rumor detection on social media with bi-directional graph convolutional networks. In: Proceedings of the AAAI Conference on Artificial Intelligence 34(01), pp. 549–556 (2020)

4. Lu, Y., Li, C.: GCAN: graph-aware co-attention networks for explainable fake news detection on social media. In: Proceedings of the 58th Annual Meeting of the Association for Computational Linguistics, pp. 505–514. Association for Computational Linguistics, Vancouver (2020). https://doi.org/10.18653/v1/2020.acl-main.48

5. Ma, J., Gao, W., Wong, K.: Rumor detection on twitter with tree-structured recursive neural networks. In: Proceedings of the 56th Annual Meeting of the Association for Computational Linguistics, pp. 1980–1989. Association for Computational Linguistics, Melbourne (2018). https://doi.org/10.18653/v1/P18-1184

6. Soroush, V., Deb, R., Sinan, A.: The spread of true and false news online. Science **359**, 1146–1151 (2018)

7. Wu, K., Yang, S., Zhu, K.: False rumors detection on sina weibo by propagation structures. In: 2015 IEEE 31st International Conference on Data Engineering, pp. 651–662. IEEE, Rio de Janeiro (2015). https://doi.org/10.1109/icde.2015.7113322

8. David, V.: An Introduction to the Theory of Point Processes: Volume I: Elementary Theory and Methods. Springer, Heidelberg (2003)

9. Hawkes, A.: Spectra of some self-exciting and mutually exciting point processes. Biometrika **58**(01), 83–90 (1971)

10. Naumzik, C., Feuerriegel, S.: Detecting false rumors from retweet dynamics on social media. In: Proceedings of the ACM Web Conference 2022, pp. 2798–2809. Association for Computing Machinery, Lyon (2022). https://doi.org/10.1145/3485447.3512000

11. Zeng, F., Gao, W.: Early Rumor Detection Using Neural Hawkes Process with a New Benchmark Dataset. In: Proceedings of the 2022 Conference of the North American Chapter of the Association for Computational Linguistics: Human Language Technologies, pp. 4105–4117. Association for Computational Linguistics, Seattle (2022). https://doi.org/10.18653/v1/2022.naacl-main.302

12. Mei, H., Eisner, J.: The neural hawkes process: a neurally self-modulating multivariate point process. In: Proceedings of the 31st International Conference on Neural Information Processing Systems, pp. 6757–6767. Curran Associates Inc., California (2017). https://doi.org/10.5555/3295222.3295420

13. Song, C., Yang, C., Chen, H., Tu, C., Liu, Z., Sun, M.: CED: credible early detection of social media rumors. IEEE Trans. Knowl. Data Eng. **33**(8), 3035–3047 (2021)

14. Tan, L., Wang, G., Jia, F., Lian, X.: Research Status of Deep Learning Methods for Rumor Detection. Kluwer Academic Publishers **82**(2), 2941–2982 (2022)

15. Zhang, X., Cao, J., Li, X., Sheng, Q., Zhong, L., Shu, K.: Mining dual emotion for fake news detection. In: Proceedings of the Web Conference 2021, pp. 3465–3476. Association for Computing Machinery, Ljubljana (2021). https://doi.org/10.1145/3442381.3450004

16. Zhao, Q., Erdogdu, M., He, H., Rajaraman, A., Leskovec, J.: SEISMIC: a self-exciting point process model for predicting tweet popularity. In: Proceedings of the 21th ACM SIGKDD International Conference on Knowledge Discovery and Data Mining, pp. 1513–1522. Association for Computing Machinery, Sydney (2015). https://doi.org/10.1145/2783258.2783401

17. Omi, T., Ueda, N., Aihara, K.: Fully neural network based model for general temporal point processes. In: Proceedings of the 33rd International Conference on Neural Information Processing Systems, pp. 2122–2132. Curran Associates Inc., New York (2019). https://doi.org/10.5555/3454287.3454477

18. Zhang, Q., Lipani, A., Kirnap, O., Yilmaz, E.: Self-attentive hawkes process. In: Proceedings of the 37th International Conference on Machine Learning, pp. 11183–11193. PMLR, Vienna (2020). https://doi.org/10.5555/3524938.3525975

19. Shchur, O., Türkmen, A., Januschowski, T., Günnemann, S.: Neural temporal point processes: a review. In: Proceedings of the Thirtieth International Joint Conference on Artificial Intelligence, pp. 4585–4593. International Joint Conferences on Artificial Intelligence Organization, Montreal (2021). https://doi.org/10.24963/ijcai.2021/623

20. Rasmussen, J.: Bayesian inference for Hawkes processes. Methodol. Comput. Appl. Probab. **15**(3), 623–642 (2013)

21. Du, N., Dai, H., Trivedi, R., Upadhyay, U., Gomez-Rodriguez, M., Song, L.: Recurrent marked temporal point processes: embedding event history to vector. In: Proceedings of the 22nd ACM SIGKDD International Conference on Knowledge Discovery and Data Mining, pp. 1555–1564. Association for Computing Machinery, California (2016). https://doi.org/10.1145/2939672.2939875

22. Jeffrey, P., Richard, S., Christopher, M.: GloVe: global vectors for word representation. In: Proceedings of the 2014 Conference on Empirical Methods in Natural Language Processing, pp. 1532–1543. Association for Computational Linguistics, Doha (2014). https://doi.org/10.3115/v1/D14-1162

23. Ma, J., Gao, W., Wong, K.: Detect rumors in microblog posts using propagation structure via kernel learning. In: Proceedings of the 55th Annual Meeting of the Association for Computational Linguistics (Volume 1: Long Papers), pp. 708–717. Association for Computational Linguistics, Vancouver (2017). https://doi.org/10.18653/v1/P17-1066

24. Ma, J., Gao, W., Wei, Z., Lu, Y., Wong, K.: Detect rumors using time series of social sontext information on microblogging websites. In: Proceedings of the 24th ACM International on Conference on Information and Knowledge Management, pp. 1751–1754. Association for Computing Machinery, Melbourne (2015). https://doi.org/10.1145/2806416.2806607

25. Ruchansky, N., Seo, S., Liu, Y.: CSI: a hybrid deep model for fake news detection. In: Proceedings of the 2017 ACM on Conference on Information and Knowledge Management, pp. 797–806. Association for Computing Machinery, Singapore (2017). https://doi.org/10.1145/3132847.3132877

26. Shu, K., Cui, L., Wang, S., Lee, D., Liu, H.: DEFEND: explainable fake news detection. In: Proceedings of the 25th ACM SIGKDD International Conference on Knowledge Discovery and Data Mining, pp. 395–405. Association for Computing Machinery, Anchorage (2019). https://doi.org/10.1145/3292500.3330935

Improving Contraction Hierarchies by Combining with All-Pairs Shortest Paths Problem Algorithms

Xinyu Song, Zhipeng Jiang$^{(\boxtimes)}$, Wenguo Yang, and Suixiang Gao

School of Mathematical Sciences, University of Chinese Academy of Sciences, Beijing 100049, China
`songxinyu21@mails.ucas.ac.cn`, {`jiangzhipeng,yangwg,sxgao`}`@ucas.ac.cn`

Abstract. Contraction hierarchies (CH) is a two-phase effective shortest path algorithm for large-scale road networks based on node contraction. However, the remaining graphs tend to be complete near the end of preprocessing, which slows down the preprocessing speed. We combine CH with the all-pairs shortest paths (APSP) problem algorithms which are efficient on complete graphs, to propose a new method. Near the end of the CH preprocessing phase, we use the APSP algorithm to obtain a distance table that contains the shortest path between all remaining nodes. Query performs a bidirectional Dijkstra search combined with a table lookup. Experimental results show that our method achieves both preprocessing and query acceleration compared to the raw CH algorithm. It allows for parameter adjustment based on spatial requirements and can be interpreted as an interpolation between CH and APSP algorithms.

Keywords: Shortest path · Contraction hierarchies · All-pairs shortest paths problem

1 Introduction

Given a graph, the shortest path problem is finding a path with the shortest "distance" between a source and a destination. In an actual network, "distance" can be the length, time, cost, etc. This problem is not only one of the core problems of graph theory but also the foundation of complex problems such as transportation, logistics and facility location. The classic methods to calculate the shortest path include the Dijkstra algorithm [7], the Floyd algorithm [5], and so on.

However, with the increase in network scale, it is often required to answer queries in a very short time. Traditional methods can no longer meet the requirements. Therefore, many researchers have turned their attention to preprocessing-based algorithms. The key idea is to pre-calculate some auxiliary information in the preprocessing phase so that the subsequent query can be answered faster than the classic methods while ensuring optimality.

W. Wu and J. Guo (Eds.): COCOA 2023, LNCS 14462, pp. 431–442, 2024.
https://doi.org/10.1007/978-3-031-49614-1_32

In recent years, preprocessing-based algorithms can be roughly categorized into goal-directed and hierarchical techniques, as well as a combination of both. The former includes the ALT algorithm [10], based on $A*$ search, Landmarks, and Triangle inequality, the Arc Flags [12] algorithm that precomputes "signposts" for each edge, and so on. The latter includes Highway Hierarchies(HHs) [13], Highway-Node Routing(HNR) [14], Reach-Based Routing [11], Contraction Hierarchies [9], and so on.

Among them, the CH algorithm has shown good performance on large-scale road networks. CH heuristically contracts nodes and adds shortcuts. A road network with one million nodes and edges can be preprocessed in 30 seconds, while queries run in about 2 milliseconds.

CH can also be combined with other technologies, such as the CHASE algorithm [4] (combined with Arc Flags), the CH-TNR algorithm [2] (combined with TNR), and the CH-HL algorithm [8] (combined with Hub Labels [1]).

CH-TNR and CH-HL are currently two competitive algorithms. They all use table lookup or vector lookup techniques for query acceleration. CH-HL stores the distance between access nodes in advance according to the level, and the query algorithm firstly searches and then uses a two-hops vector lookup. CH-TNR stores the distance to access nodes and between access nodes in advance but uses a locality filter to determine whether to perform local searches. Although the query speed of the above two algorithms is faster than CH, they require more preprocessing time. In addition, the time required to contract a node increases at the end of CH, and the graph generated by the remaining nodes tends to be complete.

APSP is one of the most famous problems in algorithm design. It refers to the problem of distances and shortest paths between all pairs of nodes in a graph.

In this paper, we present a new shortest path algorithm combining CH and APSP algorithm, which reduce both preprocessing and query time to less than CH. Towards the end of the preprocessing phase of the CH algorithm, we substitute the time-consuming preprocessing step with the computation of a complete pairwise distance table for the remaining nodes using APSP algorithm.

In Sect. 2, we outline the basics of CH and APSP problems. Section 3 shows how to combine CH and APSP and gives schemes for preprocessing and query. In Sect. 4, we conduct an experimental evaluation on real road networks and analyze the practical effect and the impact of parameters on preprocessing and query.

2 Preliminaries

Given a directed graph $G = (V, E)$ with $|E| = m$ edges and $|V| = n$ nodes. Each edge $(u, v) \in E$ has a non-negative weight $w(u, v)$, where $w : E \rightarrow \mathbb{R}^+$. A path $P = \langle v_1, v_2, ...v_p \rangle$ is an ordered sequence of nodes such that $s = v_1, t = v_p, (v_i, v_{i+1}) \in E$, and its length is defined as $w(P) := \sum_{i=1}^{p-1} w(v_i, v_{i+1})$. The path with the minimum length between $s, t \in V$ is called the shortest path $SP_{s,t}$ and its length is denoted as $d(s, t)$.

2.1 Contraction Hierarchies

Contraction Hierarchies is an efficient two-phase shortest path algorithm based on shortcuts, which is commonly used in large road networks. It has a trade-off between preprocessing and query time.

The first phase is systematic preprocessing based on node contraction and shortcut addition. CH heuristically sorts nodes by some measures of importance and prioritizes removing the least important node v. In order not to affect the shortest paths between the remaining nodes, the above removal process is achieved by replacing the path $\langle u, v, x \rangle$ with a shortcut $\langle u, x \rangle$. However, $\langle u, x \rangle$ is required if and only if $\langle u, v, x \rangle$ is the only shortest path between u and x. A simple way to achieve this is to perform a shortest path search called *witness path* search that ignores v from each source node u. Note that if a shortcut $\langle u, x \rangle$ is added when $\langle u, v, x \rangle$ is not the shortest path between u and x, the correctness of the shortest path is not affected. However, it may influence the speed of preprocessing and querying. The length of the new shortcut $\langle u, x \rangle$ is $w(u, x) = w(u, v) + w(v, x)$.

The second phase is the query phase using a variant of the bidirectional Dijkstra algorithm. After preprocessing, let $G^+ = (V, E^+)$ denotes the new graph, where E^+ contains all edges of E and added shortcuts, and $\alpha : V \to \{1, ..., n\}$ denotes the contraction order. Split G^+ into an *upward graph* $G_\uparrow := (V, E_\uparrow)$ with $E_\uparrow := \{(u, v) \in E^+ : \alpha(u) < \alpha(v)\}$ and a *downward graph* $G_\downarrow := (V, E_\downarrow)$ with $E_\downarrow := \{(u, v) \in E^+ : \alpha(u) > \alpha(v)\}$. For the shortest path query from s to t, CH performs a bidirectional Dijkstra search, including a forward search in $G_\uparrow$ and a backward search in $G_\downarrow$. If and only if there exists a shortest $s - t$ path in the original graph, the two search scopes eventually meet at node v with the highest order among all nodes included in the shortest $s - t$ path.

2.2 All-Pairs Shortest Paths Problem

Given a graph $G = (V, E)$, APSP obtains the shortest path between all pairs of nodes on G. APSP problems can be solved by running a *single-source shortest paths problem(SSSP)* algorithm $|V|$ times. The simple Dijkstra algorithm calculates SSSP in $O(n^2)$ time for non-negative real-weighted graphs. If the binary min-heap implementation of the min-priority queue is used, then it runs in $O((m + n) \log n)$ time. Combining the Fibonacci heap, Dijkstra can compute SSSP problem in $O(m + n \log n)$ time and, by repeated application, APSP in $O(mn + n^2 \log n)$ time.

When some edges have negative weight, the Dijkstra algorithm cannot be used directly, and it is more suitable to use the Floyd algorithm, a dynamic programming algorithm for solving the APSP problem and performing well on dense graphs. Floyd algorithm uses an adjacency-matrix representation instead of an adjacency-list representation of the graph and can solve the APSP problem in $O(n^3)$ time.

3   CH-APSP

As the preprocessing approach of CH proceeds, the remaining graph will gradually become dense, and it will take more time to contract a node. Take the road network of the Great Lakes provided for the 9th DIMACS challenge [6], which consists of around 2.7 million nodes and 6.9 million edges, as an example. Only use lazy updates and do not periodically reevaluate all priorities or recompute the priority of the neighbors. In Fig. 1, the time difference between two contractions gradually increases, and the increasing speed is relatively fast.

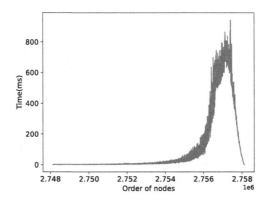

Fig. 1. Line chart of the time difference between two contractions of the last 10000 nodes of the road network of the Great Lakes.

The above results may be due to the significant impact of a node's contraction on its neighbors. As the density of the remaining graph increases, it tends to be a complete graph, and the probability of edges between any two nodes increases. So each time a node is contracted, the number of nodes affected increases, and the importance evaluation of nodes changes significantly. Therefore, more lazy updates are required, and the time difference between two node contractions increases, resulting in slower preprocessing speed.

In addition, according to the experimental results in CH-TNR, the time of the table query is less than that of the Dijkstra search, and as the number of transit nodes increases, the number of access nodes may decrease.

Therefore, based on the above observations, we propose the CH-APSP algorithm by combining the APSP problem with the CH algorithm.

3.1   Preprocessing

At the beginning of the preprocessing of CH-APSP, it is the same as the original CH algorithm. When the remaining graph becomes dense near the end of the preprocessing phase, then turn to use the APSP algorithm.

Given a node contraction order $\alpha : V \to \{1, ..., n\}$, let $G(i) = (V_i, E_i)$ denotes the graph obtained from G after i contractions and $G(0) = G$. Use $V(P) = \{v_1, ..., v_p\}$ and $E(P) = \{(v_i, v_{i+1})|i = 1, ..., p-1\}$ to denote the nodes and edges of path P, respectively.

Let M be the number of remaining non-contracted nodes. When the number of remaining nodes is M, stop the preprocessing of CH, and the graph $G(n - M) = (E_{n-M}, V_{n-M})$ is obtained, with $V_{n-M} = \{v|v \in V$ and v is not contracted$\}$ and $E_{n-M} = \{(u, v)|u, v$ are not contracted $, (u, v) \in E$ or (u, v) is a shortcut$\}$. Solve the APSP problem on $G(n - M)$ and obtain the distance table D. After that, D and a graph $G'^+ = (V, E'^+)$ are obtained, where E'^+ contains edge (u, v) which belongs to E or is an added shortcut, but v and u cannot both belong to V_{n-M}. The process is shown in Algorithm 1.

Algorithm 1: Preprocessing of CH-APSP algorithm

Input: Non-negative weighted graph $G = (V, E)$ and $M \in \mathbb{Z}$, $0 \le M \le |V| = n$.
Output: Graph $G'^+ = (V, E'^+)$ and a distance table D.
1 Initialization: $n' \leftarrow n$.
2 **while** $n' > M$ **do**
3   Heuristically sort nodes based on measures of importance.
4   Select a node v which is the least important node.
5   Contract v.
6      $n' \leftarrow n' - 1$.
7 **end**
8 Obtain graph $G(n - M)$.
9 Solve the APSP problem on $G(n - M)$.

3.2 Query

The query of the CH-APSP algorithm performs a modified bidirectional Dijkstra shortest path search combined with a table lookup.

After preprocessing, the graph $G'^+ = (V, E'^+)$ and the distance table D are obtained. Split G'^+ into two parts as is said in Sect. 2.1, the upward graph $G'_\uparrow$ and the downward graph $G'_\downarrow$. Given a source node s and a target node t, perform a forward Dijkstra search on $G'_\uparrow$ starting from s and a backward Dijkstra search on $G'_\downarrow$ starting from t. When v is settled and belongs to V_{n-M}, do not relax the edges incident to v. That is, do not continue the search from v. This node is called an *access node*.

Let $E^\wedge$ and $E^\vee$ denote the set of all access nodes found in the forward and backward search respectively. Let P_{st} denote the $s - t$ path obtained when there is no reached but not settled node. $P_{st} = P^\wedge + P^\vee$, where $P^\wedge$ and $P^\vee$ are the paths obtained from the upward and downward search. If there is no $s - t$ path, then P_{st} is ∞. The shortest $s - t$ path is obtained by the following:

$$d(s,t) = \min \left\{ w(P_{st}), \min_{v \in E^{\wedge}, v' \in E^{\vee}} \{d_{\uparrow}(s,v) + D(v,v') + d_{\downarrow}(v',t)\} \right\} \quad (1)$$

$D(v,v')$ is obtained by looking up table D. $d_{\uparrow}$ is obtained by forward search in $G'_{\uparrow}$, while $d_{\downarrow}$ is obtained by downward search in $G'_{\downarrow}$.

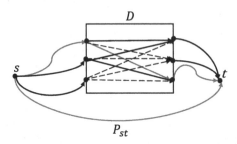

Fig. 2. A graph of the query of the two-stage CH-APSP algorithm search process.

In fact, the query can be split into two parts, one with and one without the table lookup. (See Fig. 2). Both parts can obtain the possible shortest path of their respective parts, and the shortest path between the two paths is the final $s - t$ shortest path. The query process is shown in Algorithm 2.

Algorithm 2: Query of the CH-APSP algorithm.

Input: $G'^{+} := (V, E'^{+})$, V_{n-M}, distance table D and $s, t \in V$.

Output: The shortest $s - t$ path.

1 Initialization: $d^{\wedge}[s] \leftarrow 0$, $d^{\wedge}[v] \leftarrow \infty$ for all $v \neq s$, $d^{\vee}[t] \leftarrow 0$, $d^{\vee}[v] \leftarrow \infty$ for all $v \neq t$, $Q^{\wedge} \leftarrow \{s\}$, $Q^{\vee} \leftarrow \{t\}$, $w(P_{st}) \leftarrow \infty$, $w(SP_{st}) \leftarrow \infty$;

2 **while** $Q^{\wedge} \neq \emptyset \parallel Q^{\vee} \neq \emptyset$ **do**

3   Find a node $v \in Q^{\wedge}$ such that $d^{\wedge}[v] = \min \{d^{\wedge}[u] | u \in Q^{\wedge}\}$;

4       $Q^{\wedge} \leftarrow Q^{\wedge} \setminus \{v\}$;

5   **if** $w(P_{st}) < d^{\wedge}[v]$ **then**

6 | Continue;

7   **end**

8   **if** $v \in V_{n-M}$ **then**

9 | $E^{\wedge} \leftarrow E^{\wedge} \cup \{v\}$

10   **else**

11 | Relax edges of v;

12 | (Add nodes to $Q^{\wedge}$ and change $d^{\wedge}$, just like Dijkstra algorithm.)

13   **end**

14      $w(P_{st}) = \min \{w(P_{st}), d^{\wedge}[v] + d^{\vee}[v]\}$;

15   Repeat steps 3 to 14 on $G'_{\downarrow}$;

16 **end**

17 **for** $v \in E^{\wedge}$ **do**

18   **for** $v' \in E^{\vee}$ **do**

19 | $w(SP_{st}) \leftarrow \min \{w(P_{st}), d^{\wedge}[v] + D(v,v') + d^{\vee}[v']\}$;

20   **end**

21 **end**

In Algorithm 2, steps 5 to 7 prune the search space. We can also use P_{st} to reduce the number of access nodes between steps 16 and 17. If $d^{\wedge}[v]$(or $d^{\vee}[v]$)> $w(P_{st})$, then remove v from $E^{\wedge}$ (or $E^{\vee}$).

Before proving correctness, we present the following lemma given and proved by Bauer et al. [3]. We restate the lemma here with the definition in this paper.

Lemma 1. *Given $s, t \in V$, let $k = \min\{\alpha(s), \alpha(t)\} - 1$.*

(a): For any s-t path P in $G(k)$, there exists a sequence of s-t paths $P = P_k$, $P_{k-1}, ..., P_0$, such that P_i is a path in $G(i)$ and $w(P_i) = w(P)$.

(b): For any shortest s-t path P in G, there exists a sequence of shortest s-t paths $P = P_0, P_1, ..., P_k$, such that P_i is a path in $G(i)$, $V(P) \cap V_i \subseteq V(P_i)$, and $w(P_i) = w(P)$.

Proof. Same as the proof of Lemma 1 in [3]. $\qquad\square$

The correctness of CH has been proved by Geisberger et al. in [9]. A fundamental property of CH is that we can assume that a path P consists of upward edges from s towards u followed by downward edges from u towards t. Since a shortcut may only exist between two nodes when there is a path between them in G, the path in CH corresponds to a path in G. Every shortest $s - t$ path in G still exists in CH.

The sub-path of any two nodes in the shortest path in CH is also the shortest. Otherwise, a shorter path can be obtained by replacing it with a shorter one in G. So the forward search of the CH query can find the shortest path to u and to the first transit node v on P.

Since the preprocessing of CH-APSP is the same as that of CH in the early stage, the shortcuts before the remaining M nodes are the same. Consider the upward search on $G_\uparrow$. When the path to settle u does not contain any access node except u, it can be also used to settle u on $G_\uparrow'$, and the length is the same because the pruning only occurs at the access node.

Theorem 1. *Given $s, t \in V$, $d(s, t) \leq w(P_{st})$ and $d(s, t) \leq d_\uparrow(s, v) + D(v, v') + d_\downarrow(v', t)$ for all $v \in E^{\wedge}, v' \in E^{\vee}$.*

Proof. Since P_{st} corresponds to a $s - t$ path in G, $d(s, t) \leq w(P_{st})$ obviously. In the same way, for all $v \in E^{\wedge}$ and $v' \in E^{\vee}$, we have $d(s, v) \leq d_\uparrow(s, v)$ and $d(v', t) \leq d_\downarrow(v', t)$. According to the definition of APSP, $D(v, v') = w(P_{vv'}^{(n-M)})$, where $P_{vv'}^{(n-M)}$ is the shortest $v - v'$ path in $G(n - M)$. According to (a) of Lemma 1, there is a $v - v'$ path $P_{vv'}$ in G such that $w(P_{vv'}) = w(P_{vv'}^{(n-M)})$. $w(P_{vv'}) = d(v, v')$, or otherwise there is a $P_{vv'}^{\prime(n-M)} \in G(n - M)$ such that $d(v, v') = w(P_{vv'}^{\prime(n-M)}) < w(P_{vv'}^{(n-M)}) < w(P_{vv'}^{\prime(n-M)})$. Therefore, $d(s, t) = d(s, v) + d(v, v') + d(v', t) \leq d_\uparrow(s, v) + D(v, v') + d_\downarrow(v', t)$ for all $v \in E^{\wedge}, v' \in E^{\vee}$. $\qquad\square$

Theorem 2. *The query of the two-stage CH-APSP algorithm is correct.*

Proof. Given a node contraction order $\alpha : V \rightarrow \{1, ..., n\}$. Suppose $SP_{st} = \langle s = V_1, ..., V_a, ..., V_u, ..., V_b, ..., V_p = t \rangle$, $\alpha(V_{i+1}) > \alpha(V_i)$, $i \in [1, u)$, $\alpha(V_{i+1}) < \alpha(V_i)$, $i \in [u, p)$.

(i) Suppose $M \geq \alpha(V_u)$. V_u will be settled before pruning in both upward and downward searching and the length is the same as that in CH. So we have $d(s,t) = d_\uparrow(s, V_u) + d_\downarrow(V_u, t)$.

(ii) Suppose $\alpha(V_a), \alpha(V_b) \geq M$, $\alpha(V_{a-1}), \alpha(V_{b+1}) < M$. According to (i) and the property of CH, we have $d(s, V_a) = d_\uparrow(s, V_a)$, $d(V_b, t) = d_\downarrow(V_b, t)$ and $D(V_a, V_b) = d(V_a, V_b)$. Therefore, $d(s,t) = d(s, V_a) + d(V_a, V_b) + d(V_b, t) = d_\uparrow(s, V_a) + D(V_a, V_b) + d_\downarrow(V_b, t)$.

Combined with Theorem 1, the query of the two-stage CH-APSP algorithm is correct.   □

4   Experiments and Analysis

The algorithms are implemented by C++. We used two machines, one laptop with Intel Core i7-11800H and 16 GB of RAM, and a server with Intel Xeon Platinum 8280 CPU and 1TB of RAM.

The experiment works on the road networks provided for the 9th DIMACS challenge on shortest paths [6]. The number of nodes and edges of the networks are shown in Table 1. We randomly query 100,000 times. Experiment on graphs of USA is performed on the server, and experiments on the rest are performed on the laptop.

Table 1. The number of nodes and edges of the networks

	Nodes	Edges
New York City(NY)	0.26M	0.73M
Great Lakes(LKS)	2.76M	6.89M
Full USA(USA)	23.95M	58.33M

Our implementation is not parallel. We only use lazy updates and do not periodically reevaluate all priorities or recompute the priority of the neighbors. Contract a node when the difference between its new and old importance is less than 10. The priority function is:

$$2 * \text{edgeDifference} + \text{contractedNeighbors} + \text{nodeDepth}$$

During the simulated node contraction, we do not perform the witness path search but use the maximum possible number of shortcuts to calculate the importance. However, Dijkstra witness path search with binary min-heap is performed while actually contracting the node.

4.1   Impact of APSP Algorithms and Remaining Nodes(M) on Preprocessing

Here we only consider the basic algorithms for solving APSP problems: Floyd algorithm and Dijkstra with binary min-heap. Table 2 shows some results of the preprocessing time of different M and APSP algorithms. Whether Dijkstra or Floyd algorithm is used, the preprocessing time is less than the CH algorithm in a specific range of M.

We take NY as an example to further show the influence of M on preprocessing. Figure 3 shows the variation of preprocessing time with M. Regardless of whether Dijkstra or Floyd is used, the preprocessing time of the CH-APSP algorithm decreases first and then increases. The results of using Dijkstra are significantly better than Floyd, especially on larger graphs.

Table 2. Total preprocessing time of CH-APSP algorithm using different M and APSP algorithms

	CH(s)	M	CH-APSP algorithms		$|V_{n-M}|$
			CH-Dijkstra (s)	CH-Floyd (s)	
NY	23.081	100	22.874	22.866	9122
		900	18.216	18.374	131147
		1700	16.965	20.465	137050
LKS	1327.15	1000	978.959	985.238	634293
		1500	669.427	669.347	927239
		7500	395.439	804.066	356581
USA	23398.9	14000	3933.87	18469.9	6091641
		20000	3050.68	20109.9	2977571
		25000	3048.57	40182.5	2460052

The graph gradually becomes sparse as M increases, but the number of nodes and edges also increases. When the graph in the early stage of the CH algorithm is sparse, the preprocessing speed is fast, and it slows down when the graph tends to be complete. The APSP algorithm could be faster in processing the complete graph, but the speed will slow as the number of nodes and edges increases.

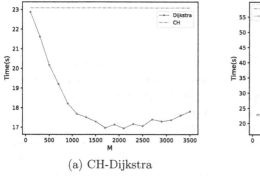

(a) CH-Dijkstra

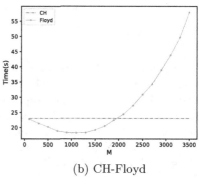

(b) CH-Floyd

Fig. 3. The variation of preprocessing time with M. The red dotted line indicates the preprocessing time of the original CH algorithm. (Color figure online)

Therefore, CH-APSP can be seen as an interpolation of APSP and CH. When $M = 0$, CH-APSP is the original CH algorithm; when $M = |V|$, CH-APSP is the APSP algorithm. The time in both extreme cases is relatively slow, and the suitable combination can be faster than both. When the time changes continuously with M, there exists a minimum value.

In addition, although Floyd performs better than Dijkstra on dense graphs when calculating the shortest path, its time has a cubic relationship with the number of nodes. Hence, the suitable range of M using CH-Dijkstra is more extensive than CH-Floyd. Take NY as an example, when the preprocessing time using CH-Floyd is more than CH, the corresponding M is 1900, while CH-Dijkstra is about 7000.

4.2   Impact of M on Query

The different algorithms for solving the APSP problem will only affect the preprocessing speed. They will not affect $G'^+ = (V, E'^+)$ and distance table D. Therefore, only the preprocessing algorithm using Dijkstra will be studied here. Table 3 shows the impact of different M on query.

Table 3. Impact of different M on query. The last column represents the proportion of the shortest path obtained through table lookup.

	CH(ms)	M	CH-APSP		
			Time(ms)	$|E^\wedge| + |E^\vee|$	Table lookup(%)
NY	0.784	100	0.762	52	17.7
		1000	0.597	109	70.7
		2000	0.537	144	93.6
LKS	6.974	1000	6.492	80	29.8
		1500	6.142	113	40.8
		7500	6.131	406	95.8
USA	164.718	20000	162.845	526	80.4
		30000	153.949	471	85.4
		40000	158.270	378	88.6

The results on all three graphs show that the query time of CH-APSP is shorter than that of CH. CH-APSP accelerates approximately 31% compared to CH when $M = 2000$ on the graph of NY. While on the graph of the USA, it is 3.9% when $M = 20000$. The performance of CH-APSP on extensive graphs could not as good as on slightly smaller graphs.

We also take NY as an example. Figure 4(a) shows the variation of the number of access nodes with M. As M increases, the number of access nodes increases rapidly, but there are fluctuations. However, it decreases at $M = 2000$, and the descent speed gradually slows. Figure 4(b) shows the proportion of the shortest path obtained by table lookup. At around $M = 1500$, the rate of increase slows down and eventually stabilizes, approaching but not reaching 1.

Figure 5 shows the variation of query time with M. As M increases, the query time gradually decreases. Initially, the descent speed is faster, but after $M = 3500$, it is slower and tends to stabilize. Based on the results in Fig. 4, the reason may be that as M increases, the number of access nodes gradually stabilizes, resulting in a relatively stable number of table lookups. In addition, most of the shortest paths come from table lookups, so the query time also tends to stabilize.

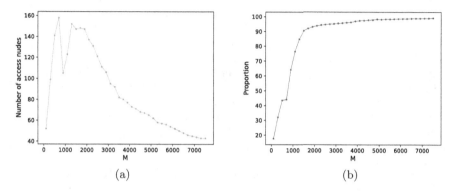

(a)   (b)

Fig. 4. (a) The variation of number of access nodes with M. (b) The proportion of the shortest path obtained by table lookup.

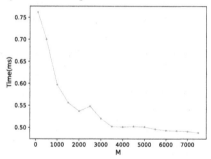

Fig. 5. The variation of query time with M.

5   Conclusions and Future Work

We have presented an algorithm that can both reduce preprocessing and query time to less than CH, allowing for parameter adjustment based on spatial requirements. It can be regarded as an interpolation between CH and APSP. Though CH-APSP can utilize the graph's properties to achieve preprocessing and query acceleration, there is still room for improvement. Firstly, CH-APSP stores the complete distance table. Consider using the properties of the graph to pure the distance table and improve its structure. Secondly, we simply used P_{st} to pure the access nodes without considering more effective methods. Finally, our algorithm does not perform well enough on large graphs like the USA, so we are considering further improvements.

References

1. Abraham, I., Delling, D., Goldberg, A.V., Werneck, R.F.: A hub-based labeling algorithm for shortest paths in road networks. In: Pardalos, P.M., Rebennack, S. (eds.) Experimental Algorithms, pp. 230–241. Springer, Berlin Heidelberg, Berlin, Heidelberg (2011). https://doi.org/10.1007/978-3-642-20662-7_20
2. Arz, J., Luxen, D., Sanders, P.: Transit node routing reconsidered. In: Bonifaci, V., Demetrescu, C., Marchetti-Spaccamela, A. (eds.) Experimental Algorithms, pp. 55–66. Springer, Berlin Heidelberg, Berlin, Heidelberg (2013). https://doi.org/10.1007/978-3-642-38527-8_7
3. Bauer, R., Columbus, T., Rutter, I., Wagner, D.: Search-space size in contraction hierarchies. Theoret. Comput. Sci. **645**, 112–127 (2016)
4. Bauer, R., Delling, D., Sanders, P., Schieferdecker, D., Schultes, D., Wagner, D.: Combining hierarchical and goal-directed speed-up techniques for dijkstra's algorithm. ACM J. Exp. Algorithmics **15** (2010)
5. Cormen, T.H., Leiserson, C.E., Rivest, R.L., Stein, C.: Introduction to Algorithms, 2nd edn. MIT Press, Cambridge, MA, USA (2001)
6. Demetrescu, C., Johnson, D.S., Goldberg, A.V.: The Shortest Path Problem: Ninth Dimacs Implementation Challenge (2009)
7. Dijkstra, E.: A note on two problems in Connexion with graphs. Numer. Math. **1**, 269–271 (1959)
8. Funke, S.: Seamless interpolation between contraction hierarchies and hub labels for fast and space-efficient shortest path queries in road networks. In: Kim, D., Uma, R.N., Cai, Z., Lee, D.H. (eds.) Computing and Combinatorics, pp. 123–135. Springer International Publishing, Cham (2020). https://doi.org/10.1007/978-3-030-58150-3_10
9. Geisberger, R., Sanders, P., Schultes, D., Delling, D.: Contraction hierarchies: faster and simpler hierarchical routing in road networks. In: McGeoch, C.C. (ed.) Experimental Algorithms, pp. 319–333. Springer, Berlin Heidelberg, Berlin, Heidelberg (2008). https://doi.org/10.1007/978-3-540-68552-4_24
10. Goldberg, A., Harrelson, C.: Computing the shortest path: a* search meets graph theory, Tech. Rep., MSR-TR-2004-24 (2004). https://www.microsoft.com/en-us/research/publication/computing-the-shortest-path-a-search-meets-graph-theory/
11. Gutman, R.J.: Reach-based routing: a new approach to shortest path algorithms optimized for road networks. In: ALENEX/ANALC, pp. 100–111 (2004)
12. Lauther, U.: An extremely fast, exact algorithm for finding shortest paths in static networks with geographical background. In: Geoinformation und Mobilität-von der Forschung zur praktischen Anwendung, vol. 22, pp. 219–230. IfGI prints, Institut für Geoinformatik, Münster (2004)
13. Sanders, P., Schultes, D.: Highway hierarchies hasten exact shortest path queries. In: Brodal, G.S., Leonardi, S. (eds.) Algorithms - ESA 2005, pp. 568–579. Springer, Berlin Heidelberg, Berlin, Heidelberg (2005). https://doi.org/10.1007/11561071_51
14. Schultes, D., Sanders, P.: Dynamic highway-node routing. In: Demetrescu, C. (ed.) Experimental Algorithms, pp. 66–79. Springer, Berlin Heidelberg, Berlin, Heidelberg (2007). https://doi.org/10.1007/978-3-540-72845-0_6

Information Theory of Blockchain Systems

Quan-Lin Li[1], Yaqian Ma[1], Jing-Yu Ma[2]([✉]) [iD], and Yan-Xia Chang[1]

[1] School of Economics and Management, Beijing University of Technology,
Beijing 100124, China
[2] Business School, Xuzhou University of Technology, Xuzhou 221018, China
mjy0501@126.com

Abstract. In this paper, we apply the information theory to provide an approximate expression of the steady-state probability distribution for blockchain systems. We achieve this goal by maximizing an entropy function subject to specific constraints. These constraints are based on some prior information, including the average numbers of transactions in the block and the transaction pool, respectively. Furthermore, we use some numerical experiments to analyze how the key factors in this approximate expression depend on the crucial parameters of the blockchain system. As a result, this approximate expression has important theoretical significance in promoting practical applications of blockchain technology. At the same time, not only do the method and results given in this paper provide a new line in the study of blockchain queueing systems, but they also provide the theoretical basis and technical support for how to apply the information theory to the investigation of blockchain queueing networks and stochastic models more broadly.

Keywords: Blockchain · Information theory · Maximum entropy principle · Steady-state probability distribution

1 Introduction

Blockchain has become a prominent topic of discussion in recent years, revolutionizing various aspects of life through its significant impact on many practical application fields. For example, finance by Kowalski et al. [8]; the Internet of Things by Torky and Hassanein [24]; healthcare by Sudeep et al. [23]; and others. The active participation of miners in the mining process is fundamental to ensuring the secure and stable operation of blockchain systems, as well as guaranteeing its sustainable development. However, the inner workings of blockchain mining are extremely obscure and challenging to examine. Conducting direct measurements on mining networks is highly complex due to the miners' privacy

Supported by the National Natural Science Foundation of China under grant No. 71932002 and the Social Science Foundation of Jiangsu Province under grant No. 23GLB018.

concerns, whereas blockchain data provides a method of direct measurement. Consequently, it is essential to develop statistical techniques using accessible blockchain data for investigating the blockchain systems.

So far blockchain research has obtained many important advances, readers may refer to a book by Swan [22]; a key research framework shown by Daneshgar et al. [3], Lindman et al. [13] and Risius and Spohrer [19]; decision in blockchain mining by Ma and Li [16] and Chen et al. [2]; and others by Lu et al. [15] and Yang et al. [25].

Applying queueing theory and Markov processes to analyze the blockchain systems is interesting and challenging, since each blockchain system not only is a complicated stochastic system but also has multiple key factors and a physical structure with different levels. Li et al. [10] provided a two-stage queueing model of the PoW blockchain system, clearly described and expressed the physical structure with multiple key factors, furthermore the matrix geometric solution was applied to give a complete solution such that the performance evaluation of the PoW blockchain system was established in a simple form. Seol et al. [20] proposed an $M(1, n)/M_n/1$ queueing model to analyze the blockchain system in Ethereum; Zhao et al. [26] established a non-exhaustive queueing model with a limited batch service and a possible zero-transaction service, derived the average number of transactions and the average confirmation time of a transaction; Mišić et al. [17] applied the Jackson network to analyze the blockchain network.

Compared with the queueing theory, the Markov process is mainly used to evaluate the throughput, confirmation time, security and privacy protection of the blockchain systems. Huang et al. [4] proposed the Markov process with an absorption state and conducted an analysis on the performance of the Raft consensus algorithm in private blockchains. Srivastava [21] calculated the transaction confirmation time in blockchain systems. Li et al. [11] discussed block access control mechanisms in wireless blockchain networks. Nguyen et al. [18] investigated the task offloading problem in mobile blockchain with privacy protection using the Markov processes and deep reinforcement learning.

The traditional reluctance of miners to share insider information regarding their competitive advantages, leading to great difficulties for these two approaches when dealing with more complex blockchain systems, such as those involving multiple mining pools. The purpose of this paper is to apply the maximum entropy principle to provide an approximate expression for blockchain systems. In information theory, entropy serves as a probabilistic measure to quantify the uncertainty of information associated with random variables. In recent years, the information entropy has been implemented in various practical domains of blockchain technology. For example, industrial Internet of Things by Khan and Byun [7]; renewable energy by Liu et al. [14]; fake news prevention by Chen et al. [1]; and medical data sharing by Liang et al. [12].

The degree of randomness in a random variable can be measured by applying the maximum entropy when its information is most uncertain. For example, a large amount of information can only be partially obtained and utilized. Jaynes [5,6] initially proposed the maximum entropy principle, which offered an approximate computational approach for unknown probability distributions. Such an approach provided a uniquely correct self-consistent method of inference for estimating probability distributions based on the available information.

The main contributions of this paper are twofold. The first one is to apply the maximum entropy principle to study blockchain queueing systems for the first time. This approach can be applied to the performance analysis of queueing systems, since expected values of various distributions can be obtained by making measurements in an operational sense. Unlike previous works for the applying queueing theory or the Markov processes, we just need to take statistical techniques by simple observation on miners. The second contribution of this paper is to provide the approximate expression of the steady-state probability distribution for the blockchain systems. So far, numerous categories of blockchain systems have yet to be thoroughly analyzed using the queueing theory or the Markov processes due to difficulties in the expression of the steady-state probability distributions. For example, the PoW blockchain systems with multiple mining pools, the PBFT blockchain systems of dynamic nodes, the DAG-based blockchain systems, the Ethereum, and the large-scale blockchain systems with either cross-chain, side-chain, or off-chain. Therefore, the results of this paper give new insights into the application of the maximum entropy principle to more complex blockchain systems, and partially solve a challenging problem in blockchain technology.

The rest of this paper is organized as follows. Section 2 introduces the blockchain queueing model briefly. In Sect. 3, we apply the maximum entropy principle to give the approximate expression of the steady-state probability distribution for the blockchain system. We also conduct numerical experiments to analyze how the key factors of the approximate expression depend on some crucial parameters in Sect. 4. Finally, the whole work is concluded in the last section.

2 Model Description

In this section, we describe a blockchain system as two stages of asynchronous processes: block-generation and blockchain-building, which is depicted in Fig. 1. To ensure clarity, we review the blockchain queueing model and adopt the notations of Li et al. [10] briefly.

Arrival Processes: Transactions arrive at the blockchain system according to a Poisson process with arrival rate λ. Each transaction must first enter and queue up in a transaction pool with infinite size.

Block-generation Processes: Each arrival transaction first queues up in the transaction pool and then waits to be mined into a block successfully. We assume that the block-generation times are i.i.d. and exponential with service rate μ_1. The transactions are chosen into the block, but they are not completely based on the First Come First Service (FCFS) from the order of transaction arrivals.

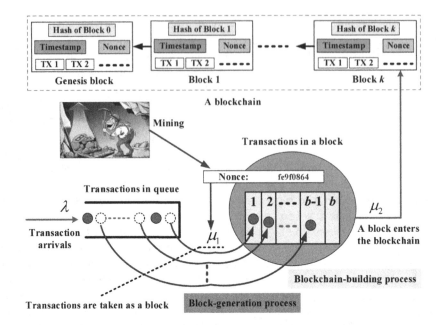

Fig. 1. A blockchain queueing system.

Block Capacity: To avoid spam attacks, we assume that the maximum size of each block is limited to b transactions. If there are more than b transactions in the transaction pool, then the b transactions are selected to form a full block while the rest of transactions are still waiting in the transaction pool and may be used to construct another block.

Blockchain-building Processes: The block with a group of transactions will be pegged to a blockchain. We assume that the blockchain-building times are i.i.d. and exponential with the service rate μ_2.

Independence: We assume that all the random variables defined above are independent of each other.

Let $I(t)$ and $J(t)$ be the numbers of transactions in the block and in the transaction pool at time t, respectively. Then, $(I(t), J(t))$ may be regarded as a state of the blockchain system at time t. The state space of this blockchain system is

$$\Omega = \{(i,j), 0 \leq i \leq b, 0 \leq j \leq \infty\}.$$

The following lemma provides a necessary and sufficient condition under which the blockchain system is stable. Here, we only restate it without proof, while readers may refer to Chap. 3 of Li [9] and Li et al. [10] for more details.

Lemma 1. *The blockchain system is stable if and only if*

$$\frac{b\mu_1\mu_2}{\mu_1 + \mu_2} > \lambda. \tag{1}$$

In what follows, we assume that the stable condition (1) is satisfied, then this blockchain system is stable. The limit

$$\lim_{t \to +\infty} p\{I(t) = i, J(t) = j\}$$

exists and is unique. Let

$$p(i, j) = \lim_{t \to +\infty} p\{I(t) = i, J(t) = j\}.$$

Then, $p(i, j), (i, j) \in \Omega$ is the steady-state probability distribution of the blockchain system.

By using the matrix-geometric solution, we can write the steady-state probability distribution under the stable condition (1), see Li et al. [10]. In the next section, we will introduce the maximum entropy principle to provide the approximate expression of the steady-state probability distribution for the blockchain system.

3 Maximum Entropy in Blockchain Systems

In this section, we provide an entropy function and some prior information, and use Lagrange method of undetermined multipliers to give the approximate expression of the steady-state probability distribution.

3.1 Entropy Function

Based on the steady-state probability distribution $p(i, j)$, we introduce the entropy function

$$H(p) = - \sum_{(i,j) \in \Omega} p(i, j) \ln p(i, j)$$

or

$$H(p) = - \sum_{j=0}^{\infty} \sum_{i=0}^{b} p(i, j) \ln p(i, j). \tag{2}$$

The maximum entropy principle states that of all distributions satisfying the constraints supplied by the given information, the minimally prejudiced distribution $p(i, j), (i, j) \in \Omega$ is the one that maximizes the entropy function of the blockchain queueing system.

3.2 Prior Information

To approximate the steady-state probability distribution $p(i, j), (i, j) \in \Omega$ using the maximum entropy principle by maximizing (2), we need to provide some prior information as follows:

(i) The normalisation:

$$\sum_{(i,j) \in \Omega} p(i, j) = 1. \tag{3}$$

(ii) The average number of transactions in the block:

$$\sum_{i=0}^{b} i \sum_{j=0}^{\infty} p(i,j) = I. \tag{4}$$

(iii) The average number of transactions in the transaction pool:

$$\sum_{j=0}^{\infty} j \sum_{i=0}^{b} p(i,j) = J. \tag{5}$$

Remark 1. *Note that the statistics of prior information selected always may be known numerically via system measurements during finite observation periods or can be determined symbolically via known analytic formulae based on operational or stochastic assumptions. For example, blockchain data has the advantage of providing direct measurements, as the fields of a block are filled by the miner of that block.*

3.3   The Maximum Entropy Principle

The steady-state probability distribution $p(i,j), (i,j) \in \Omega$ is considered as an independent variable. We maximize the entropy function (2) subject to constrains (3)–(5), the optimization model of the maximum entropy principle can be written as

$$\max \ H(p) = - \sum_{j=0}^{\infty} \sum_{i=0}^{b} p(i,j) \ln p(i,j),$$

$$s.t. \quad \begin{cases} \sum_{j=0}^{\infty} \sum_{i=0}^{b} p(i,j) = 1, \\ \sum_{i=0}^{b} i \sum_{j=0}^{\infty} p(i,j) = I, \\ \sum_{j=0}^{\infty} j \sum_{i=0}^{b} p(i,j) = J. \end{cases}$$

The following theorem provides the approximate expression of the steady-state probability distribution for the blockchain system by the maximum entropy principle.

Theorem 1. *For the steady-state probability distribution of the blockchain systems $p(i,j), (i,j) \in \Omega$, there exists a tuple of positive numbers x, y and z such that the approximate expression $\tilde{p}(i,j)$ has the following form*

$$\tilde{p}(i,j) = x y^i z^j.$$

Proof: By introducing β_0, β_1 and β_2 to equations (3)–(5), we write Lagrangian function as

$$L(p, \beta_0, \beta_1, \beta_2) = -\sum_{j=0}^{\infty}\sum_{i=0}^{b} p(i,j) \ln p(i,j) + \beta_0 \left(1 - \sum_{j=0}^{\infty}\sum_{i=0}^{b} p(i,j) \right)$$

$$+ \beta_1 \left(I - \sum_{i=0}^{b} i \sum_{j=0}^{\infty} p(i,j) \right) + \beta_2 \left(J - \sum_{j=0}^{\infty} j \sum_{i=0}^{b} p(i,j) \right), \quad (6)$$

where β_0, β_1 and β_2 are the Lagrange multipliers corresponding to constraints (3)–(5), respectively.

To find the maximum entropy solution $p(i,j)$, maximizing (2) subject to constraints (3)–(5) is equivalent to maximizing (6).

The Lagrangian function $L(p, \beta_0, \beta_1, \beta_2)$ is a multivariate function with respect to variables $p(i,j)$, β_0, β_1 and β_2. To obtain the maximum entropy solutions, we take the partial derivatives of $L(p, \beta_0, \beta_1, \beta_2)$ with respect to $p(i,j)$ and then set the results equal to zero, i.e., $\partial L/\partial p(i,j) = 0$.

If (i,j) is determined, then

$$\frac{\partial}{\partial p(i,j)} \left[-\sum_{j=0}^{\infty}\sum_{i=0}^{b} p(i,j) \ln p(i,j) \right] = -\ln p(i,j) - 1.$$

It is clear that for all $(\bar{i}, \bar{j})$, $\bar{i} \neq i$ and $\bar{j} \neq j$,

$$\frac{\partial}{\partial p(i,j)} p(\bar{i}, \bar{j}) \ln p(\bar{i}, \bar{j}) = 0.$$

Thus, we obtain

$$\frac{\partial L}{\partial p(i,j)} = [-\ln p(i,j) - 1] - \beta_0 - \beta_1 i - \beta_2 j = 0,$$

which indicates

$$\ln p(i,j) = -1 - \beta_0 - \beta_1 i - \beta_2 j. \quad (7)$$

It follows from (7) that

$$p(i,j) = \exp\left[-(1+\beta_0)\right] \exp\left(-\beta_1 i\right) \exp\left(-\beta_2 j\right). \quad (8)$$

Let

$$x = \exp\left[-(1+\beta_0)\right], y = \exp\left(-\beta_1\right) \text{ and } z = \exp\left(-\beta_2\right).$$

Then, we rewrite (8) as

$$p(i,j) = x y^i z^j. \quad (9)$$

Substituting (9) into (3) and utilizing algebraic knowledge, we have

$$x = \frac{(1-y)(1-z)}{1-y^{b+1}}. \quad (10)$$

Similarly, substituting (9) into (4) and (5), respectively, we have

$$y^{b+1} - \sum_{n=1}^{b} \frac{1}{b-I} y^n + \frac{I}{b-I} = 0 \tag{11}$$

and

$$z = \frac{J}{1+J}. \tag{12}$$

Therefore, if the average number of transactions in the block and the transaction pool can be provided, respectively, the positive numbers x, y and z exist to give the approximate expression for $\tilde{p}(i,j)$. This completes the proof.   □

Remark 2. *The theoretical expression of the mean values I and J given by Li et al. [10] are restricted to Poisson arrival processes and exponential service times, meaning that these expression are only theoretically applicable in this particular case. Nevertheless, the maximum entropy principle is not dependent on this assumption of the Poisson arrival processes and the exponential service times. It can be applied to non-Poisson arrival processes and non-exponential service times, as long as I and J can be provided, the non-linear equations can be solved to derive the approximate expression of the steady-state probability distribution for the blockchain queueing system. Therefore, the approximate expression derived in Sect. 3.3 has broad applicability.*

4   Numerical Experiments

In this section, we provide some numerical examples to verify computability of our theoretical results and analyze how the key factors y and z of the approximate expression depend on some crucial parameters of the blockchain queueing system.

Taking the situation of the Poisson arrival processes and the exponential service times in Li et al. [10] as an example, since the theoretical expression of the mean values I and J are composed of the crucial parameters λ, μ_1, μ_2 and b, we can observe the relation between the key factors and crucial parameters. Note that x is represented by y and z according to equations (10)–(12), we just need to focus on how y and z depend on these crucial parameters through numerical examples.

In the Examples 1 and 2, we take some common parameters: the maximum block size $b = 80$, the block-generation service rate $\mu_1 = 6, 7.5, 10$, the blockchain-building service rate $\mu_2 = 2$ and the arrival rate $\lambda \in (1, 3.5)$.

Example 1. We analyze how y depends on λ and μ_1. From Fig. 2, it is seen that y decreases as λ increases, while it also decreases as μ_1 increases.

Example 2. We analyze how z depends on λ and μ_1. From Fig. 3, it is seen that z increases as λ increases, while it increases as μ_1 decreases.

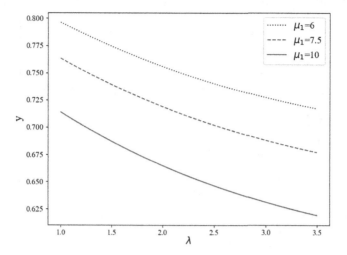

Fig. 2. y vs. λ for three different values of μ_1.

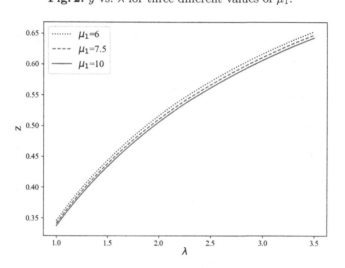

Fig. 3. z vs. λ for three different values of μ_1.

Example 3. We specifically observe how y and z depend on the maximal block size b, respectively. We take some common parameters: the arrival rate $\lambda = 1.5$, the blockchain-building service rate $\mu_2 = 2$, the maximum block size $b = 40, 80, 160$ and the block-generation service rate $\mu_1 \in (5, 7.5)$. From Fig. 4 and Fig. 5, it is seen that y and z decrease as μ_1 increases, while they increase as b increases.

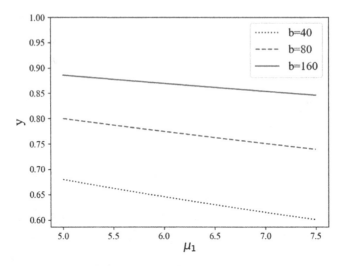

Fig. 4. y vs. μ_1 for three different values of b.

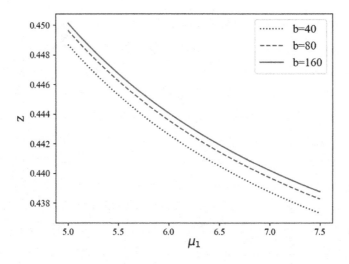

Fig. 5. z vs. μ_1 for three different values of b.

5   Concluding Remarks

In this paper, we apply the maximum entropy principle of the information theory to study the blockchain queueing system, and provide an approximate expression of its steady-state probability distribution. By obtaining this approximation, we have partially resolved a challenging issue in the blockchain technology, i.e., how to directly express the steady-state probability distributions of some large-scale and complex blockchain queueing systems. On the other hand, we use numerical examples to verify the computability of our theoretical results and analyze how the key factors of the approximate expression depend on some crucial parameters. Along these lines, we will continue our future research in the following directions:

- Investigating blockchain queueing systems with multiple mining pools, different consensus mechanisms and so on.
- Extending the information theory to blockchain queueing networks or stochastic models.
- Applying the information theory to provide a more accurate approximate expression by utilizing more prior information, such as the second and the third moments.

References

1. Chen, C.C., Du, Y., Peter, R., et al.: An implementation of fake news prevention by blockchain and entropy-based incentive mechanism. Soc. Netw. Anal. Min. **12**(1), 114 (2022)
2. Chen, J., Cheng, Y., Xu, Z., et al.: Decision on block size in blockchain systems by evolutionary equilibrium analysis. Theor. Comput. Sci. **942**, 93–106 (2023)
3. Daneshgar, F., Ameri Sianaki, O., Guruwacharya, P.: Blockchain: a research framework for data security and privacy. In: Barolli, L., Takizawa, M., Xhafa, F., Enokido, T. (eds.) WAINA 2019. AISC, vol. 927, pp. 966–974. Springer, Cham (2019). https://doi.org/10.1007/978-3-030-15035-8_95
4. Huang, D., Ma, X., Zhang, S.: Performance analysis of the Raft consensus algorithm for private blockchains. IEEE Trans. Syst. Man. Cybern. Syst. **50**(1), 172–181 (2019)
5. Jaynes, E.T.: Information theory and statistical mechanics. Phys. Rev. **106**(4), 620–630 (1957)
6. Jaynes, E.T.: Information theory and statistical mechanics II. Phys. Rev. **108**(2), 171–190 (1957)
7. Khan, P.W., Byun, Y.: A blockchain-based secure image encryption scheme for the industrial Internet of Things. Entropy **22**(2), 175 (2020)
8. Kowalski, M., Lee, Z.W., Chan, T.K.: Blockchain technology and trust relationships in trade finance. Technol. Forecast. Soc. **166**, 120641 (2021)
9. Li, Q.L.: Constructive Computation in Stochastic Models with Applications: The RG-Factorizations. Springer, Heidelberg (2010). https://doi.org/10.1007/978-3-642-11492-2_2
10. Li, Q.L., Ma, J.Y., Chang, Y.X.: Blockchain queue theory. In: Chen, X., Sen, A., Li, W.W., Thai, M.T. (eds.) CSoNet 2018. LNCS, vol. 11280, pp. 25–40. Springer, Cham (2018). https://doi.org/10.1007/978-3-030-04648-4_3
11. Li, Y., Cao, B., Liang, L., et al.: Block access control in wireless blockchain network: design, modeling and analysis. IEEE Trans. Veh. Technol. **70**(9), 9258–9272 (2021)
12. Liang, X., Chen, W., Li, J., et al.: Incentive mechanism of medical data sharing based on information entropy in blockchain environment. J. Phys: Conf. Ser. **1302**(2), 022056 (2019)
13. Lindman, J., Tuunainen, V.K., Rossi, M.: Opportunities and risks of blockchain technologies - a research agenda. In: Proceedings of the 50th Hawaii International Conference on System Sciences, Hawaii, pp. 1533–1542 (2017)

14. Liu, Z., Huang, B., Hu, X., et al.: Blockchain-based renewable energy trading using information entropy theory. IEEE T. Netw. Sci. Eng. 1–12 (2023)

15. Lu, Y., Huang, X., Zhang, K., et al.: Blockchain empowered asynchronous federated learning for secure data sharing in internet of vehicles. IEEE Trans. Veh. Technol. **69**(4), 4298–4311 (2020)

16. Ma, J.Y., Li, Q.L.: Optimal dynamic mining policy of blockchain selfish mining through sensitivity-based optimization. J. Comb. Optim. **44**(5), 3663–3700 (2022)

17. Mišić, J., Mišić, V.B., Chang, X.: Performance of bitcoin network with synchronizing nodes and a mix of regular and compact blocks. IEEE T. Netw. Sci. Eng. **7**(4), 3135–3147 (2020)

18. Nguyen, D.C., Pathirana, P.N., Ding, M., et al.: Privacy-preserved task offloading in mobile blockchain with deep reinforcement learning. IEEE Trans. Netw. Service Manag. **17**(4), 2536–2549 (2020)

19. Risius, M., Spohrer, K.: A blockchain research framework. Bus. Inform. Syst. Eng. **59**(6), 385–409 (2017)

20. Seol, J., Kancharla, A., Ke, Z., et al.: A variable bulk arrival and static bulk service queueing model for blockchain. In: The 2nd ACM International Symposium on Blockchain and Secure Critical Infrastructure, Taipei, pp. 63–72. Association for Computing Machinery (2020)

21. Srivastava, R.: Mathematical assessment of blocks acceptance in blockchain using Markov model. Int. J. Blockchains Cryptocurrencies **1**(1), 42–53 (2019)

22. Swan, M.: Blockchain: Blueprint for a new economy. O'Reilly Media, Inc. (2015)

23. Sudeep, T., Karan, P., Richard, E.: Blockchain-based electronic healthcare record system for healthcare 4.0 applications. J. Inf. Secur. Appl. **50**, 102407 (2020)

24. Torky, M., Hassanein, A.E.: Integrating blockchain and the internet of things in precision agriculture: analysis, opportunities, and challenges. Comput. Electron. Agr. **178**, 105476 (2020)

25. Yang, L., Li, M., Si, P., et al.: Energy-efficient resource allocation for blockchain-enabled industrial Internet of Things with deep reinforcement learning. IEEE Internet Things J. **8**(4), 2318–2329 (2020)

26. Zhao, W., Jin, S., Yue, W.: Analysis of the average confirmation time of transactions in a blockchain system. In: Phung-Duc, T., Kasahara, S., Wittevrongel, S. (eds.) QTNA 2019. LNCS, vol. 11688, pp. 379–388. Springer, Cham (2019). https://doi.org/10.1007/978-3-030-27181-7_23

Machine Learning with Low-Resource Data from Psychiatric Clinics

Hongmin W. Du[1], Neil De Chen[2], Xiao Li[3(✉)], and Miklos A. Vasarhelyi[1]

[1] Accounting and Information Systems Department, Rutgers University, Piscataway, NJ 08854, USA
hd255@scarletmail.rutgers.edu, miklosv@business.rutgers.edu
[2] School of Medicine, Saint Luis University, St. Louis, MO 63103, USA
[3] Department of Computer Science, University of Texas at Dallas, Richardson, TX, USA
xiao.li@utdallas.edu

Abstract. Amidst the rapid growth of big data, the success of machine learning is critically tethered to the availability and quality of training data. A pertinent challenge faced by this symbiotic relationship is the issue of "low-resource data," characterized by insufficient data volume, diversity, and representativeness, and exacerbated by class imbalances within datasets. This study delves into the intersection of machine learning and big data, exploring innovative methodologies to counteract the challenges of data scarcity. Focusing on psychiatric clinic data, marked by subjectivity and inconsistency, we outline the unique challenges posed by the nature of data in this domain. To address these challenges, we explore the potential of data augmentation-using transformations or operations on available data-and transfer learning, where knowledge from a pretrained model on a large dataset is transferred to a smaller one. Through a comprehensive exploration of these methodologies, this research aims to bolster the effectiveness of machine learning in low-resource environments, with a vision of advancing the digital landscape while navigating inherent data constraints.

Keywords: Small Data · Medical Data · Machine Learning · Low Resource Data · Data Augmentation

1 Introduction

In the era of exponential data growth, the role of machine learning has emerged as a pivotal force in extracting valuable insights and knowledge from vast and complex datasets, commonly referred to as big data. This symbiotic relationship between machine learning and big data has led to significant advancements across numerous domains, ranging from healthcare and finance to marketing and autonomous systems. However, amidst the remarkable progress achieved, a persistent challenge looms large, casting a shadow on the efficacy of machine

W. Wu and J. Guo (Eds.): COCOA 2023, LNCS 14462, pp. 455–465, 2024.
https://doi.org/10.1007/978-3-031-49614-1_34

learning algorithms-the scarcity of adequate training data and the pervasive issue of class imbalance within datasets. This challenge, synonymous with the term "low-resource data," has garnered substantial attention and constitutes a paramount concern in the field of machine learning.

Despite the undeniable potential of machine learning to revolutionize decision-making and predictive modeling, its effectiveness crucially hinges upon the quality and quantity of training data available. While the digital landscape is inundated with data, a considerable portion of it remains inadequate in terms of volume, diversity, and representativeness. This limitation not only impedes the development of accurate and robust machine learning models but also diminishes their potential to generalize to new, unseen instances. Concurrently, the uneven distribution of class labels within datasets, commonly referred to as class imbalance, further compounds the predicament by skewing the learning process in favor of the majority class, often at the expense of the minority classes. Consequently, these inherent challenges collectively underscore the pressing need to devise innovative strategies and methodologies that address the intricacies of low-resource data and mitigate the associated adversities.

In this research endeavor, we delve deep into the multifaceted domain of machine learning and big data, centering our focus on the intersection between model performance and the scarcity of training data. By analyzing the manifold implications of low-resource data scenarios and the intricate nature of class imbalance, we aim to unearth novel approaches that enable the harnessing of invaluable insights even from data-scarce environments. Through a comprehensive exploration of methodologies ranging from transfer learning and active learning to data augmentation and synthetic data generation, we strive to not only elucidate the intricacies of these strategies but also evaluate their effectiveness in ameliorating the performance of machine learning models when confronted with the challenges of low-resource data.

As we navigate through the unknown areas of this research, our ultimate goal is to contribute to the arsenal of tools and techniques that empower machine learning practitioners to transcend the constraints imposed by data scarcity and class imbalance. By doing so, we envision a future where the promise of machine learning remains steadfast, unswayed by the inherent limitations of data availability, and continues to pave the way for unprecedented advancements in our ever-evolving digital landscape.

For example, in the analysis of psychiatric clinics data, we face a difficult situation that the available data is not large enough due to the following.

– Psychiatry is a highly subjective field when it comes to gathering patient data, characterized by a scarcity of numerical information and marked inconsistency in the documentation of symptoms, signs, concerns, and the progression of cases across different psychiatrists. The lack of standardized language, apart from diagnostic criteria, presents a significant challenge for Natural Language Processing (NLP) to effectively capture and analyze the data in a coherent manner.

- Diagnosis does not always dictate the treatment plan; more frequently, it serves as a coding tool. However, diagnoses can undergo changes quite easily. Moreover, many medications are employed in off-label capacities, deviating from algorithmic guidelines. As a consequence, this diversity leads to a wide array of distinct approaches.
- Medical records in the field of psychiatry are often unreliable, with numerous patients receiving incorrect diagnoses that do not align with current diagnostic criteria. There is a prevalence of inaccurate or misleading information that may not be rectified by subsequent psychiatrists, leading to a lack of proper sanitization of the records.
- There is no evidence indicating that superior documentation or clustering of patient data directly results in improved patient outcomes. This raises an ethical dilemma, as it could be perceived that such practices primarily benefit hospital administration financially, potentially exacerbating the challenges patients face in accessing proper care.

The aforementioned facts underscore the challenging nature of researching psychiatric clinic data. Given this context, what methodologies can be employed? In this concise article, we explore potential techniques that could be applied to the analysis of psychiatric clinic data. By comparing these techniques, our aim is to identify a more effective approach. Subsequently, in future research endeavors, we aspire to extract valuable insights from psychiatric clinic data, benefiting both psychiatrists and their patients.

2 Data Augmentation

One of the ways for dealing with low-resource data is so called data augmentation. What is data augmentation? This involves artificially increasing the size of the training set by creating new data instances through transformations or operations on available data. The following are three examples.

In [9], data augmentation is employed to deal with the task of image classification. They compared and analyzed multiple methods of data augmentation, including classical image transformations like rotating, cropping, zooming, histogram based methods, and operations at Style Transfer and Generative Adversarial Networks, together with the representative examples. They also presented their own method of data augmentation based on image style transfer. Those methods generate the new images of high perceptual quality, which can be used to pre-train the given neural network in order to improve the training process efficiency. Finally, the three medical case studies are carried out to validate proposed method. They are skin melanomas diagnosis, histopathological images and breast magnetic resonance imaging (MRI) scans analysis. The image classification is utilized to provide helpful information for a diagnose. In such medical case studies, the data deficiency is a very important relevant issue. Moreover, they discussed the advantages and disadvantages of discussed methods.

In the study of computer vision, augmented data are still images. Considered operations include horizontally flipping, random cropping, tilting, and altering the color channels of the original images. Since the content of the new image is still the same, the label of the original image is preserved. This situation is changed in training networks for NLP tasks such as Machine Translation. Given a source and target sentence pair (S, T), one may need to alter it in a way that preserves the semantic equivalence between S and T. For low-resource language pairs, this is difficult to do. In [4], a novel data augmentation approach is proposed to target low-frequency words by generating new sentence pairs containing rare words in new, synthetically created contexts. Their method is found to have improved translation quality.

For text classification tasks, four simple but powerful operations are synonym replacement, random insertion, random swap, and random deletion. In [20], data augmentation techniques with those operations are employed in the study of five text classification tasks. They demonstrate strong results for smaller datasets and improve performance for both convolutional and recurrent neural networks. Across five datasets, with mentioned data augmentation techniques, using only 50% of the available training set achieved the same accuracy as normal training with all available data. Moreover, they give suggested parameters for practical use.

All above three examples contain helpful information for the analysis of psychiatric clinics data.

- Psychiatric clinics data is in text format. The task of identifying proper diagnosis is equivalent to doing text classification. Therefore, those operations mentioned in [20] may be useful.
- However, we have to select operations carefully because label (i.e., diagnosis) should be preserved, such as summarization and rearrange ordering of sentences and consider those in the case of skin melanomas diagnosis in [9].
- To preserve the label, we may treat psychiatric clinics data as data pairs of patient's suboptimal statement and doctor's diagnosis with treatment. Then, employ techniques in [4] to augment data pairs.

3   Transfer Learning

Transfer learning is a simple and powerful method. It can be used to boost model performance of low-resource neural machine translation. This technique includes taking a pre-trained model (usually trained on a large dataset) and fine-tuning it on a smaller dataset. The idea is that the pre-trained model has already learned useful features from the larger dataset, which can be applied to the smaller dataset.

Existing transfer learning methods for neural machine translation are simply transfer knowledge from a parent model to a child model once via parameter initialization. It has been showed that the encoder-decoder framework for neural machine translation is very effective in large data scenarios, but much less effective for low-resource languages. In [21], a transfer learning method is presented. This method significantly improves BLEU (Bilingual Evaluation Understudy) scores across a range of low-resource languages. The key idea is to first train a high-resource language pair of encoder-decoder (the parent model), then transfer the learned parameters to the low-resource pair (the child model) to initialize and constrain training.

In [8], a novel transfer learning method, namely ConsistTL, for neural machine translation is proposed. ConsistTL is able to continuously transfer knowledge from the parent model to the child model during the training of the child model. Specifically, the child model learning each instance under the guidance of the parent model, that is, for each training instance of the child model, ConsistTL constructs the semantically-equivalent instance for the parent model and encourages prediction consistency between the parent and child for this instance. Experimental results demonstrate that ConsistTL results gives significant improvements over strong transfer learning baselines.

A unified framework is introduced in [14] that converts all text-based language problems into a text-to-text format, which explores the landscape of transfer learning techniques for NLP. With this framework, a systematic study is made in the same paper to compare retraining objectives, architectures, unlabeled data sets, transfer approaches, and other factors on dozens of language understanding tasks.

A new idea for unsupervised domain adaptation via a remold of Prototypical Networks is introduced in [11]. The goal is to learn an embedding space and perform classification via a remold of the distances to the prototype of each class. They present Transferrable Prototypical Networks for adaptation such that the prototypes for each class in source and target domains are close in the embedding space and the score distributions predicted by prototypes separately on source and target data are similar.

If we want to use this technique to analyze psychiatric clinics data, then we may need to do the following.

- Identify a machine learning model for analysis of psychiatric clinics data.
- Find a medical data with high resource which can have the same machine learning model. This seems very hard task since psychiatry is so different from other medical fields.

4   Few-Shot/Zero-Shot Learning

This is a technique where the model is designed to make accurate predictions given only a few or no examples. This approach often involves meta-learning where the model is learning the structure or meta-knowledge across different tasks, so it learns a prior over models that is useful for new tasks. Let us first look at a few examples.

In [16], Prototypical Networks are proposed for the problem of few-shot classification. Given only a small number of examples of each new class, a classifier must generalize to new classes not seen in the training set. By computing distances to prototype representations of each class, Prototypical Networks learn a metric space in which classification can be performed. Compared to recent approaches for few-shot learning, their approach is simpler and achieve excellent results. Actually, they provide an analysis showing that some simple design decisions can yield substantial improvements over recent approaches; those approaches involve complicated architectural choices and meta-learning.

Meta-learning is a framework to address the challenging few-shot learning setting. It leverages a large number of similar few-shot tasks in order to learn how to adapt a base-learner for a new task when for the new task, only a few labeled samples are available. In [17], a novel few-shot learning method, called meta-transfer learning, is proposed. This method learns to adapt a deep neural networks for few shot learning tasks. Here, by meta, it means to train multiple tasks, and by transfer it means that learning is achieved by scaling and shifting functions of deep neural network weights for each task.

Deep neural networks is successful in the large data domain, but perform poorly on few-shot learning tasks if a classifier has to quickly generalize after seeing very few examples for the class. Generally speaking, gradient-based optimization in high capacity classifiers needs many iterative steps over many examples in order to perform well. In [15], a meta-learner model LSTMbased is proposed to learn the exact optimization algorithm used to train another learner neural network classifier in the few-shot regime. This meta-learning model is competitive with deep metric-learning techniques for few-shot learning.

In [7], an effort is made on prompts for pre-trained language models. It has shown great performance in bridging the gap between pre-training tasks and various downstream tasks. Especially, prompt tuning freezes pre-trained language models and only tunes soft prompts, gives an efficient and effective solution for adapting large scale pre-trained language models to downstream tasks.

This technique is similar to transfer learning. They transfer information obtained from high-resource data to low-resource data. Therefore, we may see it as a variation of transfer learning. The difference is that the parent's model and the child model have closer relationship.

5   Active Learning

This is a special case of machine learning where a learning algorithm can actively choose the data it wants to learn from. It's particularly useful when unlabeled data may be abundant or easy to collect, but labeling data is costly, time-consuming, or requires expert knowledge. Thus, it looks like another variation of transfer learning. We may find three examples in [3,5,6].

In text classification, labels are usually expensive and the data is often characterized by class imbalance. This gives a challenge in Real world scenarios for active learning. In [3], a large-scale empirical study is presented on active learning techniques for BERT-based classification, and a diverse set of AL strategies and datasets is addressed.

In active learning, a small subset of data is selected for annotation such that a classifier learned on the data is highly accurate. Usually, selection is done by using heuristics. To improve the effectiveness of such methods, an effort in [5] is made by introducing a novel formulation which reframes the active learning as a reinforcement learning problem and explicitly learning a data selection policy. Here, the policy takes the role of the active learning heuristic.

Active learning methods rely on being able to learn and update models from small amounts of data. Recent advances in deep learning are notorious for their dependence on large amounts of data. This difference makes deep learning is difficult to be used in active learning. However, in [6], authors combine recent advances in Bayesian deep learning into the active learning framework in a practical way and develop an active learning framework for high dimensional data, which is a task extremely challenging.

6   Self-supervised Learning

This is a type of machine learning where the model generates its own supervised learning signals from the input data itself. It is a method of training where the labels for the training data are automatically generated from the data itself, without any human annotation. For example, a model might be trained to predict the next word in a sentence, and the learned word embeddings can then be used for a task like sentiment analysis. One paradigm for self-supervised learning is from few labeled examples while making best use of a large amount of unlabeled data, that is, unsupervised pre-training followed by supervised fine-tuning.

For self-supervised learning from images, the goal is to construct image representations. They are semantically meaningful via pretext tasks that do not require semantic annotations. A lot of pretext tasks lead to representations that are covariant with image transformations. However, in [10], authors argue that semantic representations should be invariant under such transformations. Moreover, they develop Pretext-Invariant Representation Learning that learns invariant representations based on pretext tasks.

Self-supervised learning is learning from few labeled examples while making best use of a large amount of unlabeled data. [1], authors proposed an approach by using big (deep and wide) networks during pretraining and fine-tuning. They found that for their approach, the fewer the labels, the more this approach (task-agnostic use of unlabeled data) benefits from a bigger network.

In [2], a new language representation model is introduced in [2]. This new model is designed to pretrain deep bidirectional representations from unlabeled text by jointly conditioning on both left and right context in all layers. It is a pretrained model, which can be finetuned with just one additional output layer to create state-of-the-art models for a wide range of tasks

In [12], a new global logbilinear regression model is proposed. This model combines the advantages of the two major model families in the literature, global matrix factorization and local context window methods. It efficiently leverages statistical information by training only on the nonzero elements in a word-word co occurrence matrix.

7 Multi-task Learning

This involves training a model on multiple related tasks at the same time, with the aim that learning from each task with high-resource data can help improve performance on the other task with low-resource data. Examples can be found in [13, 18, 19].

8 Conclusion

Summarizing the above six techniques, we found that there are only two ways to deal with low-resource data for our particular use case.

- Data Augmentation: Generate more data by data operations.
- Transfer Learning in Wide Sense: Transfer the information or parameters from analysis of high-resource data to low-resource data. In Transfer Learning, it transfers from a parent model to a child model. In Few-Shot/Zero-Shot Learning, it transfer from previous task to new task. In Active Learning, it transfers from unlabeled data to labeled data. In Self-Supervised Learning, it transfers from own unlabeled data to labeled data. In Multi-Task Learning, it transfers one to another data where they are processed together.

How to use these techniques in analysis of psychiatric clinic data? We have discussed at the ends of Sects. 2 and 3, respectively.

Other than selection of machine learning techniques, selection of research goal is also important for analysis of psychiatric clinics data. The following suggestions stem from via the point of a psychiatrist.

If we are interested in creating a tool, then we may consider the following:

- Develop a tool capable of analyzing a patient's historical records, past medical information, and the patient's current daily note (encompassing both Subjective and Objective aspects of the visit). Utilize this information to generate a "recommended diagnosis" primarily for coding purposes-assigning a corresponding code to a specific diagnosis for billing purposes. The goal is to alleviate the documentation burden on physicians. While similar tools already exist in certain documentation systems, there's potential to enhance and refine this approach in comparison to the existing solutions. However, it remains uncertain whether surpassing the capabilities of systems like Epic is achievable.

If we are interested in clustering, then we may consider the following:

- Would recommend limiting scope to a single type of diagnosis and looking for clusters within that diagnosis, ie possible clusters of major depressive disorder.
- A lot of psychiatry involves trial-and-error for final medication selection and titration; one hypothesis is that there are subtypes which are as yet invisible to our diagnostic criteria. Identifying these subtypes and predicting them

would result in fewer trials of medications before achieving optimal control. In this case, if you have enough data, I would try repeating your analysis but only on patients with a single psychiatric diagnosis (ie, major depressive disorder); but many patients will have enough diagnoses to make me cry
- Remember that all pilots should start with small scope, your overall dataset is small for NLP but covers a massive diagnostic space in psychiatry

If you want to be spicy, then we may consider the following:

- Look at patient outcome differences between patients seen by NPs (nurse practitioners) versus MD/DO (fully trained physician)
- There is a lot of "bad psychiatry" being practiced in the US because of NPs being significantly cheaper to hire - lot of existing data on poorer outcomes but not limited to psychiatry
- Can also look at telehealth vs. in-person outpatient outcomes I am assuming you can not obtain a dataset from another medical specialty; if possible, would look into datasets with more standardized vocabulary (if you are fixed on the use of NLP) - i.e., radiology, pathology, results of endoscopies, etc. For example, in radiology, it might be very nice to take an image's READ and try to predict the IMPRESSION (the radiologist's tldr), which would save some time, and you already essentially have the training data if you have a bunch of these

In conclusion, the invaluable insights provided by the psychiatric expert offer a significant opportunity to elevate the trajectory of future research focused on psychiatric patient datasets. The suggested approach not only sheds light on the complexities of mental health data but also presents a framework to enhance the quality and depth of these datasets.

By incorporating the guidance of an up and coming psychiatric, future research endeavors can adopt a more holistic and clinically informed perspective. The emphasis on capturing the subjective experiences of patients, alongside objective observations, serves to paint a comprehensive picture of their mental health journey. This nuanced approach not only humanizes the data but also enables a more accurate representation of the intricate interplay between symptoms, concerns, and treatment progress.

Furthermore, the call for standardized language and categorization aligns with the broader objective of creating cohesive and interoperable datasets. This standardization not only facilitates meaningful comparisons across different patient cases but also streamlines data integration, thus bolstering the potential for advanced analyses and insights.

However, these suggestions come with the recognition that the ethical dimension remains of paramount importance. Ensuring patient privacy, informed consent, and protection against biases must remain at the forefront of research endeavors. By weaving these considerations into the fabric of future studies, researchers can pave the way for responsible innovation in psychiatric healthcare data.

Incorporating the recommended approach into future research not only has the potential to advance our understanding of mental health but also contributes to the ongoing dialogue between the medical and research communities. The collaborative integration of clinical expertise and data-driven insights promises to yield datasets that are not only robust but also imbued with empathy-a vital combination in the pursuit of meaningful breakthroughs in psychiatric care.

Acknowledgement. We appreciate very much to Guanghua Wang for his help in collecting references.

References

1. Chen, T., Kornblith, S., Swersky, K., Norouzi, M., Hinton, G.E.: Big self-supervised models are strong semi-supervised learners. In: Advances in Neural Information Processing Systems, vol. 33. NeurIPS (2020)
2. Devlin, J., Chang, M.-W., Lee, K., Toutanova, K.: BERT: pre-training of deep bidirectional transformers for language understanding. In: Proceedings of NAACL-HLT 2019, pp. 4171–4186. Minneapolis, Minnesota (2019)
3. Ein-Dor, L., et al.: Active learning for BERT: an empirical study. In: Proceedings of the 2020 Conference on Empirical Methods in Natural Language Processing (EMNLP), pp. 7949–7962. Association for Computational Linguistics (2020)
4. Fadaee, M., Bisazza, A., Monz, C.: Data augmentation for low-resource neural machine translation. In: Proceedings of the 55th Annual Meeting of the Association for Computational Linguistics (Short Papers), pp. 567–573 Vancouver, Canada (2017)
5. Fang, M., Li, Y., Cohn, T.: Learning how to active learn: a deep reinforcement learning approach. In: Proceedings of the 2017 Conference on Empirical Methods in Natural Language Processing, pp. 595–605, Copenhagen, Denmark (2017)
6. Gal, Y., Islam, R., Ghahramani, Z.: Deep bayesian active learning with image data. In: Proceedings of the 34th International Conference on Machine Learning, vol. 70, pp. 1183–1192. PMLR (2017)
7. Gu, Y., Han, X., Liu, Z., Huang, M.: PPT: pre-trained prompt tuning for few-shot learning. In: Proceedings of the 60th Annual Meeting of the Association for Computational Linguistics, Volume 1: Long Papers, pp. 8410–8423 (2022)
8. Li, Z., Liu, X., Wong, D.F., Chao, L.S., Zhang, M.: ConsistTL: modeling consistency in transfer learning for low-resource neural machine translation. In: Proceedings of the 2022 Conference on Empirical Methods in Natural Language Processing, pp. 8383–8394 (2022)
9. Mikołajczyk, A., Grochowski, M.: Data augmentation for improving deep learning in image classification problem. In: 2018 International Interdisciplinary PhD Workshop (IIPhDW) (2018). https://doi.org/10.1109/IIPHDW.2018.8388338
10. Misra, I., van der Maaten, L.: Self-supervised learning of pretext-invariant representations. In: Proceedings of the IEEE/CVF Conference on Computer Vision and Pattern Recognition (CVPR), pp. 6707–6717 (2020)
11. Pan, Y., Yao, T., Li, Y., Wang, Y., Ngo, C.-W., Mei, T.: Transferrable prototypical networks for unsupervised domain adaptation. In: Proceedings of the IEEE/CVF Conference on Computer Vision and Pattern Recognition (CVPR), pp. 2239–2247 (2019)

12. Pennington, J., Socher, R., Manning, C.D.: GloVe: global vectors for word representation. Proceedings of the 2014 Conference on Empirical Methods in Natural Language Processing (EMNLP), pp. 1532–1543. Doha, Qatar (2014)
13. Radford, A., Wu, J., Child, R., Luan, D., Amodei, D., Sutskever, I.: Language models are unsupervised multitask learners. OpenAI blog **1**, 9 (2019)
14. Raffel, C., et al.: Exploring the limits of transfer learning with a unified text-to-text transformer. J. Mach. Learn. Res. **21**(1), 1–67 (2020)
15. Ravi, S., Larochelle, H.: Optimization as a Model for Few-Shot Learning, ICLR (2017)
16. Snell, J., Swersky, K., Zemel, R.,: Prototypical networks for few-shot learning. In: Advances in Neural Information Processing Systems, vol. 30. NIPS (2017)
17. Sun, Q., Liu, Y., Chua, T.-S., Schiele, B.: Meta-transfer learning for few-shot learning. In: Proceedings of the IEEE/CVF Conference on Computer Vision and Pattern Recognition (CVPR), pp. 403–412 (2019)
18. Thoppilan, R., et al.: LaMDA: language models for dialog applications. arXiv:2201.08239 (2022)
19. Touvron, H., et al.: LLaMA: Open and Efficient Foundation Language Models. arXiv:2302.13971 (2023)
20. Wei, J., Zou, K.: EDA: easy data augmentation techniques for boosting performance on text classification tasks. In: Proceedings of the 2019 Conference on Empirical Methods in Natural Language Processing and the 9th International Joint Conference on Natural Language Processing, pp. 6382–6388. Hong Kong, China (2019)
21. Zoph, B., Yuret, D., May, J., Knight, K.: Transfer learning for low-resource neural machine translation. In: Proceedings of the 2016 Conference on Empirical Methods in Natural Language Processing, pp. 1568–1575. Austin, Texas (2016)

Single Image Dehazing Based on Dynamic Convolution and Transformer

Quancheng Ning and Nan Zhang[✉]

Institute of Computing Theory and Technology, Xidian University,
Xi'an 710071, China
21181214419@stu.xidian.edu.cn, nanzhang@xidian.edu.cn

Abstract. In this paper, an end-to-end multi-stage dehazing network based on convolution and Transformer is proposed. The network design is divided into three parts: encoding network, feature fusion network and decoding network. The encoding network extracts primary features of the haze image, the feature fusion network uses the serial Transformer module and the dynamic convolution module to make the information extracted from the features richer, and the decoding network is used to recover the image resolution. In the resolution restoration stage of the decoding module, the MixUp module is used to restore the high resolution image by combining the extracted primary features to reduce the loss of information. Extensive experiments were conducted on synthetic and real datasets to validate the role of Transformer module and Dynamic Convolution module in dehazing respectively. The results show that the proposed method achieves a good objective evaluation score and reconstructs a subjectively better dehazing image with a PSNR of 35.37 and a SSIM of 0.9849 on the SOTS test dataset.

Keywords: Image Dehazing · Dynamic Convolution · Gate Aggregation · Transformer

1 Introduction

In a haze scene, the presence of a large number of tiny suspended particles in the outdoor air will produce refraction and scattering of light. The refracted and scattered light will be mixed with the light reflected by the object to be observed. It will greatly reduce the clarity and contrast of the collected outdoor image, and even cause the image color to shift and a lot of details to be lost. As a result, the real image information cannot be obtained. In order to solve such problems, many image defogging algorithms have been proposed. Image dehazing is generally the first step of image processing since using dehazed images can improve the accuracy of target detection tasks.

This research is supported by National Natural Science Foundation of China under Grant Nos. 62272359 and 62172322; Natural Science Basic Research Program of Shaanxi Province under Grant Nos. 2023JC-XJ-13 and 2022JM-367.

Atmospheric Scattering Model (ASM) proposed by McCartney [1] is a famous mathematical model in computer vision and computer graphics used to describe the formation of haze images. This model is expressed as

$$I(x) = t(x)J(x) + (1 - t(x))A$$

where I is the haze image, x is the pixel location of the image, $t(x)$ is the medium transmission map, $J(x)$ is the scene radiance (dehazed image), and A is the atmospheric light vector in RGB domain. The formula expresses that the formation process of haze images is the combination of light reflected by objects and atmospheric light.

Many dehazing methods based on atmospheric scattering models estimate a priori the medium transmission map $t(x)$ and atmospheric light values as a prerequisite for computing a haze-free image, as shown in the above equation. For example, the dark channel prior [2] is the most successful a priori method and many subsequent methods have been inspired by it. The bounded context regularisation method [3] estimates the value of $t(x)$ by imposing a constraint on $t(x)$, and then estimates the atmospheric light A via the dark channel prior, which is used to obtain the dehazing image via an atmospheric scattering model. Several features related to dehazing in image dehazing are investigated in [4], including dark channel, contrast, chroma, and saturation, and it is experimentally verified that the dark channel is most relevant to the haze concentration. The first network to use deep learning for image dehazing is DehazeNet [5], which estimates the transmittance map t by means of a neural network, which sorts the network's output of the transmittance map from smallest to largest, and finds the pixels in the smaller first 0.01 of the transmittance as candidate pixels for the atmospheric light, and finds the brightest pixel among these candidate pixels that is the estimate of the atmospheric light. The idea of generative adversarial is applied to image dehazing in the dense pyramid dehazing network [6], which uses two sub-networks to estimate the medium transmission map t and the atmospheric light A, respectively, and then dehaze the image based on the atmospheric scattering model. The dehaze image and medium transmission image are spliced and input to the discriminative network, which determines whether the medium transmission map estimation of the defogged image is close to the real medium transmission map, which reflects the important role of the medium transmission map.

All-in-One Network [7] adopts the above idea, but adds a bias coefficient to the atmospheric scattering model, and then uses neural network to learn the above ratio, that is, the output of the network is multiplied by the foggy image to get the defogged image. GridDehazeNet [8] illustrates their conjecture through experiments, and thinks that it is less difficult for the network to learn the mapping from haze to dehazing directly than to learn the transmittance and atmospheric light and then restore the clear image through the atmospheric scattering model, so a pixel-to-pixel image dehazing method is proposed. GCANet [9] realizes the idea of end-to-end dehazing, and uses the pair of haze and clear images to make the model learn the mapping relationship between haze images and clear images. The above models all use haze and clear image pairs to train the network, which is a supervised learning mode. You Only Look Yourself [10] uses the idea of de-entanglement to decompose the haze image into a sub-network of haze-free images, a sub-network of medium transmission map and a sub-network of atmospheric light estimation, and then applies an atmospheric scattering model to reduce the original haze image to the output of the three sub-networks. The closer the reduced haze image is to the input haze image, the better the decomposition effect of each subnetwork on the haze image is, i.e., the better the dehazing effect is, which is an unsupervised learning model. FFA [14] is an important model in the field of dehazing, which proposes channel attention and pixel attention that can apply attention to features in the channel direction and pixel space, resulting in a large improvement in network dehazing performance. With the excellent performance of VisonTransformer in various visual tasks, DehazeFormer [21] was proposed specifically for dehazing tasks.

Image dehazing algorithms based on deep learning currently achieve optimal performance by modelling neural networks trained on haze datasets. In the process of real image dehazing, as the distribution of haze in the image is not the same, the parameters of the convolution kernel of the ordinary convolutional layer are fixed after the network training is completed, and the value of the convolution kernel parameters can not be modified according to the input, but in the process of image dehazing if it is possible to modify the parameters of the convolution kernel according to the haze features, the expressive power of the network will be improved. Dynamic convolutional layer [11] is not using only one convolutional kernel per layer, but each layer is based on multiple convolutional kernels, the convolutional kernel undergoes softmax to generate an attention score, and then a linear combination of the score and convolutional kernel to get the fused convolutional kernel, this convolutional kernel is the one that is finally used for convolutional operations. The dynamic convolution operation can improve the performance of the model on the basis of only a little increase in computational complexity. At the same time, the neural network used for dehazing also has the problem of gradient vanishing, in order to alleviate the problem of gradient vanishing, the gradient flow of the network can flow from the deep layer to the shallow layer through the residual connection. In the process of image dehazing the extracted haze feature map is very important, in order to make full use of the features using a combination of dense network and residual network to form a dense residual network. Specifically, a skip structure is added between different convolutional layers, i.e., the input of each layer is the output of the previous layers, making full use of the extracted feature maps. The ordinary convolutional layers in the dense residual network are replaced with dynamic convolution to obtain a dynamic dense residual structure. The convolution module can capture the local associations of features, but it has limited ability to capture the associations between features. In order to fully capture the associations between features, it is proposed to place the transformer module before the dynamic convolution to obtain the associations between features, and then input the feature to the dynamic convolution module. Considering that the low-level features may be lost in the process of dehazing, this paper designs the gated aggregation module to fuse the features of different levels, so that the information of the obtained features is richer.

The main contributions of our work are as follows: (1) A new image dehazing algorithm based on dynamic convolution and Transformer is proposed. The network can be divided into three parts:encoding network, feature fusion network and decoding network. (2) In order to make the feature fusion network able to fuse the global information of the features as well as the local information of the features, the T-DRC module is designed in this paper, which firstly extracts the global information by using the transformer module and then fuses the local information by using the dynamic residual component. (3) In this paper, a gated aggregation module is designed to weight different layers of features to form a feature containing more information, which is decoded to recover a haze-free image. (4) A large number of experiments have been carried out on synthetic dataset and real world dataset. The subjective and objective evaluations show that our image dehazing algorithm can achieve better dehazing results than traditional algorithms, and outperforms some current advanced algorithms interms of Peak Signal to Noise Rate (PSNR).

2   Our Method

In this paper, we propose an image dehazing algorithm based on convolution and Transformer, which can implement the dehazing process end-to-end, and the overall structure of the network is shown in Fig. 1. The network mainly contains three parts: Encoding Network, Feature Fusion Network, and Decoding Network. Specifically, the input haze image first goes through the encoding network for preliminary feature extraction and quadruple downsampling. The encoding network consists of three depth-separable convolutional modules, each of which consists of a depth-separable convolutional layer (consisting of two convolutional layers, one for group convolution and one for point-wise convolution), a batch normalization layer, and ReLU as the activation function. The first of these models is used will be used to increase the original image channel without changing the resolution. After the encoding network the primary features of the haze image are extracted and used to input the feature fusion network for feature fusion. The feature fusion network is the feature fusion of the extracted primary features using serial transformer module and dynamic residual component. The decoding network consists of two transposed convolution modules and one ordinary convolution layer, the transposed convolution module consists of a transposed convolution layer, a batch normalization layer and a ReLU activation function. When recovering the resolution of the features again, the MixUp module is used to fuse the primary features extracted from the coding network with the features that have gone through the fusion network. The last convolutional layer is used for dimensionality reduction. The haze image is skip-connection into the network output to form a residual structure to get the final haze-free image. The components of the model and the loss function are described separately next.

2.1   T-DRC Module

The module consists of two serial transformer modules and two dynamic residual components. The module structure is shown in Fig. 1.

2.2   Improved Dynamic Residual Module

The dynamic residual component of [12] consists of a dynamic dense residual block, a dynamic convolution layer, and channel attention and pixel attention modules. In my experiments, I found that dynamic convolution with this structure is difficult to train, in this paper, we use a boosted dynamic convolution module [22] to replace the original dynamic convolution module, and the overall architecture of the improved module is shown in Fig. 2.

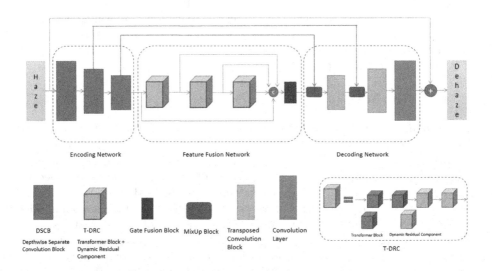

Fig. 1. Overall Architecture of the Network

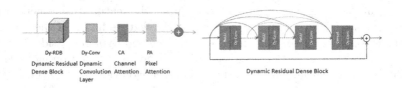

Fig. 2. Improved Dynamic Residual Module

RDB (Residual Dense Block) in [13] connects the residual structure, dense feature fusion across layers to better ensure the flow of information. The dense connection structure can reduce the shallow feature loss problem to some extent with the gradual deepening of the network, and at the same time lose a little computational complexity to improve the network performance without increasing the depth of the network. The residual structure can alleviate the gradient vanishing or gradient explosion problem of this module, allowing for better convergence. The channel attention and pixel attention modules are added to enable the network to notice different haze concentrations. Specifically, the dynamic dense residual module is shown in Fig. 2.

2.3 Dual Attention Module

The distribution of foggy in haze images taken in the real world is usually uneven, in order to enable the model to notice the difference in the distribution of haze, inspired by the literature [14], adding the channel attention mechanism and pixel attention mechanism in the feature extraction module can enhance the effect of haze removal. In this paper, this kind of module is also used to enhance the effect of haze removal, because the haze has different effects on different channels, the channel attention can apply different weights to different channels; the effects on different pixels in the features are also different, the pixel attention can apply different weights to different pixels. The structure of the channel attention and pixel attention module is shown in Fig. 3.

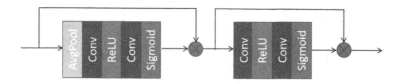

Fig. 3. Channel Attention and Pixel Attention

Specifically, the features first after global average pooling to obtain the average value of each channel, that is, in the spatial dimension of the compression of features, to get 1 * 1 * C features, after the convolution layer, ReLU activation, convolution layer, sigmoid to produce each channel in the weight, and then get to the attention and the corresponding channel multiplication that is achieved to the attention of the channel. Channel Attention Formula:

$$W_c = \sigma(Conv(\delta(Conv(AvgPool(x)))))$$

After the channel attention feature map calculation formula:

$$F_c = W_c \bigotimes F$$

Pixel Attention Formula:

$$W_p = \sigma(Conv(\delta(Conv(F_c))))$$

After the pixel attention feature map formula:

$$F_p = W_p \bigotimes F_c$$

where AvgPool represents the global average pooling operation, Conv represents the convolution operation, δ represents ReLU activation function, σ represents Sigmoid activation function, $\otimes$ represents dot product operation, F is the input feature, which represents the feature that passes through the channel's attention and the feature that passes through the pixel's attention.

2.4   Transformer Module

Convolution operation has a weak ability to capture the correlation between features. Inspired by [15], this paper designs a transformer module, the structure of which is shown in Fig. 4. Specifically, the input features are first passed through RescaleNorm [22], through MDTA (Multi-Dconv Head Transposed Attention) [15] module, and then the output features are affine transformed, skip-connextion and summed with the input features, and input to the next RescaleNorm normalisation layer, through the GDFN (Gated Dconv Feed-Forward Network) [15] network, and then after affine transformation, skip-connection summation to get the output of Transformer module.

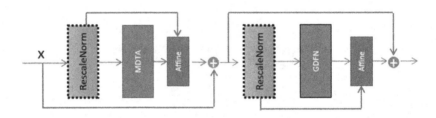

Fig. 4. Transformer Module

2.5   Gated Fusion Block

As shown in [16], fusing features from different levels is usually beneficial for both low-level and high-level tasks. To implement this idea, feature pyramids are used in [16] to fuse high-level semantic feature maps at all scales. In this paper, the gated aggregation block is designed. The structure of the block is shown in Fig. 5.

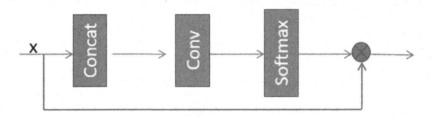

Fig. 5. Gated Aggregation Module

Assume that the output features of the three serial modules are FL, FM, and FH, which are fed into the gated aggregation block. The gated aggregation block outputs three different importance weights, and the fusion features are obtained by linearly combining the features of different levels and the corresponding weights. The calculation formula is as follows:

$$(W_l, W_m, W_h) = Softmax(Conv(F_l, F_m, F_h))$$

$$F = W_l * F_l + W_m * F_m + W_h * F_h$$

where Conv stands for convolutional layer, Softmax stands for softmax layer. F_l stands for low level features, F_m stands for middle level features, F_h stands for high level features and F stands for fused features. In the process of implementation, the number of input features of the gating module can also be flexibly changed to fuse multiple features of different layers.

2.6 MixUp Module

The MixUp module is borrowed from the module in AECR-Net [22], and the way the module aggregates features is shown in Fig. 6 below. The formula for fusing features is shown below:

$$f_\uparrow 2 = Mix(f_\downarrow 1, f_\uparrow 1) = \sigma(\theta_1) * f_\downarrow 1 + (1 - \sigma(\theta_1)) * f_\uparrow 1$$

$$f_\uparrow = Mix(f_\downarrow 2, f_\uparrow 2) = \sigma(\theta_2) * f_\downarrow 2 + (1 - \sigma(\theta_2)) * f_\uparrow 2$$

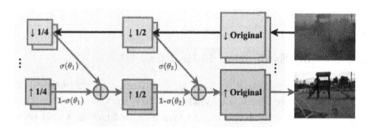

Fig. 6. MixUp Module. The first and second rows are downsampling and upsampling operations, respectively

where f↓i and f↑i are feature maps from the i-th downsampling and upsampling layer, respectively. f↑ is the final output. $\sigma(\theta i)$, i = 1, 2 is the i-th learnable factor to fuse the inputs from the i-th downsampling layer and the i-th upsampling one, whose value is determined by the sigmoid operator σ on parameter θi. During training, we can effectively learn these two learnable factors, which achieves better performance than the constant factors.

2.7   Loss Function

This paper uses $L1$ Loss function and contrast Loss function as the Loss function of model training. $L1$ Loss function calculation formula:

$$L_1 = \| \phi(I, w) - J \|_1$$

where ϕ refers to the dehazing network, w the network parameters, I the haze image, and J refers to clear image.

Reference [17] introduces contrastive learning into the field of image dehazing, as traditional image dehazing only applies clear images and does not fully utilize haze images. The author views the dehazing image as an anchor box, the clear image as a positive sample, and the haze image as a negative sample. Therefore, the process of image dehazing can be seen as a process of making the anchor box close to the positive sample and away from the negative sample. Calculation formula of contrast loss function in [17]:

$$L_c = \sum_{i=1}^{n} w_i \frac{D(G_i(J), G_i(\phi(I,w)))}{D(G_i(I), G_i(\phi(I,w)))}$$

where D refers to distance measurement, Gi, i=1,2,..., N represents the feature map extracted from the i-th layer of the fixed network, w_i representing the weight of the output feature map of the i-th layer of the fixed network.

The total loss function is composed of the above two loss functions, and the calculation formula of the total loss function is:

$$L = L_1 + \gamma L_c$$

where γ refers to the weight of L_c in the total loss L.

3   Experiments

3.1   Datasets and Evaluation Indicators

In order to verify the network's effect of dehazing in synthetic and real haze images, experiments are done on synthetic haze images and real haze images respectively. The synthetic haze image training dataset is the ITS dataset in Reside, and the test set is the SOTS-indoors dataset in Reside [18]. There are 1,399 clear haze-free images in the ITS dataset, and each haze-free image corresponds to 10 haze images with different concentrations synthesised according to the atmospheric scattering model. The SOTS indoors dataset was chosen as a test set in order to verify the effectiveness of the model. The real haze image training dataset is the NH-HAZE [19], which consists of 55 pairs of haze-free images and real haze images. In this paper, the first 50 pairs of images are used as the training set and the last five pairs of images are used as the test dataset. Since the number of haze-free and haze image pairs in the NH-HAZE dataset is too small, this paper adopts random cropping to crop image blocks with a resolution of 256*256 in the original images, and after censoring, the dataset is finally expanded to 3,877 haze-free and haze image pairs. In this paper, a general approach is adopted, and in order to evaluate the dehazing performance of this method, Peak Signal-To-Noise Ratio (PSNR) and Structural Similarity (SSIM) are used for quantitative and objective evaluation on two datasets.

3.2 Experimental Setup

This experiment is based on Pytorch 1.11.0 framework, and the graphics card NVIDIA GeForce RTX 3080 is used for the experiment. Randomly cropped image blocks of size 256*256 on the ITS dataset in Reside were used as input to the network during training, while data enhancement was performed by operations such as horizontal inversion and rotation of the data, with a batch size of 6. The Adam optimiser was used for training, with momentum decay exponential sums of 0.9 and 0.999, respectively, and an initial learning rate of 0.0002. The weights of the loss function were 1. 200 epochs were trained in this experiment, the first 100 epochs every 10 epochs learning rate decay half, the last 100 epochs can use cosine learning rate decay for learning, or set the appropriate initialisation learning rate for exponential decay learning. In this paper, several classical methods in the field of image defogging are selected for quantitative and qualitative comparison, including DCP [2], DehazeNet [5], AOD-Net [7], GridDehazeNet [8], KDDN [20], FFA [14], and DehazeFormer-T [21].

Table 1. Quantitative comparison of dehazing results on SOTS indoor synthetic dataset and NH-HAZE real dataset

Method	SOTS		NH-Haze		Pram
	PSNR	SSIM	PSNR	SSIM	Mb
DCP	15.09	0.7649	10.57	0.5196	–
DehazeNet	20.64	0.7995	16.62	0.5238	0.01
AOD-Net	19.82	0.8178	15.40	0.5693	0.002
GridDehazeNet	32.16	0.9836	13.80	0.5370	0.96
KDDN	34.72	0.9845	17.39	0.5897	5.99
FFA	36.39	0.9886	–	–	4.46
DehazeFormer-T	35.15	0.989	–	–	0.69
Our	35.37	0.9849	20.03	0.6710	5.31

3.3 Synthetic Haze Image Dataset Experiment

After training on the ITS training set, the training model is saved. Tested using SOTS-indoors test set. The defogging results are analyzed qualitatively and quantitatively. Table 1 lists the quantitative comparison results of this paper's method with representative methods.

Compared to deep learning algorithms, DCP is based on a priori dehazing algorithm, so the PSNR and SSIM are lower; DehazeNet improves the PSNR and SSIM to some extent compared to DCP; AOD-Net reduces the cumulative error by estimating the transmittance and atmospheric light at the same time to improve the SSIM to 0.8178; GridDehazeNet based on attention and multi-scale makes the PSNR and SSIM reach 32.16dB and 0.9836, respectively; KDDN further improves the PSNR, compared with GridDehazeNet, the PSNR is improved by 2.56dB, and the SSIM is improved by 0.0009. The method in this paper achieves a PSNR of 35.37 and a SSIM of 0.9849. With close number of model parameters, the peak SNR increases by 0.65 compared to KDDN network. Compared to FFA the PSNR and SSIM are low. Compared to DehazeFormer-T model PSNR is close and SSIM is low. Which is visualised for the chosen dehaze method as shown in Fig. 7 and Fig. 8.

3.4   Real Haze Image Dataset Experiment

Real haze scenes have problems such as uneven haze distribution, incomplete haze removal and colour distortion. In order to verify the dehaze effect of the method proposed in this paper in real haze scenes. It is trained and tested in NH-HAZE. The training dataset is an expanded 3877 pairs of haze and haze-free image pairs, and the test set is selected from five haze and haze-free image pairs after NH-HAZE. From Table 1, it can be seen that the PSNR and SSIM of the DCP method are low, and DehazeNet is improved to some extent compared to the DCP method; AOD-Net is improved by 0.0455 in SSIM compared

Fig. 7. There is haze image, DCP, DehazeNet, AOD-Net and haze-free image from left to right

Fig. 8. There is haze image, GridDehaze, FFA, Our, and haze-free image from left to right

to DehazeNet; and the performance of GridDehazeNet is not further improved compared to AOD-Net; KDDN achieves PSNR and SSIM of 17.39 and 0.5879, respectively. The method in this paper achieves SSIM of 0.6710 and PSNR of 20.03. However, there is still a problem of colour bias for the dehazing of the real scene, and it is guessed that the reason is that most of the 3,877 pairs of haze and haze-free image pairs that form the dataset are grassy scenes, and so the model tends to restore the fogged scenes to grass scenes. The chosen dehazing method is visualised as shown in Fig. 9 below.

Fig. 9. There is haze image, DCP, DehazeNet, AOD-Net, Our, and clear image from left to right

3.5   Ablation Experiments

In order to verify the role of dynamic convolution and the effectiveness of the transformer module, in this paper, four ablation experiments are done on the SOTS dataset, and the following four models are designed:

1) Without adding the Transformer module, the convolution module is a static convolution as Baseline.
2) Add Transformer module on top of 1).
3) Replace static convolution with dynamic convolution based on 1).
4) Add Transformer module to 3).

The experimental results are shown in Table 2.

Table 2. Performance comparison of four different model architectures

	PSNR	SSIM
Baseline	30.88	0.9622
Baseline+Transformer	33.79	0.9799
Basline+Dy	30.27	0.9624
Baseline+Dy+Transformer	35.37	0.9849

From the experimental results, it can be seen that without the transformer module, the PSNR of baseline model is on the low side; when the Transformer module is added, the PSNR and the SSIM of the model are greatly improved; After changing the static convolution to dynamic convolution in Baseline, the PSNR decreases slightly, and the SSIM and baseline are close to each other, which may be due to the initialisation problem of the model; after combining the dynamic convolution and transformer module, the dehazing performance of the model is improved, and the PSNR reaches 35.37, and the SSIM reaches 0.9849, which demonstrates that combining the dynamic convolution and transformer block can improve the dehazing performance of the network.

4 Conclusion

In this paper, we propose an image dehazing algorithm based on the serial transformer module and the dynamic residual model component module, which can recover haze and haze-free images end-to-end. In order to utilize the global information between the features and extract the local information between the features, this paper serialises the transformer module and the dynamic residual model component module to accomplish this goal. At the same time, a gating aggregation module is designed for context aggregation, which improves the performance of dehaze to a certain extent compared with other dehaze methods. Next, we will explore whether meta-learning, domain generalisation, and large language models can be applied to image dehazing, or use other attentions to replace the attention in the Transformer module to reduce the complexity of generating attention in this module.

References

1. McCartney, E.J.: Optics of the Atmosphere: Scattering by Molecules and Particles, New York, John Wiley and Sons Inc, p. 421 (1976)
2. He, K., et al.: Single image haze removal using dark channel prior. IEEE Trans. Pattern Anal. Mach. Intell. **33**(12), 2341–2353 (2011)
3. Meng, G., et al.: Efficient image dehazing with boundary constraint and contextual regularization. In: Proceedings of the 2013 IEEE International Conference on Computer Vision IEEE (2013)
4. Tang, K., Yang, J., Wang, J.: Investigating haze-relevant features in a learning framework for image dehazing. In: Proceedings of the IEEE Conference on Computer Vision and Pattern Recognition (2014)
5. Cai, B., et al.: DehazeNet: an end-to-end system for single image haze removal. IEEE Trans. Image Process. **25**(11), 5187–5198 (2016)
6. Zhang, H., Patel, V.M.: Densely connected pyramid dehazing network. In: Proceedings of the IEEE Conference on Computer Vision and Pattern Recognition (2018)
7. Li, B., et al.: AOD-Net: all-in-one dehazing network. In: 2017 IEEE International Conference on Computer Vision (ICCV) IEEE (2017)
8. Liu, X., et al.: GridDehazeNet: attention-based multi-scale network for image dehazing (2019)

9. Chen, D., et al.: Gated context aggregation network for image dehazing and deraining. In: 2019 IEEE Winter Conference on Applications of Computer Vision (WACV). IEEE (2019)

10. Li, B., et al.: You only look yourself: unsupervised and untrained single image dehazing neural network. Int. J. Comput. Vis. **129**, 1754–1767 (2021)

11. Chen, Y., et al.: Dynamic convolution: attention over convolution kernels. In: Proceedings of the IEEE/CVF Conference on Computer Vision and Pattern Recognition (2020)

12. Zhe, L., Yudong, L., Jiaying, L.: Adaptive image defogging algorithm based on dynamic convolutional kernel. Comput. Sci. **50**(06), 200–208 (2023)

13. Zhang, Y., et al.: Residual dense network for image super-resolution. In: Proceedings of the IEEE Conference on Computer Vision and Pattern Recognition (2018)

14. Qin, X., et al.: FFA-Net: feature fusion attention network for single image dehazing. In: National Conference on Artificial Intelligence Association for the Advancement of Artificial Intelligence (AAAI) (2020)

15. Zamir, S.W., et al.: Restormer: efficient transformer for high-resolution image restoration. In: Proceedings of the IEEE/CVF Conference on Computer Vision and Pattern Recognition (2022)

16. Lin, T.-Y., et al.: Feature pyramid networks for object detection. In: Proceedings of the IEEE Conference on Computer Vision and Pattern Recognition (2017)

17. Wu, H., et al.: Contrastive learning for compact single image dehazing. In: Proceedings of the IEEE/CVF Conference on Computer Vision and Pattern Recognition (2021)

18. Li, B., et al.: Benchmarking single-image dehazing and beyond. IEEE Trans. Image Process. **28**(1), 492–505 (2018)

19. Ancuti, C.O., Ancuti, C., Timofte, R.: NH-HAZE: an image dehazing benchmark with non-homogeneous hazy and haze-free images. In: Proceedings of the IEEE/CVF Conference on Computer Vision and Pattern Recognition Workshops (2020)

20. Hong, M., et al.: Distilling image dehazing with heterogeneous task imitation. In: Proceedings of the IEEE/CVF Conference on Computer Vision and Pattern Recognition (2020)

21. Song, Y., et al.: Vision transformers for single image dehazing. IEEE Trans. Image Process. **32**, 1927–1941 (2023)

22. Li, Y., et al.: Revisiting Dynamic Convolution via Matrix Decomposition (2021)

Reinforcement Learning for Combating Cyberbullying in Online Social Networks

Wenting Wang[✉], Tiantian Chen, and Weili Wu

Department of Computer Science, University of Texas at Dallas,
Richardson, TX 75080, USA
{wenting.wang,tiantian.chen,weiliwu}@utdallas.edu

Abstract. In recent decades, social network holds a pivotal role for people's communication, but it is also a particularly susceptible to cyberbullying due to their rapid information dissemination capabilities. The state-of-art research in cyberbullying mainly focus on cyberbullying detection and literature discover, which encounters significant challenges in aspects like theoretical assurances, time effectiveness, and adaptability to broad contexts. In this paper, we present a resilient framework leveraging deep reinforcement learning (DRL) to tackle the problem of cyberbullying in online social networks. Our approach leverages dynamic graph neural networks to perform network embedding and the double deep Q-network (DDQN) for the parameter learning. To evaluate the effectiveness of our proposed approach, we conducted a comprehensive set of experiments using realistic datasets. The experimental findings demonstrate that our approach outperforms the comparison methods, even we train our model with small randomly generated ER graphs. This shows the strong generalization ability of our proposed model.

Keywords: Graph neural networks · Deep reinforcement learning · Social network · Cyberbullying · Target nodes · Seed selection

1 Introduction

In today's world, the rapid growth of the internet has had a profound impact on our lives, the remarkable development of social networks have reshaped the way people connect and communicate. Concurrently, social platforms like Facebook, Twitter, LinkedIn and WhatsApp have been supplanting traditional media outlets, becoming the primary means of disseminating information and facilitating communication. These platforms have provided unprecedented opportunities for people to interact, share ideas, and build communities. However, along with their tremendous success, social networks have also given rise to a pressing and detrimental issue - cyberbullying. This harmful phenomenon has had serious and far-reaching negative consequences, particularly within the social networks.

Cyberbullying, often referred to as online bullying, is a malicious form of harassment carried out through modern electronics, such as mobile phones or computers. Online bullying or harassment has affected 59% of teenagers in the

W. Wu and J. Guo (Eds.): COCOA 2023, LNCS 14462, pp. 480–493, 2024.
https://doi.org/10.1007/978-3-031-49614-1_36

United States [20]. Cyberbullying situations often involve complexities that may not be immediately apparent. Just as in face-to-face bullying scenarios, these situations typically encompass three distinct roles: the victim, the bully, and the bystander [1].

- **Bully:** original attacker who made offensive statement to attack others.
- **Victim:** specific users who are targeted and attacked by bullies.
- **Bystander:** individuals who observe instances of cyberbullying, and these can be categorized into three groups based on their behavioral reactions: defenders, reinforcers, and outsiders.

Social networks are particularly vulnerable to cyberbullying due to convenience of information dissemination. Bully or attackers may rally a group of followers to support and perpetuate their harmful behaviors. As the influence of bullies grows, it can lead to the activation of more individuals who become attackers, exacerbating the problem of cyberbullying. Naturally, the challenge lies in devising effective measures to combat cyberbullying, with the aim of minimizing the number of victims users activated by bullying source people. This issue is commonly referred to as the problem of combating cyberbullying within social networks.

In order to protect the target victim from the attacking of bullies, we need research into cyberbullying to better understand its evolving nature and the impact of preventive measures, and use data-driven insights to refine and improve protective strategies. Inspired by the existing rumor blocking problem research in social networks, which goal is to find an optimal or near-optimal set of nodes (the blockers) to block or immunize in such a way that it minimizes the spread of the rumor or information [2], we can address the combating cyberbullying problem with analogous methods for bullying blocking for target victims. But it's important to note that while the problems of combating cyberbullying and rumor blocking share some similarities, they differ in one significant aspect - Targeted Recipients. Typically, cyberbullying is aimed at particular individuals or groups, necessitating a concentrated effort on safeguarding and aiding these directly affected users. In contrast, the objective of rumor blocking isn't tied to specific recipients; instead, it seeks to limit the dissemination of false information without predefining any particular victims.

In this paper, we illustrate that the presence of some bystanders who take a stand against cyberbullying or actively promote positive information to support cyberbullying victims can lead to a substantial reduction in the negative impact experienced by these targeted victims. An illustrative example of combatting cyberbullying is presented in Fig. 1, the directed edges denote the flow of influence propagation, and each edge is annotated with a numerical value indicating the probability of propagation. The nodes are categorized into four distinct groups: bully node, victim node, bystander node, and defender node which is selected to be the defender to spread positive influence in the network. In Fig. 1(b) shows the situation that vistims will be influenced by bullies without any defender's protection, in (c) we choose node {6} as the defender node from bystander nodes and take a stand by spreading positive information to the victim nodes, this collective action can effectively halt the attacks.

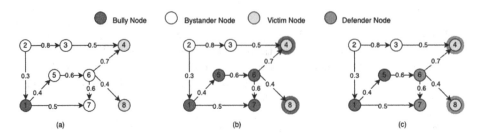

Fig. 1. An example to illustrate combating cyberbullying. (a) The initial network; (b) Cyberbullying propagation without control; (c) Combating cyberbullying with defender node 6.

In the realm of combatting cyberbullying (CCB), adopting a comprehensive and multifaceted approach that integrates various strategies and technologies is imperative. We fond reinforcement learning (RL) methods hold significant promise in the fighting against cyberbullying due to their adaptability, personalized approach, and ability to balance trade-offs. Given the remarkable advancements of DRL in various domains, including artificial intelligence, there has been a growing body of research dedicated to applying DRL techniques to tackle complex combinatorial optimization problems. In light of this progress, we introduce a pioneering end-to-end DRL framework. Our framework is designed to address the combating cyberbullying problem with a novel architecture featuring interactive graph neural networks for network embedding and leverages the DDQN for parameter learning. In summary, our contributions can be outlined as follows:

- To our knowledge, we are pioneering the introduction of an end-to-end deep reinforcement learning framework designed to efficiently tackle the problem of combating cyberbullying.
- We design a customized Random Walk method for initial node embeddings by considering both local and global nodes centrality and influences.
- To grasp the critical ripple effects of information propagation and network structure, we've devised an innovative framework that utilizes interconnected graph neural networks to acquire network embeddings.
- To evaluate our model, we conducted experiments using three real-world datasets in social networks. The comprehensive simulations substantiate the superiority of our approach over the comparison methods.

Organizations: In Sect. 2, we delve into a review of related works. Section 3 defines the preliminaries and outlines the problem statement. Section 4 is dedicated to presenting our proposed GNN and RL framework. Section 5 is dedicated to experiments and results. Section 6 concludes the paper.

2   Background and Related Work

Contemporary research on cyberbullying frequently draws from disciplines like sociology and psychology. These studies commonly adopt methods such as case studies and statistical analyses to gain a deeper understanding of the issue and provide pertinent recommendations [3,4,21].

The exploration of cyberbullying dates back to [5], a study that investigates the phenomenon and considers its potential to escalate into a challenge on par with traditional bullying. This concern is amplified by the escalating reliance on technology within society. Recently, most researchers focus on the cyberbullying identification to detect the bully events happend in social networks, Reynolds et al. [6] demonstrates how machine learning algorithms can be trained to identify specific linguistic cues used by both cyberbullies and their victims, enabling automated detection of cyberbullying content. In [10], R. Yan et al. demonstrate that the problem of combating cyberbullying is NP-hard and its objective is submodular. They propose a stochastic algorithm using reverse sampling techniques to tackle this issue. However, this approach has limitations in terms of computational overhead and scalability in large networks. The study by Dadvar et al. [7] builds upon existing literature by not only validating earlier findings using the same datasets but also extending the work to include new data. They found that deep learning models were more effective in the task at hand compared to traditional machine learning models previously applied to the same dataset. This suggests that deep learning could offer improvements in performance for similar applications. Hinduja et al. [8,9] compare cyberbullying to traditional bullying, offering specific prevention strategies for schools to implement.

The rapid advancements in DRL have led to a growing body of research that employs learning-based methods to tackle the combinatorial optimization (CO) problem. These new approaches offer promising avenues for addressing the complexities and computational challenges traditionally associated with such problems. As established in [10], the challenge of combating cyberbullying can be framed as a combinatorial optimization (CO) problem. Given this, it's logical to explore solutions for the cyberbullying challenge using deep learning (DL) and reinforcement learning (RL) strategies. Fan et al. [11]introduced a Deep Reinforcement Learning (DRL) algorithm called FINDER to address the issue of dismantling networks. The model's objective is to pinpoint influential nodes in intricate network structures. To achieve this, they employed GraphSAGE as the function approximator within their Deep Q-Network (DQN) framework. Kamarthi et al. [12]used Deep Q-Learning to detect key subgraphs, tackled the Influence Maximization problem there, and then select the higher influential nodes as seeds in the whole graph. Z. Li et al. [14] point out the existing works' limitation in the field of multi-field features as an unstructured combination and introduce a novel model called Feature Interaction Graph Neural Networks (Fi-GNN) designed to capture complex interactions among feature fields within graph-structured data in a more flexible and explicit manner.

3   Problems Formulation

A social network is commonly represented by a directed graph $G = (V, E)$, where V denotes the set of users or nodes, and E represents the collection of relationships or edges between nodes. For a given edge $(u, v) \in E$, u is termed the in-neighbor of v, while v is designated the out-neighbor of u. The in-neighbors and out-neighbors sets for a particular node v are labeled as N_v^{in} and N_v^{out}, respectively. For the edge $e = (u, v) \in E$, each has a value $p_{uv} \in p$ denotes the probability that node u will active node v. A node becomes activated when it assimilates information from other nodes, it remains inactive otherwise.

3.1   Diffusion Model

To propagate a concept across a social network, an initial set of seed nodes is activated to kick-start the influence spread. The diffusion stops when there are no more nodes left to activate. It's worth noting that the Independent Cascade (IC) model is initially conceived for single-cascade diffusion processes, but for our CCB problem, we need to consider the influence from both bullies(negative influence) and defenders(positive influence). Thus, in our research we need the Competitive Independent Cascade (CIC) model which is a variant of the Independent Cascade (IC) model that is extended to handle competition between multiple pieces of information or multiple campaigns. In the Competitive Independent Cascade model, multiple types of information or influence campaigns can compete for the chance to activate a node. In this paper, We use the CIC model to exemplify a scenario in which two competing cascades, such as negative influence and positive influence, spread simultaneously within a social network.

3.2   Problem Statement

Given a social network denoted as $G = (V, E, P)$, with the node set V, edge set E, and the propagation probability set P. Let B denote the set of initial bully nodes, T denote the target victim set and k be a positive integer for a limited budget. The **Combatting CyberBullying (CCB)** problem is to find a small subset $S \subseteq V \setminus (B \cup T)$ as seed defenders to spread positive influence and active the maximum number of victim nodes in target set T.

We can calculate the marginal influence gain as adding a node v into a seed set S with $\sigma(v; S) = \sigma(S \cup \{v\}) - \sigma(S)$. Let S_t stand for the present choice of seed nodes. Our goal is to pick the node that maximizes the value of $\sigma(v; S_t)$ to be the subsequent seed node. Calculating the impact dissemination for a set of seed nodes is a #P-hard problem, making computation of marginal gain particularly challenging. Rather than producing numerous Reverse Reachable (RR) sets as done in cutting-edge approximation methods, our paper treats CCB problem as a Reinforcement Learning problem. The objective is to discover an optimal policy for selecting k nodes to a sequence of k actions to get the maximum impact dissemination in the target victim nodes set. In the RL environment, we can denote the marginal gain as the reward for an action. Then we can utilize the DRL framework to estimate the Q-value using parameterized function.

4   Methodology

In this section, we will begin by introducing the GNN for the network embedding and then use the RL framework for the combating cyberbullying problem. Inspired [15], the CCB problems include two stages, the first involves determining the embedding of nodes, and the second employs an RL greedy method to select the k defenders.

4.1   Initial Node Embedding

Graph models generally encompass nodes, edges, global context, and connectivity features. Given their proficiency in tasks such as node classification, Graph Neural Networks (GNN) serve as an effective tool for systematically representing graphs, where edges encode node dependencies.

Algorithm 1. Customized DeepWalk (CDW)

Input: $G = (V, E)$, length threshold L, component weight α, learning rate η, embedding dimension l

Output: The embedding for each node $u \in V$, $X_u \in \mathbb{R}$, $I_u \in \mathbb{R}^l$, $R_u \in \mathbb{R}^l$

1: Initialize influence contexts $C \leftarrow \emptyset$, and X_u, I_u, R_u with $N(0, 0.01)$
2: **for** each $u \in V$ **do**
3:      $C_u^l \leftarrow \emptyset$, $C_u^g \leftarrow \emptyset$, $C_u \leftarrow \emptyset$
4:      $C_u^l \leftarrow (L \cdot \alpha)$ random walk with restart probability 0.15
5:      $C_u^g \leftarrow (1 - \alpha) \cdot L$ nodes uniformly sample from N_u^{out}
6:      $C_u \leftarrow L_u \cup G_u$
7:      Insert (u, C_u) into C
8: **end for**
9: **for** each $(u, C_u) \in C$ **do**
10:       **for** each $v \in C_u$ **do**
11:           Update X_u, I_u, X_v, R_v
12:           Use negative sampling technique to calculate:
13:               $\log \Pr(v|u) \approx \log \sigma(z_v) + \sum_{w \in N} \log \sigma(-z_w)$
                  $z_v = X_u \cdot I_u \cdot R_v + X_v$, and $\sigma(x)$ is sigmoid function
14:           **for** each $w \in N$ **do**
15:               Update X_u, I_u, X_v, R_v with stochastic gradient descent
16:                   $\Phi \leftarrow \Phi + \eta \frac{\partial}{\partial \Phi}(\log \Pr(v|u))$
17:           **end for**
18:       **end for**
19: **end for**
20: **return** X_u, I_u, R_u for each node u

For the CCB problem, we need to consider both negative influence from bullies and positive from defenders, Rather than relying on random initialization for embeddings, we design the Customized DeepWalk (CDW) method to generate the initial nodes embeddings for the input of the subsequent GNN layer.

Inspired by [13] we design the CDW to collect the nodes contexts for both local and global, and then predict the context with skip-gram method. For each node u denote C_u as the node context which includes both local which conclude the a sample nodes set of $u's$ neighbors that could be influenced from node u and the part of global context which is selected from $u's$ r-hop out-neighbors. Because cascading effect essentially involves dynamic interaction between node states, their influence potential, and their propensity to be impacted. In our research node embedding, which represent nodes as vectors to encapsulate the network's structural topology, should encapsulate three key components: the states of node itself, the influence ability to out-neighbors, and the node's own propensity to be influenced by its in-neighbors. Hence, for each node u, its embedding incorporates three features: X_u, I_u and R_u, where $X_u \in \mathbb{R}$ stands for the node u's activation state, $I_u \in \mathbb{R}^l$ is the influence potential of u to its out-neighbor nodes and $R_u \in \mathbb{R}^l$ is the receptivity propensity to be influenced by it in-neighbors. As shown in Algorithm 1, we calculate the local context C_u^l and global context C_u^g to build the context C_u of node u, then use the softmax functions to calculate the probability of a node v being influenced by node u.

The CDW approach, outlined in Algorithm 1, encompasses two phases: influence Context Creation which forms the contextual environment for each node and embedding which segment refines the parameters based on the generated context. The algorithm operates with a time complexity of $O(|V| \cdot |E|)$ (Fig. 2).

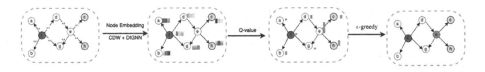

Fig. 2. Reinforcement learning framework

4.2 Dynamic Interactive Graph Neural Network

Recognizing the mutual influence between the initial embeddings in $X, I and R$, we introduce an Dynamic Interactive Graph Neural Network (DIGNN). This model can effectively captures the ongoing interplay between node features. In DIGNN, each node u is associated with a feature embedding vector $F_u = (X_u, I_u, R_u)$ and the state of graph is composed of these nodes embedding.

For state X_u of node u is designed to represent a node's activated state. In the influence process, initial bully nodes are given a constant value of $X = -1$. If node v is part of the current seed set S_t, its activation status becomes 1. Otherwise, the value of X can get cascade from both negatively and positively within the range of $[-1, 1]$, influenced by the in-neighboring nodes. The refreshed activation status of the node v at next layer $k+1$ is determined with the following formula that takes into account the influence from its in-neighbors:

$$X_v^{(k+1)} = \begin{cases} 1, & \text{if } v \in S_t \\ -1, & \text{if } v \in B \\ \tanh(\xi_X^{(k)} X_v^{(k)} + \xi_q^{(k)} q_v^{(k)}), & \text{otherwise} \end{cases} \quad (1)$$

Here $\xi_X^{(k)}, \xi_q^{(k)} \in \mathbb{R}$ are coefficient weights, $\sigma(\cdot)$ is sigmoid function and $q_v^{(k)}$ represents the anticipated influence for the node v accumulates from its incoming neighbors.

$$q_v^{(k)} = \sum_{u \in N_v^{in}} (\delta_1^{(k)} p_{uv} + \delta_2^{(k)} \text{InfluGate}(I_u^{(k)}, R_v^{(k)})) \cdot X_u^{(k)}. \quad (2)$$

We dynamically capture the diffuse weight between node v's in-neighbors' influence embedding and v's embedding of receptivity propensity. We define $e_{uv}^{(k)} = \eta^{(k)}[W^{(k)} I_u^{(k)}, W^{(k)} R_v^{(k)}]$ to assess the evolving significance of node u in relation to v, where $\eta^{(k)} \in \mathbb{R}^{2h^{(k+1)}}$ is weight parameter, $W^{(k)} \in \mathbb{R}^{h^{(k+1)} \times h^{(k)}}$ is the matrix for transformation from $h^{(k)}$ dimension to $h^{(k+1)}$ between two layers, here $[\cdot, \cdot]$ represents the concatenation. To ensure the comparability of coefficients across nodes, we employ a softmax function in combination with LeakyReLU [16] for normalization the attention coefficients.

$$\text{InfluGate}(I_u^{(k)}, R_v^{(k)}) = \frac{\exp(\text{LeakyReLU}(e_{uv}^{(k)}))}{\sum_{u \in N_v^{in}} \exp(\text{LeakyReLU}(e_{uv}^{(k)}))}, \quad (3)$$

For the influence potential feature $I_u \in \mathbb{R}^l$ of u to other nodes and the receptivity propensity $R_u \in \mathbb{R}^l$ to be influenced by others, we design the similer dynamic interactive GNN to illustrate the interactive influence as following:

$$I_v^{(k+1)} = \tanh(\gamma_I^{(k)} I_v^{(k)} + \gamma_b^{(k)} b_v^{(k)} + \gamma_X^{(k)} X_v^{(k)}), \quad (4)$$

$$R_v^{(k+1)} = \tanh(\mu_I^{(k)} R_v^{(k)} + \mu_c^{(k)} c_v^{(k)} + \mu_X^{(k)} X_v^{(k)}), \quad (5)$$

Here $\gamma_I^{(k)}, \gamma_b^{(k)}, \gamma_X^{(k)} \in \mathbb{R}$ and $\mu_S^{(k)}, \mu_c^{(k)}, \mu_X^{(k)} \in \mathbb{R}$ are weight parameters.

During DIGNN, both data propagation and structural attributes are communicated among nodes. After K cycles, the node embeddings will have assimilated insights from their neighbors. The embedding of node $u's$ could be generated with concatenating all the three components: $[X_u^{(K)}, I_u^{(K)}, R_u^{(K)}]$. Using acquired node embeddings, we define the marginal gain of node $u \in \bar{S}_t = V \setminus S_t$ concerning the present seeds set S_t as follows:

$$\hat{Q}(u, S_t; \Theta) = \theta_1^\top \text{ReLU}\left(\left[\theta_2 I_u^{(K)}, \theta_3 \sum_{v \in S_t} I_v^{(K)}, \theta_4 \sum_{w \in T \setminus (S_t \cup \{u\})} R_w^{(K)} \right] \right) \quad (6)$$

$\hat{Q}(u, S_t; \Theta)$ will be influenced by parameters denoted $\theta_1 \in \mathbb{R}^{2l}, \theta_2, \theta_3, \theta_4 \in \mathbb{R}^{l \times l}$ which will be computed in DIGNN, We train these parameters collectively using Reinforcement Learning.

4.3   Reinforcement Learning

Reinforcement Learning revolves the idea of an intelligent agent determining which actions to take in a given state to maximize cumulative rewards when interacting with its environment. Based on the characteristics of the CCB problem, we formulate the associated RL environment to address its unique challenges and constraints. RL allows intelligent agents to make decisions in complex environments to maximize cumulative rewards, making it suitable for problems like cyberbullying detection and defense, where decision-making involves dynamic and adaptive strategies to combat evolving threats. RL's ability to handle sequential decision-making and adapt to changing conditions makes it a valuable choice for tackling such problems effectively. To elaborate, we specify the components of the reinforcement learning setup as follows: states, actions, rewards, and the decision-making strategy:

- State: use vector $S_t \in \mathbb{R}^{|V|}$ to represent network state, here the value associated with node u is -1 if $u \in B$, 1 if $u \in S_t$, and 0 otherwise.
- Action: select a node $v \in \bar{S}_t$ as the defender, and we represent the action by using node embedding $X_u = 1$ for node u.
- Reward: the reward in RL environment $r(S_t, u)$ is calculate as the difference in reward when node u is added to current seeds set S_t, expressed as $r(S_t, u) = \sigma(S_t \cup \{u\}) - \sigma(S_t)$, and for the initial state $r(\emptyset) = 0$. This ensures that the cumulative reward when get S_b will correspond to the diffusion of seed set S_b, such that $\sigma(S_b) = \sum_{i=0}^{b-1} r(S_i, u_i)$.

If we represent the optimal value of Q-function for this Reinforcement Learning problem as Q^*, then the function $\hat{Q}(u, S_t; \Theta)$ parameterized with embedding will be the approximator for Q^*. This function will be learned through the Double Deep Q-Network approach. Based on $\hat{Q}(u, S_t; \Theta)$, a deterministic greedy policy will takes the optimal action $\arg\max_{u \in \bar{S}_t} \hat{Q}(u, S_t; \Theta)$. Using the node embeddings we've acquired, the score function for evaluating the incremental benefit of including a node $u \in \bar{S}_t = V \setminus S_t$ in the seed set S_t is calculated with function(7).

We use Double DQN to improve the stability and performance of RL algorithms. Double DQN use two separate neural networks: target network and online network during the learning process. It uses the online network to select the action and the target network to estimate the Q-value of that action. This helps reduce the overestimation bias and leads to more accurate Q-value estimates.

5   Experiments

In this section, we assess our proposed model using three real-world networks. First, we outline the datasets and experimental configurations. Then, we delve into the analysis and interpretation of results from various viewpoints and compare our approach with other well-established methods in the field.

5.1 Experiment Setup

For the training phase, we generate 50 Erdős-Renyi (ER) graphs with an edge probability of 0.15 and node sizes ranging between 500 and 1000 for training purposes. ER graphs are commonly used for training because their random structure offers a generalized, unbiased representation. This ensures that models trained on them are versatile and not overfitted to specific network types, making them suitable for various real-world applications. The proposed framework and baseline models are evaluated on both synthetic graph and three realistic datasets from SNAP and KONECT, with their statistics detailed in Table 1. The Wiki Vote dataset encompasses all Wikipedia voting activities from its inception until January 2008, with 7,115 nodes representing users and 103,663 edges indicating user-to-user voting interactions. Google+ directed network comprises links between Google+ users, nodes correspond to individual users, and directed edges signify that one user includes the other in their circles. Epinions is a trust network derived from the Epinions online social platform, nodes represent Epinions users, and directed edges represent trust relationships between users.

Table 1. Evaluation datasets statistics.

Dataset	# of nodes	# of edges	Type
Synthetic	2000	299501	Connection Relationship
Wiki Vote	7115	103663	Voting Relationship
Google+	23628	39242	Friendship
Epinions	75897	508837	Trust Relationship

Setup. For the directed graph $G = (V, E, P)$, we uniformly select 3% of nodes from V as the initial bully nodes set B and select various size of nodes in $V \setminus B$ as the victim nodes. For the evaluation experiments, we choose to use CIC model as the diffusion model and set propagation probability $p = 0.5$ for each edge.

Contrastive Approaches. We compare the performance for our dynamic interative GNN RL (DI-RL) with the following advanced approximation algorithm for CCB problem:

- **Reverse Influence Sampling(RIS)** [10]. Utilizes a stochastic approach through reverse sampling techniques to maximize the activation of target nodes by defenders.
- **Out-Degree (OD)** [17]. Identifies the top k nodes with the highest out-degree as the seed set.
- **Betweenness Centrality(BC)** [19]. Selects the top k nodes with the highest betweenness centrality as the seed set for positive defenders.
- **PageRank (PR)** [18]. Chooses the top k nodes with the highest PageRank scores as the seed set for defenders.

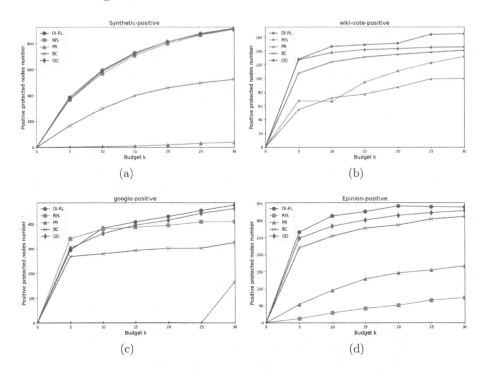

Fig. 3. Positive protected performance comparisons among different methods.

5.2   Results

We examine the association between budget k and the quantity of victim nodes triggered by both bully and defender nodes. Here we set the source bully nodes with 3% and budget $k \in [0, 30]$. For Wiki-Vote we set the diffusion probability p = 0.5 and choose the size of target victim set $|T| = 500$, the size of target victim set $|T| = 1000$ on Google+ and Epinions social networks.

In Fig. 3, the vertical axis is used for the count of nodes in T that have been activated by defenders exhibiting positive influence. The horizontal axis represents budget k. The results indicate that as the budget k increases, the positive impact on our target victim nodes also increases, leading to a reduction in the impact of cyberbullying. Based on the data gathered from our experiments, our proposed algorithm demonstrates superior results when compared to the other heuristic algorithms and works very well on big realworld data sets.

Figure 4 proves that our method by active more victim nodes with positive influence from defenders would definitely reduce the number activated by negative bully nodes. The vertical axis is used for the count of nodes in T that have been activated by cyberbully nodes exhibiting negative influence. The horizontal axis represents budget k. As we increase the budget for defender nodes,

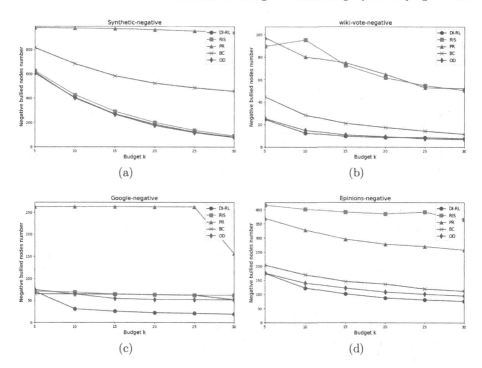

Fig. 4. Negative bullying performance comparisons among different methods.

the number of bullied victim nodes decreases. Our observations show that our method performs the best across all budgets for both synthetic and real-world networks.

6 Conclusions

In this paper, we present a framework for tackling the issue of cyberbullying through a deep reinforcement learning approach, anchored in dynamic interactive graph neural network embeddings. Our model aims to optimize the activation of defender nodes in a way that inherently reduces the negative bullying impact of bully nodes within the target victim node set. And we have rigorously tested our proposed solution through a comprehensive set of experiments. Our findings reveal that our approach excels at identifying high-quality seed defenders across various networks. Additionally, when benchmarked against several cutting-edge techniques, our method consistently outperforms the alternatives.

Acknowledgements. This work was supported in part by NSF under Grant No. 1907472 and No. 1822985.

References

1. Reynolds, K., Kontostathis, A., Edwards, L.: Using machine learning to detect cyberbullying. In: Proceedings of the 2011 10th International Conference on Machine Learning and Applications and Workshops (ICMLA'11), vol. 2, pp. 241–244. IEEE (2011)
2. Budak, C., Agrawal, D., ElAbbadi, A.: Limiting the spread of misinformation in social networks. In: Proceedings of the 20th International Conference on World Wide Web. ACM, pp. 665–674 (2011)
3. Hinduja, S., Patchin, J.W.: Cyberbullying: an exploratory analysis of factors related to offending and victimization. Deviant Behav. **29**(2), 129–156 (2008)
4. Walrave, M., Heirman, W.: Cyberbullying: predicting victimisation and perpetration. Child. Soc. **25**(1), 59–72 (2011)
5. Patchin, J.W., Hinduja, S.: Bullies move beyond the schoolyard: a preliminary look at cyberbullying. Youth Violence Juvenile Justice **4**(2), 148–169 (2006)
6. Reynolds, K., Kontostathis, A., Edwards, L.: Using machine learning to detect cyberbullying. In: Proceedings of the 2011 10th International Conference on Machine Learning and Applications and Workshops (ICMLA 2011), vol. 2, pp. 241–244. IEEE (2011)
7. Dadvar, M., Eckert, K.: Cyberbullying detection in social networks using deep learning based models. In: Song, M., Song, I.-Y., Kotsis, G., Tjoa, A.M., Khalil, I. (eds.) DaWaK 2020. LNCS, vol. 12393, pp. 245–255. Springer, Cham (2020). https://doi.org/10.1007/978-3-030-59065-9_20
8. Hinduja, S., Patchin, J.W.: Cyberbullying. Cyberbullying Research Center (2014). Retrieved 7 September, 2015
9. Slonje, R., Smith, P.K., Frisén, A.: The nature of cyberbullying, and strategies for prevention. Comput. Hum. Behav. **29**(1), 26–32 (2013)
10. Yan, R., Li, Y., Li, D., Wang, Y., Zhu, Y., Wu, W.: A stochastic algorithm based on reverse sampling technique to fight against the cyberbullying. ACM Trans. Knowl. Discov. Data (TKDD) **15**(4), 1–22 (2021)
11. Fan, C., Zeng, L., Sun, Y., Liu, Y.-Y.: Finding key players in complex networks through deep reinforcement learning. Nature Mach. Intell. **2**(6), 317–324 (2020)
12. Kamarthi, H., Vijayan, P., Wilder, B., Ravindran, B., Tambe, M.: Influence maximization in unknown social networks: Learning policies for effective graph sampling. In: Proceedings of the 19th International Conference on Autonomous Agents and MultiAgent Systems, pp. 575–583 (2020)
13. Feng, S., Cong, G., Khan, A., Li, X., Liu, Y., Chee, Y.M.: Inf2vec: latent representation model for social influence embedding. In: 2018 IEEE 34th International Conference on Data Engineering (ICDE). IEEE, pp. 941–952 (2018)
14. Li, Z., Cui, Z., Wu, S., Zhang, X., Wang, L.: Fi-gnn: modeling feature interactions via graph neural networks for CTR prediction. In: Proceedings of the 28th ACM International Conference on Information and Knowledge Management, pp. 539–548 (2019)
15. Chen, T., Yan, S., Guo, J., Wu, W.: ToupleGDD: a fine-designed solution of influence maximization by deep reinforcement learning. IEEE Trans. Comput. Soc. Syst. (2023)
16. Maas, A.L., Hannun, A.Y., Ng, A.Y., et al.: Rectifier nonlinearities improve neural network acoustic models. In: Proc. icml, vol. 30, no. 1. Citeseer, p. 3 (2013)
17. Kempe, D., Kleinberg, J., Tardos, É.: Maximizing the spread of influence through a social network. In: Proceedings of the 9th ACMSIGKDD International Conference on Knowledge Discovery and Data Mining, pp. 137–146. ACM (2003)

18. Page, L., Brin, S., Motwani, R., Winograd, T.: The PageRank citation ranking: bringing order to the web. In: Proceedings of ASIS, pp. 161–172 (1998)
19. Brandes, U.: On variants of shortest-path betweenness centrality and their generic computation. Soc. Networks **30**(2), 136–145 (2008)
20. Pew Research center. https://www.pewresearch.org/internet/2018/09/27/a-majo rity-of-teens-have-experienced-some-form-of-cyberbullying/
21. https://www.tandfonline.com/loi/wjsv20

Author Index

Printed in the United States
by Baker & Taylor Publisher Services